ELECTRON PHYSICS OF VACUUM AND GASEOUS DEVICES

ELECTRON PHYSICS OF VACUUM AND GASEOUS DEVICES

MIROSLAV SEDLAČEK
Royal Institute of Technology
Stockholm, Sweden

A Wiley-Interscience Publication
JOHN WILEY & SONS, INC.
New York / Chichester / Brisbane / Toronto / Singapore

Library of Congress Cataloging in Publication Data:

Sedláček, Miroslav.
    Electron physics of vacuum and gaseous devices / by Miroslav
Sedláček.
        p.   cm.
    Includes bibliographical references and index.
    ISBN 0-471-14527-0
    1. Electrons.   2. Electronic systems.   3. Vacuum technology.
4. Electrooptical devices.   5. Gas discharges.   I. Title.
QC793.5.E62S44   1996
537.55—dc20   95-41708
                                                        CIP

CONTENTS

APPENDICES   471

NAME INDEX   521

SUBJECT INDEX   525

PREFACE

Aus einem Buch abschreiben, gibt: ein Plagiat.
Aus zwei Büchern abschreiben, gibt: einen Essey.
Aus drei Büchern wird: eine Doktordissertation.
Aus vier Büchern: ein füntes gelehrtes Buch.
Roda Roda
(Austrian satirist, born in Croatia, 1872–1945)

Electron physics encompasses a large part of physics. It deals with everything from electrons and the ions from which they are freed or attached, the interaction of these particles with fields and matter, electron collisions with atoms and molecules, as well as electron emission and the electric conductivity in gases.

This book, an introduction to electron physics of free particles, results from notes given to my students at the Royal Institute of Technology in Stockholm over the last two decades. It is my hope that a larger group of electrical engineering and physics students, as well as professional physicists and engineers working in fields ranging from electron tubes to large accelerators, will find this book useful.

I have attempted to stress the physical principles at the occasional expense of technical completeness, especially those principles common to a group of devices. Besides giving working principles for different tubes and devices, I have included some technical data to show the state of the art and their use in possible applications.

I use the International System of Units (SI) throughout. The literature cited is the most appropriate for further reading. I begin some chapters with a historical note, and the references credit the scientists or engineers who were responsible for major breakthroughs and advancements in the field of electron physics.

I owe thanks to many colleagues who provided information and discussions of the manuscript, especially to Prof. M. Eriksson, Dr. W. B. Herrmannsfeldt, Dr. S. Martin, Dr. A. Pinelli, and Prof. S. Torvén. Dr. Staffan Rosander gave the permission to use many figures and portions of text from his student textbook *Accelerator Technology*. Special thanks go to Prof. Peter Trower, who corrected my English and made very many useful comments. Many thanks also goes to the institutes, firms, and industries who provided photographs and allowed the use of figures from their publications. Most important in the refining process were my students. My TEX written notes were scrutinized by their ever watchful eyes, identifying quite a few errors. Those that remain are mine alone, and I would appreciate any notification of their existence.

I dedicate this work to my wife Nada, for without her support and forbearance, sitting in an armchair and reading while I was typing on my Macintosh, this book would have never been produced.

Miroslav Sedlaček

Zagreb, Stockholm

INTRODUCTION

The latin word **electrica** was probably coined by Gilbert [1] at the end of the six-teenth century, from the greek word for amber ηλεκτρον, and translated into English **electricity** by Sir Thomas Browne [2]. For almost 200 years it was used in terms like "electrical fire," "electrical æther," or "electrical fluid," and "*There seems to be a Quantity of this Æther in all Bodies*," both "*originally-electric*" (insulators) and "*non-electric*" (conductors) [3]. However, the laws of electrolysis, found by Faraday [4] in 1831, showed that elementary electric charges must exist. Even if he interpreted the experiments as a proof that "*equivalent atom*" quantities correspond to equal quantity of electricity, he was "*jealous to use the expression atom*" because of difficulties in understanding their real nature. Instead of "ether" Faraday intro-duced "tubes of force" which originate and terminate on electric charges [5] and thus were the fundamental expression of electricity. Some 50 years later Stoney [6] interpreted Faraday's electolysis laws as requiring every atomic bond to carry a "*certain quantity of electricity ... which is the same in all cases ... 1/10^{20} Ampères*." Ten years later he wrote [7]: "*There may be several charges in one chem-ical atom, and there appear to be at least two in each atom. These charges will be convenient to call* **electrons**."

The first experiments that connected electricity to particles were made by Plücker [8], working with gas discharges at low pressure. He observed that his cath-ode emitted "rays" which produced a shadow on the glass tube. These "rays" could be deflected by a magnetic field. A decade later, Hittorf's [9] similar experiments showed that these "rays" move in a straight line, starting at the cathode, but in a magnetic field spiral along the field lines. Varley [10] opined that "cathode rays" are small particles torn off the cathode, and their negative charge expains the deflection in magnetic field. While bombarding insulators with cathode rays he observed also a deflection, but his contemporaries could not repeat electrostatic deflection experiments because of poor vacuum. So Hertz [11] theorized that cathode rays were electromagnetic waves and delayed the discovery of the electron for more than 25 years. J. Perrin [12] perfomed an important experiment in 1895. By introducing a Faraday cage in a discharge tube he was able to prove that cathode rays have negative electricity, however, he believed they were negative ions. Hertz considered the possibility that cathode rays were particles, and he proposed an experiment to measure their charge-to-mass ratio e/m. J. J. Thomson [13], the first to succeed with such experiments, did so in 1897, with revolutionary results. He showed that cathode rays differed from all known ions in that their mass was tiny—only 1/200 of the mass of the hydrogen ion. In the same year, Kaufmannn [14] obtained the same result and suggested that cathode rays cannot be particles torn off the cathode, while Fitzgerald [15] expressed the opinion that they are free electrons. Modern physics was thus started.

However, as early as 1873, an important experiment had already been perfomed. Guthrie [16] showed that air near a heated iron sphere becomes conductive, at low temperature for positive charges, and at higher temperatures for negative charges. Ten years later, Edison [17] demonstrated his famous incandescent lamp with a carbon filament, but with a separated electrode. He showed that a galvanometer needle was deflected when the meter was connected between the superfluous electrode and the positive side of the filament, while no deflection occured when connected to the negative side. This so-called Edison effect, investigated extensively in the next years, caused Preece [18] to opine that negatively charged particles constitute the current between the filament and the extra electrode, subsequently corroborated by the discovery of the electron by J. J. Thomson.

Important breakthroughs came about one after another. In the year of the discovery of the electron, Braun [19] constructed the first cathode ray tube, which was thereafter known as the *Braunian tube* in Europe. Richardson, starting with the classical theory of metals where the electrons were assumed to act like gas molecules, wrote down the equation for thermionic emission of electrons from metals. His temperature dependence was wrong—square root of temperature, not the temperature squared—but his exponential was correct.

In 1904, Fleming [20] developed the diode, which he called a *thermionic valve*—hence the name "valve" for an electron tube. The first diode was a simple glass bulb with tungsten filament and wire anode. Wehnelt [21] showed the same year that barium and strontium emitted electrons in great abundance. These discoveries formed the basis of electron tube technology, which was to dominate electronics for 50 years. An important step was taken in 1906 by Lee de Forest [22] who introduced a third electrode, the grid, between the cathode and the anode so that his triode could amplify electical signals.

The first World War saw the rapid development of electron tubes for their use in wireless communications. Already at the beginning of the century, Marconi and Tesla showed that electromagnetic waves could transmit information, their transmitters being interrrupted spark gaps. By 1915, transatlantic communication was made possible due in large part to the realization that de Forest's triode could amplify a carrier signal and demodulate it on the receiving side.

The years 1918 to 1960 comprised the era of electron tubes. The technology improved continuously. With the start of regular public broadcasting, electron tube production increased steadily so that the tubes became cheaper and cheaper. In the 1930s, more grids were introduced which opened new low- and high-frequency applications in communication, industry, and science. Cathode ray tubes, phototubes, indicator tubes, gas-filled rectifiers, ignitrons, and other electron tubes were designed and produced. The first experiments to transmit video signals also started in the 1930s. The frequency domain was extended; thus before the second World War it was possible to generate waves as short as a few centimeters. The first radar, of 8 meter wavelength, was constructed in 1934, but the invention of the magnetron in 1940 by Boot and Randall [23] made radar possible for wartime. Also, electron tubes became smaller and smaller so that they could be easily combined with passive circuit components on boards, the precursors of modern printed circuits. This technology was used in the MANIAC, the first programmable computer with 19,000 electron tubes whose volume required a large laboratory room.

But the future of the electron tube was doomed. In 1948 Bardeen, Brattain, and Shockley [24] invented the transistor. Within 10 years, semiconductor technology dominated the field of electronics since transistors and intregrated circuits were simpler, cheaper, and more versatile. The last electron tube factory closed in 1984.

There are, however, still fields where at present, because of inherent limitations, semiconductors cannot replace electron tubes. These are cathode ray tubes, TV tubes, high-power transmitter and microwave tubes, X-ray tubes, photomultipliers, image amplifiers, and converters, to mention a few. Fluorescent lights, other gas-filled tubes, and gas lasers are examples of essential electron tubes.

Electron optics is another field where great progress has been made since Busch, [25] in 1926, recognized the essential similarities between light rays and electron beams. The rapid development of modern biology and surface science depends to a large degree on the electron microscope, elementary particles research depends on large particle accelerators and photomultipliers, an material physics research depends on microprobes, electron microscopes, and spectometers. Using electron beams one can drill, cut, mill, melt, and weld all varieties of materials, including those hard to handle or to process. Production of integrated circuits depends ever more on electron beams and gas discharges: electron beams < 10 nm in diameter can write a pattern on a silicon chip; an ion implantation device can dope semiconductors with high precision; ion etching, sputtering, and depostion are widely used to produce integrated circuits.

Devices which depend on the interaction of free particles with electric or magnetic fields allow us to "see" individual atoms, to detect which elements exist on a surface < 1 μm^2, or, with an accelerator 27 km in circumference, to study quarks, the smallest entities of the atomic material world.

The advances in semiconductor technology not withstanding, the electron physics of vaccum and gaseous devices will remain an essential part of our technology and, therefore, our economy. A simple integrated circuit costs a fraction of a dollar, a color television monitor tube costs $100, a high power klyston costs $50,000, and an advanced electron microscope with accessories costs $150,000. Tubes and devices, based on interaction of free charges with electric and magnetic fields in vacuum or in gases, are far from a finished chapter. They will often be used in the future, and we will see their position secured in both research and production as their technology is developed. The physical background and the working principles of these tubes and devices make up the content of this book.

1

Fields and Orbits

Stimulated by Babbages articles describing an automatic calculating machine, the Difference Engine, Edvard Scheutz and his father Georg completed in 1843 the first working Difference Engine. The engine consisted of three major units: the calculating unit, the transfer mechanism and the printing unit. Three machines were built, and used to print mathematical tables. (Technical Museum, Stockholm).

1.1. ELECTROMAGNETIC FIELD

1.1.1. Introduction

Electron tubes and devices which use free electrons and ions in vacuum or gas at low pressure must include a source of charged particles and electrodes or magnets to generate electric, magnetic, and electromagnetic field. To understand the properties of these devices it is necessary to know the field distributions which, with the particles equation of motion, describe the physics behind these devices and their behavior.

Field distributions are the solutions, for specific boundary conditions, of Maxwell's equations [26]

$$\nabla \times \mathbf{H} = \mathbf{J} + \frac{\partial \mathbf{D}}{\partial t}, \qquad \nabla \cdot \mathbf{D} = \rho,$$
$$\nabla \times \mathbf{E} = -\frac{\partial \mathbf{B}}{\partial t}, \qquad \nabla \cdot \mathbf{B} = 0, \tag{1.1}$$

where

$$\mathbf{D} = \varepsilon \mathbf{E}, \qquad \mathbf{B} = \mu \mathbf{H}, \qquad \mathbf{J} = \sigma \mathbf{E}.$$

Field distributions and Newton's second law determine the orbits taken by the particles, the force acting on the particles is the Lorentz force

$$\mathbf{F}_{\text{Lorentz}} = \frac{d(m\mathbf{v})}{dt} = q\mathbf{E} + q\mathbf{v} \times \mathbf{B}. \tag{1.2}$$

In these equations $\mathbf{E}$ is the electric, $\mathbf{H}$ the magnetic field, $\mathbf{D}$ the dielectric displacement vector, $\mathbf{B}$ the magnetic induction, and $\mathbf{v}$ the velocity of the particle. σ is the conductivity, $\varepsilon = \varepsilon_r \varepsilon_0$, and $\mu = \mu_r \mu_0$, where ε is the dielectric constant, μ the permeability, and $\varepsilon_0 \mu_0 = 1/c^2$ with c the velocity of light.

1.1.2. Electrostatics

Many electron physics applications require only static fields, that is,

$$\partial \mathbf{E}/\partial t = \dot{\mathbf{E}} = \partial \mathbf{B}/\partial t = \dot{\mathbf{B}} = 0.$$

For an electrostatic field, $\mathbf{J} = 0$, but Gauss' law holds. Thus we can uniquely associate the electric potential V with the electric field at every point in space

$$\mathbf{E} = -\nabla V, \tag{1.3}$$

if the electric field is nonlossy — that is conserves energy. Outside the electrodes this electric potential is described by Poisson's equation

$$\nabla \cdot (\nabla V) = \nabla^2 V = -\frac{\rho}{\varepsilon}, \tag{1.4}$$

while inside the region where the space-charge can be neglected the potential is determined by Laplace's equation

$$\nabla^2 V = 0. \tag{1.5}$$

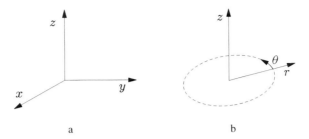

Fig. 1.1. Coordinate system definition. (**a**) Rectangular and (**b**) cylindrical.

The solution to these two equations must satisfy certain boundary conditions:
1. On the surface of an electrode, i, the potential must be constant:

$$V(x, y, z) = V_i = \text{const.} \tag{1.6}$$

2. The electric field at the boundary between two media, i and j, must satisfy the condition

$$\mathbf{n} \times (\mathbf{E}_j - \mathbf{E}_i) = 0, \qquad \mathbf{n} \cdot (\varepsilon_j \mathbf{E}_j - \varepsilon_i \mathbf{E}_i) = \sigma, \tag{1.7}$$

where σ is the surface charge density and $\mathbf{n}$ the unit outward pointing vector (directed form i to j).

To apply these ideas to practical problems we must choose a coordinate system. The symmetry of the situation encountered indicates for almost all applications either a rectangular (Cartesian) or a cylindrical system, Fig. 1.1. In electron optics the axially symmetric lenses suggest the use of cylindrical coordinates. Thus the left side of Eqs. 1.4 and 1.5 expressed in rectangular coordinates is

$$\nabla^2 V = \frac{\partial^2 V}{\partial x^2} + \frac{\partial^2 V}{\partial y^2} + \frac{\partial^2 V}{\partial z^2}, \tag{1.8}$$

and in cylindrical coordinates

$$\nabla^2 V = \frac{1}{r} \frac{\partial}{\partial r} \left(r \frac{\partial V}{\partial r} \right) + \frac{1}{r^2} \frac{\partial^2 V}{\partial \theta^2} + \frac{\partial^2 V}{\partial z^2} = \frac{\partial^2 V}{\partial r^2} + \frac{1}{r} \frac{\partial V}{\partial r} + \frac{1}{r^2} \frac{\partial^2 V}{\partial \theta^2} + \frac{\partial^2 V}{\partial z^2}. \tag{1.9}$$

In a rotationally symmetrical system the azimuthal term is zero, $\partial V / \partial \theta = 0$.

Analytic solution of Laplace's or Poisson's equation can only be obtained for some special and simple geometries. In other cases, conformal mapping can be fruitfully used. However, in most cases of practical interest, numerical methods will be required (Section 1.2).

With the determination of the electric potential the electric field components can be found in rectangular coordinates by

$$E_x = -\frac{\partial V}{\partial x}, \qquad E_y = -\frac{\partial V}{\partial y}, \qquad E_z = -\frac{\partial V}{\partial z}, \tag{1.10}$$

and in cylindrical coordinates

$$E_r = -\frac{\partial V}{\partial r}, \qquad E_\theta = -\frac{1}{r} \frac{\partial V}{\partial \theta}, \qquad E_z = -\frac{\partial V}{\partial z}. \tag{1.11}$$

The electrostatic energy stored in the volume dv, $\mathbf{D}^2 dv/(2\varepsilon)$, is in a closed system, where the electrostatic potential is given as a solution to Eq. 1.4:

$$W = \int\int\int [\frac{\varepsilon}{2}(\nabla V)^2 - \rho V]dv. \tag{1.12}$$

1.1.3. Magnetostatic

The procedure outlined above for electrostatics, by introducing a scalar magnetic potential V_m, is of limited use for magnetostatic, since such solutions can be found only in the region where $\mathbf{J} = 0$ and $\mu = \mu_0$. This is only the case when the magnetic field is low and the surface between the magnetic material and vacuum is approximately equipotential. This is true for weak permanent magnets where $\nabla \times \mathbf{H} = 0$.

The requirement of stationary magnetic field on Maxwell's equations results in

$$\nabla \times (\mathbf{B}/\mu) = \mathbf{J}, \qquad \nabla \cdot \mathbf{B} = 0,$$

which can be written as

$$\frac{1}{\mu}\nabla \times \mathbf{B} + \nabla\frac{1}{\mu} \times \mathbf{B} = \mathbf{J}.$$

Introducing vector potential, $\mathbf{A}$, such that $\mathbf{B} = \nabla \times \mathbf{A}$ yields

$$\frac{1}{\mu}\nabla \times \nabla \times \mathbf{A} + \nabla\frac{1}{\mu} \times \nabla \times \mathbf{A} = \mathbf{J}.$$

Magnetic potential thus can be expressed by

$$\nabla^2\mathbf{A} + \frac{1}{\mu}[\nabla\mu \times (\nabla \times \mathbf{A})] = -\mu\mathbf{J}. \tag{1.13}$$

Boundary conditions for $\mathbf{B}$, using the same conventions as for electrostatic, are

$$\mathbf{n} \times [\mathbf{B}_j/\mu_j - \mathbf{B}_i/\mu_i] = \mathbf{K}_{ij}, \qquad \mathbf{n} \cdot (\mathbf{B}_j - \mathbf{B}_i) = 0, \tag{1.14}$$

where $\mathbf{K}_{ij}$ is the surface current, which flows at the boundary between i and j.

It is sometimes convenient to introduce a scalar potential. This can, without limitations, be done in a region where the current density is zero. The field distribution inside such regions can be obtained by solving Laplace's equation (Eq. 1.5). But inside a current-carrying region $\nabla \times \mathbf{H}$ is no longer zero and an auxiliary vector $\mathbf{M}$ must be defined in an appropriate way to make $\nabla \times (\mathbf{H} - \mathbf{M})$ irrotational. A magnetic scalar potential, V_m, can then be introduced defining the magnetic field vector as $\mathbf{H} = -\nabla \cdot V_m + \mathbf{M}$, where the scalar magnetic potential satisfies

$$\nabla^2 V_m = -\nabla \cdot \mathbf{M}. \tag{1.15}$$

The advantage of this method is simplicity, while its main disadvantage is that only one closed region is allowed. This restriction results from the fact that a constant can be added to a scalar potential without changing the differential equation. With two or more closed regions with iron involved, the constant remains undefined, and the solution inside iron can only be obtained by solving the vector potential equation. Another disadvantage is that special equations on the boundary between air (vacuum) and magnetic material must be introduced. Regardless of these limitations, the method is used to solve some three-dimensional problems numerically.

In most practical cases the field distribution in rectangular or cylindrical coordinates must be found. Cylindrical coordinates for problems with rotational symmetry, and rectangular coordinates for those with the same x–y cross-section for all z-values, reduce the problem to two dimensions. The vector potential and the current density then in rectangular Cartesian coordinates will only have nonzero A_z and J_z components, while in cylindrical coordinates, only the A_θ and J_θ will be different from zero.

Recalling that

$$\nabla^2 \mathbf{A} = \nabla(\nabla \cdot \mathbf{A}) - \nabla \times (\nabla \times \mathbf{A}),$$

the equation to be solved in rectangular coordinates is

$$\frac{\partial^2 A_z}{\partial x^2} + \frac{\partial^2 A_z}{\partial y^2} + \frac{1}{\mu}\left[\frac{d\mu}{dx}\frac{\partial A_z}{\partial x} + \frac{d\mu}{dy}\frac{\partial A_z}{\partial y}\right] = \mu J_z, \tag{1.16}$$

and the equation to be solved in cylindrical coordinates is

$$\frac{\partial^2 A_\theta}{\partial r^2} + \frac{1}{r}\frac{\partial A_\theta}{\partial r} + \frac{\partial^2 A_\theta}{\partial z^2} + \frac{A_\theta}{r^2} + \frac{1}{\mu}\left[\frac{\partial \mu}{\partial z}\frac{\partial A_\theta}{\partial z} + \frac{\partial \mu}{\partial r}(\frac{A_\theta}{r} + \frac{\partial A_\theta}{\partial r})\right] = \mu J_\theta. \tag{1.17}$$

When the solution is known the field components can be obtained in rectangular coordinates by

$$B_x = \frac{\partial A_z}{\partial y}, \qquad B_y = \frac{\partial A_z}{\partial x}, \tag{1.18}$$

and in cylindrical coordinates

$$B_r = -\frac{\partial A_\theta}{\partial z}, \qquad B_z = \frac{1}{r}\frac{\partial}{\partial r}(r A_\theta). \tag{1.19}$$

The magnetostatic energy stored in the volume element dv is $B^2 dv/(2\mu)$, and in a closed system, where the vector potential is given by Eq. 1.13, the magnetic energy is given by

$$W = \int\int\int [\mathbf{J} \cdot \mathbf{A} - \frac{1}{2\mu}(\nabla \times \mathbf{A})^2]dv. \tag{1.20}$$

1.1.4. Wave Equation

High-power microwaves are generated in microwave tubes, where an electron beam interacts with the high-frequency electromagnetic field. Similar conditions exist in some accelerators. To describe the interaction and to compute particle orbits in this field, the wave equation must be solved. In most practical cases the interaction takes place inside microwave cavities, especially those with rotational symmetry, so both the electric and the magnetic field can be described by a sine wave function with constant frequency,

$$\mathcal{E} = \Re \mathbf{E}(x, y, z)\, e^{j\omega t}, \qquad \mathcal{H} = \Re \mathbf{H}(x, y, z)\, e^{j\omega t}. \tag{1.21}$$

Thus

$$\frac{\partial \mathcal{E}}{\partial t} = \Re j\omega \mathbf{E}\, e^{j\omega t}, \qquad \frac{\partial \mathcal{H}}{\partial t} = \Re j\omega \mathbf{H}\, e^{j\omega t}. \tag{1.22}$$

Inside the cavity $\mathbf{J} = 0$ and $\rho = 0$ and Maxwell's equations are

$$\begin{aligned}
\nabla \times \mathbf{H} = j\omega\mathbf{D}, && \nabla \cdot \mathbf{D} = 0, \\
\nabla \times \mathbf{E} = -j\omega\mathbf{B}, && \nabla \cdot \mathbf{B} = 0,
\end{aligned} \tag{1.23}$$

so taking into account the expressions for $\nabla \times \nabla \times \mathbf{E}$ and $\nabla \times \nabla \times \mathbf{H}$, respectively, we write

$$\nabla^2 \mathbf{E} + k^2 \mathbf{E} = 0, \qquad\qquad \nabla^2 \mathbf{H} + k^2 \mathbf{H} = 0, \tag{1.24}$$

where $k^2 = \omega^2 \mu_r \mu_0 \varepsilon_r \varepsilon_0$.

Often only the solution for lowest frequency mode is needed. In the rotationally symmetric case with $\mu_0 \varepsilon_0 = 1/c^2$ and $\mu_r = \varepsilon_r = 1$, the magnetic field has only an azimuthal component,

$$H = H_\theta(r, z),$$

inside the cavity and the wave equation becomes

$$\frac{\partial^2 H_\theta}{\partial r^2} + \frac{1}{r}\frac{\partial H_\theta}{\partial r} + \frac{\partial^2 H_\theta}{\partial z^2} + \left(k^2 - \frac{1}{r^2}\right) H_\theta = 0. \tag{1.25}$$

By substituting a potential which decreases with distance from the symmetry axis, $H_\theta = U/r$, Eq. 1.25 becomes

$$\frac{\partial^2 U}{\partial r^2} - \frac{1}{r}\frac{\partial U}{\partial r} + \frac{\partial^2 U}{\partial z^2} + k^2 U = 0. \tag{1.26}$$

While neglecting the small losses to the cavity walls, to which the electric field must be normal

$$\mathbf{n} \times \mathbf{E} = 0,$$

we now have an eigenvalue problem. The value of k can be determined from the energy stored in the cavity,

$$k^2 = \min \frac{\int_S \frac{1}{r}\left(\frac{\partial U^2}{\partial r} + \frac{\partial U^2}{\partial z}\right) dS}{\int_V \frac{U^2}{r} dV}. \tag{1.27}$$

Guessing the first value of k, we solve numerically Eq. 1.26, using the solved value of k as the input for the next iteration until convergence is obtained.

Knowing the solution, $U = U(r, z)$, the field components are obtained from

$$H_\theta = \frac{U}{r}, \qquad E_z = -\frac{j}{\omega\varepsilon_0}\frac{1}{r}\frac{\partial U}{\partial r}, \qquad E_r = \frac{j}{\omega\varepsilon_0}\frac{1}{r}\frac{\partial U}{\partial z}. \tag{1.28}$$

In most problems dealing with the interaction of charged particles and electromagnetic fields in cavities the Q-value and shunt impedance of the cavity are important parameters. The Q-value is defined as the ratio of stored power and losses to the cavity walls and possible inner dielectric or conductive materials,

$$Q = \frac{2\pi f_0 W}{\sum P_{wi}}, \tag{1.29}$$

while the shunt impedance can be defined in many different ways. A commonly used definition in both accelerator technology and in the design of microwave tubes is

$$R_{sh} = \frac{VV^*}{2\sum P_{wi}}. \tag{1.30}$$

Here W is the total energy stored in the cavity, P_{wi} the losses, f_0 the frequency of the appropriate, often the lowest, mode, and V the complex voltage over the gap as seen by a particle traversing the cavity. The voltage is defined by

$$V = \int\limits_0^L (E_z)_{r=0} \, e^{j\omega z/v} dz,$$

that is, the integral of the electric field along the axis of the cavity considering the particle velocity v. In accelerator technology and when the particles are electrons v can usually be replaced by the velocity of light c.

Cavities with coupling holes or unsymmetrical design use three-dimensional programs.

1.2. NUMERICAL METHODS

HISTORICAL NOTES. The method of least squares, one of the most useful tools in numerical analysis, was first attempted by Johann Tobias Mayer and Ruđer Bošković in the second half of the 18th century. It was left to Gauss [27] to develop the method. He observed that it involves the solution of a set of linear equations with a large number of unknowns and thus becomes difficult to calculate. He suggested dividing the system of equations into a few subsystems by neglecting the "undesired" variables. After solving the subsystems the resulting approximations could be used to solve the complete system in a first approximation. The process could be repeated as many times as necessary until the variation of the unknowns become less than the desired accuracy. It was extended by Seidel [28] and is called the *Gauss-Seidel method.*

Gauss also worked with interpolation processes. He was aware of the possibility of replacing the partial derivatives with corresponding differences and of using the type of solution he proposed for the least square problem, to numerically solve partial differential equations.

Between Gauss' work and the first world war, attempts were made to solve some mechanical and electromagnetic problems using these iteration methods. However, for most practical cases this formidable task involved so many calculations that it almost exceeded the human abilities.

Analog methods were therefore developed to solve Laplace's equation: current-flow model, elastic-membrane model, and resistor analog. These methods, confined to two dimensions, could simulate even rotationally symmetric problems. The accuracy was limited, but an approximate solution could be obtained rather easily.

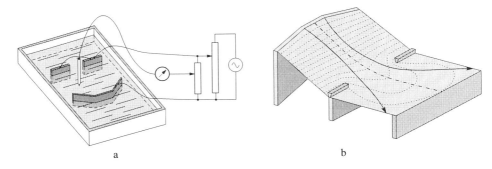

a   b

Fig. 1.2. (a) Current-flow and **(b)** elastic-membrane analog.

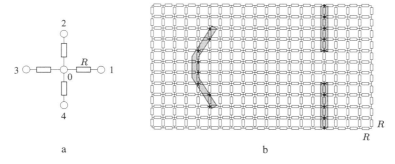

Fig. 1.3. (a) A node in a resistor analog and (b) resistor mesh.

The current-flow analog used a uniform conducting medium, like tap water, in a trough, where metallic electrodes with the desired shape were used to simulate the electrodes, as seen in Fig. 1.2a. An alternating current source (of ~1000 Hz) was connected to the electrodes with a relative amplitude corresponding to the design potentials. With a thin wire probe the potential at any point was measured using a zero current method.

For an elastic uniformly stretched membrane and deformed by a displacement in height proportional to the electrode potentials, as seen in Fig. 1.2b, the vertical displacement is proportional to the potential of the studied electrode configuration. This analogy can be extended by letting small spheres, say ball-bearing, roll over the membrane. Their motion corresponds to the motion of charged particles in the corresponding electric field. Elastic-membranes were used to model many electron tubes and lenses during the 1930 and 1940s.

A mesh with equal resistors connected between the nodes, seen in Fig. 1.3, also simulated the solution of Laplace's equation. In any node Kirchhoff's law,

$$\sum I_i = 0,$$

is valid, so

$$\frac{V_1 - V_0}{R} + \frac{V_2 - V_0}{R} + \frac{V_3 - V_0}{R} + \frac{V_4 - V_0}{R} = 0,$$

or

$$V_0 = \frac{1}{4}(V_1 + V_2 + V_3 + V_4).$$

Later we will see that this last equation is a five-node approximation to the solution of Laplace's equation. By choosing the resistor values in correspondence to an axis of rotational symmetry, these problems could also be solved. By sending a current of correct amplitude into some nodes, it was possible to simulate the solution of Poison's equation.

These early efforts gave rise to the development of the analog computer, which was decisively displaced by the digital computer with its speed, memory capacity, and cost, when these become widely available.

There was no proof that the numerical solution to the partial differential equations existed, numerical stability was not very well understood. In a 1928 paper, Courant, Friedrichs, and Lewy [29] showed that if the original differential equations were replaced with a set of equations corresponding to the numerical approximation, arbitrarily close in a well-understood sense, the numerical solution might not have any relationship to the true solution. This astonishing result was forgotten until von Neumann took it up again in the mid-1940s [30]. He developed a heuristic procedure, generalizing the work of Courant, Friedrichs and Lewy, which opened the possibility to test numerical stability. His work was later expanded, and rigorous proofs were given as to which conditions must be fulfilled for a system of equations to be solved by iteration methods. Young [31] formulated a set of conditions which, if fulfilled, make certain that the solution of the linear system will converge towards a true solution. However, for nonlinear systems there is no proof that such a solution corresponds to the solution of the physical problem.

Not waiting for the mathematical niceties, the scientific community used iteration processes to solve field problems in mechanics and electromagnetism. The work, done by hand and using mechanical computing machines, caused some ingenious methods to be devised to check these computations. Tens or hundreds of thousands of additions and multiplications had to be done. Such computations were major parts of many a doctoral thesis in the 1940s and involved a year or more of painstaking work.

The computers completely changed the situation. Computational errors were no longer a problem. The number of arithmetic operations increased a thousandfold, while the time needed to obtain the solution decreased by a similar factor. By the end of 1950s the first programs were written which constitute the stem for modern versions. Today our programs solve in three dimensions electric and magnetic fields and the wave equation. These programs had completely and forever changed the conditions for constructing modern electron devices. The same computation, which previously took a year, can today be made in a few seconds on a work station.

The motion of charged particles is determined by electric and magnetic field distributions; these fields are dependent on the boundary conditions on electrodes, on electrical currents, and on space-charge. Only in some simple cases is it possible to compute field distribution analytically. Generally the boundary conditions are such that it is all but impossible to obtain an analytical solution.

Numerical methods differ in important respects to the analytical ones. First the problem must be solved in a finite *"universe,"* inside the limits of a mesh of crossing lines, as seen in Fig. 1.4. These lines cross each other in points called *nodes*. The universe is thus divided in small mesh elements of different form, triangular, rectangular, or other. The nodes can either lie on the boundary or be defined only inside the universe. In the later case, the nodes directly outside the universe are classified separately to store the information about the boundary conditions. Inside every mesh element, different parameters such as potential, space-charge, and permeability are assumed to be constant or vary linearly, quadratically, or whatsoever.

Instead of a continuous function, as in the analytical solution, a discrete function is obtained and its numerical values have the following properties:

- The value is known only at the nodes.
- The boundary conditions determine the value near the boundary.
- The value at any node is a function of the values at its nearest neighbors. The type of differential equation decides the functional form.

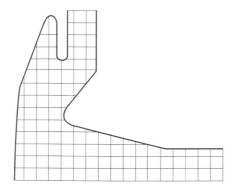

Fig. 1.4. "Universe" with nodes and meshes.

The third property shows that any unknown potential can be computed by solving a system of simultaneous equations. This solution is unique and converges to the analytical solution when the distance between the nodes approaches zero if some conditions are fulfilled [31].

Numerical methods to solve partial differential equations have been developed, of which the **Finite Difference Method** and the **Finite Element Method** are the most widely used.

In the first method, differentials are replaced by differences à la C. F. Gauss. The constructors of accelerators in 1950s were the first to use the method and wrote the first programs.

In the second, the finite element method, the universe is divided into small mesh elements. For each mesh element a physical quantity, for example the stored energy, is computed. The variational principle, along with the requirement that the total energy should be minimum, results in a system of discrete equations, which have to be satisfied for every mesh element. The method was proposed by Courant [32] but was all but forgotten until the mid-1960s, when it was used in the computation of mechanical structures.

Both methods have in common that they require the simultaneous solution of a system of equations.

The computer programs for FDM and FEM are often divided into three parts, as seen in Fig. 1.5. In the first part, the geometry of the problem and the boundary conditions are treated. In the second part the solution of the system of simultaneous equations is made. In the third part a graphic presentation of the computed potential, field, or other relevant parameters is presented.

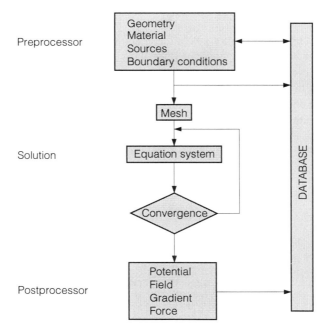

Fig. 1.5. Simplified flow diagram for FDM and FEM programs.

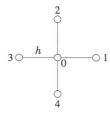

Fig. 1.6. Definition of a node and its four neighbors.

1.2.1. Finite Difference Method

The finite difference method (FDM) is suited to solve partial differential equations with scalar potentials. Our discussion will be restricted to two-dimensional problems, either planar or rotationally symmetric. In both cases the universe is divided into small rectangular, but not necessarily quadratic, mesh elements, as seen in Fig. 1.6.

We start with Cartesian coordinates; for simplicity a quadratic mesh with $\Delta x = \Delta y = h$ is assumed.

A node in the mesh has four neighboring nodes, as seen in Fig. 1.6, at each of which the potential can be expanded in a power series,

$$V_1 = V_0 + h\frac{\partial V_0}{\partial x} + \frac{h^2}{2}\frac{\partial^2 V_0}{\partial x^2} + \cdots.$$

Neglecting all terms of higher than first order we obtain

$$\frac{\partial V}{\partial x} \approx \frac{V_1 - V_0}{h} \simeq \frac{V_0 - V_3}{h}$$

and

$$\frac{\partial^2 V}{\partial x^2} \approx \frac{1}{h}\left(\frac{V_1 - V_0}{h} - \frac{V_0 - V_3}{h}\right) = \frac{1}{h^2}(V_1 - 2V_0 + V_3). \qquad (1.31)$$

Similar expression is obtained for the y derivative.

These central-difference approximations are introduced in Laplace's equation, Poisson's equation, or the wave equation. For example, in Poisson's equation we obtain

$$\frac{\partial^2 V}{\partial x^2} + \frac{\partial^2 V}{\partial y^2} \approx \frac{1}{h^2}(V_1 + V_2 + V_3 + V_4 - 4V_0) = -\frac{\rho_0}{\varepsilon},$$

giving

$$V_0 = \frac{1}{4}(V_1 + V_2 + V_3 + V_4 + h^2\frac{\rho_0}{\varepsilon}). \qquad (1.32)$$

Here the potential at the central node is the arithmetic mean of the potentials in the four neighboring nodes increased by the space-charge term.

In cylindrical coordinates the $1/r$ term,

$$\frac{1}{r}\frac{\partial}{\partial r}\left(r\frac{\partial V}{\partial r}\right) \approx \frac{1}{r_0 h}\left[\frac{r_0}{h}(V_2 - 2V_0 + V_4) + \frac{V_2 - V_4}{2}\right],$$

causes, for example, Laplace's equation to become

$$V_0 = \frac{1}{4}\{V_1 + (1 + \frac{h}{2r_0})V_2 + V_3 + (1 - \frac{h}{2r_0})V_4\}. \qquad (1.33)$$

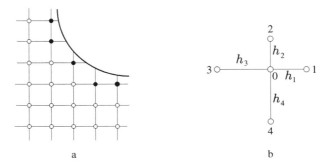

Fig. 1.7. (a) Mesh near the boundary. **(b)** Definition of distances to neighboring nodes. o, regular node; •, irregular node.

These expressions are valid for all the so-called *regular* nodes, those in which all neighboring nodes are located at a distance h. Near the edge of the universe this requirement is violated, as some *irregular* nodes have a smaller distance to the mesh lines, as seen in Fig. 1.7a. Besides, in one case the boundary, for example an electrode, has constant potential, $V(x, y) = V_0$; in another case the value of the derivative of the potential, $\partial V / \partial n$, is known. The first is called a *Dirichlet boundary condition*, whereas the second is referred to as a *Neumann boundary condition*.

According to Fig. 1.7b, the Dirichlet boundary condition gives the following partial derivatives in rectangular coordinates:

$$\frac{\partial^2 V}{\partial x^2} \approx \frac{2}{h_1 + h_3}\left(\frac{V_1 - V_0}{h_1} - \frac{V_0 - V_3}{h_3}\right),$$
$$\frac{\partial^2 V}{\partial y^2} \approx \frac{2}{h_2 + h_4}\left(\frac{V_2 - V_0}{h_2} - \frac{V_0 - V_4}{h_4}\right). \tag{1.34}$$

Laplace's equation for an irregular node then takes the form

$$V_0 = \frac{1}{D}(a_1 V_1 + a_2 V_2 + a_3 V_3 + a_4 V_4), \tag{1.35}$$

with

$$a_1 = \frac{h_2 + h_4}{h_1}, \qquad a_2 = \frac{h_1 + h_3}{h_2}, \qquad a_3 = \frac{h_2 + h_4}{h_3}, \qquad a_4 = \frac{h_1 + h_3}{h_4},$$

and

$$D = a_1 + a_2 + a_3 + a_4.$$

Similar expressions can be obtained for cylindrical coordinates, allowing for the $1/r$ term. Instead of a_2 and a_4 introduce

$$b_2 = a_2\left(1 + \frac{1}{r_0 a_2}\right) \quad \text{and} \quad b_4 = a_4\left(1 - \frac{1}{r_0 a_4}\right)$$

in an equation similar to Eq. 1.35.

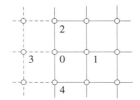

Fig. 1.8. Reflection at Neumann boundary.

For the Neumann boundary condition the value of the derivative for the stationary electric and magnetic fields, $\partial V/\partial n$, is zero, and thus can simply be obtained by reflection, as seen in Fig. 1.8:

$$V_3 - V_1 = 0, \qquad V_3 = V_1.$$

We therefore choose the Neumann boundary to coincide or at least to be parallel with the node lines. When solving the wave equation inside a cavity (Neumann boundary condition apply to all walls), the expressions for the coefficients of the irregular nodes become rather complicated. Note that the symmetry axis in planar and rotationally symmetric problems is a Neumann boundary condition.

Note also that the differential equation with the central-difference approximation, valid for the central node and its four neighboring nodes, is not the only possibility. Sometimes the central node and its eight neighboring nodes are used. The local error is reduced by one third, with a corresponding increase in computation time. In three dimensions the central-difference approximation usually takes into account the six neighboring nodes.

In most modern FDM programs it is enough to specify the boundary conditions as geometrical coordinates by declaring the type, Dirichlet or Neumann, and the mesh size. The code then lays out the mesh and computes the coefficients $a_{n1} \ldots a_{n4}$ and D_n for all irregular nodes and puts, if quadratic mesh is used, all the a_{nn} coefficients equal to 1.

For each node one obtains one equation with constant coefficients, a_{ij}, for the whole system a matrix equation

$$\mathbf{A}\,\mathbf{V} - \mathbf{B} = 0. \qquad (1.36)$$

Here $\mathbf{A}$ is a matrix of coefficients a_{ij} with $N \times N$ elements, N being the number of nodes in the universe, $\mathbf{V}$ and $\mathbf{B}$ are vectors with length N. In Laplace's equation all B_i are equal to zero and are given by the space-charge part in Poisson's equation. In any row of the $\mathbf{A}$ matrix, only five elements differ from zero. Such matrices are called *sparse* and are suitable for iterative solution. When actually solving the Laplace's equation, it is advisable to normalize the matrix by dividing all five coefficients by the diagonal element of the row, a_{ii}. For any row with k columns the ith row will have the form

$$0 \ldots 0 \quad a_{i,i-k} \quad 0 \ldots 0 \quad a_{i,i-1} \quad 1 \quad a_{i,i+1} \quad 0 \ldots 0 \quad a_{i,i+k} \quad 0 \ldots 0.$$

Storing the complete $\mathbf{A}$ matrix with $N \times N$ elements would require an unnecessarily large part of computer memory, so only those coefficients, which are $\neq 0$ or $\neq 1$, are stored — that is, those coefficients associated with irregular nodes. For all regular nodes, only four coefficients need be stored in Cartesian coordinates, or four coefficients per row in cylindrical coordinates, while other vectors store the information on the ordinal numbers of irregular nodes, for every row the first and last regular node in a sequence of regular nodes,

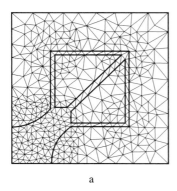
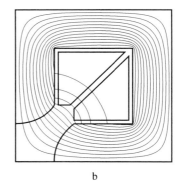

a   b

Fig. 1.9. (a) Triangular mesh for a FEM solution of a magnetic quadrupole quadrant and **(b)** the computed magnetic field lines.

and the information on potential values of the boundary nodes. This is all that is needed to obtain an iterative solution and requires much less space in the computer memory.

1.2.2. Finite Element Method

The problem to be solved, although mathematically continuous, must be made discrete for the finite element method by dividing the universe into elements, usually triangular. In contrast to the FDM, the mesh elements are deformed near the boundary so that at least one of the triangle corners lies on the boundary, as seen in Fig. 1.9a.

For any mesh element the stored energy (electric, magnetic, elastic, thermal, or gravitational) can be computed. Since energy is a scalar quantity, the sum of the energy in all mesh elements is equal to the total energy in the universe. It is therefore possible to obtain the differential equations, which describe the system using the variational principle. This means that Eq. 1.12 or 1.20 must be applied to each mesh element.

For a magnetic field with a rotational symmetry ($|\mathbf{A}| = A_\theta = A$, $|\mathbf{J}| = J_\theta = J$), for example, we obtain

$$W = \int \int \int [\mathbf{J} \cdot \mathbf{A} - \frac{1}{2\mu}(\nabla \times \mathbf{A})^2]dv, \tag{1.20}$$

which in equilibrium for an isolated system must be minimal. Small variations in A change W by $\sim (dA)^2$, so it can be assumed that A varies nearly linearly inside each triangle,

$$A = u + vr + wz. \tag{1.37}$$

As seen in Fig. 1.10, the derivatives of the potential for the node 0 can be computed from the values of the potential in the six neighboring nodes. For example, by using the triangle 012 we obtain

$$\begin{aligned}
\frac{\partial A}{\partial r} &= v = \frac{(A_1 - A_0)(z_2 - z_0) - (A_2 - A_0)(z_1 - z_0)}{(r_1 - r_0)(z_2 - z_0) - (r_2 - r_0)(z_1 - z_0)}, \\
\frac{\partial A}{\partial z} &= w = \frac{(A_1 - A_0)(r_2 - r_0) - (A_2 - A_0)(r_1 - r_0)}{(r_1 - r_0)(z_2 - z_0) - (r_2 - r_0)(z_1 - z_0)},
\end{aligned} \tag{1.38}$$

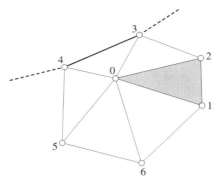

Fig. 1.10. Triangles around the node 0.

while the coefficient u is simply

$$u = \frac{(A_0 + A_1 + A_2)}{3}.$$

By integrating W over triangle 012 we obtain

$$W_{012} = \left[Ju - \frac{1}{2\mu}(v^2 + w^2) \right] S, \tag{1.39}$$

where S is the area of the triangle. Similar expressions can be obtained for the other five triangles with common corners to node 0.

According to the variational principle the energy of the system does not change if a small deviation, dA_0, is introduced at the node 0 and the potentials in the neighboring nodes are kept constant,

$$\frac{\partial W}{\partial A_0} = \frac{\partial}{\partial A_0}(W_{012} + W_{023} + \cdots + W_{061}) = 0. \tag{1.40}$$

If Eq. 1.39 and corresponding equations for other triangles are introduced in Eq. 1.40, a differential equation is obtained which, in the same form, is valid for every node in the universe. In this differential equation, enter as parameters u, v, and w and the corner coordinates of all triangles.

Accordingly, for every ith node an equation with constant coefficients, c_{ij}, is obtained, and for the whole universe a matrix is obtained:

$$\mathbf{CA} - \mathbf{B} = 0. \tag{1.41}$$

The matrix of the coefficients is $\mathbf{C}$, $\mathbf{A}$ and $\mathbf{B}$ are vectors of a length N, and B_i includes term with current density in the ith node. In any row of the matrix $\mathbf{C}$, only seven elements differ from zero.

1.2.3. Solution of Simultaneous Systems

In both FDM and FEM a system of simultaneous equations is obtained, the solution of which results in a potential value in each node. These equations are linear for a Laplace equation, while for the space-charge, magnetic field problem and wave equation they are nonlinear (variable ρ, ε, or μ, or unknown k). The system has a great number of equations;

in modern applications there are up to a few hundred thousands nodes. A direct solution of this system by known methods (Gauss, Cholewsky, QR-solve) cannot be obtained because of computer time and memory limitations. So it is advisable to limit direct solutions to about 2000–3000 nodes. One must not forget that the matrix of the coefficients is sparse; such systems can be solved faster and with a smaller amount of computer memory by iterative methods.

The theory of iterative solutions of linear system of equations is well developed. Nonlinear systems require caution, especially when using a direct method — for example, that of Newton (danger for oscillations) — and so it is customary to linearize them by iteratively solving for the potential while holding ρ, μ, or k constant. Only when the value of the potential has converged are new values of ρ, μ, or k computed, and then the next value of the potential is calculated.

1.2.3.1. Linear Problems

Iterative methods successively solve for the potential node by node. The solution, **Y**, of the system is obtained from the matrix equation,

$$\mathbf{C\,Y} = \mathbf{B}, \tag{1.42}$$

where, from the beginning, **Y** differs from zero only in certain nodes, those with a Dirichlet boundary condition in the electrostatic or magnetostatic case, and nodes with nonvanishing current density in the case of stationary magnetic field. For all other nodes, it is customary to put a zero at the start. In turn, node by node, a corrected value is computed. The ith equation

$$\sum_j c_{ij} y_j = b_i, \qquad 1 \leq j \leq N,$$

is solved for y_i. It was proved that the equations must be solved either node by node (Gauss-Seidel method) or row by row for the iterative method to remain stable.

When in the first iteration the potential is computed for all nodes, the process continues to the next iteration. In each kth iteration and for each node, the residual, $\Delta_i = y_i^{(k+1)} - y_i^{(k)}$, is added to $y_i^{(k)}$. This iteration procedure is very slow. To speed up the convergence, the residual can be multiplied by a so-called overrelaxation factor, ω, such that

$$y_i^{(k+1)} = y_i^{(k)} + \omega \Delta_i, \tag{1.43}$$

$$\Delta_i = [\frac{b_i}{c_{ii}} - \sum_1^{i-1} \frac{c_{ij}}{c_{ii}} y_j^{(k+1)} - \sum_{i+1}^{n} \frac{c_{ij}}{c_{ii}} y_j^{(k)} - y_i^{(k)}], \tag{1.44}$$

where in the first sum the already corrected values are included, while in the second the old values from the previous iteration appear. In an FDM process the **C** matrix can already be normalized so that all diagonal elements are unity, $c_{ii} = 1$. The method is known as Successive **O**ver**R**elaxation (SOR) method.

For each problem there is an optimal overrelaxation factor, ω_0, where

$$1 \leq \omega_0 < 2.$$

Figure 1.11 shows the number of iterations as a function of the overrelaxation factor when solving a Laplace problem with $\sim$10,000 nodes, with ω_0 typically between 1.75 and 1.85.

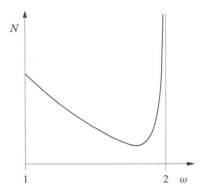

Fig. 1.11. Variation of the number of iterations, N, with the overrelaxation factor, ω.

It is important to make the correct choice of ω_0. That is the weak point of SOR. There is no way to find the optimal value directly. Various schemes, described in the computing literature, have been adopted to find the optimum.

Considering Eq. 1.42 the matrix $\mathbf{C}$ can be split in

$$\mathbf{C} = \mathbf{L} + \mathbf{D} + \mathbf{U},$$

where $\mathbf{D}$ is the diagonal part of $\mathbf{C}$, $\mathbf{L}$ the lower triangular part and $\mathbf{U}$ the upper triangular part of $\mathbf{C}$. Both $\mathbf{L}$ and $\mathbf{U}$ have zeros on the diagonal. In FDM process the matrix $\mathbf{D}$ is an identity matrix. During the iteration the matrix $-\mathbf{C}^{-1}(\mathbf{L} + \mathbf{U})$ maps one set of values of $\mathbf{Y}$ into the next one. This iteration matrix has eigenvalues, which show how the residuals are suppressed during one iteration. The convergence is determined by the decaying of the largest eigenvalue. Its modulus is called the *spectral radius*, ρ_j. The optimal overrelaxation factor can be chosen as [33]

$$\omega_0 = \frac{2}{1 + \sqrt{1 - \rho_j^2}}.$$

Spectral radius can be computed for some special cases analytically — for example, on a square with Dirichlet boundary conditions. In practical computations an approximation

$$\omega_m = \frac{2}{1 + \sqrt{1 - \dfrac{(x_M + \omega - 1)^2}{\omega^2 x_M}}}, \tag{1.45}$$

is often used, where ω is an arbitrary initial value ($1 \leq \omega < 2$) and x_M is a characteristic quantity for the problem. x_M can be found by saving from iteration to iteration
(a) the maximum absolute value of the residual vector Δ,
(b) the sum of the absolute values of all Δ_i components, or
(c) the length of Δ.
x_M can then be computed by dividing the computed value a, b, or c from the iteration $k+1$ by that from iteration k. ω_m must be kept constant for a certain number of iterations to avoid oscillations. Usually alternative b gives the fastest convergence. When ω_m no longer changes appreciably, ω_0 is set equal to ω_m and the iteration is continued until convergence is obtained.

Brandt [34] introduced in 1977 a new method, the multigrid method. The method solves the problem successively on finer and finer mesh. It starts again with the linear system

$$\mathbf{C}_h \, \mathbf{Y}_h = \mathbf{B}_h, \tag{1.46}$$

where subscript h denotes that the system represents a discretization with the mesh size h. Assume that an approximation to the solution is $\overline{\mathbf{Y}}_h$. If $\mathbf{Y}_h$ denotes the exact solution of Eq. 1.46, the error is

$$\varepsilon_h = \mathbf{Y}_h - \overline{\mathbf{Y}}_h. \tag{1.47}$$

The residual, often called *defect,* is defined as

$$\mathbf{d}_h = \mathbf{C}_h \, \overline{\mathbf{Y}}_h - \mathbf{B}_h. \tag{1.48}$$

Because $\mathbf{C}_h$ is a linear operator the error can be expressed by

$$\mathbf{C}_h \, \varepsilon_h = -\mathbf{d}_h. \tag{1.49}$$

In a coarser mesh — for example, $H = 2h$ — Eq. 1.49 becomes

$$\mathbf{C}_H \, \varepsilon_H = -\mathbf{d}_H. \tag{1.50}$$

The important point is that $\mathbf{C}_H$ has only 1/4 nodes as compared to the initial problem. To pass from the finer to the coarser mesh a transformation is needed. Somehow, every node in the finer mesh must influence the coarser mesh. This is done by a restriction operator, $\mathbf{R}$, which connects the residuals on both meshes

$$\mathbf{d}_H = \mathbf{R} \, \mathbf{d}_h. \tag{1.51}$$

Reciprocally, when going from the coarse to the fine mesh, another operator, called *prolongation operator* or *interpolation operator,* $\mathbf{P}$, is necessary to obtain the error distribution

$$\overline{\varepsilon}_h = \overline{\mathbf{P}} \, \overline{\varepsilon}_H. \tag{1.52}$$

Both operators can be linear; for example,

$$\mathbf{R} = \begin{pmatrix} \frac{1}{16} & \frac{1}{8} & \frac{1}{16} \\ \frac{1}{8} & \frac{1}{4} & \frac{1}{8} \\ \frac{1}{16} & \frac{1}{8} & \frac{1}{16} \end{pmatrix} \tag{1.53}$$

and

$$\mathbf{P} = \begin{pmatrix} \frac{1}{4} & \frac{1}{2} & \frac{1}{4} \\ \frac{1}{2} & 1 & \frac{1}{2} \\ \frac{1}{4} & \frac{1}{2} & \frac{1}{4} \end{pmatrix}. \tag{1.54}$$

Generally, whenever $H = 2h$, $\mathbf{R}$ should be adjoint operator to $\mathbf{P}$. When $\overline{\varepsilon}_h$ is known the solution can be updated

$$\mathbf{Y}_h^{new} = \overline{\mathbf{Y}}_h + \overline{\varepsilon}_h. \tag{1.55}$$

One step of this coarse-mesh correction scheme is:
- Compute the error on the fine mesh from Eq. 1.47.
- Use the restriction operator to obtain the residua! on the coarse mesh, Eq. 1.51.

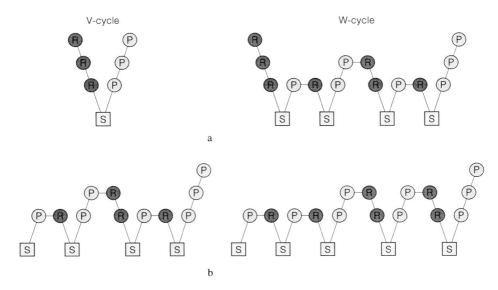

Fig. 1.12. Patterns of multigrid cycles for a solution with 4 mesh sizes. R and the descending line (\) denote restriction, P and the ascending lines (/) denote prolongation, and S denotes the exact solution. **(a)** Original multigrid method. **(b)** Full multigrid algorithm.

- Compute the solution on the coarse mesh by solving Eq. 1.50.
- Use the interpolation operator to compute the correction on the fine mesh using Eq. 1.52.
- Obtain the next approximation by Eq. 1.55.

The process can easily be widened to include more than two mesh sizes. The exact solution on the coarsest mesh can then be obtained by some of the direct matrix inversion methods. By calling the process from the finest to the coarsest mesh and back a cycle, two computational patterns evolved: the V-cycle and the W-cycle. The V-cycle is faster, but the W-cycle gives a better solution. Figure 1.12a shows the order of the computation processes for those cycles when four mesh sizes are used.

After the introduction in the 1970s, the method was further developed into the **Full MultiGrid** Algorithm. The process starts by an exact solution on the coarsest mesh, and it continues with cycles with finer and finer mesh (Fig. 1.12b). The important difference is that the right-hand side of Eq. 1.42 must be known for all mesh sizes.

SOR must not be used in multigrid methods. The error during the relaxation process can be analyzed by the Fourier series, and it will show both low- and high-frequency components. In a Gauss-Seidel process ($\omega = 1$) with a small mesh size (i.e., with a large number of nodes), the low-frequency components will only slightly be changed in each iteration, while the high-frequency components are smoothed by a large factor. All amplitudes of the Fourier series are amplified in SOR, which explains the faster convergence. In multigrid methods, error components which have a wavelength less than $2H$ cannot be observed and thus not reduced on the coarser grid. On the coarser grid only the errors, which correspond to the low frequency components on the fine grid, are actually reduced. Upon the transition to the finer grid the high-frequency components are smoothed so that both the high- and the low-frequency components become small. As the solution on the coarse grid takes a much shorter computing time, multigrid methods have an advantage in speed (in a problem with 100,000 nodes the solution can be obtained about 10 times faster). However, the presentation of irregular boundaries is more difficult, because the restriction and prolongation operators must be treated in a more complicated manner.

1.2.3.2. Accuracy Considerations

The accuracy of the solution depends on many factors. If the simultaneous system of equations could be solved exactly, the solution would only be approximate because of many reasons — for example, mesh size. It is desirable to work with small meshes, especially when the results will be used in orbit computations. However, decreasing the mesh size results in a rapid increase in computer memory and computation time. In the first approximation the computation time increases with the number of iterations N as $N \ln N$ if SOR is used or approximately as N in FMG. Thus reducing the mesh size by a factor of two will increase the computation time either sixfold or twofold. Also the shape of the mesh elements is important, especially for the triangular elements in FEM. If the program generates many highly nonequilateral triangles, the solution in these regions will loose accuracy. Also somewhat different solutions result when triangular, rectangular, or quadratic mesh elements are used.

If different methods are used in different parts of the universe, upon the transition boundary the solutions will not be continuous, and thus errors must be introduced. This is the case with some three-dimensional codes, which compute the magnetic field and use vector potential inside the iron and scalar potential outside of it.

Even the solution of the linear system of equations introduces errors. In the iteration process some characteristic quantity changes — for example, the residual vector's largest absolute value, its length, or the sum of the absolute values of its components. This quantity must be used to decide when the process should be stopped. In practice the process converges when the potential changes in the sixth or seventh digit. More prolonged iterations produce only oscillation of the potential values. But even this is uncertain for if the derivative of the solution in a node is small, the residual value will also be small. Experience, test runs, and finally comparison with measurements must be the guide when using these programs.

Finally, the computer noise influences the results. The solution gives the potential, but it is the field strength that is needed. Since the field strength is the derivative of the potential, the accuracy of the potential solution must be great. Generally double precision in floating point operations is required.

1.2.3.3. Nonlinear Problems

The space-charge inside an electron beam region, the permeability inside iron, the dielectric constant inside an isolator, and the resonant frequency of a cavity depends on the field distribution. Moreover, they are nonlinear functions of the field strength. Thus the coefficients of the C matrix, or equivalently B, will vary as the iterative solution is computed because they are functions of y_i. There is no mathematical proof that an iterative solution to a nonlinear system exists, but the experience has shown that it can be found with some restrictions. One restriction is that the mesh elements must be similarly shaped, a condition automatically fulfilled in FDM. However, in FEM the individual mesh elements, usually triangles, must be checked near the boundary. Many FEM programs build in a feature to optimize the shape of the mesh elements.

Two alternate procedures can be used to solve nonlinear problems. In the first and most popular, the problem is linearized, keeping the space-charge, permeability, dielectric constant, or cavity frequency constant while solving Eq. 1.42. Here the convergence is more dependable and oscillation tendency is less pronounced. In the second, in each iteration Newton's method is used to compute the corresponding quantity.

Multigrid methods can also be used to solve nonlinear problems. Both procedures can be used, namely linearization and Newton's method. But a great strength is that FMG can directly be applied, if a suitable nonlinear relaxation to smooth the errors, and a restriction procedure to make correction on the coarser mesh is made. A FAS algorithm (**F**ull **A**pproximation **S**torage algorithm) was developed in which no nonlinear equations must be solved except on the coarsest mesh.

(a) The Linearized Underrelaxation

The procedure employed depends on the type of problem to be solved. In Poisson's equation with space-charge, nothing is initially known about the particle positions, which dictate the space-charge distribution. With Laplace's equation and zero space-charge the potential is computed first, and then the electric field and particle orbits. This gives the space-charge at all nodes. Next, this linear system is solved, but this time with the right-hand side different from zero — that is, Poisson's equation. For magnetic fields a constant μ is initially assumed, the linear system is solved, and new μ value is computed at every node. When the μ values are corrected and when the permeability does not appreciably change, the computation is finished. In computing a resonant cavity a guessed value for the resonant frequency, f_0, is used followed by a solution of the linear system, which adjusts the frequency. The computation can be stopped when the change becomes sufficiently small.

With this method of solving a nonlinear system the solution sometimes oscillates so strongly that it diverges. An underrelaxation of the coefficients c_{ij} or b_i, depending on the problem, serves to keep these oscillations under control. Experience shows that in computing the solution of Poisson's equation with space-charge, the underrelaxation factor, β, should be kept between 0.5 and 0.8. In every node the new space-charge value should be computed according to

$$\rho_i^{(k+1)} = \rho_i^{(k)} + \beta[\rho_i'^{(k+1)} - \rho_i^{(k)}],$$

where $\rho_i'^{(k+1)}$ is the value of ρ_i at the end of the iteration k, when the solution of the linear system converges, and $\rho_i^{(k)}$ is the value at the beginning of the iteration. When solving the magnetic field, μ is underrelaxed with β between 0.1 and 0.2 since the B–H curve is strongly nonlinear in some parts, and

$$\mu_i^{(k+1)} = \mu_i^{(k)} + \beta[\mu_i'^{(k+1)} - \mu_i^{(k)}].$$

For the cavity resonant frequency a suitable underrelaxation factor is 0.5.

(b) The Nonlinear Overrelaxation Method

In nonlinear overrelaxation procedure, Newton's method is used to compute the change of the coefficients c_{ij} or b_i, depending on the type of problem, in every node in each iteration. The more stable *regula falsi* method is generally not used because of large demands on computer memory.

As an example we take a magnetic field problem where in a node

$$c_0 y_0 = \sum c_i y_i + b, \tag{1.56}$$

and y_i is assumed to be constant, while c_i is changing with y_0. This equation can be solved using Newton's method:

$$(y_0 + dy_0) \cdot (c_0 + \frac{\partial c_0}{\partial y_0} dy_0) = \sum c_i y_i + \sum \frac{\partial c_i}{\partial y_0} y_i dy_0.$$

We can compute dy_0 as follows:

$$dy_0 = \frac{\sum (c_i y_i)/c_0 - y_0}{1 + \frac{\partial c_0}{\partial y_0} \frac{y_0}{c_0} - \frac{(\sum \partial c_i / \partial y_0) y_i}{c_0}}. \tag{1.57}$$

The numerator shows the linear change in y_0, and the denominator improves this value by accounting for the nonlinearity. y_0 is first computed using Eq. 1.56 and dy_0 from Eq. 1.57. Next $y' = y_0 + dy_0$ is computed, and this result is overrelaxed with

$$y_0^{(k+1)} = y_0^{(k)} + \omega [y' - y_0^{(k)}],$$

where the overrelaxation factor, ω, usually lies between 1.3 and 1.6.

1.2.4. Comparison Between FDM and FEM

With both FDM and FEM programs, which can solve the problem, the choice depends upon the application. Memory and time requirements do not differ much, so the available computer resources do not decide the choice. A comparison of the solutions shows that they do not differ much. Some problems, like Laplace's, Poisson's, or the wave equation, can be solved with either method. However, magnetic field problems are more suited to FEM solution because FDM uses scalar potentials and permits only one closed region.

The rectangular mesh of FDM compared with the commonly used triangular of FEM, shows the advantage of the latter because an increase in the mesh elements density where the best accuracy is needed is readily possible. The boundary conditions can even be better approximated since triangles are easily deformed near the boundary, so that all nodes become regular. This deformation, however, must not be too large, since oblong triangles with small angles worsen the interpolation. Most FEM codes therefore first build the mesh, then use an iterative procedure, which optimizes the triangle shape. The FDM rectangular meshes are suitable for geometries which have many straight lines, so their solutions are more accurate than those of the FEM. Curved boundaries, however, result in many irregular nodes, which require separate defining equations. If h is the mesh size, then the local error in regular nodes is proportional to h^4, and at irregular nodes it is proportional only to h^3. Matrix coefficients c_{ij} can sometimes violate the condition of diagonal dominance [31], needed to guarantee that the solution of a linear simultaneous system of equations is convergent.

We emphasize that both the FDM and the FEM solve for potential while when computing particle orbits the field strength is needed. Thus, numeric derivatives must be computed and interpolated, which in each process decreases the accuracy of the results.

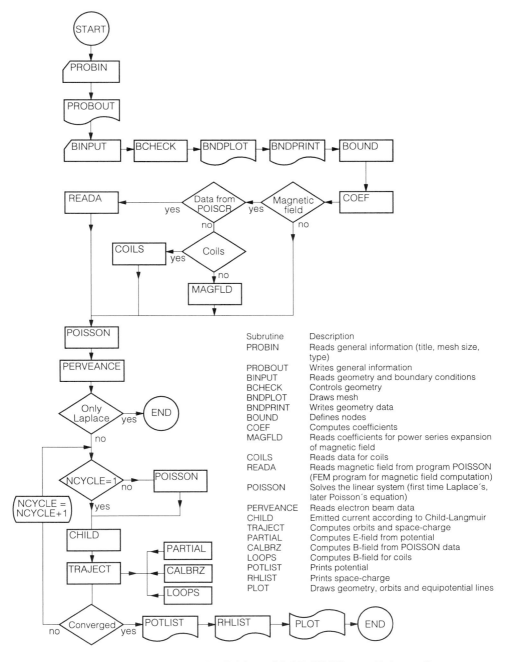

Fig. 1.13. Simplified flow diagram of a slightly modified E-GUN Electron Trajectory Program. The program has ~14,000 FORTRAN statements and calls 63 subroutines. The memory requirement for 50,000 nodes is ~1.5 Mbyte. A simple problem with ~10,000 nodes takes less than 1 minute to compute on a work station.

Numeric derivatives on a mesh with equal size of the quadratic or rectangular elements produce a better approximation than the mesh with triangles of different size and shape. Therefore the FEM uses different smoothing methods. For example, to compute the field components in equidistant intervals, harmonic analysis with numerical damping of high frequency terms is used. From these smoothed field components the particle orbits are computed. This process requires much computer time. When an electron optical lens or an electron gun is computed, and space-charge must be allowed for, the smoothing must be done repeatedly. The use of the FDM methods is therefore recommended,* with a FEM solution to the stationary magnetic field distribution, which needs to be computed only once.

The FDM and FEM programs are large and complex, and the language is usually FORTRAN or C. A two-dimensional (2-D) code has some 10,000 statements, and a 3-D program has still more. Figure 1.13 shows a simplified flow diagram of the E-GUN program,** a typical 2-D program, which computes the electrostatic field and electron or ion orbits, considering space-charge and magnetic field.

Today many different programs exist which solve stationary electric and magnetic field problems. Some, usually 2-D programs, can be obtained free of charge from, for example, CERN or other large public institutions, while others are proprietary and can only be rented or purchased. Industries have developed their own programs to solve complicated 3-D geometries. An example is CERN's quadrupole magnet for their antiproton collector as seen in Fig. 1.14.

With the increasingly ready access to large supercomputers the optimization in 3-D of a magnet, a lens, an electron gun, or a cavity is possible, even when the computations must be repeated with a variety of parameters.

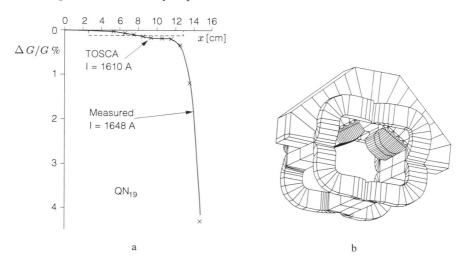

a   b

Fig. 1.14. (a) Normalized magnetic field gradient. CERN's quadrupole magnet for antiproton collector. (b) 3-D prisms for the numerical solution; only the upper half of the iron is shown. Program TOSCA. (After C. W. Trowbridge, *Status of Electromagnetic Field Computation,* Rutherford Appleton Lab, with permission.)

---

* Kasper, E., "Numerical design of electron lenses," *Nucl. Instr. Meth.*, **175**, 187 (1981).
** Herrmannsfeldt, W. B., "Electron Trajectory Program," *SLAC Report 331*, Stanford, 1988.

Other methods have been developed which can solve partial differential equations, examples of which are integral equation systems and the Monte-Carlo methods. The latter can be suitable for special cases, but neither has achieved the extension that the FDM and the FEM have.

With the advent of massively parallel processors and new methods in the software, problems of large complexity will be solved, problems where simultaneously electrical, mechanical, and thermal properties will be considered. It seems, however, that the solution of these problems will rely on the discretization of the field equations with known techniques of the FDM and the FEM.

1.3. PARTICLE ORBITS

1.3.1. Equation of Motion

The position of a charged particle is completely determined by three coordinates — for example, in the rectangular coordinates with x, y, z. The particle velocity, $\mathbf{v}$, also has three components, $\dot{x}$, $\dot{y}$, $\dot{z}$.

The particles motion in electric and magnetic fields is described by the relativistic equation of motion

$$\frac{d}{dt}(m\mathbf{v}) = \frac{d}{dt}\left[\frac{m_0\mathbf{v}}{\sqrt{1-v^2/c^2}}\right] = q\,\mathbf{E} + q\,\mathbf{v} \times \mathbf{B}, \tag{1.58}$$

where m_0 is the particle rest mass, q its charge, and c the velocity of light.

By scalar multiplication of Eq. 1.58 with $\mathbf{v}$ we obtain

$$\mathbf{v} \cdot \frac{d}{dt}\left[\frac{m_0\mathbf{v}}{\sqrt{1-v^2/c^2}}\right] = q\,\mathbf{E} \cdot \mathbf{v} + q\,\mathbf{v} \cdot (\mathbf{v} \times \mathbf{B}),$$

where

$$\mathbf{v} \cdot (\mathbf{v} \times \mathbf{B}) = 0,$$

because $\mathbf{v} \times \mathbf{B}$ is perpendicular to both $\mathbf{v}$ and $\mathbf{B}$. We introduce, on the right side,

$$\mathbf{v} = \frac{d\mathbf{s}}{dt},$$

with the result

$$\frac{d}{dt}\left(\frac{m_0c^2}{\sqrt{1-v^2/c^2}}\right) = q\,\mathbf{E}\,\frac{d\mathbf{s}}{dt}.$$

In terms of electric potential, V, the above equation of motion becomes

$$\frac{d}{dt}\left(\frac{m_0c^2}{\sqrt{1-v^2/c^2}} + qV\right) = 0. \tag{1.59}$$

When integrating this equation we must choose initial conditions, the simplest are

$$V = V_0 = 0, \qquad v = 0 \qquad \text{for} \qquad t = 0.$$

These initial conditions are generally used in physics and electrotechnical engineering. They mean that zero potential is chosen at the cathode, or in an ion source, where the velocity of the particles is zero (neglecting the Maxwell's velocity distribution; see Section 2.2.1).

$$\frac{m_0 c^2}{\sqrt{1 - v^2/c^2}} = -qV + m_0 c^2, \tag{1.60}$$

which can also be rewritten as

$$m_0 c^2 \left[\frac{1}{\sqrt{1 - v^2/c^2}} - 1 \right] = -qV.$$

The left side is the particle's kinetic energy, and the right side is its potential energy, which must be positive, $W_p > 0$, so that the velocity is real. Positive particles can, therefore, move only inside a region where V is negative, whereas negative particles move only where V is positive.

For charged particles the kinetic energy, if given in electronvolts, is numerically equal to the potential, V, and can be compared with the particle rest mass energy $m_0 c^2$ by

$$\xi = \frac{W_p}{m_0 c^2}. \tag{1.61}$$

Velocity can then be expressed as

$$v = c \frac{\sqrt{2\xi + \xi^2}}{1 + \xi}, \tag{1.62}$$

where, when ξ is much smaller than 1,

$$v = \sqrt{\frac{2W_p}{m_0}}, \tag{1.63}$$

and classical mechanics can be used. If, on the contrary, ξ becomes greater then 0.2, relativistic mechanics must be used to solve Eq. 1.58. In many applications, however, classical mechanics is sufficient, so *in the rest of the book, except sporadically, the equation of motion will be used in the framework of classical physics*

$$\mathbf{a} = \frac{d^2 \mathbf{s}}{dt^2} = \eta_q (\mathbf{E} + \mathbf{v} \times \mathbf{B}), \tag{1.64}$$

$$v = \frac{ds}{dt} = \sqrt{-2\eta_q V}, \tag{1.65}$$

where

$$\eta_q = q/m_0.$$

For electrons

$$v = 5.93 \cdot 10^5 \sqrt{V} \text{ m/s},$$

with $m_0 = 9.1083 \cdot 10^{-31}$ kg, which corresponds to an energy of 511 keV. Already at an energy of 7 keV the error is 1 percent if Eq. 1.65 is used for electrons. This means that we are forced to use relativistic formulae when the particles are electrons and acceleration voltages on the order of a few tens of kilovolts are used. Figure 1.15 shows

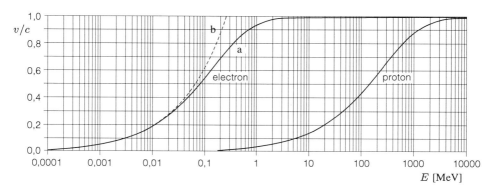

Fig. 1.15. Electron and proton velocity as a function of energy.

how the velocity of electrons changes as a function of energy (curve a). Curve b shows the result which would be obtained with the classical formula. The rest mass of protons is 1.66×10^{-27} kg, which corresponds to an energy of 938 MeV. Only in accelerator technology must we allow for the relativistic effects when the orbits of protons or heavy ions in electric or magnetic field should be computed.

In many practical applications of electron physics, two coordinate systems are mostly used: rectangular and cylindrical. In rectangular coordinates we have

$$\mathbf{v} = \dot{x}\,\mathbf{i} + \dot{y}\,\mathbf{j} + \dot{z}\,\mathbf{k}, \tag{1.66}$$

$$\mathbf{a} = \ddot{x}\,\mathbf{i} + \ddot{y}\,\mathbf{j} + \ddot{z}\,\mathbf{k}, \tag{1.67}$$

$$\mathbf{E} = -\frac{\partial V}{\partial x}\,\mathbf{i} - \frac{\partial V}{\partial y}\,\mathbf{j} - \frac{\partial V}{\partial z}\,\mathbf{k}, \tag{1.68}$$

$$\mathbf{v} \times \mathbf{B} = (v_y B_z - v_z B_y)\,\mathbf{i} + (v_z B_x - v_x B_z)\,\mathbf{j} + (v_x B_y - v_y B_x)\,\mathbf{k}, \tag{1.69}$$

and in cylindrical coordinates we obtain

$$\mathbf{v} = \dot{r}\mathbf{r}_0 + r\dot{\theta}\boldsymbol{\theta}_0 + \dot{z}\,\mathbf{z}_0, \tag{1.70}$$

$$\mathbf{a} = (\ddot{r} - r\dot{\theta}^2)\mathbf{r}_0 + (r\ddot{\theta} + 2\dot{r}\dot{\theta})\boldsymbol{\theta}_0 + \ddot{z}\,\mathbf{z}_0. \tag{1.71}$$

where $\ddot{r}$ is the radial, $r\ddot{\theta}$ the azimuthal, $\ddot{z}$ the axial, $r\dot{\theta}^2$ the centripetal, and $2\dot{r}\dot{\theta}$ the Coriolis' acceleration. For rotationally symmetric problems, $E_\theta = 0$ and $B_\theta = 0$, so

$$\mathbf{E} = -\frac{\partial V}{\partial r}\,\mathbf{r}_0 - \frac{\partial V}{\partial z}\,\mathbf{z}_0, \tag{1.72}$$

and

$$\mathbf{v} \times \mathbf{B} = v_\theta B_z \mathbf{r}_0 + (v_z B_r - v_r B_z)\boldsymbol{\theta}_0 - v_\theta B_r \mathbf{z}_0, \tag{1.73}$$

which yield

$$\ddot{r} - r\dot{\theta}^2 = \eta_q\left(-\frac{\partial V}{\partial r} + v_\theta B_z\right), \tag{1.74}$$

$$r\ddot{\theta} + 2\dot{r}\dot{\theta} = \eta_q(v_z B_r - v_r B_z), \tag{1.75}$$

$$\ddot{z} = \eta_q\left(-\frac{\partial V}{\partial z} - v_\theta B_r\right). \tag{1.76}$$

In the static field case it is only important *where* the particle is at the moment, not *when* it comes there. Thus we can eliminate the time from Eq. 1.64 by using Eq. 1.65.

Starting with the energy equation,

$$v^2 = \dot{x}^2 + \dot{y}^2 + \dot{z}^2 = -2\eta_q V,$$

and choosing z as the independent variable, we obtain

$$\dot{x} = \frac{dz}{dt}\frac{dx}{dz} = \dot{z}x' \qquad\qquad \dot{y} = \frac{dz}{dt}\frac{dy}{dz} = \dot{z}y', \tag{1.77}$$

giving

$$\dot{z}^2(1 + x'^2 + y'^2) = -2\eta_q V,$$

and

$$\dot{z} = \sqrt{\frac{-2\eta_q V}{1 + x'^2 + y'^2}}, \tag{1.78}$$

which we will use in most computations of electron lenses and electron guns.

1.3.2. Numerical Orbit Computations

Ordinary differential equations with known initial conditions describe the motion of charged particles. These equations are not stiff and are numerically solved by either single-step methods (for example, Runge–Kutta methods) or multistep methods, which can use extrapolation to zero step length (Bulirsch–Stoer method). Among the Runge–Kutta methods preference should be given to the Runge–Kutta–Fehlberg variant (or other similar variants), which allows estimating the local error.

Irrespective of which method is used, the second-order equation 1.58 or 1.64 should be reduced to a system of simultaneous first-order equations,

$$\mathbf{y}' = \mathbf{f}(x, \mathbf{y}),$$

where the approximate value of $\mathbf{y}(x)$ is to be found at discrete points,

$$x_n = a + nh, \quad n = 0, 1, 2, \cdots, N, \quad Nh = b - a.$$

Here h is the step length, and a and b are the initial and end points of the computation interval.

In the classical case we obtain six ordinary differential equations, three for the velocities and three for the coordinates. In the relativistic case an extra equation must be added to describe the relativistic mass increase. Most of the programs, which compute particle orbits, do so in the relativistic formulation, because it does not increase appreciably the computing time.

In numerical computations, time t is generally used as the independent variable, transformed in a suitable way. Other variables can be transformed and normalized too, to avoid large exponents of 10. A common transformation is

$$\mu = \frac{m}{m_0}, \qquad \mathbf{w} = \frac{\mathbf{v}}{c}, \qquad \tau = ct, \qquad \mathcal{E} = \frac{\mathbf{E}}{c},$$

and for electrons we have

$$\kappa = \frac{q}{m_0 c} = 586,674 \ [\text{A/N}].$$

The relativistic equations of motion and their transformed correspondences are

| Relativistic equation | Transformed equation |

$$\frac{d}{dt}(mc^2) = q\mathbf{E}\mathbf{v}, \qquad\qquad \frac{d\mu}{d\tau} = \kappa\mathcal{E}\mathbf{w},$$

$$\frac{d(m\mathbf{v})}{dt} = q\mathbf{E} + q\mathbf{v} \times \mathbf{B}, \qquad \frac{d\mathbf{w}}{d\tau} = \frac{1}{\mu}\left[\kappa(\mathcal{E} + \mathbf{w} \times \mathbf{B}) - \mathbf{w}\frac{d\mu}{d\tau}\right], \qquad (1.79)$$

$$\frac{d\mathbf{s}}{dt} = \mathbf{v}, \qquad\qquad\qquad \frac{d\mathbf{s}}{d\tau} = \mathbf{w}.$$

In high-frequency field, $\varphi = \omega t$ can be introduced, where φ is the phase angle.

Whether relativistic or classical equation of motion, energy conservation should be used to verify and improve the velocity components, after every few steps. By interpolating the potential corresponding to a point in the orbit, the velocity can be computed from the energy equation. Since the potential value is more reliable than the electric field strength, which is obtained as a numerical derivative, the velocity vector can be corrected by adjusting its absolute value while keeping its direction unchanged.

1.4. STEP-BY-STEP SIMULATION
OF TIME-VARIABLE PROCESSES

During the past generation, new methods have been developed by plasma physicists, which made possible the simulation of motion of thousands of particles in electric and magnetic fields. These methods, which work step by step, have acquired the popular name "leap-frog" methods because the position and forces which influence each particle are computed at equidistant time intervals, Δt, while the particle velocities are computed in the middle of each time interval, $t + \Delta t/2$. A statistically distributed group of particles is chosen and the space and time derivatives of the electric and magnetic fields are replaced by their finite differences. The relativistic equation of motion, with its corresponding difference equations, is evaluated on a mesh in a manner similar to that described in Section 1.2. Today's different step-by-step programs, from one- to three-dimensional, are large codes (typically >20,000 statements) and suitably must be run on large, fast computers.

These codes are able to:

- Use Newton's laws and the Lorentz force to move the particles.
- Solve Maxwell's equations to obtain field distributions.
- Solve nonlinear problems.
- Follow the particles from arbitrary initial conditions.
- Solve transients.
- Accept inhomogeneous and anizotropic regions,
- Use arbitrary boundary conditions, even if ideal ones are common (limited "universe" and its form).

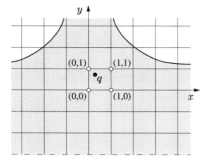

Fig. 1.16. Mathematical mesh for the step-by-step method. A particle with charge q is represented by a space-charge ρ in four neighboring nodes and by a current density, J, along the four sides of the mesh element, which includes the particle.

The programs were developed to simulate the complicated plasma processes, but are now used for simulation of microwave tubes, in electron optics, and in accelerator technology. As an important complement to theory, they now provide understanding of earlier unexplainable phenomena.

For these numerical methods the mesh is similar to the one in Fig. 1.2, in one, two, or three space dimensions, but with time steps, Δt, introduced. The benefits of using a mesh are similar to those in the FDM and FEM methods. With N particles a computation using direct interaction needs N^2 arithmetic operations, and interaction through a mesh needs only N.

Step-by-step methods start at $t = 0$ by solving Maxwell's equations including electrodes, coils, currents and all particles in their initial positions and with their velocities. The particles are then allowed to move step by step under the influence of the computed fields and forces. Hundreds or thousands of particles (even more in three-dimensional problems) are moved for hundreds or thousands of time steps. The computation is done in phase space with the coordinates $\mathbf{x}$ and $\mathbf{v}$. The particle kind can be mixed, say electrons and ions, each named by a subscript, i. $\mathbf{x}$ and $\mathbf{v}$ can take all possible values inside the limits of the universe. Potentials and field components are known only at nodes, marked by a subscript — for example, j — which actually are one, two, or three subscripts in the case of one-, two-, or three-dimensional simulations, respectively. The position and the velocity of the particles determine the space-charge and the current density, implying charge and current distributions since the space-charge is defined only at the nodes, and the current density only along the sides, which connect nodes. After all particles have been moved one time step, the space-charge and the current density are known, and the field equations can be solved. By knowing the field distribution and the position of the particles, the forces on the particles are computed by interpolation. This process is outlined in Fig. 1.17.

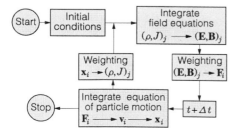

Fig. 1.17. Step-by-step simulation program. The nodes of the mesh are enumerated by j, and the particles are enumerated by i.

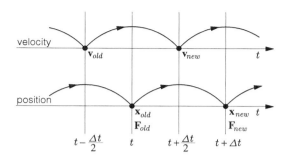

Fig. 1.18. Moving the particles in the leap-frog method. In the jump to the next position the centered velocity is used, in the computation of the velocity the centered force. (After *High-Power Microwave Sources*, edited by Victor L. Granatstein and Igor Alexeff, © 1987, Artech House, Inc., Norwood, MA, with permission.)

In the most common integration method, the so-called leap-frog method, the equation of motion is to be integrated for each particle

$$m\frac{d\mathbf{v}}{dt} = \mathbf{F} \qquad \text{and} \qquad \frac{d\mathbf{x}}{dt} = \mathbf{v}. \tag{1.80}$$

These two equations are replaced by difference equations

$$m\frac{\mathbf{v}_{new} - \mathbf{v}_{old}}{\Delta t} = \mathbf{F}_{old}, \tag{1.81}$$

and

$$\frac{\mathbf{x}_{new} - \mathbf{x}_{old}}{\Delta t} = \mathbf{v}_{new}, \tag{1.82}$$

where the subscript *old* stands for the value at the beginning, and *new* stands for the value at the end of the step. In every step the velocity, **v**, is determined at time $t + \Delta t/2$, while the space coordinates, **x**, are found at time t, as seen in Fig. 1.18. Thus, the position and velocity are not known simultaneously. Using the same step length Δt, the points of definition of each quantity jump alternate in time. So the initial values pose a problem because normally the particle position and the velocity are both known at $t = 0$. To start the integration it is necessary to compute velocity at the time $-\Delta t/2$, knowing the force **F** at $t = 0$. Care is taken that the kinetic and potential energies are computed at the same time, t.

The error of this simple integration method converges to zero when $\Delta t \to 0$, so if an appropriate Δt is chosen, the total error can be kept small. The leap-frog method is simple, uses moderate amounts of computer memory, and is fast, substantial advantages when simulating thousands of particles in thousands of steps.

The Lorentz force has two components, $\mathbf{F}_e$ and $\mathbf{F}_m$,

$$\mathbf{F} = \mathbf{F}_e + \mathbf{F}_m = q\,\mathbf{E} + q\,\mathbf{v} \times \mathbf{B}, \tag{1.83}$$

so both field strengths must be computed at the place where the particle is at the moment. This means that in the expression for the Lorentz force the centered velocity must be used

$$m\frac{\mathbf{v}_{t+\Delta t/2} - \mathbf{v}_{t-\Delta t/2}}{\Delta t} = q\left[\mathbf{E}_t + \frac{\mathbf{v}_{t+\Delta t/2} + \mathbf{v}_{t-\Delta t/2}}{2} \times \mathbf{B}_t\right], \tag{1.84}$$

which in three dimensions means three simultaneous equations. A simpler and faster method is to separate the electric and the magnetic force completely [35]. Introduction of

$$\mathbf{v}^- = \mathbf{v}_{t-\Delta t/2} + \frac{q}{m}\mathbf{E}_t\frac{\Delta t}{2}, \qquad\qquad \mathbf{v}^+ = \mathbf{v}_{t+\Delta t/2} - \frac{q}{m}\mathbf{E}_t\frac{\Delta t}{2}, \qquad (1.85)$$

separates the acceleration caused by the electric field. Because the magnetic field only can change the direction of $\mathbf{v}$ but not its value, $|\mathbf{v}^+| = |\mathbf{v}^-|$. The magnetic force is then

$$m\frac{\mathbf{v}^+ - \mathbf{v}^-}{\Delta t} = \frac{q}{2}(\mathbf{v}^+ + \mathbf{v}^-) \times \mathbf{B}_t,$$

or

$$\mathbf{v}^+ = \mathbf{v}^- + \frac{q}{2m}(\mathbf{v}^+ + \mathbf{v}^-) \times \mathbf{B}_t\Delta t.$$

This expression can be vector multiplied from the left by $\mathbf{B}_t\Delta t$, yielding

$$\mathbf{v}^+ = \mathbf{v}^- + \frac{q}{2m}\frac{2}{1 + \left(\frac{q}{2m}\mathbf{B}_t\Delta t\right)^2}\mathbf{v}^\star \times \mathbf{B}_t\Delta t, \qquad (1.86)$$

with

$$\mathbf{v}^\star = \mathbf{v}^- + \frac{q}{2m}\mathbf{v}^- \times \mathbf{B}_t\Delta t. \qquad (1.87)$$

Computation of velocity $\mathbf{v}_{t+\Delta t/2}$, known as *Boris' algorithm*, involves:
- Computing $\mathbf{v}^-$ according to Eq. 1.85
- Turning it into $\mathbf{v}^+$ by using Eqs. 1.86 and 1.87 and then
- Rewriting Eq. 1.85

$$\mathbf{v}_{t+\Delta t/2} = \mathbf{v}^+ + \frac{q}{m}\mathbf{E}_t\frac{\Delta t}{2}. \qquad (1.88)$$

For small time steps, the rotation angle, $\Delta\theta$, is approximately $\omega_c\Delta t$, where ω_c is the cyclotron frequency, Eq. 3.10:

$$\omega_c = \frac{q}{m}B_t,$$

which with the definitions of $\mathbf{v}^-$ and $\mathbf{v}^+$ gives

$$\left|\tan\frac{\Delta\theta}{2}\right| = \frac{|\mathbf{v}^+ - \mathbf{v}^-|}{|\mathbf{v}^+ + \mathbf{v}^-|} = \frac{\omega_c\Delta t}{2}. \qquad (1.89)$$

The equation can be used to obtain an estimate of the local error.

When the velocity $\mathbf{v}(t + \Delta t/2)$ is computed the position of the particle at the time $t + \Delta t$ is

$$\mathbf{x}(t + \Delta t) = \mathbf{x}(t) + \mathbf{v}(t + \Delta t/2)\Delta t. \qquad (1.90)$$

The process is repeated for each particle inside the limits of the universe. The position, $\mathbf{x}$, of each particle is known at time t, and its velocity is known at time $t - \Delta t/2$. However, the force on each particle must be known at $\mathbf{x}$, and here one runs up against a problem, because the force is a function of the electric and magnetic field strength at $\mathbf{x}$. This means that the solution of Maxwell's equations inside the whole universe must be available. One component of the electric field depends on the potentials of the electrodes in the universe, while another comes from the charge of the particles. For time-varying problems

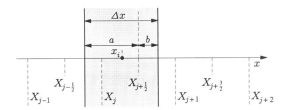

Fig. 1.19. Linear weighting of space-charge. (After *High-Power Microwave Sources,* edited by Victor L. Granatstein and Igor Alexeff, © 1987, Artech House, Inc., Norwood, MA, with permission.)

the electric field coming from $-\partial \mathbf{B}/\partial t$ must be included. Similarly, the magnetic field strength comes from coils or permanent magnets, from the currents corresponding to all particles which are moving inside the universe, and from $\partial \mathbf{E}/\partial t$, if any.

The universe is a mesh and the electric and the magnetic fields are known only in its nodes. We must define now how the nodes are coupled to the charges and currents from different particles. At time t the particle i will be found at $\mathbf{x}_i$. The position of the particle is a continuous function of the coordinates, but the solution to Maxwell's equations is only known in the nodes. Therefore, the charge of the particle at $\mathbf{x}_i$ is distributed over more than one node. The simplest procedure would be to assign the charge to the nearest node, but a significantly better process, as seen in Fig. 1.19, is to assign a weighted charge to the two, four, or six nearest nodes for the one-, two-, or three-dimensional case, respectively, by linear interpolation. Space-charge and currents will thus be equalized, and the fluctuations in the field strength and the numerical noise will be reduced. The weighting is usually linear, because higher-order weighting requires too much computer time.

Two different approaches are used in connection with weighting. In one the charge is represented by a particle situated at $\mathbf{x}_i$, and the programs using this model are usually called "particle-in-cell" (PIC) codes. In the other the charge is seen as a cloud covering a mesh element, with the center of gravity at $\mathbf{x}_i$. These programs are called "cloud-in-cell" (CIC) codes. For the one-dimensional case of Fig. 1.19 there is no difference between the PIC and the CIC codes. The particle, or the center of gravity of the cloud, is at x_i. The part of charge of the cloud, as in Fig. 1.19, which is inside the region with the length $\Delta x/2$ to the right of the node X_j is added to the space-charge in this node. What remains, b in Fig. 1.19, lies between the node X_{j+1} and $X_{j+1} - \Delta x/2$ is added to node X_{j+1}. Mathematically this adds to the node X_j the charge

$$q(X_j) = \frac{\Delta x - (x_i - X_j)}{\Delta x} q_i = \frac{X_{j+1} - x_i}{\Delta x} q_i, \qquad (1.91)$$

and to the node X_{j+1}

$$q(X_{j+1}) = \frac{x_i - X_j}{\Delta x} q_i, \qquad (1.92)$$

respectively.

It is also advisable to use weighting when field strengths are computed. For linear weighting, the electric field, in the one-dimensional case, would be

$$E(x_i) = \left[\frac{X_{j+1} - x_i}{\Delta x}\right] E_j + \left[\frac{x_i - X_j}{\Delta x}\right] E_{j+1}. \qquad (1.93)$$

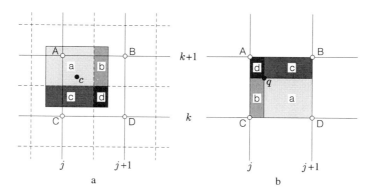

Fig. 1.20. Weighting of space-charge in a two-dimensional case. The shaded areas are proportional to the space-charge. (**a**) CIC code, where c shows the center of the cloud. (**b**) PIC code, where q shows the position of the particle. In both codes the area a is added to the node A, and so on.

The field strength is computed by a simple approximation, for example the x component of the electric field according to

$$E_j = -\frac{V_{j+1} - V_{j-1}}{2\Delta x},$$

with a similar expression for E_{j+1}. The weighted value is used at time t to obtain the force.

The difference between the CIC code and the PIC code is evident in the two-dimensional case, seen in Fig. 1.20. Weighting with the central node and its 8 nearest neighbors is also used for highly inhomogeneous situations.

When the space-charge and currents are computed at all nodes, Maxwell's equations can be solved. There the FDM, the FEM, or other procedures can be used as described in Section 1.2. The FDM is often preferred, since the rest of the computation is also done using finite differences. However, both the FDM and the FEM require large computer resources both in memory and in speed. Recall that the computations described in Section 1.2 were made only once, but for time-varying problems, with hundreds of time steps, Maxwell's equations must be solved hundreds of times. It is therefore necessary to use a smaller number of mesh elements than for steady-state solutions. With a reduced number of mesh elements, direct methods using matrix inversion may be faster than iteration. However, it is often possible to be clever. Both the PIC and CIC codes are normally used to solve time-variable problems where natural solution often has an oscillating character, like for plasmas or in microwave tubes. For the latter even the boundary conditions exhibit periodicity. If a periodic approximation can adequately represent the boundary conditions it is possible to apply fast Fourier transform (FFT), which can result in an appreciable saving in computation time.

With the solution of Maxwell's equations obtained, the circle is closed. All forces are known, and in the next time step new positions and velocities can be computed. From these data new space-charge and current distributions are obtained and Maxwell's equation is again solved. The scale and complexity of problems that can be simulated depend only on the computer resources available. In Section 7.3.2 and 7.4.2 two examples of simulations of microwave tubes using the step-by-step method are shown, Figs. 7.19 and 7.46.

In conclusion, the numeric computations of the field distribution and of particle orbits require that the users of even the well-tested and well-documented codes possess the analytical and programming skills to understand their construction and functioning. Otherwise, the uninitiated often use the program in a manner its author never intended, with unpleasant results. As one expert* in the field puts it: *"In the last resort, though, trajectory tracing* (and computation of the potential distribution) *can only be learned by writing programs for oneself, not by reading the accounts, however instructive, of others."*

---

* P. H. Hawkes in *Applied Charged Particle Optics.*

2

Statistical Mechanics and Electron Emission

Lee de Forest triode from 1908. The tube was used in audion detectors, precursors of radio receivers. It had two tungsten wires as cathode, a meander-shaped grid, and an ~ 1 cm^2 plate as anode (Telemuseum, Stockholm).

2.1. STATISTICAL MECHANICS

HISTORICAL NOTES. Newton's [36] *Philosophiae Naturalis Principia Mathematica* was published in 1687. Soon the scientific community accepted Newton's three fundamental postulates, which determine the motion of all particles, if their position and velocity are known at any time. In 1660s, Boyle [37] and Mariotte [38], independently, found experimentally that the product of the pressure, p, and volume, V, for a gas is constant. For more than a hundred years no contradiction was found to the causal mechanics of Newton or the phenomenological mean values of Boyle's law.

Things began to change at the turn of the 19th century. In 1802 Gay-Lussac [39] extended Boyle's law to show that the constant was proportional to the temperature, T. In 1811 Avogadro [40] found that certain weight of every gas occupied the same volume at the same pressure, thus giving rise to the number we know today by his name. An atomic theory, where every gas atom had a given mass, explained this fact. If all gases were build of atoms, these particles behaved according to Newton's mechanics, and thus all processes must be reversible if the direction of time is reversed. But in 1824 Carnot [41] proposed the second law of thermodynamic, which made phenomenological thermodynamic processes irreversible. Clausius [42] then in 1865 introduced the concept of entropy and "thermal death."

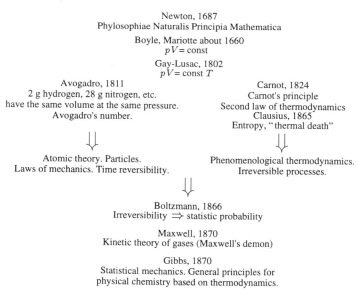

Newton, 1687
Phylosophiae Naturalis Principia Mathematica

Boyle, Mariotte about 1660
$pV = $ const

Gay-Lusac, 1802
$p\bar{V} = $ const T

Avogadro, 1811
2 g hydrogen, 28 g nitrogen, etc.
have the same volume at the same pressure.
Avogadro's number.

Carnot, 1824
Carnot's principle
Second law of thermodynamics
Clausius, 1865
Entropy, "thermal death"

Atomic theory. Particles.
Laws of mechanics. Time reversibility.

Phenomenological thermodynamics.
Irreversible processes.

Boltzmann, 1866
Irreversibility $\Rightarrow$ statistic probability

Maxwell, 1870
Kinetic theory of gases (Maxwell's demon)

Gibbs, 1870
Statistical mechanics. General principles for
physical chemistry based on thermodynamics.

Fig. 2.1. Historical overview.

Atoms and phenomenological thermodynamics remained incompatible until in 1866 Boltzmann [43] suggested that instead of talking about irreversibility, one should use the concept of statistic probability. His idea was not readily accepted until 1870 when Maxwell [44] produced his kinetic theory of gases based on statistical mechanics. With his now famous demon, Maxwell showed that even the most improbable event is allowed, but unlikely. In the same year Gibbs [45] founded the modern version of statistical mechanics and showed that physical chemistry is based on thermodynamics.

At the turn of the century all physics problems appeared to be solved [46]. Nothing could be less true. In 1901 Planck [47] abandoned a 2000-year-old principle, which said that the nature does not make jumps — *"Natura non facit saltus"*, in his explanation of the radiation of a black body, thus beginning the quantum theory. In 1905 Einstein [48], using Planck's theory, explained the photoelectric effect, and with it he established the particle–wave duality.

After Edison's [17] discovery of the current emerging from the negative side of a heater in an incandescent lamp, and J. J. Thomson's [13] explanation, Fleming [20] constructed the first diode. Richardson [49] concluded that the temperature dependence of the emission of electrons was dominated by an exponential term of the form $-W/kT$. The cathode in the first diodes was tungsten, but in 1904 Wehnelt [21] introduced the oxide cathode. The first oxide cathodes were very unreliable. Pierce* described later the activation as *"a wishful process consisting of glowing the filament brightly, applying voltage, and drawing current to the plate* (US name for anode) *until, in happy circumstances the plate got red-hot, giving off enough gas to produce a blue glow and further increase the activity and current."* In 1914 Langmuir [50] experimented with thoriated tungsten, and these cathodes were prevalently used until 1920s, when industry learned how to make reliable oxide cathodes.

The first diodes were used as detectors in radio receivers, which had only a tuned circuit and a rectifying diode. The advantage of having a third electrode, the grid, was first noticed by de Forest [22], he called the tube "audion." From the beginning, audion was implemented only as a detector, but Armstrong [51] used the triode as an amplifier with feedback and negative grid bias; the circuit remained the basis of electronics. More electrodes were added to form a tetrode (Schottky [52]) and pentode (Holst and Tellegen [53]), the latter being the most used tube in radio receivers until the invention of the transistor.

That selenium changes resistance when exposed to light was discovered by Smith [54]. Soon, selenium photocells played an important role in photometry. A few years later Hertz [55] detected that a spark gap ignited much easier when illuminated by the light of a spark from another gap. Thorough investigation showed that the UV light is responsible for the breakdown, and the effect was strong when the incident light hit the negative terminal. Hallwachs [56] continued the experiments by illuminating a zinc sphere by UV light and found that *"the negative electricity leaves a body and follows electrostatic lines of force in its passage under the influence of ultraviolet radiation."* Photoemission was for a long time called the *Hallwachs effect*. Elster and Geitel [57] observed that sodium and potassium amalgam emit electrons even when irradiated by visible light. They enclosed the cathode covered by amalgam in an evacuated glass envelope and made the first phototube. Experiments with photoemission showed soon that the number of electrons released is proportional to the intensity of the incident light, and that the maximum energy of emitted electrons increases linearly with the frequency of the light. These facts were explained by Einstein [48] in his paper on photoemission which gave him the Nobel Prize.

In Newton's mechanics, forces can be superposed and thus the interaction of two or more particles can be computed. However, if the number of particles is large — for example, in a gas — it is practically impossible to follow the motion of each particle by applying the laws of mechanics. Thus with large numbers of particles only their *collective properties* can be observed. Such a viewpoint is called *statistical*; and like all other descriptions of nature, statistical methods result from experience. Statistical assertions about a gas only describe the behavior of the gas in its entirety and say nothing about individual particles. They predict the probable number of particles, which have a certain common property, but not which ones will posses it.

The motion of the particles and their collisions are completely determined by Newton's laws. Knowing the conditions of a gas at any time, these laws allow the position and velocity of each particle to be computed for the rest of eternity. In contrast, statistical mechanics admits no possibility whatsoever of determining the initial conditions. They are, in principle, undefined. Thus, deprived of any practical possibility to find the initial position and velocity of each molecule in a gas, we are reduced to talking about the probability of finding a gas in a certain state.

---

* Under the pseudonym J. J. Coupling, *Science Fiction*, November 1946.

From every initial condition a course of events develops. Observing similar systems, but with a large number of different initial conditions, a common behavior will be exhibited. In statistical mechanics the mean values of all physical quantities in all possible states which a system can occupy are determined. Those states, which have the largest probability, are found in nature.

In the phenomenological description of matter used in thermodynamics, the mean values (e.g., pressure, temperature, etc.) are the elements, which the laws of nature connect to create thermodynamic quantities. A specific property of thermodynamics is the irreversibility of physical processes: Thermodynamical processes proceed in one temporal direction. Heat flows from a hot to a cold body until both reach the same temperature. However, if we film a mechanical motion, and run the film backwards, both the original motion and the film obey the laws of mechanics, and no contradiction will be revealed. In classical mechanics we do not differentiate between the past and the future, yesterday could as well be tomorrow, and no harm to the prediction power would result. Irreversibility appears to contradict the laws of mechanics.

In 1866 Boltzmann [43] resolved this conundrum. He asserted that we can only speak about more or less probable states, and he introduced the concept of statistical probability to replace irreversibility. A transition to more probable states is to be expected, but is not an unconditional consequence of the laws of mechanics.

Thus, every phenomenological state is composed of a great number of microstates, each with differing density distributions, pressures, and temperatures. Each microstate is *a priori* equally probable. The more microstates which exist, the larger the probability of finding the most probable state in the nature. Experiments — for example, the diffusion of light — show that the reversibility is not complete. All gases, as a consequence of refraction and reflection, diffuse light because of small, local density deviations caused by statistical fluctuations. These fluctuations can be computed using the methods of statistical mechanics and agree with measurements. Thus statistical mechanics explains why the sky is blue.

Let us observe a closed system with N interacting particles:

$$N = \sum N_j = \text{const}, \tag{2.1}$$

where N_j is the number of particles in a small energy interval, j. We assume that the number of particles does not change and that the interaction between these particles is sufficiently weak so that the total system energy is determined only by totality of the individual particles energies*:

$$U = \sum N_j E_j = \text{const}. \tag{2.2}$$

These two equations must be satisfied for each particle distribution in each energy state.

With so many particles in a state, after a few collisions the difference from the initial conditions is appreciable, and the particles "forget" their past. Assume we observe the particles at equal time intervals, which are long enough for the particles to have made numerous collisions. A similar result would be obtained if the particles were randomly distributed on different energy states. We assume only that Eqs. 2.1 and 2.2 apply. If the experiment is repeated many times, large differences between distributions are observed,

---

* The symbol E denotes energy in this chapter.

but some are often repeated. The mean value over all distributions gives the mean of the distribution. The less probable distributions can be neglected as compared to the enormously dominating, most frequently occurring distribution. This conclusion opens the possibility to study which particle distributions along the energy axis are statistically possible, and to compute their probability, W, for a specific distribution. Then even the most probable distribution can be found.

There exist, however, some fundamental conditions that define which distributions are physically possible. They depend on the properties of the three types of particles found in the nature, which are:

1. Particles can be identified and named. An exchange of the particle a with the energy E_a with the particle b with the energy E_b ($E_a \neq E_b$) gives a new distribution. (e.g., gas particles).
2. Particles do not have identity and exchange does not give a new distribution (e.g., photons).
3. Particles do not have identity but in any state only one particle is allowed to exist (e.g., electrons).

For distinguishable particles: The particles have identity, different energy states are distinguishable, many particles can share the same energy state, the total number of particles is constant (Eq. 2.1), and the total energy in the system is constant (Eq. 2.2). The particles are distributed in different energy states so that all these conditions are fulfilled. Let us divide the energy into intervals of equal size, each containing S_j energy states which are usually different for different energy intervals, as seen in Fig. 2.2.

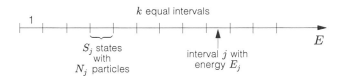

Fig. 2.2. Energy axis with equal size energy intervals.

Let us take three particles with the same energy:

$$\text{——— a — b — c ———} \qquad \text{——— a — c — b ———}$$

These are identical distributions since an exchange of particles with the same energy does not produce a new distribution. If we have six particles and three energy intervals (E, $2E$ and $3E$), one possible distribution with a total energy of $11E$ is

$$
\begin{array}{ll}
3E & \text{——— e — f ———} \\
2E & \text{————— d ————} \\
E & \text{— a — b — c —}
\end{array}
$$

After a time we can find any particle in any energy interval as long as Eq. 2.2 is satisfied, so

$$
\begin{array}{lll}
3E & \text{——— b — d ———} & \text{————— d —————} \\
2E & \text{————— f —————} & \text{— e — b — c —} \\
E & \text{— c — e — a —} & \text{——— a — f ———}
\end{array}
$$

are possible distributions.

All particles can be permuted and for the six particles we have 6! or 720 possibilities. But all permutations in the same energy interval are identical, so this is reduced to

$$\frac{6!}{1!\,2!\,3!} = 60$$

distribution variants. Thus for N particles, which have at their disposal k energy intervals, the number of permutations will be

$$\frac{N!}{N_1!\,N_2!\cdots N_k!} = \frac{N!}{\Pi N_j!}. \tag{2.3}$$

Equation 2.3 describes the number of possibilities to distribute N particles on k energy intervals under the condition that the first interval contains N_1 particles, and so on. In the energy interval j there are N_j particles, which can be distributed on S_j energy states. Since every state can have more than one particle, the first particle has S_j states to choose, but that is true even for the second, third, and all other particles in the same energy interval. We can therefore distribute N_j particles on

$$S_j \cdot S_j \cdot S_j \cdots S_j = S_j^{N_j}$$

different ways. The total number of ways to distribute N particles on k energy intervals is

$$\mathcal{W}_{MB} = N! \prod_k \frac{S_j^{N_j}}{N_j!}. \tag{2.4}$$

For indistinguishable particles, like electrons (case 3): The particles do not have identity, different energy states are distinguishable, only one particle can be in one energy state, the total number of particles is constant, and the total energy in the system is constant.

We repeat the reasoning of the random distribution of particles in different energy intervals, but we must take care of Pauli's principle [58] which says that only one particle can occupy an energy state:

$$N_j \le S_j.$$

For three particles, which can share six different energy states, two of the possible distributions are

The first particle can take any of the six energy states, the next particle must choose from the remaining five, and so on. In total there are $6 \cdot 5 \cdot 4 = 120$ possibilities. Mutual permutation of particles again does not mean a new distribution, so we have 120/3! or 20 variants of this distribution.

N_j particles can thus in an energy interval be distributed on S_j states on

$$\frac{S_j(S_j - 1)(S_j - 2)\cdots(S_j - N_j + 1)}{N_j!}$$

different ways. Using binomial coefficients this can be written as

$$\frac{1}{N_j!}\frac{S_j!}{(S_j - N_j)!} = \binom{S_j}{N_j}. \tag{2.5}$$

The total number of ways to distribute N electrons on k energy intervals is

$$\mathcal{W}_{FD} = \prod_k \binom{S_j}{N_j}. \tag{2.6}$$

Which distribution is the most probable in each of the two cases? A closed system, left to itself, shows a tendency toward to distribute itself to attain the highest value $\mathcal{W}_{max}$. To see how $\mathcal{W}$ changes when the distribution changes, we use as an example electrons and assume 50 particles in the lowest energy states, 10 in each energy interval, as seen in Fig. 2.3.

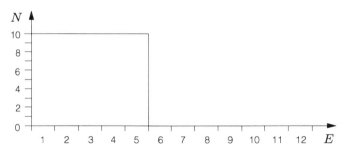

Fig. 2.3.

Here we have

$$\Sigma N_j = 50, \qquad \Sigma N_j E_j = 150,$$

and

$$\mathcal{W}_0 = \binom{10}{10}\binom{10}{10}\binom{10}{10}\binom{10}{10}\binom{10}{10} = 1.$$

We now add 20 energy units to our system by heating, so that

$$\Sigma N_j = 50, \qquad \Sigma N_j E_j = 170,$$

and assume that this distribution is obtained by moving four particles from the first to the sixth energy interval, as seen in Fig. 2.4. Thus

$$\mathcal{W}_1 = \binom{10}{6} \cdot 1 \cdot 1 \cdot 1 \cdot 1 \cdot \binom{10}{4} = 44100.$$

Is this near the maximum?

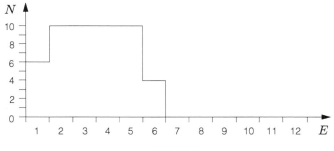

Fig. 2.4.

Another distribution, seen in Fig. 2.5. gives

$$W_2 = \binom{10}{7} \cdot 1 \cdot 1 \cdot 1 \cdot 1 \cdot \binom{10}{1}\binom{10}{1}\binom{10}{1} = 120000,$$

$W_2 \approx W_1 \gg W_0.$

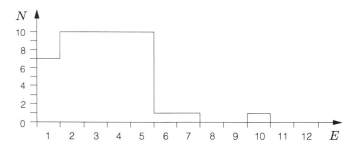

Fig. 2.5.

The high-energy "tail" of the distribution seems to have a large influence on W. Maybe it would be wise to try a smoother transition to the "tail," seen in Fig. 2.6, which gives

$$W_3 = 1 \cdot 1 \cdot \binom{10}{9}\binom{10}{8}\binom{10}{5}\binom{10}{3}\binom{10}{3}\binom{10}{1}\binom{10}{1} = 163296000000 \approx 10^{11}.$$

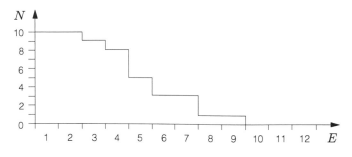

Fig. 2.6.

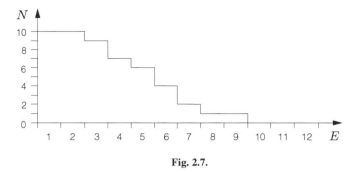

Fig. 2.7.

We can try another small change and also make a rounding of the transition, seen in Fig. 2.7, and obtain

$$W_4 = 1 \cdot 1 \cdot \binom{10}{9}\binom{10}{7}\binom{10}{6}\binom{10}{4}\binom{10}{2}\binom{10}{1}\binom{10}{1} = 238140000000 \approx 10^{11}.$$

We see that $W_3 \approx W_4$. Both are approximately equal.

All 7827 distributions in this example can be computed by computer simulation and are displayed below.

W	(each $*$ marks 50 distributions)	number
10^0–10^1		0
10^1–10^2		0
10^2–10^3		3
10^3–10^4		10
10^4–10^5	$*$	70
10^5–10^6	$***$	173
10^6–10^7	$**************$	717
10^7–10^8	$**********************$	1109
10^8–10^9	$********************************$	1642
10^9–10^{10}	$***$	2063
10^{10}–10^{11}	$**$	2006
10^{11}–10^{12}	$*$	33

To find the mean W we add all possible distributions

$$44100 + 120000 + 1.6 \cdot 10^{11} + 2.38 \cdot 10^{11} + \text{all other distributions,}$$

which, after dividing by the number of distributions, gives a mean value of $W \sim 10^{11}$, because W_{max} and the distributions thereabout dominate.

Thus, we can find any W, if at a certain moment we observe the system, but it is most probable that we will find a distribution near W_{max}. It is important to remember that in the nature we will not observe 50 particles, but 10^{20} to 10^{30} particles per cubic meter. For in plasma there can be 10^{20} free electrons per cubic meter, in a crystal 10^{29} per cubic meter. This will make W_{max} still more dominant. With such a large number the behavior of an average particle can be computed by starting from rather simple principles: W_{max} is very insensitive to small changes in the neighborhood of W_{max} (e.g., a rounding of the "tail" gives a high W value).

This means that a simple rearrangement of two particles in the "tail" of two distributions, W_1 and W_2, both of which are near W_{max}, will only slightly influence the most probable distribution W_{max}, because W_{max} includes many distributions, which are almost equal. Such a rearrangement, seen in Fig. 2.8, can be accomplished by moving

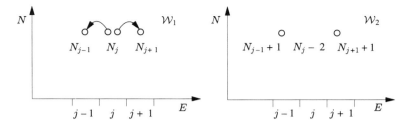

Fig. 2.8. Rearrangement of two particles in the "tail" of two distributions, $\mathcal{W}_1$ and $\mathcal{W}_2$.

one particle up one energy interval and another particle one energy interval down from some interval, "j ," in the "tail," and it satisfies all our conditions, especially Eqs. 2.1 and 2.2.

$$\frac{\mathcal{W}_2}{\mathcal{W}_1} = \frac{\binom{S_1}{N_1}\binom{S_2}{N_2}\cdots\binom{S_{j-1}}{N_{j-1}+1}\binom{S_j}{N_j-2}\binom{S_{j+1}}{N_{j+1}+1}\cdots\binom{S_n}{N_n}}{\binom{S_1}{N_1}\binom{S_2}{N_2}\cdots\binom{S_{j-1}}{N_{j-1}}\binom{S_j}{N_j}\binom{S_{j+1}}{N_{j+1}}\cdots\binom{S_n}{N_n}} =$$

$$= \frac{S_{j-1}-N_{j-1}}{N_{j-1}+1}\cdot\frac{N_j(N_j-1)}{(S_j-N_j+2)(S_j-N_j+1)}\cdot\frac{S_{j+1}-N_{j+1}}{N_{j+1}+1}\approx$$

$$\approx \frac{\left(\dfrac{S_{j-1}}{N_{j-1}}-1\right)\left(\dfrac{S_{j+1}}{N_{j+1}}-1\right)}{\left(\dfrac{S_j}{N_j}-1\right)^2}\approx 1,$$

which also can be written as

$$\frac{\dfrac{S_j}{N_j}-1}{\dfrac{S_{j-1}}{N_{j-1}}-1} = \frac{\dfrac{S_{j+1}}{N_{j+1}}-1}{\dfrac{S_j}{N_j}-1}.$$

$S_1/N_1 - 1, S_2/N_2 - 1, \ldots S_j/N_j - 1, \ldots$ is a sequence of numbers in "j "

$$\frac{S_j}{N_j} - 1 = \text{const } a^j = e^\alpha e^\nu = e^{\alpha+\nu},$$

where ν only is a numbering which shows where on the energy axis the energy interval "j " is situated. We can therefore put

$$\nu = \beta E_j$$

and obtain

$$\frac{S_j}{N_j} - 1 = e^{\alpha+\beta E_j} \tag{2.7}$$

or

$$\frac{N_j}{S_j} = \frac{1}{1 + e^{\alpha+\beta E_j}},$$

where α and β are two parameters.

If we let the width of the energy intervals go to zero, the limit is

$$P_{FD} = \frac{dN}{dS} = \frac{1}{1 + e^{\alpha + \beta E}}. \tag{2.8}$$

We can use the same procedure for distinguishable particles. Having found $\mathcal{W}_{max}$ we move two particles to obtain, from Eq. 2.4,

$$\frac{\mathcal{W}_2}{\mathcal{W}_1} = \frac{N! \, \dfrac{S_1^{N_1}}{N_1!} \dfrac{S_2^{N_2}}{N_2!} \cdots \dfrac{S_{j-1}^{N_{j-1}+1}}{(N_{j-1}+1)!} \dfrac{S_j^{N_j-2}}{(N_j-2)!} \dfrac{S_{j+1}^{N_{j+1}+1}}{(N_{j+1}+1)!} \cdots \dfrac{S_n^{N_n}}{N_n!}}{N! \, \dfrac{S_1^{N_1}}{N_1!} \dfrac{S_2^{N_2}}{N_2!} \cdots \dfrac{S_{j-1}^{N_{j-1}}}{(N_{j-1})!} \dfrac{S_j^{N_j}}{(N_j)!} \dfrac{S_{j+1}^{N_{j+1}}}{(N_{j+1})!} \cdots \dfrac{S_n^{N_n}}{N_n!}} =$$

$$= \frac{S_{j-1}}{N_{j-1}+1} \cdot \frac{N_j(N_j-1)}{S_j^2} \cdot \frac{S_{j+1}}{N_{j+1}+1} \approx 1,$$

or

$$\frac{\dfrac{S_j}{N_j}}{\dfrac{S_{j-1}}{N_{j-1}}} = \frac{\dfrac{S_{j+1}}{N_{j+1}}}{\dfrac{S_j}{N_j}},$$

which can be written as

$$\frac{N_j}{S_j} = e^{-(\alpha + \beta E_j)},$$

or at the limit when the width of the energy interval goes to zero:

$$P_{MB} = \frac{dN}{dS} = e^{-(\alpha + \beta E)}. \tag{2.9}$$

Figures 2.9a and 2.9b show each distribution, the first is called the **Fermi–Dirac** and the other the **Maxwell–Boltzmann** distribution.

In Eqs. 2.8 and 2.9 the parameters α and β do not depend on E but on the total energy, U, of the system and the number of particles N. What is the physical meaning of these parameters? If β is defined for any part of the curve, it will be valid for the whole curve. Observe the "tail" in Fig. 2.9a and compare it with the same region in Fig. 2.9b. In any energy interval E_j, Eq. 2.6 is valid:

$$\mathcal{W}_j = \binom{S_j}{N_j} = \frac{S_j(S_j-1)(S_j-2)\ldots(S_j-N_j+1)}{N_j!}.$$

But in the "tail" $S_j \gg N_j$; that is, there are very many empty states compared to the number of particles. Therefore we neglect all numbers in the numeratorr, which we subtract from S_j and obtain

$$\mathcal{W}_j \approx \frac{S_j^{N_j}}{N_j!}.$$

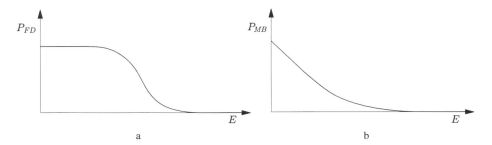

Fig. 2.9. (a) Fermi–Dirac distribution. **(b)** Maxwell–Boltzmann distribution.

This implies that the "tails" in both distributions are identical, those of the FD distribution and the MB distribution. Thus for a low-density electron gas the FD distribution can be approximated by the MB distribution with good accuracy, and this can be realized as follows: Keep the energy constant and observe an energy interval E_j in the "tail" of the distribution so that $S_j \gg N_j$. For both distributions, β is a constant, does not depend on E, and thus is the same for both distributions. Consequently, in the "tail" the FD and the MB distributions are both simple decaying exponentials and we can calculate β for a thin gas, which follows the MB distribution. As the curves diverge at lower energy the same β is still valid for both.

To find out the physical meaning of β let us realize that, since β is common for both distributions, we can use classical mechanics. For a gas the equation of state is

$$pV = N_M R T, \qquad R = N_A k, \tag{2.10}$$

where p is the pressure, V the volume, N_M the number of kilomoles in the gas, R the gas constant, N_A Avogadro's number, and k Boltzmann's constant.

Imagine a closed cubical box containing an ideal thin gas with $N = N_M N_A$ particles, which can be approximated as free particles. The pressure which the gas exerts on the walls of the box can be computed using the principle of virtual work as

$$dA = F\,ds = pS\,ds = p\,dV = 2dN < \frac{1}{2}mv_x^2 >,$$

where dA is the virtual work, F the force on the box wall, S the wall area, ds the virtual displacement, and dN the number of particles in the volume element dV. Factor 2 comes from the momentum change when the particle bounces of the wall, and its velocity is changed from $+v$ to $-v$. We denote the mean value of a physical quantity by bracketing $< >$. With the number of particles per unit volume

$$n = \frac{dN}{dV}$$

the pressure on the wall caused by the perpendicular component of the velocity is

$$p = 2\,n < \frac{1}{2}mv_x^2 >.$$

But the three velocity components are equivalent

$$< v_x^2 > = < v_y^2 > = < v_z^2 >,$$

so the pressure is

$$p = \frac{2}{3} n < \frac{1}{2} m v^2 > = \frac{2}{3} n < E >.$$

Multiplying by the box volume, V, and using the equation of state (Eq. 2.10) we obtain

$$\frac{2}{3} < E > = kT. \tag{2.11}$$

MB distribution determines the probability, P, to find a particle in a certain volume element of the six-dimensional phase space (three space and three momentum coordinates)

$$dP = \text{const } e^{-\beta E} dV_6 = \text{const } e^{-\beta E} dp_x dp_y dp_z dx dy dz,$$

where p_x, p_y, and p_z are momentum components mv_x, mv_y, and mv_z. Since the particle must be inside the volume we obtain

$$P = \text{const} \int e^{-\beta E} dp_x dp_y dp_z dx dy dz = 1. \tag{2.12}$$

Gas particles inside the box have a mean energy given by the total energy and by the number of particles

$$< E > = \frac{U}{N}.$$

The product NdP gives the number of particles in a volume element of the phase space, so the mean energy is

$$< E > = \frac{\int E N dP}{\int N dP}.$$

Any particle position is *a priori* equally probable, while the kinetic energy depends only on the momentum. The mean kinetic energy of a particle is, therefore,

$$< E > = \frac{\int \frac{1}{(2m)} (p_x^2 + p_y^2 + p_z^2) e^{-\beta(p_x^2 + p_y^2 + p_z^2)/(2m)} dp_x dp_y dp_z}{\int e^{-\beta(p_x^2 + p_y^2 + p_z^2)/(2m)} dp_x dp_y dp_z}.$$

By substitution the triple integral is transformed into a sum of three integrals with the form

$$\frac{1}{\beta} \frac{x^2 e^{-x^2} dx}{e^{-x^2} dx}.$$

Since

$$\frac{d}{dx} (x e^{-x^2}) = e^{-x^2} - 2x^2 e^{-x^2},$$

the integral is

$$x e^{-x^2} \Big|_{-\infty}^{\infty} = \int_{-\infty}^{\infty} e^{-x^2} dx - 2 \int_{-\infty}^{\infty} x^2 e^{-x^2} dx = 0,$$

or

$$\frac{\int\limits_{-\infty}^{\infty} x^2\, e^{-x^2} dx}{\int\limits_{-\infty}^{\infty} e^{-x^2} dx} = \frac{1}{2}.$$

Considering all components we obtain

$$<E> = <E(p_x)> + <E(p_y)> + <E(p_z)> = \frac{3}{2}\frac{1}{\beta}.$$

Introducing this expression into Eq. 2.11 we obtain

$$\beta = \frac{1}{kT}.$$

This gives us the known form of the Maxwell–Boltzmann distribution

$$P_{MB} = \text{const}\, e^{-E/(kT)}, \tag{2.13}$$

where the constant can be calculated using Eq. 2.12 and integrating,

$$\text{const} = \left(\frac{1}{2\pi mkT}\right)^{3/2}.$$

Now we find α in the Fermi–Dirac distribution by introducing

$$E_F = -\frac{\alpha}{\beta},$$

so that the Fermi–Dirac distribution is

$$P_{FD} = \frac{1}{1 + e^{(E-E_F)/(kT)}}, \tag{2.14}$$

where E_F is called *Fermi energy*, corresponding to the energy of a state, which is half filled, as seen in Fig. 2.10a.

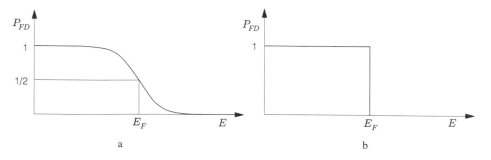

Fig. 2.10. **(a)** Fermi–Dirac distribution and the definition of Fermi energy. **(b)** Fermi–Dirac distribution at 0 K.

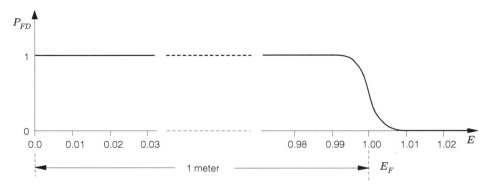

Fig. 2.11. Fermi–Dirac distribution for aluminum at 300 K.

At $T = 0$ (Fig. 2.10b), all states are filled to the Fermi energy, E_F, as all particles are in the lowest possible energy state. The FD distribution of a crystal (10^{29} particles per cubic meter) at room temperature, seen in Fig. 2.11, shows how few electrons are in higher energy states.

The derivative of P_{FD} with respect to E at E_F is

$$\frac{dP_{FD}}{dE} = -\frac{1}{4kT},$$

and so the "tail" of the FD distribution, seen in Fig. 2.12, starts at

$$E \approx E_F + 2kT.$$

From there out in energy the FD distribution can be approximated by the MB distribution. The shaded areas are equal. The number of lower-energy states which are vacated is equal to the number of particles which are moved to higher-energy states, so the FD distribution is symmetric in respect to Fermi energy.

$$P_{FD}(E_F + dE) = 1 - P_{FD}(E_F - dE).$$

An important consequence is observed in semiconductors where electrons and holes behave identically and thus can be treated in the same way.

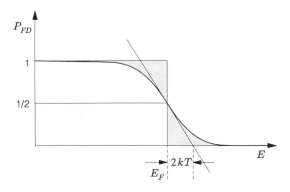

Fig. 2.12. Transition to the "tail" in the Fermi–Dirac distribution.

In the special case of $T = 0$ the Fermi energy can easily be computed. The number of states in the energy interval $E, E + dE$ for electrons, which have 2 spin directions, is given by (see Appendix A)

$$dS(E) = \frac{V}{h^3} 4\pi \sqrt{(2m)^3} \sqrt{E} dE, \qquad (2.15)$$

and the total number of electrons is

$$N = \int \frac{dN}{dS} dS = \int P_{FD} dS = \frac{V}{h^3} 4\pi \sqrt{(2m)^3} \int \frac{\sqrt{E} dE}{1 + e^{(E-E_F)/(kT)}}. \qquad (2.16)$$

At absolute zero temperature, $P(E) = 1$ for $E < E_F$ and $P(E) = 0$ for $E > E_F$, so

$$N = \frac{V}{h^3} 4\pi \sqrt{(2m)^3} \int_0^{E_F} \sqrt{E} dE$$

and

$$E_{F(T=0)} = \frac{h^2}{2m} \left(\frac{3}{8\pi} \right)^{2/3} n^{2/3}, \qquad (2.17)$$

where $n = N/V$ is the number of particles per unit volume.

At higher temperatures, but as long as $kT \ll E_F$, by expanding the denominator of Eq. 2.16 in a power series we obtain

$$E_F = E_{F(T=0)} \left[1 - \frac{\pi^2}{12} \left(\frac{kT}{E_{F(T=0)}} \right)^2 \right].$$

Fermi energy decreases slowly with the temperature.

For crystals ($n = N/V \sim 10^{29}$ particles per cubic meter), the Fermi energy at $T = 0$ is ~ 10 eV, which corresponds to a temperature of $\sim 100,000$ K. In plasma (10^{18} or 10^{19} ion-electron pairs per cubic meter) at $T = 0$ the Fermi energy is only $\sim 10^{-6}$ eV. So we only have to consider the "tail" of the distribution and thus can apply the MB distribution to plasma along the entire energy axis.

To see better the difference between the FD and the MB distribution we introduce de Broglie [59] wavelength for particles and look at them as wave packets. A particle with momentum p has a wavelength of

$$\lambda = \frac{h}{p},$$

which, if it has exactly the Fermi energy, is

$$\lambda = \frac{h}{\sqrt{2mE_F}}.$$

We introduce Eq. 2.17 for Fermi energy at $T = 0$ so that

$$\lambda = \sqrt[3]{\frac{8\pi}{3}} \sqrt[3]{\frac{V}{N}}.$$

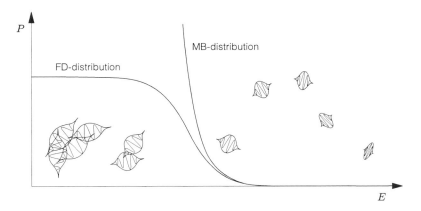

Fig. 2.13. Illustration of wave packets and Fermi–Dirac distribution.

The distance between two particles in a box with the sides of length L and volume $V = L^3$ is

$$d = \sqrt[3]{V/N}.$$

Thus a particle whose energy is equal to Fermi energy at $T = 0$ is represented by a wave packet with a wavelength $\sim 2d$. The distance between the particles is smaller, or for lower energies much smaller, than de Broglie wavelength, so the wave packets overlap. We cannot, therefore, represent these wave packets by particles in a classical sense, because they interfere. When $E \gg E_F$, as in the FD distribution "tail," the wave length becomes shorter and smaller than the distance between the particles. The probability of interference between these wave packets is small, so the classical model can be used and the particles can be viewed as material points, as shown in Fig. 2.13.

The behavior of the electrons in the FD distribution "tail" can be treated as analogous to gas particles. Even before the advent of quantum physics, the physicists were speaking of "electron gas" in crystals, and especially in metals, a name proposed by Drude [60] when he tried to explain the high conductivity of metals as an "electron gas." "Electricity particles" (= electrons) must be able to move freely inside a metal (= gas). Lorentz [61] used these ideas when he developed his "classical" theory of metals where he considered so-called binary gases (i.e., gases with two types of molecules, with one of a much larger mass than the other) from which the "electron gas" idea resulted. Lorentz explained, for example, electrical (σ) and thermal (κ) conductivity of metals. The Lorentz number, $L = \kappa/\sigma T$, is in a first approximation a constant for all metals. A problem for Lorentz' theory was that the "electron gas" energy could not be observed. Since according to the kinetic theory of gases the mean energy of gas particles does not depend on the particle mass, all particles would have the same mean energy. Light electrons with the same energy as the other particles in a metal would have such large velocities that would easily be observed, which is not true.

This difficulty disappears when Pauli's principle is considered. In the classical theory all electrons should have zero energy at 0 K. Pauli's principle says that the energy of the "electron gas" at $T = 0$ is exceedingly high. Why can't we observe this energy in the heat capacity?

A crystal can be heated by bringing it in contact with a gas at high temperature. The gas atoms collide with the crystals atoms and in each of these collisions an energy of kT can be imported to the crystal and converted into oscillations of the atoms in the crystal lattice. Even the electrons are heated. But this is true only for the electrons, which have an empty state inside a step kT in energy available, as seen in Fig. 2.11. Only electrons near the highest occupied energy state can increase their energy. In the small interval in the "tail" of the FD distribution the electrons can move to higher energy states. So for all other electrons it has no effect that the metal is in contact with a warm gas.

The "electron gas" heat capacity can be approximated by drawing a straight line in a figure similar to Fig. 2.12, but with such a slope in E_F that it cuts the abscissa at $E_F + 4kT$. This line approximates a triangle with an area corresponding to the number of electrons, n, which at the temperature T have moved to a higher energy state. The triangle area is

$$\frac{1}{2}\frac{4kT}{2} = kT,$$

and

$$\frac{n}{N} = \frac{kT}{E_F}, \qquad \text{or} \qquad n = N\frac{kT}{E_F}.$$

For aluminum (trivalent) at 300 K, $kT \sim 0.026$ eV and $E_F \sim 12$ eV, thus $kT/E_F \sim 0.002$ and $n \sim 0.002N$. The average electron energy increases as $\frac{8}{3}kT$ (the center of gravity of the triangle) and the total energy increase, ΔE, is

$$\Delta E \approx n\frac{8}{3}kT = \frac{8}{3}\frac{N}{E_F}(kT)^2.$$

Thus the electron gas heat capacity is

$$c_e = \frac{\Delta E}{\Delta T} \approx \frac{16k^2}{3}\frac{N}{E_F}T.$$

Exact result (Sommerfeld):

$$c_e = \frac{\pi^2 k^2}{2}\frac{N}{E_F}T.$$

$c_e \sim 3 \cdot 10^4$ J/m^{3}K is almost two orders of magnitude lower than the heat capacity of aluminum, $c_{Al} = 2.1 \cdot 10^6$ J/m^3 K.

If a crystal could be heated to 100,000 K, kT would be of the same order as the Fermi energy, all electrons could move to a higher energy state, and the "electron gas" would increase the crystal heat capacity. It would be like a classical gas, but at this temperature the crystal has ceased to exist because it melted and evaporated.

2.2. ELECTRON EMISSION

A lone atom can be in a number of different energy levels, each described by a wave function $\psi(x, y, z, t)$ that is a solution to the Schrödinger [62] wave equation. Consequently, ideal crystal energy levels can be described by similar wave functions where the potential is a periodic function of the spatial coordinates, the period being determined by the crystal lattice dimensions.

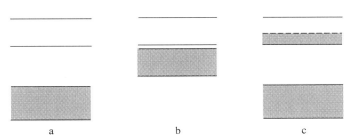

Fig. 2.14. Energy bands. (**a**) Isolator. (**b**) Intrinsic semiconductor. (**c**) Metal.

Solutions only exist for certain energies and energy levels assemble in broad bands, corresponding to individual lattice ions that lie close together. This can be seen as a concurrence between all crystal ions and their influence on the electrons. In metals, any electron in the uppermost energy (conduction) band does belong to the entire crystal.

Electrons occupy some of the possible states, while others are electron-free, with the number of occupied states in an energy band depending on the number of electrons and the temperature. Without an external electric field the electrons move in the crystal randomly, so that the net flow in any direction is zero. If all states in the uppermost energy band are occupied, a small external electric field will not change the electron energy. An external electric field may increase the energy of some electrons, if there are empty states in the energy band with three different outcomes:

(a) The highest energy, the *valence band*, is filled with electrons, and it lies at large distance from the next-higher energy empty band, as shown in Fig. 2.14a. Even at a high temperature, only a negligible number of electrons can be promoted across the gap, so the electrical conductivity is extremely low and the crystal is an isolator.

(b) The filled valence band lies near the next empty band, as seen in Fig. 2.14b. At low temperature the crystal is an isolator, but with increasing temperature some electrons cross the gap and the crystal begins to conduct current when an external electric field is applied, with the conductivity depending directly on the proximity to the next band and temperature. The crystal is an intrinsic semiconductor.

(c) The valence band is only partially filled with electrons because there are either unoccupied states or two bands overlap, as seen in Fig. 2.14c. The crystal is a good conductor, a metal, and the valence band is called the *conduction band*.

Electron emission uses the later two cases. In the intrinsic semiconductors the gap between the valence band and next empty band is so large at room temperature (300 K) that only few electrons bridge the gap. The conductivity is considerably increased by introducing foreign atoms in the crystal lattice in a process called *doping*. If such atoms at 0 K have one, at least partially filled band, which lies near the empty conduction band, electrons will be promoted into the conduction band and the crystal will conduct current. Such foreign atoms are called *donors* and the crystal becomes an n-type semiconductor, as shown in Fig. 2.15a. If the foreign atoms have an empty energy band near the crystals valence band at 0 K, as seen in Fig. 2.15b, the electrons at higher temperature can cross from the crystal valence band to the foreign atom empty band and leave a hole in the valence band. Such crystal also becomes conductive, the foreign atoms are called *acceptors*, and the crystal is a p-type semiconductor.

Donor and acceptor atoms are ionized, and the conductivity becomes acceptable when the ionization potential is $\sim kT$, which at 300 K is 0.026 eV.

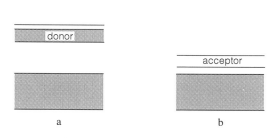

Fig. 2.15. Energy bands in a doped semi-conductor.

The electrons (holes) in the conduction (valence) band move inside the crystal in all direction with the velocities determined by the Fermi–Dirac distribution. Without an external electric field the net electron flow is zero. However, near the crystal surface there exists a potential gradient even when no current is flowing across the surface, because also in the best vacuum foreign atoms are adsorbed on the surface. These atoms, when ionized, generate an electric field, which dies out inside the crystal. Surface irregularities generate also surface donors and acceptors.

Surface charges screen opposite sign charges inside the crystal. On a metal surface these charges form a very thin layer because of the many free charges in the conductivity band. On a semiconductor surface the situation is more complicated, and since the number of charges is much smaller, the layer becomes thicker, 10^{-6} m. The potential varies with distance from the surface.

We now compare a conduction band electron deep inside the crystal, with one near the surface. If both have the same kinetic energy, the distance to the bottom of the conduction band must be the same. The distance to the Fermi level must be different because the electric field near the surface gives that electron a potential energy. Since the Fermi level must be constant inside the whole crystal (Eq. 2.23), the conduction band must be bent near the surface, with the bending generated by surface charges. This macroscopic change of potential bends the band level and strongly influences the ability of the crystal to emit electrons.

Depending on the crystal type, metal, or type of semiconductor, the conduction band bending differs. For example, there can be an excess of surface charges if the adsorbed

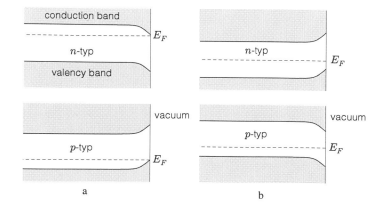

Fig. 2.16. Bending of energy bands near the surface of the crystal.

atoms are donors and the crystal the n-type, or if the adsorbed atoms are acceptors and the crystal is p-type. The bending differs in each case; that is, the field near the surface has a different sign, as seen in Fig. 2.16a. When the adsorbed atoms are of the opposite type, there is a deficit of charges, as seen in Fig. 2.16b.

An external electric field introduces a complication. Such a field enters the crystal and produces an inner space-charge. Even if such a field cannot penetrate deep into the crystal (deeper into a semiconductor than in a metal), it can generate so many charges that the layer changes character. An n-type crystal can behave on the surface as a p-type crystal and vice versa.

An electron, a few interatomic distances from the surface of a metal (or a few tens for a semiconductor), moves under the influence of the described fields. When it comes within one to three interatomic distances of the surface, it meets all the irregularities in the microscopic field, caused by the termination of the lattice.

Atoms on the surface do not have any neighbors on one side, and thus the electric field is different. When an electron leaves the crystal, a dipole is formed. For metals, this dipole layer is very thin, but for all crystals the effect depends on the orientation of the crystalline planes and is influenced by the adsorbed atoms. These surface fields prevent many electrons with low kinetic energy from leaving the crystal.

An electron with sufficient kinetic energy to penetrate the potential barrier at the surface feels another force, which acts at long distances, the mirror force, as seen in Fig. 2.17. This force is a consequence of the charges induced when the electron leaves the crystal. Neglecting all surface irregularities, the mirror force is

$$F = \frac{e^2}{16\pi \epsilon_0 x^2}.$$

The electron is thus moving in an electric potential

$$\varphi = \frac{e}{16\pi \epsilon_0 x} + \text{const.}$$

For the electron to move from $x = x_0$ to infinity, as seen in Fig. 2.18, its energy must exceed the Fermi energy, E_F, by the work function

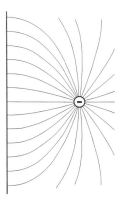

Fig. 2.17. Electric field caused by mirror charges.

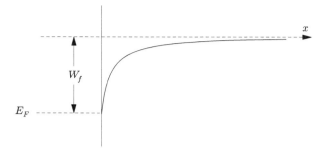

Fig. 2.18. Potential distribution caused by mirror force.

$$W_f = e\varphi_f = \frac{e^2}{16\pi\,\epsilon_0 x_0} + \text{energy due to all aforementioned effects.} \qquad (2.18)$$

W_f differs for different crystal planes because of unequal surface dipole moments.

W_f is also temperature-dependent because the Fermi energy varies with temperature, and the surface dipole layer thickness changes. The work function has a negative temperature coefficient because the Fermi energy decreases with increasing temperature for both metals and semiconductors.

The situation at the surface of a crystal is very complicated, so we can question the validity of the thermionic emission equation derived by Sommerfeld [63], which assumes completely free electrons inside the crystal and no interaction between the electrons and the lattice.

Turning the problem around, if the crystal and its surrounding (vacuum) are in thermal equilibrium, the same number of electrons will cross the boundary layer in each direction in a given time interval. If true, we can look upon the electrons outside the crystal as free electrons. These electrons lie in the "tail" of the Fermi–Dirac distribution, so we can use classical mechanics in our computations and need not know much about the complicated electric field distribution across the boundary. The crystal and its surrounding will be in thermal equilibrium if the Fermi energy is equal on both sides.

Any change in an isolated system increases the system entropy, S. Thermal equilibrium of an isolated system must, therefore, correspond to its highest entropy. Otherwise, spontaneous processes should occur and the entropy would increase further. At equilibrium,

$$dS = 0 \qquad \text{and} \qquad S = \text{maximum.}$$

While p and T are constant, any change of the state of the system results in

$$dS \geq \frac{dQ}{T} = \frac{dU + p\,dV}{T},$$

or

$$dU - T\,dS + p\,dV = d(U - TS + pV) = dG \leq 0,$$

where Q is the heat and U the internal energy. The expression inside the parentheses is called the *thermodynamic potential* or *Gibbs free energy*, and it plays an important role in the equilibrium of chemical reactions and between different phases in solid state.

We can, as an example, take melting. At temperature above the melting point the thermodynamic potential is higher for the crystal than for the melt. The crystal melts spontaneously in an irreversible process which decreases G. At the temperature for which G has the same value both for the melt and for the crystal, the mixture is in equilibrium, both phases coexist, and the temperature is called the *melting point*. It is similar with chemical reactions. A generalized chemical reaction can be written as

$$\sum N_{Ma} A \rightarrow \sum N_{Mb} B,$$

where A and B represent chemical substances and N_M must satisfy the condition that the total number of atoms on both sides of the equation must be the same. If the quantities of A and B are the molar weights, N_M is the number of kilomoles entering the reaction. The change in G can thus be written as

$$dG = \left(-\sum N_{Ma} \frac{dG_a}{dN_{Ma}} + \sum N_{Mb} \frac{dG_b}{dN_{Mb}} \right) dN_M,$$

where, for example, $-N_{Ma} dN_{Ma}$ means the decrease in kilomoles of the substance A.

$$\mu = \left(\frac{dG}{dN_M} \right)_{p,T=\text{const}}$$

is called the *chemical potential* and thus defines the change in the thermodynamic potential per kilomole of a substance in a chemical reaction with p and T kept constant. In equilibrium, G obtains the lowest possible value, but this does not imply that the internal energy, U, is also a minimum. If, for example, by increasing U the entropy is also increased, at constant temperature and pressure, the increase of S can be greater than the increase of U.

For charged particles the electric potential energy is added so that

$$G_e = U - TS + pV + q\varphi.$$

In electrochemical reactions — for example, an electrolytic solution — the charge involved is $q = zFN_M$, where z is the ion valence, N_M the number of kilomoles, and F the Faraday constant (9.64846×10^7 As/kmol). If dN_M kilomoles of charged particles enter the system, the thermodynamic potential will, if temperature, pressure, and electric potential are kept constant, increase by

$$dG_e = \mu dN_M + zF\varphi dN_M,$$

and the *electrochemical potential* will be

$$\left(\frac{dG_e}{dN_M} \right)_{p,T,\varphi=\text{const}} = \mu + zF\varphi = \mu_e.$$

It is easy now to define the change in the electrochemical potential per charged particle. For electrons $z = -1$ and introducing $dN_M = dN/N_A$, recalling that F/N_A is equal to electron charge, we obtain

$$\left(\frac{dG_e}{dN} \right)_{p,T,\varphi=\text{const}} = \mu' - e\varphi = \mu'_e,$$

where μ' and μ'_e are the chemical and electrochemical potential per electron, respectively. μ'_e is the energy added when an electron enters the system. The equation is a consequence of our defining φ ambiguously so that only changes of entropy are important and a term in form of $S_0 T$ can be added without changing the result.

For a system of charged particles electrochemical potential equals Fermi energy:

$$\mu'_e = E_F.$$

This can be proved recalling the relation between the entropy, S, and the probability, W, of the state of the system

$$S = k \, ln W. \tag{2.19}$$

By taking the logarithm of Eq. 2.6,

$$\ln W = \sum [\ln S_j! - \ln N_j! - \ln (S_j - N_j)!],$$

and using Stirling's formula

$$d(\ln N_j!) = \ln N_j \, dN_j,$$

we obtain

$$d(\ln W) = \sum [-\ln N_j + \ln(S_j - N_j)] dN_j = \sum \ln\left(\frac{S_j}{N_j} - 1\right) dN_j. \tag{2.20}$$

By comparing the natural logarithm argument here with that of Eq. 2.7 we obtain

$$\sum \ln\left(\frac{S_j}{N_j} - 1\right) dN_j = \frac{1}{kT} \sum E_j dN_j - \frac{E_F}{kT} \sum dN_j,$$

which, after multiplication by kT, is

$$T d(k \ln W) = T dS = \sum E_j dN_j - E_F \sum dN_j, \tag{2.21}$$

where the first term is the change of the system total energy (Eq. 2.2) while the second term, $\sum dN_j$, which is equal to dN, is the change of the number of system particles (Eq 2.1). We can write

$$dG_e - E_F dN = 0,$$

which gives

$$\left(\frac{dG_e}{dN}\right)_{p,T,\varphi=\text{const}} = E_F. \tag{2.22}$$

Two systems with charged particles which come into contact experience charge transfer so that when the equilibrium is achieved we obtain

$$dG = dG_1 + dG_2 = 0.$$

But

$$dN_1 = -dN_2,$$

so

$$\mu_{e_1} = \mu_{e_2}.$$

Thus the two systems will be in equilibrium when

$$E_{F_1} = E_{F_2}. \tag{2.23}$$

Many physical phenomena — for example, contact potential, Seebeck's thermoeffect [64], and the way semiconductors work — are a consequence of this fact.

2.2.1. Thermionic Emission

In thermodynamic equilibrium, a crystal boundary will be crossed by the same number of electrons traveling in opposite directions and the Fermi energy will be the same on both sides of the boundary. We calculate the number of electrons per unit time that enter the crystal by assuming that outside the crystal the electrons are free in the vacuum, because they are in the "tail" of the Fermi–Dirac distribution. Thus we avoid the complications at the boundary (e.g., lattice irregularities, surface charges, adsorbed atoms, etc.), replacing the boundary by a perfect plane surface. We will assume that this idealized surface is perpendicular to, say, the x-axis. Outside the crystal the potential energy is the sum of the Fermi energy E_F and the work function W_f.

The number of electrons per unit volume of phase space is determined by the number of states per unit volume (Eq. 2.15) and by the Fermi–Dirac distribution.

$$dN = \frac{dN}{dS}dS = \frac{1}{1 + e^{(E-E_F)/(kT)}} \frac{1}{h^3} 4\pi \sqrt{(2m_e)^3}\sqrt{E_k}\, dE_k\, dx\, dy\, dz.$$

The number of electrons crossing the boundary per unit time and per unit area is

$$dn = \frac{dN}{dy\, dz\, dt} = \frac{1}{h^3}4\pi\sqrt{(2m_e)^3}\frac{v_x\sqrt{E_k}}{1 + e^{(E-E_F)/(kT)}}\, dE_k.$$

Making an energy to velocity transformation

$$E_k = \frac{1}{2}m_e v^2, \qquad dE_k = m_e v\, dv,$$

changing from spherical to rectangular coordinates

$$4\pi v^2 dv = dv_x\, dv_y\, dv_z,$$

and introducing

$$4\pi\sqrt{E_k}\, dE_k = 4\pi\sqrt{\frac{m_e}{2}}v m_e v\, dv = 4\pi\frac{\sqrt{m_e^3}}{\sqrt{2}}v^2 dv = \frac{\sqrt{m_e^3}}{\sqrt{2}}dv_x\, dv_y\, dv_z,$$

we get

$$dn = 2\frac{m_e^3}{h^3}\frac{v_x\, dv_x\, dv_y\, dv_z}{1 + e^{(E-E_F)/(kT)}}.$$

The total electron energy, the sum of its potential and kinetic energy, is at the boundary

$$E = E_F + W_f + E_K = E_F + W_f + E_x + E_y + E_z,$$

or

$$E - E_F = e\varphi_f + E_x + E_y + E_z.$$

The current density is

$$J_{sat} = \int e\, dn = 2e\frac{m_e^3}{h^3}\int\limits_0^\infty\int\limits_{-\infty}^\infty\int\limits_{-\infty}^\infty \frac{v_x\, dv_x\, dv_y\, dv_z}{1 + e^{e\varphi_f/(kT)+m_e(v_x^2+v_y^2+v_z^2)/(2kT)}}.$$

In the y and z directions parallel to the boundary, the electrons can have all possible velocities. To simplify the integration, transform the rectangular coordinates into polar ones

$$v_y^2 + v_z^2 = v_r^2, \qquad\qquad dv_y dv_z = v_r dv_r d\theta,$$

and make the substitution

$$\frac{m_e}{2kT} v_r^2 = u, \qquad v_r dv_r = \frac{kT}{m_e} du, \qquad a = e^{-(2e\varphi_f + m_e v_x^2)/(2kT)}.$$

Using

$$\int\limits_0^\infty \frac{du}{1 + \frac{e^u}{a}} = \ln e^u - \ln\left(e^u + a\right)\Big|_0^\infty = \ln(1 + a),$$

the integral over the y–z plane becomes

$$\frac{kT}{m_e} \int\limits_0^{2\pi} d\theta \int\limits_0^\infty \frac{du}{1 + e^u\, e^{e\varphi_f/(kT) + m_e v_x^2/(2kT)}} = 2\pi\frac{kT}{m_e} \ln\left(1 + e^{-(2e\varphi_f + m_e v_x^2)/(2kT)}\right).$$

We now use a quantum mechanical fact that electrons have a wave nature, and waves are reflected at discontinuities. If the electron reflection probability is r, then the boundary will be crossed by $1 - r$ electrons, where r is a function of electron energy. The current density becomes

$$J_{sat} = 4\pi e \frac{m_e^2 kT}{h^3} \int\limits_0^\infty (1 - r) \ln\left(1 + e^{-(2e\varphi_f + m_e v_x^2)/(2kT)}\right) v_x dv_x. \qquad (2.24)$$

We evaluate this integral under certain assumptions. Thermionic emission is observed above ~ 1000 K. The measured work function for most metals and semiconductors is between 1 and 6 eV, the numerical value of the work function divided by kT is ~ 10, and the exponential term becomes $\sim 10^{-5}$. Expanding the natural logarithm

$$\ln(1 + x) = x - \frac{x^2}{2} + \frac{x^3}{3} - \frac{x^4}{4} + \cdots,$$

we need keep only the first term. Replacing the energy-dependent reflection coefficient r by its mean value, we obtain the emitted current density

$$J_{sat} = 4\pi e \frac{m_e k^2 T^2}{h^3} (1 - <r>) e^{-e\varphi_f/(kT)} = A(1 - <r>) T^2 e^{-e\varphi_f/(kT)}. \qquad (2.25)$$

This is the *Richardson equation* [49], sometimes also called *Richardson–Dushman* equation [65]. The equation constant

$$A = 4\pi e \frac{m_e k^2}{h^3} = 1.2 \cdot 10^6 \frac{A}{m^2 K^2}$$

is called the *universal thermionic constant* and is valid for all crystals.

The exponential term is responsible for the largest part of the increase in current density as temperature is increased. The increase due to T^2 is small and not confirmed experimentally. Taking as an example tungsten at 2500 K, a 1 percent change in temperature results in a 2 percent change in J_{sat} because of the T^2 factor, compared to a 20 percent increase in the exponential term. In nature, few effects change so fast as the electron emission, where doubling of temperature increases the current density 100-fold.

That the reflection coefficient must be small has been confirmed both by quantum mechanical studies and experimentally. Although not unambiguous, experimental results indicate that r can increase for slow electrons and may be strongly energy-dependent. Generally, the reflection coefficient is neglected and J_{sat} is usually written

$$J_{sat} = AT^2 e^{-e\varphi_f/(kT)}.$$

Equation 2.24, with the natural logarithm argument replaced by the first term of the series, shows the number of emitted electrons per unit time and unit area in the velocity interval dv_x:

$$\frac{dn(v_x)}{dv_x} = 4\pi \frac{m_e^2 kT}{h^3}(1 - r)v_x e^{-(2e\varphi_f + m_e v_x^2)/(2kT)}.$$

Dividing by the number of emitted electrons

$$n_e = \frac{J_{sat}}{e},$$

the velocity distribution is

$$\frac{1}{n_e}\frac{dn(v_x)}{dv_x} = \frac{m_e}{kT}\frac{1-r}{1-<r>}v_x e^{-m_e v_x^2/(2kT)}, \tag{2.26}$$

which is *Maxwell's velocity distribution*. This is expected since the electrons outside the crystal are in the "tail" of the Fermi–Dirac distribution, the region that agrees with the Maxwell–Boltzmann distribution. The mean kinetic energy of the electrons leaving the crystal in the x direction is

$$<E(p_x)> = kT,$$

and in the y and z directions

$$<E(p_y)> = <E(p_z)> = \frac{1}{2}kT,$$

which together gives

$$<E> = 2kT. \tag{2.27}$$

The cathode cools when it emits electrons because of the work function and mean kinetic energy of the emitted electrons. The power loss per unit area is

$$P_{loss} = J(\varphi_f + \frac{2kT}{e}).$$

The largest power loss, however, comes from thermal radiation from the cathode surface — roughly proportional to T^4 — and from the conduction to its supports.

The thermionic constant, A, and the work function, W_f, for a material can be determined by measurements, but they are extremely difficult to perform. Only by working in ultra-high vacuum and with monocrystals have consistent results been obtained for thermionic constant of tungsten, and they are still only a half of the theoretical value. The work function in all practical cathodes is also difficult to measure. The emission never comes from the whole cathode surface. The orientation of the crystal lattice planes, impurities (like on the surface adsorbed atoms), and materials which come from other electrodes (primarily anode by sputtering) greatly influence the measurements. Semiconductors are often porous and thus have reduced emitting area, while surface irregularities, which make, on the microscopic scale, the cathode surface resemble an Alpine landscape, increase the local electric field and thus the emission. Thus cathodes made of the same material and with exactly the same dimensions give different current densities. We can imagine dividing the cathode surface into small regions with area ΔS_i, across which we assume a constant work function. Produce a mean value

$$<W_c> = \frac{1}{S}\Sigma W_i \Delta S_i$$

and the current density

$$(J_{sat})_{eff} = \frac{1}{S}\Sigma J_{sat\,i}\Delta S_i.$$

Therefore, we write Richardson's equation as

$$(J_{sat})_{eff} = A_R T^2 \, e^{-W_R/(kT)},$$

where A_R and W_R are the measured cathode constants.

2.2.2. The Influence of Electric Field

The foregoing not withstanding, we now assume that the emitting surface is without any irregularities and that the work function is constant across the whole surface. To be able to measure or use the emitted current, we must introduce at least one more electrode into the vacuum system, which collects all or a part of the emitted electrons.

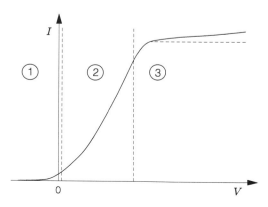

Fig. 2.19. Current–voltage relationship for a diode.

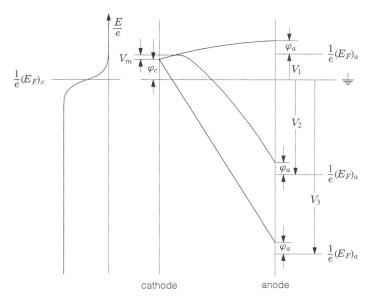

Fig. 2.20. Potential distribution in a diode with the Fermi–Dirac distribution showed along the potential axis.

Such a system with two electrodes is called a *diode*. The electron-emitting, negative electrode is the *cathode*, and the electron-collecting, positive electrode the *anode*.* We assume that the cathode is grounded. The current–voltage relationship for a diode, shown in Fig. 2.19, can be divided in three regions: the initial current region, the space-charge region, and the saturation current region.

The potential variation between the cathode and the anode for all three regions is shown schematically in Fig. 2.20. Observe that the positive direction of the anode voltage is directed downward. We assume that the reflection coefficient, r, is zero for both the cathode and the anode, and that the electrodes are closely spaced, plane, and parallel.

1. Initial Current Region

Here the anode potential is negative with respect to the ground, so the electrons on their way from the cathode to the anode must move against the potential difference

$$\Delta V = V_1 - \varphi_a, \qquad V_1 < 0.$$

From Richardson's equation we obtain the current density

$$J_b = AT^2 \, e^{e(V_1 - \varphi_a)/(kT)} = J_{sat} \, e^{e(V_1 - \varphi_a + \varphi_c)/(kT)}, \tag{2.28}$$

where V_1 is the anode voltage (i.e., the potential difference between the Fermi levels of the cathode and the anode), $e\varphi_c$ is the cathode work function, and $e\varphi_a$ is the anode work function. If I_b is the current emitted by the cathode a curve $\ln(I_b)$ as a function of V_1 can be drawn which, according to Eq. 2.28 should be a straight line, suggesting that the emitted electrons follow a Maxwellian velocity distribution.

---

* The terms "electrode," "cathode," "anode," and "ion" were coined by Faraday ["Experimental Researches in Electricity, VI, §662 ff", *Phil. Trans. Roy. Soc.*, 77 (1834).]

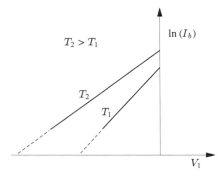

Fig. 2.21. Current-voltage relationship in the initial current region.

Measurements which show agreement with the theory in large current limits are very difficult to perform at low currents (below 10^{-15} A), where the reflection coefficient dependence on current was reported [66].

2. Space-Charge Region

Inspection of Eq. 2.28 could call forth the false believe that when $V_1 - \varphi_a + \varphi_c = 0$, the anode current becomes equal to saturation current. This is not true because there is created in front of the cathode a region with negative space-charge. The energy of emitted electrons is MB-distributed, and so there are more slow electrons than fast ones, which produce a cloud repelling the slowest ones. The corresponding potential minimum in front of the cathode is called the *virtual cathode*. The fraction of the electrons which pass the potential minimum, V_m, seen in Fig. 2.20, is determined primarily by the voltage between the cathode and anode. The current density is

$$J_r = AT^2 \, e^{-e(\varphi_c + V_m)/(kT)} = J_{sat} \, e^{-eV_m/(kT)}, \tag{2.29}$$

where the potential minimum, V_m, depends on both J_r, V_a, and initial velocity distribution. Therefore, an analytical expression for V_m can only be obtained for some special geometries, such as plane [67], cylindrical [68], and spherical [69].

A good practical solution can be obtained by assuming that the initial electron velocities at the cathode are zero. All electrons have the same velocity at the same distance from the cathode, and so V_m must be zero. Mostly the anode voltage in a diode is greater than a hundred volts. If the electrode near the cathode is a control grid (for example, the triode of Section 3.3), the voltage between them can be much smaller. However, the anode field penetrates the grid and produces an equivalent voltage of a few volts at least. The initial kinetic energy of thermionic electrons is $\sim kT$, which corresponds to 0.1–0.25 eV at a cathode temperature of 1000–2500 K. Thus V_m must also be of similar magnitude and can be neglected.

The existence of space-charge, ρ, implies that Poisson's equation must be used to derive the current-voltage relationship, in addition to the continuity and energy conservation equations. The continuity equation holds the current density to be independent of the x coordinate. We solve this problem only for the planar geometry, seen in Fig. 2.22. Introducing

$$V = V_2 + \varphi_c - \varphi_a$$

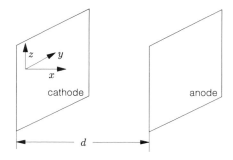

Fig. 2.22. Geometry in a planar diode.

and

$$\frac{d^2V}{dx^2} = -\frac{\rho}{\epsilon_0} \qquad \text{(Poisson's equation)},$$

$$J = \rho v \qquad \text{(Continuity equation)},$$

$$eV = \tfrac{1}{2}m_e v^2 \qquad \text{(Energy conservation)},$$

we obtain

$$\frac{d^2V}{dx^2} = -\frac{J}{\epsilon_0\sqrt{2eV/m_e}}.$$

After multiplying by $2dV/dx$ and integrating we obtain

$$\left(\frac{dV}{dx}\right)^2 = -\frac{4J\sqrt{V}}{\epsilon_0\sqrt{2e/m_e}} + C.$$

At the low limit, where at the cathode $dV/dx = 0$ because the initial velocities are neglected, we obtain

$$\frac{dV}{dx} = 2\sqrt{-\frac{J}{\epsilon_0}\sqrt{\frac{V}{2e/m_e}}},$$

which, when integrated with $V_{(x=0)} = 0$, gives

$$\frac{4}{3}V^{3/4} = 2\sqrt{-\frac{J}{\epsilon_0}\sqrt{\frac{1}{2e/m_e}}}\,x.$$

In most cases $V_2 \approx V_a \gg \varphi_c - \varphi_a$. The anode current ($x = d$) in a planar diode is

$$J_r = -\frac{4\epsilon_0}{9}\sqrt{\frac{2e}{m_e}}\frac{V_a^{3/2}}{d^2} = -2,334\cdot 10^{-6}\frac{V_a^{3/2}}{d^2}\;\frac{\text{A}}{\text{m}^2}. \qquad (2.30)$$

This expression, called *Child–Langmuir law,* shows the space-charge limited anode current to be proportional to the anode voltage raised to 3/2 power. This relationship between the current and the anode voltage is valid for all cathode–anode geometries and we can write

$$I = pV_a^{3/2}, \qquad (2.31)$$

where the constant, p, called the *perveance,* only depends on the form of the electrodes.

Note that the current density in the space-charge region is independent of the cathode temperature, since both $\varphi_c - \varphi_a$ and the initial velocity distribution vary slowly with the temperature and $\varphi_c - \varphi_a$ inherently is small. This result is valid even when the virtual cathode is considered, so *the heater current need not to be stabilized* when space-charge-limited current emission is used in an electron device.

3. Saturation Region

Current saturation is reached when the anode voltage is sufficiently high so that all the emitted electrons are collected, the space-charge region is depleted, and the virtual cathode and the corresponding potential minimum, V_m, cease to exist.

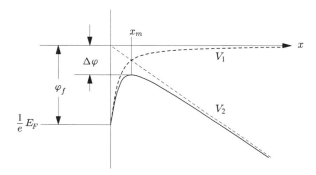

Fig. 2.23. Potential distribution near the cathode in the saturation region.

The situation in front of the cathode, shown in Fig. 2.23, is similar to that of Fig. 2.18. Because of the high anode potential, a superposition of two electric fields must be considered: the mirror charge field

$$E_1 = -\frac{e}{16\pi\epsilon_0 x^2},$$

and the constant external field E_c. At x_m in front of the cathode the field magnitudes are equal and the forces on an electron leaving the cathode exactly cancel

$$F_1 + F_2 = -\frac{e^2}{16\pi\epsilon_0 x^2} + eE_c = 0.$$

Thus

$$x_m = \sqrt{\frac{e}{16\pi\epsilon_0 E_c}}.$$

The work function decreases, the corresponding difference in the potential distribution is

$$\Delta\varphi_f = \frac{1}{e}\left[\int_{x_m}^{\infty} F_1 dx + \int_{x_m}^{0} F_2 dx\right] = -\frac{e}{16\pi\epsilon_0 x_m} - E_c x_m = -\frac{1}{2}\sqrt{\frac{eE_c}{\pi\epsilon_0}},$$

and the saturation current density is

$$J_m = J_{sat}\, e^{e\sqrt{eE_c/(\pi\epsilon_0)}/(2kT)}. \tag{2.32}$$

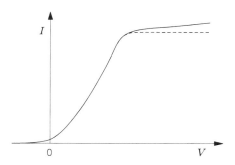

Fig. 2.24. Schottky effect.

In the saturation current region the slow raise in current density with the anode voltage, the so-called Schottky effect [70], can considerably increase the anode current at high anode voltage, as seen Fig. 2.24. Plotting ln (J_m) with square root of V_a results in a straight line.

Measurements show small periodical deviations from a straight line at high electric field. Even of no practical interest, they show a weak reflection of the electrons at the potential barrier that is measurable when the barrier becomes thin enough. This reflection occurs when the electron from inside the crystal comes near the surface and again at the border with the vacuum. The number of reflected electrons depends on the thickness of the barrier, which depends on the external electric field.

For the common oxide cathodes the Schottky effect does not give an increase of the saturation current density according to Eq. 2.32 because the very rough cathode surface and x_m depend strongly on the electric field strength.

2.2.3. Photoemission

When a metal or a semiconductor surface is illuminated by radiation of a suitable wavelength, electrons leave the surface. The photon energy, $E = hf$, can be completely transferred to an electron inside the crystal, and if it is high enough the electron will be ejected. The electron energy must be sufficiently larger than the Fermi level to overcome the boundary potential barrier (work function) and with kinetic energy to spare. The probability for this process, its quantum efficiency, expressed as the number of emitted electrons per incoming photon, is low — between 10^{-5} and 10^{-1}.

If the cathode temperature is low enough, we can use the FD distribution corresponding to 0 K (Fig. 2.10b). A photon with energy higher than the work function of the crystal can transfer all its energy to an electron, which can be emitted. Thus for each crystal there is a minimum-frequency photon, which will liberate electrons with zero kinetic energy. At higher frequencies the emitted electron energy increases as (Einstein's formula [48])

$$hf - W_f = \frac{1}{2}m_e v^2. \tag{2.33}$$

The minimum frequency is therefore $f_0 = W_f/h$. Introducing the limiting wavelength, $\lambda_0 = c/f_0$, where c is the velocity of light, we obtain

$$\lambda_0 = \frac{hc}{e\varphi_f} = \frac{1240.4}{\varphi_f} \ \text{[nm]},$$

with $\varphi_f = W_f/e$ (Eq. 2.18).

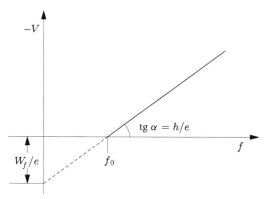

Fig. 2.25. Graphic presentation of Einstein's formula.

At higher temperature some electrons have energy greater than E_F, and so the minimum wavelength moves toward longer wavelengths and becomes less sharp. Einstein's formula fails and must be extended by [71]

$$\ln \frac{I}{T} = C + \ln\left[F\left(\frac{hf - W_f}{kT} \right) \right],\qquad(2.34)$$

where I is the anode current and C and F are constants.

The minimum wavelength at low temperature can be measured as follows: A photoemitting cathode is placed near an anode in a vacuum chamber. The cathode is illuminated by monochromatic light, whose wavelength is varied. Changing the anode voltage and measuring the emitted anode current, a negative anode voltage is found where the current becomes zero. Plotting the anode voltage at zero anode current with light frequency, a straight line of Fig. 2.25 is obtained. In this graphical presentation of Einstein's formula the slope is proportional to Planck's constant; the point where the line crosses the abscissa determines the minimum frequency, and the point where it crosses the ordinate gives the work function.

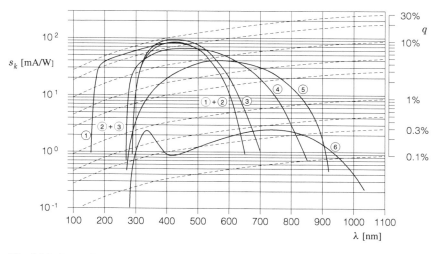

Fig. 2.26. Spectral sensitivity, s_k, and quantum efficiency, q (dashed lines), for different photocathodes (Philips). 1, UV (GaAs(Cs)); 2, blue (SbKCs); 3, green (SbCs$_3$); 4, red (SbNa$_2$KCs); 5, extended red (SbNa$_2$KCs); and 6, infrared-sensitive (AgOCs).

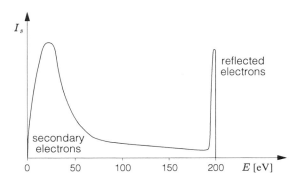

Fig. 2.27. Energy distribution of secondary electrons. The narrow peak on the right shows the reflected primary electrons. (After *Handbook of Vacuum Physics, Vol. 2, Physical Electronics,* A. H. Beck, Ed., 1968, Pergamon Press.)

For any wavelength above the minimum, the photocurrent is proportional to the intensity of the incoming light: The larger the intensity, the greater the number of photons. All photons with the same energy have the same probability of triggering the emission of electrons.

Most photocathodes show a selective photoelectric effect. Every photoemitting material has a specific emission maximum. The quantum efficiency decreases and the bandwidth increases in proportion to the atomic number. Today mostly compound photocathodes use alkaline metals, as seen in Fig. 2.26, to attain a higher sensitivity — that is, a higher quantum efficiency with a less marked maximum in the visible region.

2.2.4. Secondary Emission

A crystal bombarded by electrons or ions, metal or isolator, emits electrons in a process called *secondary emission*. This differs from the thermionic emission, because it is detected even at very low current densities, where the bombarding particles cannot increase the temperature in the target area. A classical explanation is that some incoming particles are reflected by the electrons and ions in the crystal lattice. Such particles lose only part of their energy, but calculations show that their number must be much smaller than the observed number of emitted electrons. Measured energy distribution shows that ~ 10 percent of all emitted electrons have their energy near that of the bombarding beam and these electrons are the reflected electrons. The remaining ~ 90 percent have energies ~ 15 eV and are the true secondary electrons, as seen in Fig. 2.27, for a primary beam of 200 eV.

The number of "secondary" electrons (i.e., the number of true secondary and of reflected primary electrons) produced by the incoming beam is called the *secondary emission gain, δ*. Let $n(x, E)dx$ be the number of secondary electrons per primary electron with energy E, released in a layer dx thick at a depth x. $f(x)$ is the probability that a secondary electron created at this depth will reach the surface. Thus the secondary emission gain is

$$\delta = \int_0^\infty n(x, E)\, f(x)dx.$$

In contrast with thermionic emission, the secondary emission depends on a much larger volume in the target area. The work function and the surface change the gain, but are not

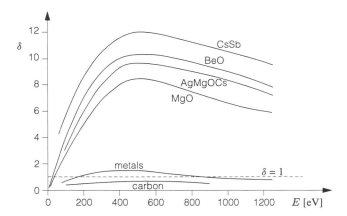

Fig. 2.28. Secondary emission gain for various materials with primary beam energy. (After *Moderne Vakuumelektronik*, J. Eichmeier, © 1981, Springer-Verlag, with permission.)

the determining factors as observed when a metal with high δ was coated by another one with low δ. The coated layer thickness had to be appreciable before a substantial decrease of the gain was observed. This implies that the secondary electrons are generated over much of the primary electron trajectory in the material. At the same time there must be internal absorption, because after a maximum is reached, the gain decreases with the primary particle energy. This is confirmed by an increased gain at an oblique angle of incidence, since the secondary electrons have a shorter path to the surface.

The theory of the secondary emission, still under development, can be divided in two parts: (a) computation of the number and energy distribution of the secondary electrons produced in the target area and (b) analysis of the transport processes which determine the movement of the secondary electrons and their collisions with the electron gas and ions in the crystal lattice. Only the electrons whose momentum component perpendicular to the surface is big enough to overcome the work function can leave the material.

Quantum physics says that secondary electrons are generated not by direct collisions but by the excitation of valence band electrons, where energy distribution has a maximum near $E = E_F$. Thus most excited electrons have approximately the same direction as the incoming primary electrons. The transport theory does not fit the measurements for metals, and it is worse for semiconductors and isolators.

Experimentally, it is difficult to measure δ because the surface strongly influences the gain. For metals the gain is slowly varying with energy and remains below 2, heavier metals (e.g., Ni, Mo, W) showing a larger δ than the alkaline metals. Metal oxides on metal base show a high gain, in some cases > 30. For high gain, complex materials are used, important among which are Ag–O–Cs, MgO on Mg and BeO on BeCu, as seen in Fig. 2.28.

The experimental results can be interpreted as follows: Every excitation uses, on average, $\sim E_F$ of the primary particle energy. A primary electron can thus only excite some electrons and the probability of excitation per unit length decreases with electron velocity. A slow electron loses its energy near the surface, so the excited electrons can, if their momentum is opportunely directed, leave the material, since they need not cover a distance. Higher-energy electrons penetrate deeper into the material and release a larger number of secondary electrons, many of which lie so deep in the material that they have no chance to reach the surface. Thus the gain decreases when the energy of the primary beam exceeds a certain value.

The high absorption power for electrons of metal surfaces is a consequence of partially filled metal conduction bands. Thin isolating metal oxide layers have no free states in the valence band and thus cannot absorb the excited electrons, which easily leave the material. To continuously emit electrons the oxide layer must be thin and a conducting material must lie below the layer to replace the emitted electrons.

Although positive ions also cause secondary emission, their gain is much lower than that for electrons, since in an ion–valence-electron collision only a small fraction of the ion's energy is transferred to the electron, a consequence of the large mass difference. Thus the bombarding ion energy must be much greater than that of electrons to give a comparable secondary emission gain. Ion-induced secondary emission plays an important role in gas discharges (see Section 9.3).

Secondary emission always appears when a beam hits a surface, and thus it is often important to avoid or to reduce the secondary emission. This is accomplished by covering the surface with a low-gain material, like graphite. Or the electrodes can be shaped (e.g., like a cup) so that the emitted secondary electrons are intercepted by another part of the electrode. Sometimes a retarding electric field is built which stops the secondary electrons in front of the electrode from which the secondary emission should be damped or avoided.

2.2.5. Field Emission

Electron emission can be observed from cold electrodes in the presence of a high electric field near the surface, and it is called *field emission*. The low temperature of the electrodes makes thermionic emission negligible, and this effect has no connection with the Schottky effect.

For field emission a field strength of $10^8 - 10^9$ V/m in front of a perfectly plane surface is required. Surface unevenness and sharp tips substantially increase the local field strength, so that field emission is observed at a much lower average field. Large current densities heat these protrusions, which further increases emission, leading to a breakdown in vacuum between two electrodes even at moderate voltages (e.g. transmitting tubes). Sharp edges, corners, and small radii of curvature must be avoided when constructing high-voltage electron tubes and devices. The system often must be conditioned to withstand high electric field strength by slowly increasing the electrode voltage causing weak flashovers, which burn down protrusions.

Field emission can be understood qualitatively by recalling the wave nature of the electrons inside the crystal. A large electric field in front of the surface reduces the potential barrier width, allowing some conduction band electrons to tunnel through the barrier, as shown in Fig. 2.29.

Although the wave function falls exponentially inside the barrier, a small, finite amplitude will be present on the other side; the shorter the barrier, the larger the wave function amplitude outside the crystal.

Field emission from metals at 0 K [72] is expressed as

$$J_{sat} = CE^2 \, e^{-D/E}, \tag{2.35}$$

where $C \sim 10^4 - 10^5$ A/V^2 and $D \sim 10^9 - 10^{11}$ V/m are two material dependent constants.

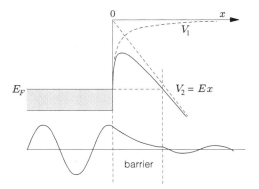

Fig. 2.29. Field emission. Potential barrier and the wave function of the tunneling electron.

Field emission conditions are changed at higher temperatures in that the FD distribution "tail" electrons see a narrower potential barrier, which increases the field emission. In field emission cathodes (e.g., electron microscopes) it is better to heat the cathode (1000–2000 K) by applying a lower field strength than by using a cold cathode.

Field emission and vacuum breakdown are often a problem in high-voltage and high-electric-field-strength devices. Besides avoiding sharp edges, polishing the electrodes and/or plating them with high-work-function materials (e.g., gold or tantalum) helps solve this problem. However, field emission is not the only cause of vacuum breakdown. Many different mechanisms seem to be involved. Small particles can be torn from electrodes, accelerated in the gaps between the electrodes, and upon impact cause local heating and cratering. Not only the field strength, but even the voltage and the choice of material, plays an important role.

2.2.6. Cathode Materials

In electron devices which use free electrons, the cathode is one of the most important parts. When thermionic emission is used, the cathode material must have a low work function and a melting point high enough to avoid significant evaporation at working temperature. In high-anode-voltage applications the cathode must be able to endure ion bombardment. Photocathodes should have a high quantum efficiency and a low work function. Cathodes used as secondary electron sources must have a high gain. All these conditions cannot be satisfied with a single material.

Alkaline metals have the lowest work function but also a low melting point (e.g., cesium evaporates at 28°C). The work function and melting temperature of some materials are given in Table 2.1.

1. Thermionic Emission Cathodes

There are five categories of materials used for thermionic cathodes: pure metals, metal coating metal surfaces, semiconductors on metal surfaces, impregnated cathodes, and rare earth hexaborides. Each material has properties suitable for different applications.

(a) Pure Metals

Tungsten is the usual pure metal used for cathodes, despite its high work function (4.52 eV), because of its high melting point (3400 K). In view of its homogeneity, it is insensitive to fast positive ion bombardment. Tungsten is used in X-ray tubes, high-power transmitting tubes, and most applications calling for high anode voltage. The tungsten cathode

Table 2.1

Work function and melting point

Material	Work function	Melting point	Material	Work function	Melting point
Ag	4.7	960.5	Nd	3.3	1016
Al	3.0	660.3	Ni	5.0	1453
Au	4.8	1064.6	Pt	6.0	1772
Ba	2.52	729	Rb	1.8	39.6
C	4.7	3827	Sr	2.1	768
Ca	3.2	839	Ta	4.1	3014
Cd	4.1	320.9	Ti	4.09	1670
Cs	1.8	28.5	Th	3.4	1755
Cu	4.1	1084.6	W	4.52	3407
Fe	4.7	1535	Zr	4.1	1852
Hg	4.5	-38.7	LaB_6	2.74	2210
Ir	5.4	2443	NdB_6	4.57	
K	1.8	63.4	TaB	2.89	
La	3.3	920	TaC	3.14	
Li	2.2	180.7	ThO_2	4.57	
Mo	4.3	2617	TiC	3.35	
Na	1.9	98	ZrB	4.48	

Source: *Vacuum Tubes*, K. R. Spangenberg, 1948, McGraw-Hill; and *American Institute of Physics Handbook*, D. E. Gray, ed., © 1967, McGraw-Hill.

working temperature depends on the current density: 2400 K, 2500 K, and 2600 K show a saturation current density of 0.4, 0.9, and 2.3 A/cm², respectively. The lifetime is limited by evaporation. These cathodes are usually designed so that their cross-section can be reduced by 10−15 percent during the lifetime. At 2600 K the tungsten evaporation rate is 3×10^{-8} g/s cm², which corresponds to ~ 0.2 g/cm² or 0.01 mm in 2000 hours. The thicker the wire, the longer its life at the same temperature. Water vapor appreciably reduces the lifetime, since in contact with a hot tungsten surface it forms free hydrogen and tungsten oxides, whose evaporation is much faster than the metal. Tungsten oxides accumulate on the cold walls of the tube, where they are reduced back into tungsten by the hydrogen. Producing water, the vapor enters again into the cycle and the process can continue, even with very little water vapor in the tube. This process cycles destructively. For glass tubes the glass darkens with time, as seen in light bulbs. In addition, at high temperature tungsten recrystallizes, becomes brittle, and breaks easily under mechanical stress caused, for example, by magnetic forces when heater voltage is connected. At room temperature the initial current surge is some 15 times larger than the working temperature current (most light bulbs burn when switched on).

Sometimes cathodes are made of tantalum, where the work function is 4.1 eV and the melting point is 3000 K. Compared to tungsten, tantalum requires lower temperature to produce the same current density. Its drawback is that oxygen strongly reduces the emission. Tantalum is used sparsely, for complicated cathode shapes, since it can be worked more easily than tungsten.

Most semiconductor and dispenser cathodes are indirectly heated by tungsten spirals. A few percent of rhenium can be added to tungsten to make it less brittle.

(b) Metal-Coated Metal Surfaces

These materials are used in high-power transmitting tubes since they show less crystallization than do pure metals. Thorium oxide added to tungsten was observed to have much higher emission than the pure metal tungsten [50]. Pure thorium is not suitable because its work function is 3.4 eV and the melting point is 1755 K.

Tungsten alloyed with ~ 1 percent thorium has increased emission, which depends on the reduction of the oxide to pure thorium, which diffuses between tungsten crystals to the surface. There thorium builds up a monoatomic electropositive layer producing a strong electric field and reducing the work function to 2.63 eV, smaller than that of pure thorium. Traces of carbon also seem to play a role. Probably, carbon and tungsten make a carbide which reduces the thorium oxide. A thorated tungsten cathode working temperature is 1800−2000 K, very near the thorium melting point. The metal is restrained by surface tension, which is higher for tungsten than for thorium. Thorium evaporation is negligible.

Thorium oxide (1−2 percent) is added to tungsten before sintering. During pumping at ~ 1600 K, a hydrocarbon compound (e.g., acetylene or naphthalene vapor) is introduced, which is reduced on the hot surface. Carbon then diffuses into tungsten to form tungsten carbide, and the cathode must then be activated. This is done by holding the cathode at 2800 K for ~ 1 minute while inside the thorium oxide is reduced to metallic thorium. The temperature is then reduced to ~ 2100 K for tens of minutes, during which time thorium diffuses to the surface and makes a monoatomic layer, which increases the emission.

Thorated tungsten cathodes can exhibit decreased emission over time or after over-loading. The damage can be reversed by repeating the second part of the activation process. Such restoration is possible until most of the thorium is consumed.

Thorated tungsten cathodes are sensitive on residual gases, and oxygen greatly reduces the emission. Ion bombardment also limits the lifetime, so thorated tungsten cathodes are not used in devices where the anode voltage exceeds 10 kV.

Besides thorated tungsten, thorated iridium and zirconized and titanized tungsten are sometimes used as cathode material.

(c) Semiconductor Cathodes

Among the most important thermionic emission materials are alkaline metal oxides like those of barium, strontium, and calcium because of their low work function (~ 1 eV), and thus at a low working temperature (900–1100 K) a high current density can be obtained.

These oxides, however, are easily poisoned, so oxide cathode production starts with the carbonates of the metals. Heating in vacuum decomposes the carbonates forming oxides and carbon dioxide, which must be pumped away. To activate the cathodes a suitable voltage is connected between the cathode and other electrodes while under vacuum. The current increases slowly from zero until full emission is reached.

Most oxide cathodes are indirectly heated by a tungsten wire inside a hollow metallic electrode whose surface is coated with the emitting material.

Oxide cathode operation and activation is not very well understood. For example, it is known that not only the cathode material but also the base material and the production methods play an important role in the manufacture. Nickel is usually used as base material for oxides because it evaporates slowly at the working temperature and has a low heat con-ductivity. Improved performance is obtained if a few hundreds of a percent of manganese, magnesium, zirconium, or silicon are added, which diffuse to the cathode surface and reduce the barium, strontium, and/or calcium oxides. Thermodynamic equilibrium is thus obtained for the pressure of especially barium, which slowly evaporates at the working temperature.

In the manufacture of oxide cathodes a barium/strontium/calcium carbonate paste with a fixing agent (like cellulose varnish) is sprayed, painted, centrifuged, or deposited by electrophoresis or doping on the base material. The cathode is heated and the paste dries into a layer between tens and hundreds of a micrometer thick, depending on the use for the cathode. It is then mounted in the tube, which is evacuated to $\sim$ 0.1 Pa. When heating starts the water vapor and adsorbed gases are driven out. At $\sim$1200 K the carbonates start to decompose into oxides and carbon dioxide, the porous oxide layer is sintered, the fixing agent decomposes, and the carbon oxidizes with carbon dioxide and is pumped off as carbon monoxide. This process must be done rapidly with a high-speed pumping system. At the end of the decomposition, when the pressure starts to drop, the temperature is increased to $\sim$1500 K for a short time. A sudden pressure drop indicates that the carbonates are decomposed and the cathode is outgassed. During this process the tube is also heated from the outside to outgas the inner surfaces. The electrodes can be outgased by high-frequency heating.

The activation process is then started by connecting a voltage between the cathode and other electrodes, keeping the temperature at $\sim$ 1250 K, a compromise between not allowing barium to evaporate too rapidly and a reasonably fast activation process. The emission current is low from the beginning but stabilizes after a while, finishing the cathode activation. This activation process is believed to be similar to electrolysis. Barium, liberated by the decomposition of its oxide, is transported to the surface by diffusion and continues so under normal operation.

There is large variation in the emission of oxide cathodes for identical cathodes, which should therefore be used under space-charge-limited conditions. Typical current densities are up to 1 A/cm^2 under DC drive and up to tens of amperes per square centimeter when used in short pulses, but the emission decreases during the pulse. Three time constants are observed. The first, $\sim$10^{-4} s, is the most important and depends probably on the changes of the inner space-charge and surface charges. It limits the pulse length in pulsed microwave tubes to some microseconds, after which an output power decrease can be observed. The second, between 0.01 and 1 s, probably results from the redistribution of the donors by the external electric field. The third, between 100 and 1000 s, is probably caused by the change in crystal lattice donor number near the surface.

The advantages of oxide cathodes are their simple manufacture, high current density, and low working temperature. The disadvantages are their sensitivity to ion bombardment and susceptibility to poisoning at working temperature caused by residual gases (e.g., oxygen, carbon monoxide, and chlorine) and water vapor at room temperature. Their lifetime is relatively short (1000−2000 hours), limited by the consumption of the emissive material, since the evaporation of barium at working temperature is rather fast. The lifetime can be extended by a lower working temperature.

(d) Impregnated Cathodes

All cathodes have a reserve of emitting material, but impregnated cathodes have a large reserve, which increases their lifetime. Thus for 10,000-hour lifetime applications, like in high power or satellite microwave tubes, these cathodes are used.

There are three main types of impregnated cathodes as seen in Fig. 2.30. All are indirectly heated from a tungsten spiral and have a base of nickel or molybdenum.

The L-cathode [73], developed around 1950, has behind the porous tungsten a storage of barium aluminate ($5BaO_2 \cdot Al_2O_3$) and tungsten. Originally barium carbonates

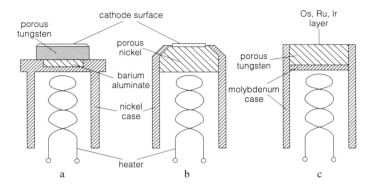

Fig. 2.30. Modern impregnated cathodes. (**a**) L-cathode. (**b**) Matrix cathode. (**c**) Dispenser cathode.

were used, but the decomposition outgassing took too much time and the porous tungsten became oxidized. When the decomposition temperature was increased, the barium transport to the surface was too fast, and the lifetime became shorter. The barium aluminate cathodes develop very little gas during activation and the whole process is fast. Here the barium aluminate is thought to react with tungsten, forming free barium, and the barium vapor is presumed to diffuse through the tungsten pores to make a thin surface layer. Some oxygen is present in the tungsten plug and some barium reaches the surface in oxide form. The work function of L-cathodes is ~ 2 eV.

The matrix cathode stores barium and strontium oxides in a nickel matrix, and its emitting surface is manufactured by sintering nickel powder, barium and strontium carbonates and an "activator," zirconium hydride. More ZrH_2 is in a thicker sintered nickel matrix below the emitting surface. All is mounted in a case of nickel. Evacuation and heating decompose the carbonates as in the oxide cathode. At working temperature, oxides are reduced to free barium and strontium, with the aid of zirconium. Barium and strontium diffuse to the surface and replace the evaporated surface layer.

Dispenser cathodes, also known as B-cathodes, are very much like the L-cathodes, the main difference being the lack of storage for barium aluminate. During manufacturing, molten aluminate penetrates the porous tungsten, filling all pores with aluminate and providing a source of barium. The B-cathode working principle is similar to that of L-cathodes; however, a mixture of barium aluminate and barium-calcium aluminate ($5BaO \cdot 2Al_2O_3 \cdot 3CaO$) improves the emission properties, where the role calcium plays is not known.

Around 1970, it was found that an ~ 500 nm thick coating of osmium, ruthenium, rhenium, or iridium increases the cathode life time appreciably. These M-cathodes work at $\sim 100°$ C lower temperature than do B-cathodes but give the same current density. The work function is 1.8 eV. In 1984 osmium or ruthenium were directly added to tungsten before sintering the porous matrix, so that these MMM-cathodes give a current density of 10 A/cm² within a lifetime exceeding 100,000 hours. M-cathodes are used in most microwave tubes, especially those for satellite transmission.

All impregnated cathodes have a relatively low working temperature: L-cathodes, $\sim 1\,300°$C; matrix cathodes, $\sim 950°$C; and M-type dispenser cathodes, $\sim 1050°$C. Current densities of a few amperes per square centimeter DC and up to 20 A/cm² in pulsed operation can be obtained. All are insensitive to residual gases — except oxygen, water vapor, and carbon oxides — and the sensitivity to these gases is much less than that of oxide cathodes. L-cathodes and dispenser cathodes can be removed from vacuum and stored in nitrogen atmosphere or dry air for some hours without harm. Tantalum, titanium, and stainless steel should not be used near the cathode when heated to working temperature. Impregnated cathodes are less sensitive to ion bombardment, because surface barium is replenished from the storage when heated and thus can be used at anode voltages up to over 100 kV.

(e) Hexaboride Cathodes

A common characteristic feature of all cathodes, pure metals excepted, is their sensitivity to residual gases, which poison them even at room temperature. Thus if they operate in devices which are opened to air, they must be replaced every time the device is vented. The hexaborides of rare earth metals, like lanthanum, gadolinium, and yttrium, at lower temperature have greater emission than does thorated tungsten and can repeatedly be exposed to atmospheric air without losing their emission properties. The hexaboride cathodes, among which lanthanum hexaboride is most widely used, have a working temperature between $1100°C$ and $1600°C$, at which the current densities over 100 A/cm^2 are obtained. At the highest temperatures the material evaporates fast, but is replaced by diffusion from inside. The cathode is impoverished at high temperature because the boron atoms diffuse into the base material, unless this is chemically inert. A sensitivity to residual gases at low working temperature usually can be cured by a slight increase in heating power.

The hexaboride cathodes are manufactured by sintering the powder in the desired shape under pressure and heating, or as monocrystals. They are easily machined, but have difficulty adhering to base materials. Tantalum and tantalum carbide, rhenium, and carbon are adequate base materials, at moderately high temperatures only rhenium and carbon are still good enough, and at the highest temperatures only carbon is useful. This complicates the construction since mechanical stability must be achieved as well as chemical isolation maintaining the lifetime. Pyrolytic graphite is a suitable base material and can be used as the heater element.

2. Photocathodes

Modern photocathode materials are CuBe oxides, different antimonides, or oxides of alkaline metals, as seen in Fig. 2.26. These materials exhibit a high quantum efficiency and thus exhibit high light sensitivity I_k/Φ, where I_k is the emitted current at a flux Φ: a few microamperes per lumen and square centimeter or radiation sensitivity of tens of milliamperes per watt and square centimeter.

Photocathodes are manufactured by coating glass (semitransparent) or metal. The latter are mounted inside a vacuum envelope and the light enters through a glass window. Semitransparent photocathodes are coated on the inside of the vacuum envelope. The choice of the window defines the desired light transmission. Borosilicate glass is usually used, except for ultraviolet applications where the window must be made of quartz glass.

Cesium photocathodes, when liquid nitrogen-cooled and illuminated with a time-constant laser light whose wavelength corresponds to the work function of the photocathode, will emit electrons with extremely low velocity spread. Such cathodes, illuminated with nanosecond pulsed laser light (gigawatts), are used as very-high-current electron sources, up to some megaamperes.

3. Secondary Emission Cathodes

The choice of materials for secondary emission cathodes depends on the desired secondary emission gain. Alkaline oxides, CuBe oxides, and antimonide-coated electrodes will emit secondary electrons with high yield. Secondary emission cathodes are used mostly in electron multipliers (e.g., photomultipliers). In channel multipliers and channel plates, the secondary emission is obtained from specially doped glass.

4. Field Emission Cathodes

Field emission cathodes are used in electron microscopy. To increase the lifetime, a thin wire is welded on a "V"-shaped, directly heated cathode. The cathode temperature can be kept much lower than for thermionic emission. Field emission is also used in vacuum microelectronic triodes (Section 5.1.1).

3

Charged-Particle Dynamics

Aurora borealis. Charged particles, mostly electrons with energy of a few kiloelectronvolts, move along the earth magnetic field lines, colliding with the molecules in the upper atmosphere and exciting their atoms which emit light. (Photograph Jan Olav Andersen, Andenes, Norway.)

3.1. INTRODUCTION

Charged particles moving in a vacuum interact with the electric and magnetic fields which determine their direction of motion. The force is given by Newton's second law, Eq. 1.2:

$$\frac{d\,(m\mathbf{v})}{dt} = q\,\mathbf{E} + q\,\mathbf{v} \times \mathbf{B}, \tag{3.1}$$

where m is the particle mass, $\mathbf{v}$ its velocity, and q its electric charge; $\mathbf{E}$ is the electric and $\mathbf{B}$ the magnetic field, the right side being the Lorentz force.

For arbitrary fields it is not possible to solve analytically Eq. 3.1, but some general conclusions can be made. Defining the electric potential of the particle emitting electrode to be zero and neglecting the emitted particle velocity distribution, the relativistic particle energy equation (Eq. 1.60) is

$$m_0 c^2 \left[\frac{1}{\sqrt{1 - v^2/c^2}} - 1 \right] + qV = 0,$$

which, for $v \ll c$ and $m = m_0 = \text{const}$, reduces to the classical

$$\frac{mv^2}{2} + qV = 0, \tag{3.2}$$

with particle velocity

$$v = \sqrt{-2\frac{q}{m}V}, \tag{3.3}$$

completely determined by the electric potential at the particle position, assuming $v = 0$ at $V_0 = 0$.

The electric field alone accounts for the particle energy change and the magnetic field only changes its direction, because the centripetal magnetic force is always perpendicular to the velocity. In a homogeneous magnetic field with no electric field, and the initial velocity perpendicular to $\mathbf{B}$, the particle moves in a circular orbit.

In a general electromagnetic field, particles move in helicoidal orbits with variable radii of curvature, for fields varying in time or space, the particles are accelerated by electric fields as well as by time-varying magnetic fields. The fields can restrict particles to a limited region of space or they can reflect them.

We start with simple stationary fields where electric field strength can include a component generated by other charged particles. For electrons, relativistic formulae must be used, even when modest acceleration voltages are present. The fields generated by the electrodes, current loops, and permanent magnets must be known to solve the equation of motion using the methods described in Chapter 1.

First we consider homogeneous static fields and nonrelativistic particles, and we neglect space-charge effects. Next, the influence of the space-charge on the motion of electrons in simple electron tubes with three electrodes are considered. Such tubes and many other electron devices induce currents in external circuits, forming electrical signals.

Therefore, we discuss simple cases of time-varying fields. In the chapters on Electron Optics, High Perveance Electron Beams, and Microwave Tubes, charged particle dynamics is treated more in details, taking into account space-charge and high-frequency effects.

3.2. UNIFORM ELECTRIC AND MAGNETIC FIELD

To examine the motion of the particles in uniform fields, we start with Eq. 3.1. Assume a source of electrons consisting of a cathode, an anode, and a system of electrodes, which generate a narrow electron beam with velocity $v_0 = \sqrt{eV_0/(2m_0)}$, where V_0 is the voltage of the last electrode. Let this low-current beam enter a homogeneous electric field region between two parallel plate electrodes (Fig. 3.1), with the field oriented perpendicular to the beam direction, v_0.

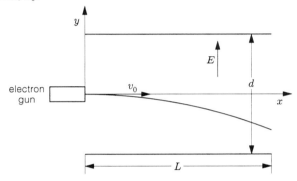

Fig. 3.1. Motion of an electron in a homogeneous electric field.

The equation of motion in rectangular coordinates is

$$m\frac{d^2x}{dt^2} = 0, \qquad m\frac{d^2y}{dt^2} = -eE,$$

and at $t = 0$ we have

$$x = 0 \qquad y = 0 \qquad v = v_x = v_0.$$

After integrating we obtain

$$v_x = v_0, \qquad v_y = -\frac{e}{m}Et,$$

$$x = v_0t, \qquad y = -\frac{1}{2}\frac{e}{m}Et^2.$$

The orbit,

$$y = -\frac{1}{2}\frac{e}{m}E\frac{x^2}{v_0^2},$$

is a parabola, like that of an ideal projectile in a homogeneous gravitational field.

The deflection in a field of length L, with distance between the electrodes d and the voltage V_d, is

$$y_L = -\frac{1}{4}\frac{L^2}{d}\frac{V_d}{V_0},$$

and the beam will leave the electric field region at the angle

$$\tan \alpha = -\frac{L}{2d}\frac{V_d}{V_0}.$$

A large deflection needs long deflecting electrodes, high deflection voltage, small distance between the electrodes, and low acceleration voltage in the electron gun. In practice, not all of the conditions can be obtained simultaneously, especially if the deflecting voltage is a high-frequency signal. The field is not homogeneous along the whole orbit, and the fringing field at both ends can be appreciable. A more realistic view is given in Section 4.7.

For combined electric and magnetic fields we divide the vectors in Eq. 3.1 into components parallel ($\parallel$) and perpendicular ($\perp$) to the magnetic field $\mathbf{B}$ such that

$$m\frac{d}{dt}(\mathbf{v}_\parallel + \mathbf{v}_\perp) = q\,[\mathbf{E}_\parallel + \mathbf{E}_\perp + (\mathbf{v}_\parallel + \mathbf{v}_\perp) \times \mathbf{B}], \qquad (3.4)$$

where $\mathbf{v}_\parallel \times \mathbf{B}$ is zero. We separate this equation into components

$$m\frac{d\mathbf{v}_\parallel}{dt} = q\,\mathbf{E}_\parallel \qquad (3.5a)$$

and

$$m\frac{d\mathbf{v}_\perp}{dt} = q\,(\mathbf{E}_\perp + \mathbf{v}_\perp \times \mathbf{B}). \qquad (3.5b)$$

Under our assumption that the electric and magnetic fields are homogeneous, Eq. 3.5a describes a uniformly accelerated motion. Taking the z axis along the magnetic field, the solution of Eq. 3.5a is

$$z = z_0 + v_{z_0}t + \frac{1}{2}\frac{q}{m}E_\parallel t^2, \qquad (3.6)$$

where v_{z_0} and z_0 are the initial particle velocity and position.

Equation 3.5b, which describes the particle motion in the plane perpendicular to $\mathbf{B}$, can be simplified by dividing the velocity $\mathbf{v}_\perp$ into a constant velocity $\mathbf{v}_D$ and a variable one, $\mathbf{u}$, both perpendicular to $\mathbf{B}$. $\mathbf{v}_D$ can be defined as

$$\mathbf{v}_\perp = \mathbf{v}_D + \mathbf{u}, \qquad \mathbf{v}_D = \frac{\mathbf{E}_\perp \times \mathbf{B}}{|\mathbf{B}|^2}. \qquad (3.7)$$

Since the time derivative of $\mathbf{v}_D$ is zero we obtain

$$m\frac{d\mathbf{u}}{dt} = q\left[\mathbf{E}_\perp + \mathbf{u}\times\mathbf{B} + \frac{(\mathbf{E}_\perp \times \mathbf{B})\times\mathbf{B}}{|\mathbf{B}|^2}\right],$$

where $\mathbf{E}_\perp \times \mathbf{B}$ is perpendicular to $\mathbf{B}$. The third term is equal to $-\mathbf{E}_\perp$, and

$$m\frac{d\mathbf{u}}{dt} = q\,(\mathbf{u}\times\mathbf{B}). \qquad (3.8)$$

The acceleration $d\mathbf{u}/dt$ is perpendicular to $\mathbf{u}$, whose magnitude does not change, only its direction. Since $d\mathbf{u}/dt$ is constant, because $\mathbf{u}$ always is perpendicular to $\mathbf{B}$, this component represents a central motion with a circular orbit, so

$$\left|\frac{d\mathbf{u}}{dt}\right| = \frac{u^2}{r},$$

and the radius of curvature is

$$r = \frac{u}{(q/m)B} = \frac{u}{\omega_c}, \tag{3.9}$$

where

$$\omega_c = \frac{q}{m} B \tag{3.10}$$

is the *cyclotron frequency*. Note that $\omega_c \sim 1/m$ and $r \sim m$, which results in large differences between electron and ion radii of curvature; for example, in a 0.1 T magnetic field a 1000 eV electron moves in a circle with radius 0.75 mm, while a proton circle has a 32 mm radius. In the earth magnetosphere, ~ 100 keV electrons, generating the aurora in a $\sim 10^{-5}$ T magnetic field, have orbits with radii ~ 100 m, while orbit radii of the protons in the solar wind measure many kilometers.

Thus, charged particle motion in homogeneous electric and magnetic fields is a superposition of three motions:

1. Uniformly accelerated motion in the $\mathbf{E}_{\parallel}$ direction with acceleration $q\mathbf{E}_{\parallel}/m$,
2. Uniform motion with the velocity $|\mathbf{E}_{\perp}|/|\mathbf{B}|$ perpendicular to both $\mathbf{B}$ and $\mathbf{E}$, and
3. Uniform circular motion in a plane perpendicular to $\mathbf{B}$. Looking in the magnetic field direction, positive particles rotate counterclockwise, whereas negative ones rotate clockwise.

Figure 3.2 shows the motion of an electron in a field where $\mathbf{E}_{\parallel}$ initially retards the electron, stops the motion in the z-direction, and then accelerates. Note that the projection of the orbit in the x–y plane is a cycloid.

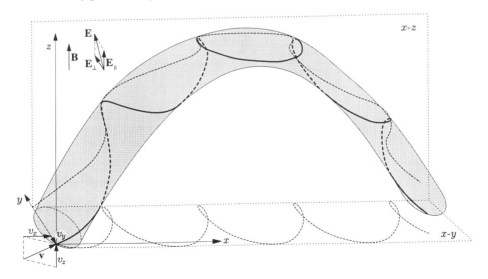

Fig. 3.2. Motion of an electron in static electric and magnetic fields.

If the magnetic field is not homogeneous, the radius of curvature will increase in a decreasing field, and vice versa. Without an electric field, particles are bound to follow the magnetic field lines even when bent. This property governs many cosmic phenomena.

A practical application of crossed electric and magnetic fields is the magnetron (Section 7.4.2). Here

$$\mathbf{E} = -E_y\,\mathbf{j}, \qquad\qquad \mathbf{B} = -B_z\,\mathbf{k},$$

where $\mathbf{j}$ and $\mathbf{k}$ are unit vectors in y and z direction. The electron equation of motion is

$$\frac{dv_x}{dt} = \eta v_y B_z = \omega_c \frac{dy}{dt}, \qquad \eta = \frac{e}{m_0}, \tag{3.11}$$

$$\frac{dv_y}{dt} = \eta E_y - \eta v_x B_z = \eta E_y - \omega_c \frac{dx}{dt}. \tag{3.12}$$

and at $t = 0$ we have

$$x = y = 0, \qquad v_x = v_y = 0.$$

The time derivative of Eq. 3.12, introducing dv_x/dt from Eq. 3.11, is

$$\frac{d^2 v_y}{dt^2} + \omega_c^2 v_y = 0,$$

with the general solution

$$v_y = C_1 \sin \omega_c t + C_2 \cos \omega_c t.$$

The initial conditions reduce this equation to

$$v_y = C_1 \sin \omega_c t. \tag{3.13}$$

This solution introduced into Eq. 3.11 gives

$$\frac{dv_x}{dt} = \omega_c v_y = \omega_c C_1 \sin \omega_c t,$$

and after integration we obtain

$$v_x = C_1(1 - \cos \omega_c t).$$

Equating the derivative of Eq. 3.13 with Eq. 3.12 and imposing initial conditions results in

$$\eta E_y - \omega_c C_1 = 0$$

and

$$C_1 = \frac{\eta E_y}{\omega_c} = \frac{E_y}{B_z}.$$

The solution is thus

$$x = \frac{E_y}{\omega_c B_z}(\omega_c t - \sin \omega_c t), \qquad v_x = \frac{E_y}{B_z}(1 - \cos \omega_c t), \tag{3.14}$$

$$y = \frac{E_y}{\omega_c B_z}(1 - \cos \omega_c t), \qquad v_y = \frac{E_y}{B_z}\sin \omega_c t. \tag{3.15}$$

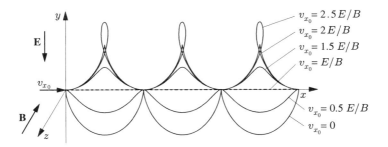

Fig. 3.3. Cycloidal orbit in perpendicular electric and magnetic fields.

The electron orbit is a cycloid, seen in Fig. 3.3. The orbits become hypocycloids or epicycloids, depending on the ratio of v_{x_0} to $|E_y|/|B_z|$. When this ratio equals one and v_{y_0} is zero, the solution degenerates, and the particle moves along the x axis with constant velocity.

3.3. ELECTRON TUBES, TRIODE

In many electron devices the current supplied by the cathode is used to generate or amplify electrical signals. The emitted current is controlled by varying the potential of an electrode near the cathode, the grid, made as a loose mesh or a helix.

The simplest of such amplifier tubes is a triode [22]. The grid potential alone does not influence the current, since the potential distribution in the tube depends on the grid–cathode and the anode–cathode voltage, but also on the current emitted by the cathode. This is a consequence of (a) the space-charge, which forms a virtual cathode in front of the physical cathode (Section 2.2.2), and (b) the distribution of charges, determined by the interelectrode capacitances and the grid and anode potentials.

As seen in Fig. 3.4a, the capacitances C_{gc}, C_{ac}, and C_{ga} form a triangle network, which can be transformed into a star network like

$$C_c = C_{ac} + C_{gc} + \frac{C_{ac}C_{gc}}{C_{ga}}.$$

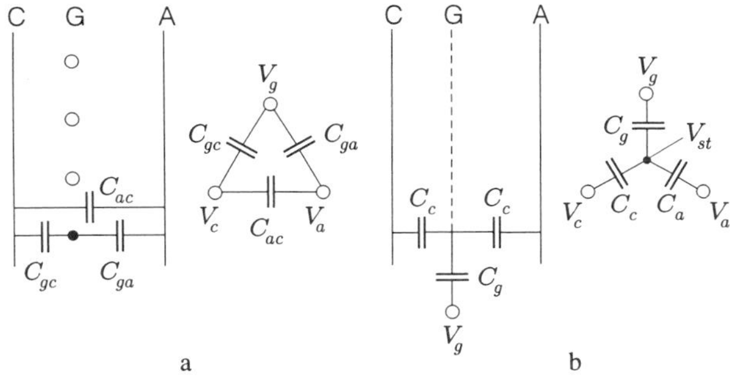

Fig. 3.4. Interelectrode capacitances in a triode.

The charges on the electrodes are

$$q_k = C_c(V_c - V_{st}),$$
$$q_g = C_g(V_g - V_{st}), \quad \text{and} \tag{3.16}$$
$$q_a = C_a(V_a - V_{st}),$$

if V_{st} is the potential at the center of the star, where the charge is

$$q_c + q_g + q_a = 0, \tag{3.17}$$

because the voltage sources, connected to the electrodes, can only influence the distribution of the charges on the electrodes, but the sum of all charges must be zero. Introducing Eq. 3.16 into Eq. 3.17 and letting the cathode potential be zero, $V_c = 0$, the potential at the center of the star is

$$V_{st} = \frac{V_g + \dfrac{C_a}{C_g}V_a}{1 + \dfrac{C_a}{C_g}\left(1 + \dfrac{C_c}{C_a}\right)} \approx V_g + DV_a, \tag{3.18}$$

where

$$D = \frac{C_a}{C_g} \tag{3.19}$$

is a measure for the field penetration of the anode through the grid, where we assume that $C_a/C_g \ll 1$. Since the potential of the star point is the triode control potential, the triode can be replaced by an equivalent diode. Equation 2.30 can be used to find the anode current variation with grid and anode potentials,

$$I_a = pV_{st}^{3/2} = p(V_g + DV_a)^{3/2}. \tag{3.20}$$

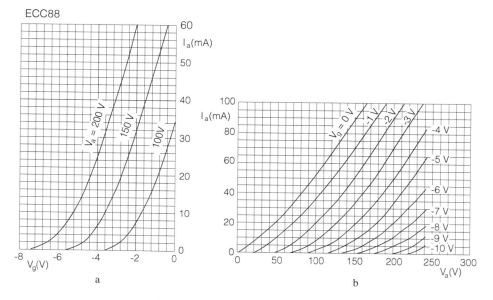

Fig. 3.5. Families of curves for the triode ECC88.

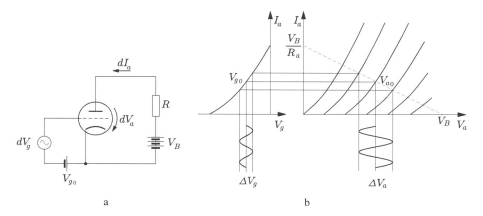

Fig. 3.6. (a) Triode circuit. (b) Graphic determination of the triode working point.

Equation 3.20 describes two families of curves: The first (Fig. 3.5a) is I_a with V_g, V_a constant, and the second (Fig. 3.5b) is I_a with V_a, V_g constant. The working point of the triode is controlled by the DC voltages V_{g0} between the cathode and the grid (grid bias, chosen so that the grid potential, V_g, is negative) and V_{a0} between the cathode and the anode. This choice of working point guarantees small losses in the control circuit, there is no DC grid current, and the current–voltage characteristic uses its linear part.

A simple triode circuit is shown in Fig. 3.6a. Superimposing a small-amplitude AC signal, dV_g, on the grid bias V_{g0} will change both the anode current and voltage,

$$dI_a = \left(\frac{\partial I_a}{\partial V_g}\right)dV_g + \left(\frac{\partial I_a}{\partial V_a}\right)dV_a, \tag{3.21}$$

which, for working points in the linear current–voltage characteristic part, have constant transconductance

$$g_m = \left(\frac{\partial I_a}{\partial V_g}\right)_{V_a=\text{const}}, \tag{3.22}$$

dynamic anode resistance

$$R_i = \left(\frac{\partial V_a}{\partial I_a}\right)_{V_g=\text{const}}, \tag{3.22}$$

and amplification factor

$$\mu = \left(\frac{\partial V_a}{\partial V_g}\right)_{I_a=\text{const}} = \frac{1}{D}. \tag{3.24}$$

For these constants, Barkhausen's formula is valid

$$g_m R_i D = \frac{g_m R_i}{\mu} = 1, \tag{3.25}$$

and Eq. 3.21 becomes

$$dI_a = g_m dV_g + \frac{1}{R_i} dV_a. \tag{3.26}$$

The assumption in Eq. 3.19 that D is a constant is simplistic since space-charge depends on the emitted current, I_a, and thus D decreases and μ increases with anode current.

In the triode circuit, the anode voltage is

$$V_a = V_B - I_a R_a.$$

When the control grid is modulated by a signal, the change in the anode voltage is

$$dV_a = -dI_a R_a.$$

Using Eqs. 3.25 and 3.26, the change in the anode current is

$$dI_a = \mu \frac{dV_g}{R_i + R_a}, \tag{3.27}$$

where the graphical solution is shown in Fig. 3.6b. The working point is given by V_{a_0} and V_{g_0}. The line from the source voltage V_B to the point defined by V_B / R_a is the load line. When the control grid is modulated by a signal with a peak-to-peak amplitude ΔV_g, the anode voltage slips along the load line, and the amplified signal divided by the modulation signal, the voltage amplification, is

$$A = \frac{\Delta V_a}{\Delta V_g}.$$

Because of the negative grid bias the grid current is generally small, and the input power is also small. The large output power comes from converting the anode voltage power supply energy into the load circuit energy by the kinetic energy of the electrons. This energy conversion is discussed more in detail in the next section.

Modern transmission tubes often have one additional electrode, the screen grid, between the control grid and the anode. The anode-control grid and anode-cathode capacitances in such tetrodes are much smaller, and the influence of the anode potential change on the current is very much reduced.

3.4. TIME-VARIABLE ELECTRIC FIELD

Electron motion in a time-variable electric field is complicated, since the field changes as the electrons travel from the cathode to the anode. The field is generated by the electrical charge distributions from power supplies connected to the electrodes. These power supplies, with passive elements like resistors, inductances, and capacitors, make the external circuit which is affected as an electron moves in vacuum between the electrodes.

A charged particle moving near a conductor induces mirror charges on its surface which arrange themselves so as to keep the electric field inside the conductor zero (Fig. 3.7a). If there are several isolated conductors in the neighborhood of the particle, their potential will change as the particle moves. If these conductors are connected by perfectly conducting wires, the potential difference will remain zero by currents in the wires which connect the conductors. Figure 3.7b shows an electron moving between electrodes A and B connected to power supply with the voltage V_B, so that B is at a positive potential. If the electron moves Δx, and the corresponding potential increase is ΔV, the electron kinetic energy increases $-e\Delta V$, which comes from the power supply. When an electron gains energy

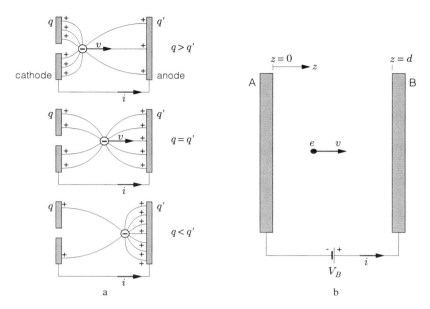

Fig. 3.7. (a) Redistribution of surface charges while an electron moves between two electrodes. (b) Induced current in the external circuit.

in the electric field, the external source which generates the field must lose the same amount of energy. Thus, when an electron is retarded by a field, the lost kinetic energy appears elsewhere in the system.

While the electron moves Δx a positive charge, Δq, moves from A to B through the power supply, which has done $\Delta q V_B$ work. Thus

$$\Delta q V_B - e \Delta V = 0,$$

or

$$\Delta q = \frac{e}{V_B} \Delta V. \tag{3.28}$$

If Δt is the time for the electron to move over Δx, then

$$\frac{\Delta q}{\Delta t} = \frac{e}{V_B} \frac{\Delta V}{\Delta t}, \tag{3.29}$$

or

$$i = \frac{dq}{dt} = \frac{e}{V_B} \left[\frac{\partial V}{\partial x} \frac{dx}{dt} + \frac{\partial V}{\partial y} \frac{dy}{dt} + \frac{\partial V}{\partial z} \frac{dz}{dt} \right] = -\frac{e}{V_B} \mathbf{E} \cdot \mathbf{v}, \tag{3.30}$$

where i is the current in the external circuit connecting A and B, $\mathbf{E}$ is the electric field, and $\mathbf{v}$ is the electron velocity. Since $\mathbf{E}$ is proportional to V_B, the current does not depend on the voltages connected to the electrodes. i is the current in the external circuit even if the power supply is taken away, and the electrodes A and B are only connected by a wire or an impedance. We define $\mathbf{E}_1(x, y, z) \equiv \mathbf{E}(x, y, z)/V_B$ [m^{-1}] as the electric field between A and B when B is at a potential 1 V higher than A.

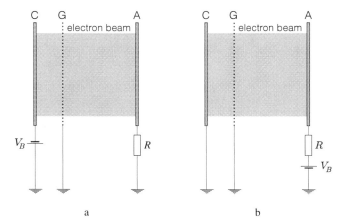

Fig. 3.8. Two modulation possibilities. (**a**) The beam is accelerated between the cathode and the control grid. (**b**) The beam is accelerated between the grid and the anode.

Instead of a single electron, consider a space-charge ρ flowing between the electrodes, so that the induced current in the external circuit is

$$i = \int \rho \mathbf{v} \mathbf{E}_1 \, dx \, dy \, dz = \int \mathbf{J} \mathbf{E}_1 \, dx \, dy \, dz, \tag{3.31}$$

and $\mathbf{J} = \rho \mathbf{v}$ is the current density in the volume $dx\,dy\,dz$, the integral extending over the space-charge region.

Thus the induced current in the external circuit *does not depend* on the impedance in the circuit, and if an electron beam can be modulated with a weak signal, it can extract power from the supply which generates the electric field and accelerates the electrons. That is exactly what happens in electron tubes.

There are two types of modulation of an electron beam which differ in the power supply connection. Figure 3.8a shows an electron beam which is accelerated between the cathode and a control grid whose mesh wire thickness and distance between wires is ideally zero. Such a grid is completely transparent to electrons and is an equipotential surface. If all electrons have the same velocity at the grid plane and if n electrons pass the grid, the current $I_0 = ne$ is induced in the resistor R, which results in a voltage drop $I_0 R$. Because of this voltage drop, the electrons see a retarding field, and each electron loses $eI_0 R$ kinetic energy between the grid and the anode, so n electrons will lose $neI_0 R = I_0^2 R$. But the induced current generates exactly the same power in the resistor, so the kinetic energy the electrons lose between the grid and the anode is converted into heat in the resistor. The remaining electron kinetic energy heats the anode.

If the number of electrons which pass the grid is varied without changing their velocity, the current will be time-dependent. Assume a sinusoidal modulation,

$$i = I_0 + I_1 \sin \omega t,$$

where the time for the electron to move from the grid to the anode is small compared to the period of the modulation voltage. The induced current will be equal to the electron current, and a voltage drop $(I_0 + I_1 \sin \omega t)R$ is generated across the resistor. The total power developed in the resistor is then

$$i^2 R = (I_0 + I_1 \sin \omega t)^2 R,$$

and its mean value over one period is

$$P = I_0^2 R + \frac{1}{2} I_1^2 R,$$

which is greater than the power generated under DC conditions. However, the number of electrons passing through the grid did not change, so the mean energy of the electrons at the anode is lower in the AC than in the DC case. During the half period when the electron current is greater than I_0, the voltage drop over the resistor is larger and thus the retarding field is stronger. More than half of the electrons lose more energy as compared to the DC case.

In the second case with the power supply and resistor connected in series (Fig. 3.8b), the conversion of DC to AC power occurs differently. Assume that the electrons pass the grid with negligible velocity and that the number of electrons varies with time. As the induced current passes the resistor, the anode voltage varies, and the power supply compensates for the power loss in the resistor. The remainder of the power from the supply is first converted into electron kinetic energy, and then into heat when the electrons hit the anode. Assuming the same current variation with time as in the previous case, the result will be the same.

Both modulation methods are used in practice. Klystron, traveling wave, and similar microwave tubes use the first method to convert DC into microwave power. The energy from the DC power supply is first converted into the kinetic energy of the electrons. By velocity modulation, the electron beam kinetic energy is converted into a current modulation, which then generates microwave power by inducing currents in the load. The remaining kinetic energy is converted into heat in the collector electrode. The second procedure describes well the conditions in triodes and similar electron amplifier tubes.

Problems arise, however, when the electron transit time between electrodes is of the same order as the period of the signal to be amplified. The induced current in the external circuit, connecting two grids, G_2 and G_3, shown in Fig. 3.9, depends on the modulated electron beam which passes between them. Let the current modulation be made by a weak signal, V, to G_1, and place a resistor, R, between G_2 and G_3. The grids are again transparent to electrons and can be approximated as equipotential surfaces.

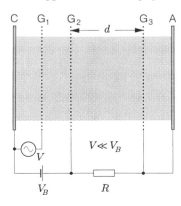

Fig. 3.9. Induced current in the outer circuit.

The induced current through the resistor depends only on the electron current between G_2 and G_3. Electrons reaching the anode neutralize only the positive charge which arrives there to meet them, and they cause no extra current through R. The current through the resistor remains the same even if the anode replaces grid G_3.

The current through R is thus determined by Eq. 3.31, with a small difference: Time variation of the current must now be taken into account, and the current density becomes a function of time, $J(x, y, z, t)$. If the grid is wider than the electron beam, the electric field from G_2 to G_3 everywhere is perpendicular to the grid planes. The normalized field strength, E_1, is then $1/d$, where d is the distance between the grids and the voltage V_B between G_1 and G_2 accelerates the electrons to a velocity v_0 when they pass G_2. In a coordinate system where the velocity is along the z axis and taking into consideration only the time-varying part we obtain

$$|\mathbf{J}| = J_1(x, y) \sin \omega(t - \frac{z}{v_0}),$$

and

$$I_1 = \int J_1(x, y)dx dy,$$

where the integral is over the electron-beam cross section. Substituting this expression, together with $|\mathbf{E}_1| = 1/d$ in Eq. 3.31, we obtain

$$i(t) = \frac{I_1}{d} \int_0^d \sin \omega(t - \frac{z}{v_0})dz = \frac{I_1}{d} \frac{v_0}{\omega}\left[\cos\left(\omega t - \frac{\omega d}{v_0}\right) - \cos \omega t\right]. \qquad (3.32)$$

Difference between cosine functions is a product of sine functions, so

$$i(t) = I_1 \frac{\sin \omega d/(2v_0)}{\omega d/(2v_0)} \sin \omega\left(t - \frac{d}{2v_0}\right),$$

and introducing the transit time $\tau = d/v_0$ we obtain

$$i(t) = I_1 \frac{\sin \omega\tau/2}{\omega\tau/2} \sin \omega\left(t - \frac{\tau}{2}\right) = I_1 M_B \sin \omega\left(t - \frac{\tau}{2}\right), \qquad (3.33)$$

where M_B is the *beam coupling coefficient*. The phase of the induced current in the external circuit is the same as that of the electron current midway between the grids.

The beam coupling coefficient determines the ratio between the induced current and the AC component of the electron current (Fig. 3.10). For $\omega\tau \sim 0$ (i.e., for the DC and for AC current when the transit time is short compared to the period of the AC signal), $M_B \sim 1$. When the modulation frequency increases, M_B decreases, and when the transit time becomes equal to the period of the modulation signal the induced AC current in the external circuit is zero. Since the induced current is the sum of all contributions in the region between the electrodes, the total induced current must always be zero when the number of periods of the modulating signal is an integer multiple of the transit time.

The interelectrode distance (e.g., in a triode between the cathode and the control grid) in high-frequency transmitting tubes is ≤ 1 mm, and a modulation voltage of ~ 100 V allows an upper frequency of ~ 100 MHz. In microwave triodes, which work up to ~ 4000 MHz, distances between the cathode and the grid are < 0.1 mm, and those between the grid and the anode are < 1 mm (see Fig. 7.1).

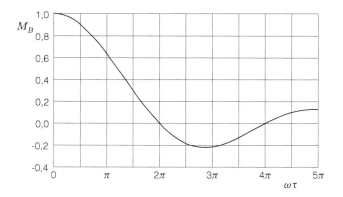

Fig. 3.10. Beam coupling coefficient vs. transit time.

Strictly speaking, the electron velocity changes a little when modulated by an AC signal. If the electrons travel a sufficient distance, the fast electrons catch up with the slow ones. In the electrode configuration of Fig. 3.11, the electrons are accelerated between the cathode and the grid G_1, and between G_1 and G_2 there is an AC signal source which they enter with the velocity v_0 moving along the z axis from the origin where the electrons cross grid G_1. The AC field accelerates the electrons according to Eq. 3.1,

$$\frac{d^2 z}{dt^2} = -\frac{e}{m} E_z.$$

In terms of the distance, d, between G_1 and G_2 and the time-varying voltage amplitude, A, the electric field strength is

$$E_z = -\frac{A}{d} \sin \omega t, \tag{3.34}$$

which we introduce into the equation of motion and integrate to get

$$v = \frac{dz}{dt} = v_0 - \frac{eA}{\omega m d} (\cos \omega t - \cos \omega t_1), \tag{3.35}$$

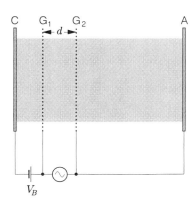

Fig. 3.11. Velocity modulation of an electron beam.

where $v_0 = \sqrt{2\,(e/m)\,V_B}$ is the electron velocity and t_1 the time when they cross G_1. Equation 3.35 describes the instantaneous velocity of the electrons between G_1 and G_2. The time, t_2, when the electrons leave G_2, when the signal amplitude, A, is small compared to V_B, (i.e., the electron velocity changes little between the grids) is

$$t_2 = t_1 + \frac{d}{v_0}.$$

Introducing t_2 into Eq. 3.35 we get

$$v_{G_2} = v_0 + \frac{2eA}{\omega m d} \sin \frac{\omega d}{2v_0} \sin \omega t_0, \qquad (3.36)$$

where

$$t_0 = t_1 + \frac{d}{2v_0}$$

is the time when the electrons pass the half-way point between G_1 and G_2. This gives the same expression for the beam coupling coefficient, M_B, as in Eq. 3.33, so

$$v_{G_2} = v_0 + \frac{eM_BA}{mv_0} \sin \omega t_0 = v_0 \left[1 + \frac{M_BA}{2V_B} \sin \omega t_0 \right] \qquad (3.37)$$

An AC signal between the grids G_1 and G_2 velocity modulates the electron beam so that in the region between G_2 and the next electrode, faster electrons catch up with the slow ones. This modulation method is used in microwave tubes.

4

Electron Optics

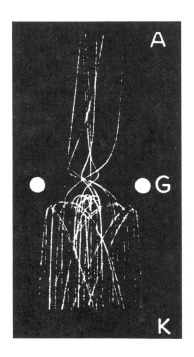

Simulation of electron orbits in a triode with bearing balls
on a streched rubber membrane. The construction of many
electron tubes in the 1930s was based on such experiments.
(From *Grundlagen der Elektronenoptik*, W. Glaser, © 1952,
Springer-Verlag, with permission.)

4.1. INTRODUCTION

Electric and magnetic fields affect electron or ion beams similarly as lenses affect light rays. A premise is that the particles move without collisions and that the space-charge can be neglected, which is satisfied, if the mean free path between two collisions with the residual gas molecules is longer than the length of the electron or ion orbit and if the beam current is low. The analogy with the light rays is so deep that the same terminology can be used, like focus and focal length, aberrations, beam intensity, lens systems, and optical instruments. There are, however, natural limitations on how far this parallelism can be driven. Two photons never repel each other, whereas two charged particles do. Only as an exception the refractive index is changing with time in light optics, and electric and magnetic field can change the direction and strength in nanoseconds. Disregarding these extremes, the same physical background lies behind both light and electron optics. It is Fermat's [74] law in the form valid for refraction on a planar boundary:

The path a light ray takes between two points is such that the optical path is a minimum.

$$s = n_1 \overline{AO} + n_2 \overline{OB} \to \text{min}, \qquad \frac{ds}{dx} = 0,$$

where s is the optical path and n_1 and n_2 the refractive indices, as seen in Fig. 4.1, so

$$n_1 x - n_2 (d - x) = n_1 \sin \alpha_1 - n_2 \sin \alpha_2 = 0.$$

Snell's [75] law follows directly

$$\frac{\sin \alpha_1}{\sin \alpha_2} = \frac{n_2}{n_1}. \tag{4.1}$$

In a medium where the refractive index varies continuously, Snell's law becomes

$$S = \int_{P_1}^{P_2} n \, ds \ \to \ \text{min}.$$

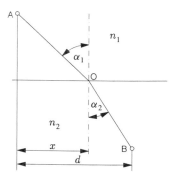

Fig. 4.1. Fermat's law.

Table 4.1

Similarities and differences between electron and light optics

Electron optics	*Geometric optics*
Analogies	
Electron beam	Light beam
Electron lens	Glass lens
Electric or magnetic deflection system	Glass prism
Electric or magnetic mirror	Mirror
Differences	
Particle velocity (v)	Refractive index (n)
$n \sim v$ (v = particle velocity)	$n \sim 1/v$ (v = velocity of light)
$(n_1/n_2)_{max} \sim 100$	$(n_1/n_2)_{max} < 3$
Electron lens: $n = \mathrm{f}(\mathbf{E},\mathbf{B})$, variable f	Glass lens: $n = $ const, $f = $ const
Variable electric or magnetic deflection	Variable deflection – prism must rotate
Continuous change of the refractive index	Finite number of refractive surfaces
If potential not equal on both sides: two different focal lengths	Often same medium of both sides of the lens, $f_1 = f_2$
Magnetic field: anisotrophy (rotation)	Normally no anisotrophy
Rotationally symmetric lenses positive	Positive and negative lenses

An analogous expression can be derived in mechanics (principle of least action):

$$s = \int\limits_{P_1}^{P_2} mv(x, y, z)ds \;\; \rightarrow \; \min, \tag{4.2}$$

where v is the velocity of the particle.

In light optics the main optical elements are mirrors, prisms, and rotationally symmetric lenses. In electron optics their equivalences are: electric and magnetic lenses, mirrors and deflection elements — prisms. Lenses and deflection elements are common, whereas electron optical mirrors occur rarely. Contrary to light optics, lenses without rotational symmetry are frequently used in electron optics. The property that particles which move in a rotationally symmetric field can image an object is a consequence of Eq. 4.2 and is valid also for neutral particles in a rotationally symmetric gravitational field.

The most important similarities and differences between electron and light optics are summarized in Table 4.1.

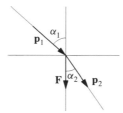

Fig. 4.2. Newton's force at the refraction boundary.

HISTORICAL NOTES. Scientists disputed the nature of the light from ancient times. In the second half of the seventeenth century, Newton [76] developed his particle theory. At about the same time, Huygens [77] assumed that light consists of waves. Newton believed that it was impossible to explain the rectilinear propagation of light by wave nature, and Huygens could not understand why light rays do not influence each other when they cross. In Newton's theory the light consists of small particles, corpuscles, emitted by the light source, which follow rectilinear orbits. Each particle has a momentum

$$\mathbf{p} = m\mathbf{v}.$$

When such a particle crosses a boundary between two media, its momentum changes according to

$$\mathbf{F} = \frac{d\mathbf{p}}{dt} = \frac{d}{dt}(m\mathbf{v}).$$

Newton supposed a force, perpendicular to the boundary, so the change in momentum occurs in the direction of the force (Fig 4.2). The momentum component perpendicular to the force does not change, and thus

$$p_1 \sin \alpha_1 = p_2 \sin \alpha_2,$$

which is Snell's law, if the index of refraction is written as

$$n = \text{const } p.$$

As Newton's force acts perpendicularly to the boundary, the incident ray, the refracted ray, and the normal to the surface lie in the same plane. No work is done on a particle which moves on a surface perpendicular to the force. Refraction surfaces are therefore equipotential surfaces.

Newton's authority was responsible for the domination of the particle theory during the whole eighteenth century. In 1802 Young [78] demonstrated dark and light bands behind two slits. He explained the phenomenon by interference of light waves but could not get attention. Fresnel [79] won a French Academy prize with his work on diffraction, based on the wave theory. Poisson pointed out that according to Fresnel's theory a bright spot should be seen in the shadow of small, round object. Arago made the experiment and found the spot. Laplace contributed with basic theoretical work and gained support for the wave theory. The theory was crowned by the work of Maxwell [26]. His equations tie the electromagnetic field and light phenomena together.

New problems emerged in 1905, when Einstein [48] explained the photoelectric effect by light particles. The particle–wave duality was established. de Broglie [59] proposed the idea that even particles must exhibit wave properties, which was confirmed experimentally by Germer and Davisson[80]. This was followed by Busch's [25] work, where he showed that a rotationally symmetric magnetic field has all properties of a lens.

In 1858 Plücker [8] and in 1879 Crookes [81] showed that "cathode rays" throw a shadow behind an obstacle in a gas discharge. Supported by Hertz [11], the scientific community believed that "cathode rays" are electromagnetic waves. Braun [19] constructed in 1897 the first cathode ray tube with magnetic deflection and made the first oscilloscope.

Busch work opened the way for electron optics. Rogowski and Flegler [82] showed that with two coils a sharp picture can be made. Davisson and Calbick [83] constructed 1931 the first electrostatic lens and computed the focal length for a rotationally symmetric opening. In the same year Ruska and Knoll [84] built the first electron microscope (Ruska Nobel prize 1986), and 3 years later Holst, Teves, and Veenmans [85] built the first image converter and image amplifier.

During the time between the two world wars the theory of electron optics was further developed by Ruska, Scherzer, Zworykin, and Glaser. Zworykin constructed the first iconoscope and initiated the best-known application of electron optics — television.

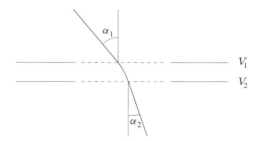

Fig. 4.3. Refraction of an electron orbit between two meshes ($V_2 > V_1$).

For a charged particle, accelerated by voltage V, the momentum is

$$p = mv = \sqrt{-2qmV},$$

where q is the charge and m is the mass of the particle. Assume that the velocity of the particle is zero where $V = 0$, in accordance with Section 1.3.1. Comparing this equation with Newton's light theory, the refraction index can be written as

$$n = \text{const} \sqrt{-2qmV} \propto \sqrt{V}. \tag{4.3}$$

Imagine two plane metal electrodes with a fine mesh over a hole in each plate, as seen in Fig. 4.3. Let the potentials on the respective plate be V_1 and V_2. The velocity of the particle is proportional to the square root of the potential, and

$$\frac{v_2}{v_1} = \sqrt{\frac{V_2}{V_1}}.$$

The potential difference between the electrodes is $\Delta V = V_2 - V_1$, so

$$\frac{\sin \alpha_1}{\sin \alpha_2} = \sqrt{1 + \frac{\Delta V}{V_1}}.$$

A double electrode with a fine mesh refracts particle orbits similarly as a boundary between two media refracts light rays. However, the mesh scatters the particles and the electric field bulges between the meshes, which results in strong absorption and bad imaging.

Instead of well-defined refraction surfaces, electron optics uses continuously varying electric or magnetic fields, so a larger variation in electron optical lenses and lens systems construction than in light optics is possible. But electron optics has an important restriction: Rotationally symmetric electron lenses cannot be made with a negative focal length. This is very inconvenient; to compensate the lens errors in light optics, combinations of positive and negative lenses are used. In electron optics other, more complicated methods must be applied.

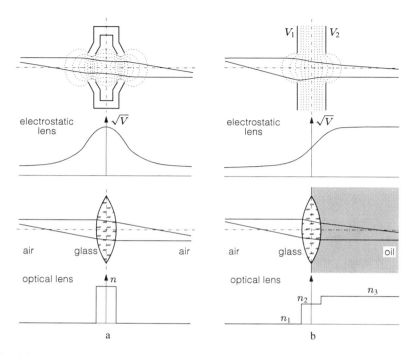

Fig. 4.4. Similarity between rotationally symmetric electron and glass lenses. The particles are electrons, and the orbits and potential distribution are computer simulations. (**a**) The same refractive index on both sides of the lens. (**b**) Different refractive indices.

4.2. PARAXIAL THEORY FOR ROTATIONALLY SYMMETRIC ELECTRON LENSES

4.2.1. Orbit Equation in Rotationally Symmetric Fields

In rotationally symmetric fields $E_\theta = 0$ and $B_\theta = 0$. Define

$$\eta = \frac{q}{m_0}, \tag{4.4}$$

where q is the charge and m_0 is the rest mass of the particle. The classical equations of motion (Eqs. 1.74 – 1.76) are

$$\ddot{r} - r\dot{\theta}^2 = -\eta\left(\frac{\partial V}{\partial r} - v_\theta B_z\right), \tag{4.5}$$

$$r\ddot{\theta} + 2\dot{r}\dot{\theta} = -\eta(v_r B_z - v_z B_r), \tag{4.6}$$

$$\ddot{z} = -\eta\left(\frac{\partial V}{\partial z} + v_\theta B_r\right). \tag{4.7}$$

4.2.1.1. Electrostatic Field

In accordance with the definition in Section 1.3.1, the potential at the initial point of the particle orbit is zero and so is the initial velocity of the particle, neglecting the Maxwellian velocity distribution. The particles can thus never obtain an azimuthal velocity component. In such a field we obtain

$$\ddot{r} = -\eta \frac{\partial V}{\partial r}, \tag{4.8}$$

$$\ddot{z} = -\eta \frac{\partial V}{\partial z}. \tag{4.9}$$

In electron optics it is important to know the radial position of the particle as a function of its axial position, so by using Eqs. 1.77 and 1.78 the independent variable, t, can be transformed into the axial position, z.

$$\dot{r} = \frac{dz}{dt}\frac{dr}{dz} = \dot{z}r', \tag{4.10}$$

$$\dot{z} = \sqrt{\frac{-2\eta V}{1 + r'^2}}. \tag{4.11}$$

For the second derivative we obtain

$$\ddot{r} = \frac{d}{dt}\left(\frac{dr}{dt}\right) = \frac{d}{dt}(\dot{z}r') = \dot{z}\frac{dz}{dt}\frac{dr'}{dz} + \ddot{z}r' = \dot{z}^2 r'' + \ddot{z}r' = -\eta \frac{\partial V}{\partial r}. \tag{4.12}$$

By introducing Eqs. 4.11 and 4.9 into Eq. 4.12 we obtain

$$r'' + \frac{1 + r'^2}{2V}\left[\frac{\partial V}{\partial z}r' - \frac{\partial V}{\partial r}\right] = 0. \tag{4.13}$$

Some important conclusions are:
- The orbit equation (Eq. 4.13) is a nonlinear second-order differential equation including partial derivatives of the potential.
- η has vanished. In an electrostatic field the solution of Eq. 4.13 has the same form for both positive and negative particles, and particles with different masses follow the same orbits.
- The partial derivatives of V are divided by V, so if the potential on all electrodes changes for a factor — for example, by connecting all electrodes to a common source through potential dividers — the orbits do not change. A common DC source need not to be stabilized. However, this is not true if there is a magnetic lens somewhere in the system.
- If the dimensions of all electrodes are increased or decreased by a factor, the orbits will change by the same factor.

There are no approximations in the derivation of Eq. 4.13; the only limitation is the assumption of rotational symmetry. Even if the equation is formally simple, the solution is governed by the two-dimensional potential function $V(r, z)$ and its partial derivatives. This complicates the analysis of the equation. By developing $V(r, z)$ in a power series as a function of the potential along the axis, the potential becomes a function of only one variable, and the differential equation is no longer partial. But only linearization of Eq. 4.13 allows conclusions on its electron optical meaning. Equation 4.13 is a suitable form for numerical computations.

4.2.1.2. Magnetostatic Field

In a magnetostatic field the force is determined not only by the strength of the field but also by the velocity of the particle, $\mathbf{F} = \mathbf{v} \times \mathbf{B}$. Different particles feel different forces at the same point, r, z, depending on the direction from which they come. If a particle has a radial velocity or the field has a radial component, an azimuthal force will act on the particle, so all three velocity components must be considered.

We will use vector potential, $\mathbf{A}$, to describe the magnetic field. Because of the assumed rotational symmetry, only its θ-component differs from zero. The axial and the radial magnetic field components are defined by Eq. 1.19:

$$B_r = -\frac{\partial A_\theta}{\partial z}, \qquad\qquad B_z = \frac{1}{r}\frac{\partial}{\partial r}(r A_\theta). \qquad (1.19)$$

Introducing Eq. 1.19 into the equations of motion (Eqs. 4.5 – 4.7) we obtain

$$\ddot{r} = -\eta\left[-\dot\theta\frac{\partial}{\partial r}(r A_\theta)\right] + r\dot\theta^2, \qquad (4.14)$$

$$r\ddot\theta + 2\dot r\dot\theta = -\eta\left[\dot r\frac{1}{r}\frac{\partial}{\partial r}(r A_\theta) + \dot z\frac{\partial A_\theta}{\partial z}\right], \qquad (4.15)$$

$$\ddot{z} = -\eta\left(-r\dot\theta\frac{\partial A_\theta}{\partial z}\right). \qquad (4.16)$$

The left side of Eq. 4.15 is

$$r\ddot\theta + 2\dot r\dot\theta = \frac{1}{r}\frac{d}{dt}(r^2\dot\theta)$$

Substituting

$$\frac{d}{dt} = \frac{dz}{dt}\frac{\partial}{\partial z} + \frac{dr}{dt}\frac{\partial}{\partial r},$$

the right side becomes a total differential

$$\eta\left[\dot r\frac{1}{r}\frac{\partial}{\partial r}(r A_\theta) + \dot z\frac{\partial A_\theta}{\partial z}\right] = \frac{\eta}{r}\left[\dot r\frac{\partial}{\partial r}(r A_\theta) + \dot z\frac{\partial}{\partial z}(r A_\theta)\right] = \eta\frac{1}{r}\frac{d}{dt}(r A_\theta).$$

Equation 4.15 can thus be written as

$$\frac{d}{dt}(r^2\dot\theta) = -\eta\frac{d}{dt}(r A_\theta). \qquad (4.17)$$

The integration gives the angular velocity

$$\dot\theta = -\eta\left(\frac{A_\theta}{r} + \frac{C}{r^2}\right), \qquad (4.18)$$

where C is an integration constant. If $\dot\theta = \dot\theta_0$, $r = r_0$, and $A_\theta = A_{\theta 0}$ at $z = 0$, we obtain

$$C = \frac{-\dot\theta_0 r_0^2}{\eta} - A_{\theta 0} r_0. \qquad (4.19)$$

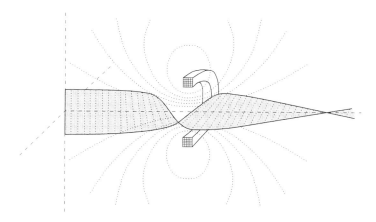

Fig. 4.5. Rotation of the electron orbit in a magnetic lens.

$\dot{\theta}$ can now be eliminated in the radial equation. Introducing Eq. 4.18 into Eq. 4.14 the orbit equation becomes

$$\ddot{r} = \eta^2\left(A_\theta + \frac{C}{r}\right)\left[\frac{C}{r^2} - \frac{\partial A_\theta}{\partial r}\right] = -\frac{\eta^2}{2}\frac{\partial}{\partial r}\left[\left(A_\theta + \frac{C}{r}\right)^2\right]. \qquad (4.20)$$

For the axial direction, Eq. 4.18 must be introduced into Eq. 4.16:

$$\ddot{z} = \eta r\dot{\theta}\left(\frac{\partial A_\theta}{\partial z}\right) = -\frac{\eta^2}{2}\frac{\partial}{\partial z}\left[\left(A_\theta + \frac{C}{r}\right)^2\right]. \qquad (4.21)$$

Equations 4.18, 4.20, and 4.21 show that the particles are accelerated in the axial and radial direction and rotate around the axis, as seen in Fig. 4.5.

Electrostatic and magnetostatic fields can be combined. Introducing

$$D^2 = -\frac{\eta}{2V}\left(A_\theta + \frac{C}{r}\right)^2, \qquad (4.22)$$

two symmetric equations are obtained

$$\ddot{r} = -\eta\frac{\partial}{\partial r}[V(1 - D^2)], \qquad \ddot{z} = -\eta\frac{\partial}{\partial z}[V(1 - D^2)].$$

In analogy to the derivation of Eq. 4.13 the orbit equation for the radial direction is

$$r'' + \frac{1 + r'^2}{2V}\left\{r'\frac{\partial}{\partial z}[V(1 - D^2)] - \frac{\partial}{\partial r}[V(1 - D^2)]\right\} = 0. \qquad (4.23)$$

4.2.2. Power Series for the Electrostatic and the Magnetostatic Field

The radial differential equation governs imaging. The equation is nonlinear and contains partial derivatives, limiting the possibilities of analyzing. The partial derivatives can be transformed into power series.

4.2.2.1. Electrostatic Field

Potential function, $V(r, z)$, is the solution of Laplace's equation and can be expanded in a power series

$$V(r, z) = \sum_{n=0}^{\infty} V_n(z) r^n. \tag{4.24}$$

The derivatives are

$$\frac{\partial V}{\partial r} = \sum n r^{n-1} V_n, \qquad \frac{\partial^2 V}{\partial r^2} = \sum n(n-1) r^{n-2} V_n, \qquad \frac{\partial^2 V}{\partial z^2} = \sum r^n V_n''.$$

Introducing these expressions into Laplace's equation in cylindrical coordinates (Eq. 1.9), we obtain

$$\sum n(n-1) r^{n-2} V_n + \sum n r^{n-2} V_n + \sum r^n V_n'' = 0.$$

This equation must be satisfied for every exponent of r

$$n(n-1) r^{n-2} V_n + n r^{n-2} V_n + r^{n-2} V_{n-2}'' = 0.$$

Hence, the recursion formula is

$$V_n = -\frac{1}{n^2} V_{n-2}''. \tag{4.25}$$

The symmetry calls for $V(r, z) = V(-r, z)$. This can be satisfied only if all $V^{2n+1}(z) = 0$, $n = 0, 1, 2, \ldots$. Accordingly, Eq. 4.24 can be written as

$$V(r, z) = \sum_{n=0}^{\infty} (-1)^n \left(\frac{r}{2}\right)^{2n} \frac{1}{(n!)^2} V_0^{(2n)}(z) = V_0(z) - \frac{r^2}{4} V_0''(z) + \frac{r^4}{64} V_0^{IV} \mp \cdots, \tag{4.26}$$

where $V_0 = V(0, z)$. The electric field components are

$$E_z = -\frac{\partial V}{\partial z} = -V_0' + \frac{r^2}{4} V_0''' - \frac{r^4}{64} V_0^{V} \pm \cdots \tag{4.27}$$

and

$$E_r = -\frac{\partial V}{\partial r} = \frac{r}{2} V_0'' - \frac{r^3}{16} V_0^{IV} \pm \cdots. \tag{4.28}$$

The series for the planar case is

$$V(x, y) = \sum_{n=0}^{\infty} y^n V_n(x) \quad \text{with} \quad V_n = -\frac{1}{n(n-1)} V_n''$$

and

$$V(x, y) = V_0(x) - \frac{y^2}{2!} V_0''(x) + \frac{y^4}{4!} V_0^{IV}(x) \mp \cdots. \tag{4.29}$$

4.2.2.2. Magnetostatic Field

The power series for the magnetostatic field can be obtained similarly. As the particles move in vacuum, where there are no coils or iron, scalar magnetic potential, V_m, can be used:

$$V_m(r, z) = \sum_{n=0}^{\infty} (-1)^n \left(\frac{r}{2}\right)^{2n} \frac{1}{(n!)^2} V_{m0}^{(2n)}(z),$$

where $V_{m0} = V_m(0, z)$. The axial component of the magnetic field is

$$B_z(r, z) = -\frac{\partial V_m}{\partial z} = B_0 - \frac{r^2}{4}B_0'' + \frac{r^4}{64}B_0^{IV} \mp \cdots, \qquad (4.30)$$

where $B_0 = B_z(0, z)$. From

$$\nabla \cdot \mathbf{B} = 0$$

follows

$$\frac{1}{r}\frac{\partial}{\partial r}(r B_r) + \frac{\partial B_z}{\partial z} = 0,$$

and by integration we obtain

$$B_r(r, z) = -\frac{r}{2}B_0' + \frac{r^3}{16}B_0''' \mp \cdots. \qquad (4.31)$$

According to Eq. 1.19, in a rotationally symmetric case we have

$$A_\theta = \frac{1}{r}\int r B_z dr = \frac{r}{2}B_0 - \frac{r^3}{16}B_0'' \pm \cdots. \qquad (4.32)$$

In practice, the magnetic field is sometimes known only along the axis from measurements. It is possible to estimate the magnitude of the radial field near the axis using Eqs. 4.31 or 4.28. However, these equations should *never be used* in connection with numerical orbit computations. In both equations, as well as in Eqs. 4.27 and 4.30, enter the derivatives, and if the field is measured *the numerically computed derivatives will oscillate vigorously*. It cannot be expected that the numerical orbit computations will give the correct answer. The only possibility to use the measured values is to smooth the data by some numerical procedure which minimizes, say, the fourth- or higher-order derivatives and then limit the series expansion of the fields to the same order.

4.2.3. Paraxial Approximation of the Orbit Equation

The differential equation (Eq. 4.23) determines the particle orbits in rotationally symmetric fields. The equation can be solved analytically only in some very special cases, for example with constant axial electric or magnetic field. In most cases of practical interest, the solution can be obtained only numerically. Numeric computations have a drawback: It is difficult to draw general conclusions and to understand how the variations of parameters influence the imaging errors.

It is similar in light optics. According to Snell's law we have

$$n_1 \sin(\alpha + \alpha_1) = n_2 \sin(\alpha - \alpha_2),$$

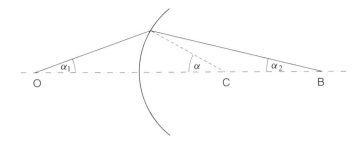

Fig. 4.6. Refraction of a light ray on a spherical surface. C is the center of curvature, O is the object, and B is the image.

as seen in Fig. 4.6. Only if $\alpha_1 \ll 1$ and $\alpha_2 \ll 1$ the sine function can be replaced by its argument, and the angle α_2 is a linear function of α_1. For small angles of incidence, all light rays coming from a point on the optical axis will join in an image point which also lies on the axis, otherwise different imaging errors appear. To obtain an error-free image, all light rays coming from the object must make small angles with the optical axis. These light rays are called *paraxial*, and the imaging is called *paraxial imaging* or *Gauss' approximation*.

In electron optics the same approximation must be made — linearization of Eq. 4.23. Two steps must be taken:

1. Expansion in power series, where the partial derivatives are replaced by the derivative of the field distribution along the optical axis, keeping only the linear terms,
2. elimination of r'^2 — that is linearization of Eq. 4.23.

Replacing $V_0(0, z)$ with V and $B_0(0, z)$ with B in Eqs. 4.27, 4.28, 4.30, and 4.31 we have

$$E_z = -V', \qquad E_r = \frac{r}{2}V'', \qquad\qquad B_z = B, \qquad B_r = -\frac{r}{2}B'. \qquad (4.33)$$

Furthermore,

$$r'^2 \ll 1. \qquad (4.34)$$

Note that the last approximation limits the slope of the orbit, but not the size of the object. Replacement of partial derivatives limits the size of the lens.

With some restriction we can assume that the particles start outside the magnetic field. The constant C in Eq. 4.18 is then zero. The paraxial equation becomes

$$r'' + r'\frac{V'}{2V} + \frac{r}{4V}\left(V'' + \frac{\eta}{2}B^2\right) = 0. \qquad (4.35)$$

Equation 4.35 is a linear and homogeneous second-order differential equation, which has two independent solutions. The general solution is a combination of two particular solutions, u and w:

$$r(z) = C_1 u(z) + C_2 w(z). \qquad (4.36)$$

The rotation of the orbit in the magnetic field, according to Eq. 4.18, occurs with the angular velocity

$$\dot{\theta} = \theta'\dot{z} = \eta\frac{A_\theta}{r} = \frac{\eta}{2}B,$$

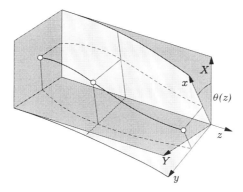

Fig. 4.7. Lipich's law.

where only the first term in the series for A_θ, Eq. 4.32, is used. For paraxial orbits $\dot{r} \ll \dot{z}$, and Eq. 4.11 is approximated by $\dot{z} \approx \sqrt{2\eta V}$. The orbit rotation is

$$\theta = \sqrt{\frac{\eta}{8}} \int \frac{B}{V} dz. \tag{4.37}$$

The important result is that in the paraxial approximation the orbit rotation does not depend either on position or on slope, if the particle starts outside the magnetic field. The rotation angle is the same for all orbits, so when computing the imaging properties of a lens, it is possible to disregard rotation. The particle orbits can be seen as projections in two perpendicular planes, which rotate with the angle $\theta(z)$, as shown in Fig. 4.7. In optics this is called *Lipich's law* [86].

The relations do not change even if $C \neq 0$ in Eq. 4.18. The only difference is the addition of a constant angle, but the differential equation (Eq. 4.35) remains the same. Furthermore, the angle of rotation inside the lens does not depend on either the initial position or the initial slope, and the paraxial imaging is not altered. In the rest of the chapter we will assume that $C = 0$.

The original differential equation (Eq. 4.23) is nonlinear and has partial derivatives. Taking into account the nonlinear term, r'^2, and higher-order terms in the series for the electric or magnetic field, different imaging errors will show up. Such errors are called *aberrations*. Often it is enough to take into consideration the third-order terms, and the errors are then called *third-order aberrations*.

4.2.4. Imaging Laws

The paraxial orbit equation (Eq. 4.35) is *homogeneous and linear*. The general solution is

$$r(z) = C_1 u(z) + C_2 w(z). \tag{4.36}$$

Choose two particular solutions, as seen in Fig. 4.8, which at $z = z_o$ satisfy

$$u(z_o) = 1, \qquad u'(z_o) = 0, \qquad r(z_o) = r_o,$$

$$w(z_o) = 0, \qquad w'(z_o) = 1, \qquad r'(z_o) = r'_o.$$

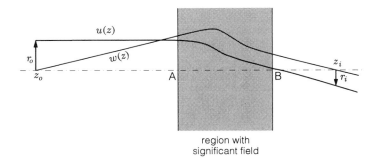

Fig. 4.8. Two particular solutions to Eq. 4.35.

In this case $C_1 = r_o$ and $C_2 = r'_o$. Assume that $w(z) = 0$ at $z = z_i$. Then

$$r(z_i) = r_o u(z_i).$$

The particle orbit $u(z)$ crosses the plane $z = z_i$ in a point whose distance from the optical axis, r_i, is proportional to r_o and *independent on the slope of the orbit* at $z = z_o$. The ratio

$$M_l = \frac{r_i}{r_o} \tag{4.38}$$

is called the lateral magnification. It is a constant for all paraxial orbits and the base of all optical imaging.

Assuming that both the object and the image are outside the region with significant field strength, the orbits there are straight lines. It is useful to choose symmetric orbits for the two particular solutions: the *first* and the *second principal orbit,* shown in Fig. 4.9. The second principal orbit enters the region with significant field from the object side with zero slope, $u'(z) = 0$. The first principal orbit starts from the same point, but with such a slope that after passing the region with significant field, its slope is zero, $w'(z) = 0$. These two orbits cross at $z = z_i$. The plane $z = z_i$, perpendicular to the optical axis, is called the *Gauss' image plane* (or Gauss' plane). Extrapolating the straight parts of the first principal orbit, they will cross at a plane $z = z_{H1}$, the first principal plane. The second principal plane is defined in a similar way — that is, at $z = z_{H2}$. The focal lengths are the distances from the point where the extensions of the principal orbits cross the optical axis to the corresponding principal plane.

$$\begin{aligned} f_1 &= z_{H1} - z_{F1}, \\ f_2 &= z_{F2} - z_{H2}. \end{aligned} \tag{4.39}$$

The object distance a and the image distance b are

$$\begin{aligned} a &= z_{H1} - z_o, \\ b &= z_i - z_{H2}. \end{aligned} \tag{4.40}$$

For the first principal orbit, we obtain

$$\begin{aligned} z < z_{H1}, \quad & w(z) = w(z_{H1}) + (z - z_{H1})w'(z_{F1}), \\ z > z_{H1}, \quad & w(z) = w(z_i) = w(z_{H1}), \qquad w'(z) = 0, \end{aligned} \tag{4.41}$$

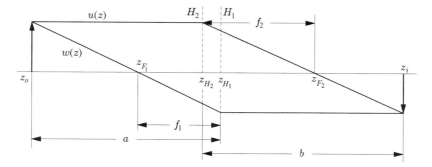

Fig. 4.9. Definition of principal orbits, focal lengths and principal planes. $w(z)$ is the first principal orbit, $u(z)$ the second principal orbit, z_{F1} the object focus, z_{F2} the image focus, f_1 the object focal length, f_2 the image focal length, and H_1 the first and H_2 the second principal plane.

and for the second we have

$$z < z_{H2}, \qquad u(z) = u(z_o) = u(z_{H2}), \qquad u'(z) = 0,$$
$$z > z_{H2}, \qquad u(z) = u(z_{H2}) + (z - z_{H2})u'(z_{F2}). \tag{4.42}$$

According to these definitions we have

$$w'(z_{F1}) = w'(z_o) = \frac{w(z_{H1})}{z_{H1} - z_{F1}} = \frac{w(z_i)}{f_1},$$

$$w(z_o) = w(z_{H1}) - \frac{w(z_{H1})}{f_1}(z_{H1} - z_o) = w(z_i)\{1 - \frac{z_{H1} - z_o}{f_1}\}$$

for the first principal orbit, and

$$u'(z_{F2}) = u'(z_i) = -\frac{u(z_{H2})}{z_{F2} - z_{H2}} = -\frac{u(z_o)}{f_2},$$

$$u(z_i) = u(z_{H2}) - \frac{u(z_{H2})}{f_2}(z_i - z_{H2}) = u(z_o)\{1 - \frac{z_i - z_{H2}}{f_2}\}$$

for the second principal orbit. The lateral magnification is then

$$M_l = \frac{u(z_i)}{u(z_o)} = \frac{w(z_i)}{w(z_o)},$$

and hence

$$1 - \frac{z_i - z_{H2}}{f_2} = \frac{1}{1 - \frac{z_{H1} - z_o}{f_1}} \qquad \text{or} \qquad (1 - \frac{b}{f_2}) \cdot (1 - \frac{a}{f_1}) = 1,$$

which is the *lens equation*

$$\frac{f_1}{a} + \frac{f_2}{b} = 1. \tag{4.43}$$

The lateral magnification can be expressed more symmetrically by a, b, f_1, and f_2:

$$M_l = \frac{b}{a}\frac{f_1}{f_2}. \tag{4.44}$$

The angular magnification is

$$M_\theta = \frac{r'(z_o)}{r'(z_i)} = \frac{b}{a}. \tag{4.45}$$

For a symmetrical lens, with $f_1 = f_2$, the lateral and the angular magnifications are equal.
   The radial equation (Eq. 4.35) can also be written as

$$\sqrt{V}\frac{d}{dz}(\sqrt{V}\frac{dr}{dz}) + \frac{1}{4}(V'' + \eta\frac{B^2}{2})r = 0. \tag{4.46}$$

Introducing the two particular solutions, $u(z)$ and $w(z)$, and multiplying the two equations with u and w, respectively,

$$w\sqrt{V}\frac{d}{dz}(\sqrt{V}\frac{du}{dz}) + \frac{1}{4}(V'' + \eta\frac{B^2}{2})uw = 0,$$

and

$$u\sqrt{V}\frac{d}{dz}(\sqrt{V}\frac{dw}{dz}) + \frac{1}{4}(V'' + \eta\frac{B^2}{2})uw = 0.$$

Subtracting the second equation from the first, we obtain

$$\frac{d}{dz}\{\sqrt{V}(u'w - uw')\} = 0.$$

This equation can be integrated

$$\sqrt{V_i}(u'_i w_i - u_i w'_i) = \sqrt{V_o}(u'_o w_o - u_o w'_o).$$

If w and u represent the first and the second principal orbit, then

$$w'_i = 0, \qquad f_1 = \frac{w_i}{w'_o}, \qquad u'_o = 0, \qquad f_2 = \frac{u_o}{u'_i}.$$

Hence

$$-\sqrt{V_i}w_i u'_i = \sqrt{V_o}u_o w'_o,$$

or

$$\frac{f_2}{f_1} = \sqrt{\frac{V_i}{V_o}}. \tag{4.47}$$

The ratio between the focal lengths is equal to the square root of the ratio between the potentials which prevail on both sides of the lens, where the fields can be assumed negligible. If the potential is equal on both sides of a lens, which is always the case in magnetic lenses, the focal lengths are equal.

In a general case it is impossible to find the focal lengths and the positions of the principal planes as a function of the axial potential distribution, $V(z)$, or the magnetic field, $B(z)$, without numerical computations. However, like in the geometrical optics, simplification by a "thin" lens is an approximation, which is acceptable, if the region with a significant field strength is short compared to focal lengths. Equation 4.35 can be transformed using a substitution proposed by Picht [87]:

$$\rho = r\sqrt[4]{V}, \tag{4.48}$$

which gives

$$\rho'' + T\rho = 0, \qquad T = \left[\frac{3}{16}\left(\frac{V'}{V}\right)^2 + \frac{\eta}{8}\frac{B^2}{V}\right] \tag{4.49}$$

and which can be integrated:

$$\rho'(z_{\mathrm{B}}) - \rho'(z_{\mathrm{A}}) = -\int_A^B T\rho\,dz.$$

It is assumed that the significant field stretches from $z = A$ to $z = B$, as seen in Fig. 4.8. A similar expression can be obtained directly from Eq. 4.35, but a detailed analysis shows that the error is proportional both to r and V'. In the above integral the error is proportional to r^2 and V'^2 and considerably smaller. Assume that ρ changes only insignificantly inside the region of integration, so that ρ can be taken out:

$$\rho'(z_{\mathrm{B}}) - \rho'(z_{\mathrm{A}}) \approx -\rho\int_A^B T\,dz. \tag{4.50}$$

Observing the first and the second principal orbit an approximation for the focal lengths is

$$\frac{1}{f_1} = \sqrt[4]{\frac{V_2}{V_1}}\int_A^B T\,dz, \qquad \frac{1}{f_2} = \sqrt[4]{\frac{V_1}{V_2}}\int_A^B T\,dz. \tag{4.51}$$

which for an electrostatic or a magnetostatic lens gives

Electrostatic lens   Magnetostatic lens

$$\frac{1}{f_2} = \frac{3}{16}\sqrt[4]{\frac{V_1}{V_2}}\int_A^B \left(\frac{V'}{V}\right)^2 dz, \qquad (4.52) \qquad \frac{1}{f_2} = \frac{\eta}{8V}\int_A^B B^2\,dz, \qquad (4.53)$$

$$f_1 = f_2 \quad \text{is valid strictly,}$$

$$\theta = \sqrt{\frac{\eta}{8V}}\int_A^B B\,dz.$$

The longer the focal length of the lens, the better the accuracy of these approximations. The error is generally below 20 percent, except for lenses with very short focal length.

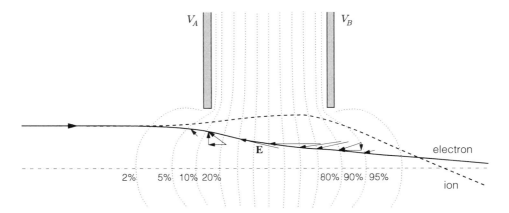

Fig. 4.10. Particle orbits in an electrostatic lens.

There is a point in using Eq. 4.46 to compute an approximate expression for the focal length, even if it *never should be used as a base for a design*. It gives an interesting illustration of the fact that almost all rotationally symmetric lenses are positive. Observe the orbit of an electron in an electrostatic lens, seen in Fig. 4.10, where $V_B > V_A$. The radial component of the field increases with the distance from the optical axis. In the left half of the lens the electron comes nearer to the optical axis, because of the outward direction of the electric field. When the electron enters the right half of the lens, the direction of the radial field has changed, but the field is weaker because the electron is nearer the axis. The outward force in the right half of the lens is therefore smaller than the inward force in the left half. The lens is positive and remains so even for a positive particle. The left side acts as a negative lens, but when the ion enters the right half, the inward acting force is stronger and the lens is positive. There is no difference if $V_A > V_B$.

Integrating Eq. 4.46 for an electrostatic lens, we obtain

$$\sqrt{V_B}\left(\frac{dr}{dz}\right)_{z=z_B} - \sqrt{V_A}\left(\frac{dr}{dz}\right)_{z=z_A} = -\frac{1}{4}\int_A^B \frac{r V''}{\sqrt{V}}dz.$$

In a short and weak lens, approximate $r(z_A) \approx r(z_B) \approx r_0$ and observe the second principal orbit. The approximate value of the image focal distance is

$$\frac{1}{f_2} \approx \frac{1}{4\sqrt{V_B}}\int_A^B \frac{V''}{\sqrt{V}}dz. \tag{4.54}$$

When integrating Eq. 4.54, the first part of the integral between A (where the significant field starts) and a point midway between the electrodes gives a focusing action. The second part, from the middle of the lens to B (where the field becomes negligible) results in a defocusing action, as seen in Fig. 4.11b. This is a consequence of the proportionality between the radial electric field and V'' for paraxial orbits (Eq. 4.28). When the second derivative of the potential along the axis is positive, the radial force is directed toward the optical axis, and vice versa. In the special case of Fig. 4.10, there are two factors which act together and make the lens positive. The first factor, radial electric field proportional

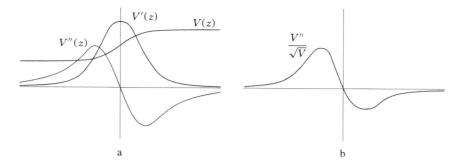

Fig. 4.11. (a) Potential distribution, $V(z)$, and its first and second derivative for an electrostatic lens with two electrodes. **(b)** The function $V'''/\sqrt{V}$ for the same lens.

to the radial position, was already mentioned. The second factor is the change in velocity. In the right half of the lens the velocity is higher, and because $v \propto \sqrt{V}$, the integrand $V''/\sqrt{V}$ in the right half of the lens must be smaller than that in the left half. Therefore, the first part of the integral to the point midway between the electrodes is considerably larger than its second part, and the lens is positive.

The integral in Eq. 4.54 can be integrated partially. With

$$u = \frac{1}{\sqrt{V}} \quad \text{and} \quad dv = V'' dz,$$

the focal length becomes

$$\frac{1}{f_2} \approx \frac{V'(z_B) - V'(z_A)}{4V(z_B)} + \frac{1}{8\sqrt{V(z_B)}} \int_A^B \frac{V'^2}{V^{3/2}} dz. \tag{4.55}$$

Assuming that $\mathbf{E} = 0$ left of A and right of B, the first term in Eq. 4.55 is zero, and only the integral remains:

$$\frac{1}{f_2} \approx \frac{1}{8\sqrt{V(z_B)}} \int_A^B \frac{V'^2}{V^{3/2}} dz. \tag{4.56}$$

The interesting fact about this extremely bad approximation is that the integrand includes V'^2. Independently if the first derivative is positive or negative, the value of the integral will be positive and so will be the lens.

If the electric field is different from zero on any side of the lens, the first term in Eq. 4.55 can outweigh the second. This can be achieved only with an electrode and a field-free region on one side, *an aperture* (Fig. 4.14),

$$\frac{1}{f_2} \approx \frac{V'(z_B) - V'(z_A)}{4V(z_B)}. \tag{4.57}$$

The lens can then be positive or negative, depending which field strength is stronger. A special case, often met in practice, is an anode with a hole in front of a planar cathode, with a field-free region behind the anode. The electric field is positive in the region between

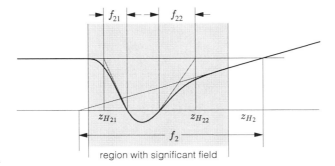

Fig. 4.12. Particle orbit in a lens where the orbit crosses the optical axis twice inside the lens.

the cathode and the anode, $V'(z_A) > 0$, but behind the anode the field is zero, $V'(z_B) = 0$. The anode opening acts as a negative, divergent lens, as can be seen in Fig. 4.14a.

Many useful electron-optical lenses cannot be described in such a simple way. We have assumed that both the object and the image, as well as both focal points, lie outside the region with significant field strength. This is not always true. For example, in the electron microscope the objective lens always has such a short focal length, so that the object lies deep inside the magnetic field, and the electron orbits cross the optical axis inside the lens. Often the object is placed very near such a crossing point to obtain a large magnification. This configuration is similar to an immersion lens in geometrical optics, where the focal length is shortened by using a special oil to change the refraction index. The only interesting electron orbit part is the part from the object to the place where the magnetic field becomes negligible. The orbit part before the object is only used for object "illumination."

The focal length can be defined in an analogous way as earlier, namely by

$$-f_{2n} = \frac{r_0}{(r')_{z=z_n}},\tag{4.58}$$

where the subscript n denotes the nth crossing of the optical axis, as seen in Fig. 4.12.

4.2.5. Matrix Formulation of the Paraxial Orbit Equation

In a region without electric or magnetic field, a drift space, the particle orbit is described by

$$r(z) = r(z_o) + r'(z_o)(z - z_o),$$
$$r'(z) = r'(z_o)$$

or in matrix form

$$\begin{pmatrix} r \\ r' \end{pmatrix} = \begin{pmatrix} 1 & z - z_o \\ 0 & 1 \end{pmatrix} \begin{pmatrix} r_o \\ r'_o \end{pmatrix}.\tag{4.59}$$

The lens equation, written in matrix form, would to a great extent simplify the computations of lens systems, especially if many lenses are coupled in series.

Observe a thin lens — that is, a lens whose focal lengths are much longer than the region with significant field strength. Assume that the radial coordinate, r, does not change much during the passage through the lens and write

$$\begin{pmatrix} r \\ r' \end{pmatrix} = \begin{pmatrix} 1 & 0 \\ a_{21} & a_{22} \end{pmatrix} \begin{pmatrix} r_o \\ r'_o \end{pmatrix}. \tag{4.60}$$

For the second principal orbit, with r_o at z_{H2}, we obtain

$$r'(z_{H2} - \epsilon) = 0, \qquad r'(z_{H2} + \epsilon) = -\frac{r_o}{f_2},$$

which gives

$$a_{21} = -\frac{1}{f_2}.$$

The first principal orbit, with r_o at z_{H1}

$$r'(z_{H1} + \epsilon) = 0, \qquad r'(z_{H1} - \epsilon) = \frac{r_o}{f_1},$$

gives

$$r' = 0 = -\frac{1}{f_2} r_o + a_{22} r'_o,$$

or

$$a_{22} = \frac{f_1}{f_2}.$$

The passage through a thin lens can therefore be described by a matrix equation

$$\begin{pmatrix} r \\ r' \end{pmatrix} = \begin{pmatrix} 1 & b \\ 0 & 1 \end{pmatrix} \begin{pmatrix} 1 & 0 \\ -\frac{1}{f_2} & \frac{f_1}{f_2} \end{pmatrix} \begin{pmatrix} 1 & a \\ 0 & 1 \end{pmatrix} \begin{pmatrix} r_o \\ r'_o \end{pmatrix}. \tag{4.61}$$

where a and b are the object and the image distance according to Fig. 4.9.

Quadrupole and multipole lenses conform naturally to matrix formalism, and are often used in accelerators and beam transport systems (see also Section 4.5 and Appendix H). There thousands of lenses are connected in series and the matrix formalism is the only method whereby these systems can be computed. The matrix equations are developed also for bending magnets, acceleration cavities, and other components of accelerators and beam transport systems.

4.3. ELECTRIC AND MAGNETIC
ROTATIONALLY SYMMETRIC LENSES

It is possible, at least in principle, to numerically compute the field distribution in any electron lens. Numerical computations give, however, rarely a deeper insight in the physical problem. Many lenses with similar geometry and different electrode potential or magnetic field must be computed to get general conclusions from a graphical representation of the results. Therefore, some approximate calculations can give a construction hint.

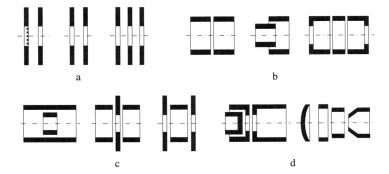

Fig. 4.13. Rotationally symmetric electrostatic lenses. (**a**) Plane electrodes with a hole. (**b**) Cylindrical lenses. (**c**) Combined lenses. (**d**) Cathode lens.

4.3.1. Electrostatic Lenses

Rotationally symmetric electron lenses, seen in Fig. 4.13, can be divided according to different features:

(a) Number of electrodes — one, two, three or more,

(b) Electrode form — plane or cylindrical,

(c) Potential on both sides of the lens — $V_1 = V_2$, $V_1 < V_2$ or $V_1 > V_2$.

4.3.1.1. Aperture

Figure 4.14 shows electron orbits with potential and field distribution for an aperture. Assume that the region on the right side of the aperture is field-free, and that the hole has a small diameter compared to the distance to the nearest electrode on the left. The described geometry can be realized in two ways: Either the left electrode is designed as a mesh or it is a cathode (the latter only according to Fig. 4.14a). The first possibility is seldom used in electron optics because of inhomogeneity in field distribution, scattering, and losses in the mesh. The other is used frequently as a cathode – anode combination.

With the above-mentioned assumptions approximate

$$0 < z < L, \qquad V'(z) = \text{const}, \qquad V''(z) = 0,$$

where L is the distance between the electrodes. In the hole V' changes the value abruptly. From being constant, it falls to zero on the right side. The second derivative can be described by a Dirac function, as indicated in Fig. 4.14. Even if the approximation looks bad, the analysis suggests how the choice of the parameters influences imaging.

Observing the second principal orbit, with initial values $r = r_0, r' = 0$, the orbit equation can be integrated between 0 and $L - \epsilon$, giving

$$r = r_0.$$

In the region around L equation 4.35 takes the form

$$r'' + r\frac{V''}{4V} = 0.$$

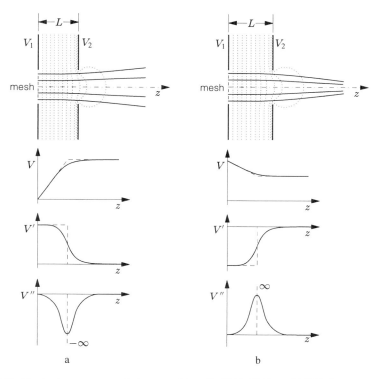

Fig. 4.14. Electrode geometry and electron orbits with $V(z)$, $V'(z)$, and $V''(z)$ for an aperture. (a) $V_2 > V_1$, negative lens. (b) $V_2 < V_1$, positive lens. - - - - approximation, ——— computed.

On the right side of the lens, at $z = L + \epsilon$ we obtain

$$\int_{L-\epsilon}^{L+\epsilon} r'' dz = - \int_{L-\epsilon}^{L+\epsilon} r \frac{V''}{4V} dz,$$

or

$$r' = \frac{V'}{4V_2} r_0.$$

Hence

$$\frac{f_2}{L} = - \frac{4V_2}{V'L} = - \frac{4V_2}{V_2 - V_1} = - \frac{4Q}{Q-1}, \qquad (4.62)$$

where

$$V' = \frac{V_2 - V_1}{L} \qquad \text{and} \qquad Q = \frac{V_2}{V_1}.$$

The focal length is independent on the diameter of the hole, d, if $d \ll L$.

Depending on the potentials, V_1 and V_2 the lens can be positive ($V_2 < V_1$) or negative ($V_2 > V_1$). Note that an aperture is the only negative rotationally symmetric lens if $V_2 > V_1$. Regretfully, such a lens needs a mesh to generate a homogeneous electric field on one side of the lens and cannot be used for imaging. Equation 4.62 is generally good enough

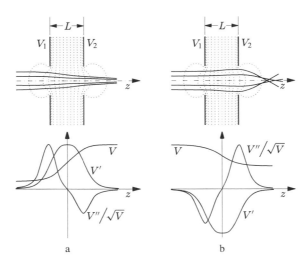

Fig. 4.15. Electrode geometry and electron orbits with $V(z)$, $V'(z)$, and $V''(z)/\sqrt{V(z)}$ for a lens with two apertures. (**a**) $V_1 < V_2$, (**b**) $V_1 > V_2$.

if used as an approximation for the focal length of a cathode – anode combination, if no other lenses are involved.

4.3.1.2. Lens with Two Apertures

This lens has two parallel plane electrodes, at a distance L, and a hole in each of them, as seen in Fig. 4.15.

With the same assumption as earlier, a similar technique of replacing the second derivative of the potential at the electrode plane by Dirac function can be used. After integration of the equation of motion, and after normalizing

$$Q = \frac{V_2}{V_1},$$

the image focal distance is

$$\frac{f_2}{L} = -\frac{8}{3}\frac{1}{(1 - 1/Q)(1 - \sqrt{Q})}. \tag{4.63}$$

The object focal distance is obtained from Eq. 4.47:

$$f_1 = \frac{f_2}{\sqrt{Q}}.$$

The principal planes, H_1 and H_2, are situated at

$$\frac{z_{H2}}{L} = \frac{1}{6}\frac{5Q + 3}{1 - Q}$$

and

$$\frac{z_{H1}}{L} = \frac{1}{6}\frac{5 + 3Q}{1 - Q}.$$

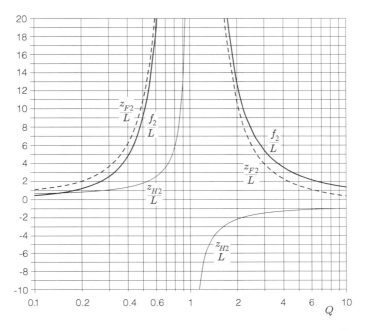

Fig. 4.16. Normalized focal length, f_2/L, the position of the image focal point, z_{F2}/L, and the principal plane z_{H2}/L, for a lens with two apertures, as a function of the ratio between the potentials on both sides of the lens, $Q = V_2/V_1$.

Equation 4.63 shows that the lens always is positive. An accelerating lens ($Q > 1$) is stronger than a retarding one ($Q < 1$). This is valid for all lenses with two apertures and does not depend on the electrode shape.

Figure 4.16 shows the normalized focal length, the position of the image focus, and the position of the principal plane:

1. Observe that the electron orbit crosses the optical axis inside the lens field if $Q > 9$ ($z_{F2}/L < 0.5$).
2. The second principal plane lies always outside the lens, on the left of the left electrode if $Q > 1$ and to the right if $Q < 1$.
3. The weaker the lens the further away from the left electrode ($z = -L/2$) is the principal plane. The position of the focal point, $F_2 = f_2 + z_{H2}$, differs for $Q > 1$ by not more than 30 percent, even for $Q = 9$ from the focal distance f_2. However, for $Q = 0.2$ the value of F_2 is more than the twice f_2.
4. For $Q > 1$ the lens can be replaced by a thin lens aproximation without large errors. This is not true if $Q < 1$. Already at $Q = 0.4$ the error becomes unacceptable. The explanation is the position of the second principal plane. To take a lens as thin, the distance between the principal planes must be small compared to the focal length. For this lens $z_{H1} - z_{H2} = L/3$ for all values of Q. For example, at $Q = 5$, $L/3 \ll f_2 = 2.7\,L$; at $Q = 0.2$, f_2 is only $1.2L$ and thus almost three times smaller.

4.3.1.3. Lens with Three Apertures ("Einzellens")

It is often desirable to focus particle beams without changing the beam energy. Einzellens does it, and the central electrode is often connected to cathode potential. According to

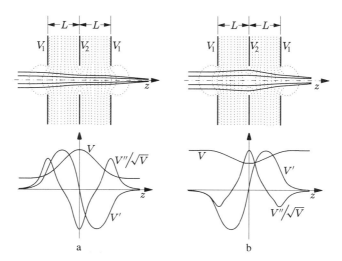

Fig. 4.17. Electrode geometry and electron orbits with $V(z)$, $V'(z)$, and $V''(z)/\sqrt{V(z)}$ for an einzellens. **(a)** $V_1 < V_2$, **(b)** $V_1 > V_2$.

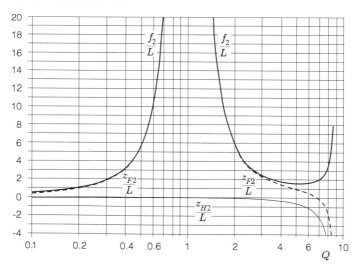

Fig. 4.18. Normalized focal length, f_2/L, the position of the image focal point, z_{F2}/L, and the position of the second principal plane, z_{H2}/L, for an einzellens as a function of the ratio between the potentials, $Q = V_2/V_1$.

Eq. 4.47, both focal distances are equal and both principal planes are the same distance from the symmetry plane:

$$f_1 = f_2, \qquad z_{H1} = -z_{H2}.$$

An approximate value for the focal distance and the principal plane positions can be computed similarly as for an aperture. If the hole diameter $d \ll L$, then

$$\frac{f_2}{L} = -\frac{8}{3} \frac{1}{(Q-1)(4 - 3/\sqrt{Q} - \sqrt{Q})}, \qquad (4.64)$$

$$\frac{z_{H2}}{L} = 1 - \frac{4}{(3 - \sqrt{Q})(1 + \sqrt{Q})}.$$

Figure 4.18 shows the normalized focal length and the position of the second principal plane. Note that einzellens is stronger than a lens with two apertures (Fig. 4.16). For a week lens, $Q \approx 1$, both principal planes are near the central electrode, and the einzellens can be approximated by a thin lens, but Eq. 4.51, valid for a week lens, can only be used if $0.2 < Q < 2$. For $Q > 4.91$ and $Q < 0.2$ the electron orbit crosses the optical axis inside the lens, and the approximation is no longer valid.

4.3.1.4. Cylindrical Lens

Cylindrical lenses are made of coaxial tubes, with or without apertures. These lenses are often used in focusing low-perveance electron beams, the most important applications being cathode ray tubes, TV and computer screens, and so on.

With equal inner radii, and the distance between the tubes small compared to the radius, the potential distribution along the axis can be approximated by

$$V \approx \frac{1}{2}(V_1 + V_2) + \frac{1}{2}(V_2 - V_1) \tanh(1.32\frac{z}{R}). \tag{4.65}$$

An approximate value for the focal distance can be obtained from Eq. 4.54

$$\frac{f_2}{R} = \frac{\sqrt{Q}}{0.44} \frac{\sqrt{Q} + 1}{(\sqrt{Q} - 1)^2} \tag{4.66}$$

In many textbooks on electron optics,* diagrams similar to Figs. 4.16 and 4.18 can be found, valid for different tube diameters and distances, which can be used as a first approximation in numerical design of cylindrical lenses.

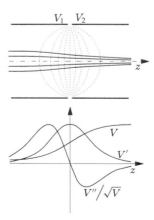

Fig. 4.19. Cylindrical lens. $V_2 > V_1$.

---

*See, for example, K. R. Spangenberg, *Vacuum Tubes.*

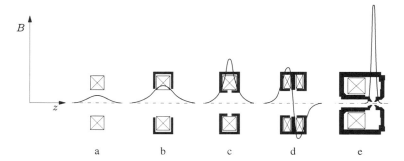

Fig. 4.20. Rotationally symmetric magnetostatic lenses. (**a**) Coil without iron. (**b**) Open coil with iron. (**c**) Coil with iron and a short gap. (**d**) Two coils with inverted fields. (**e**) Coil with iron and poles.

4.3.2. Magnetostatic Lenses

Rotationally symmetric magnetostatic lenses can be divided in:
(a) Lenses with or without iron,
(b) Lenses with or without alternation of the magnetic field direction along the axis.

All rotationally symmetric magnetic lenses are *positive* in accordance with Eq. 4.53. Also, pure magnetic lenses, which always have the same potential on both sides of the lens, must have *equal focal lengths*

$$f_1 = f_2.$$

They differ from the electrostatic lenses in an important point: The focal length depends on the charge and the mass of the particles. The orbits rotate in accordance with Eq. 4.37.

Magnetic lenses are used more often than electrostatic lenses. The most important reason is the easier construction of strong magnetic lenses. By comparing the electric and the magnetic force

$$q\mathbf{E} = q\,\mathbf{v} \times \mathbf{B},$$

and taking as an example an electron accelerated to an energy of 10,000 eV, it is easy to see that an electric field of 5900 V/mm must be used to obtain the same effect as with a magnetic field of 0.1 T. To generate a magnetic field of 0.1 T, using coils with iron is easy. At higher energies, magnetic lenses are still more favorable.

In electrostatic lenses, approximate calculations of the focal length for simple aperture lenses give reasonable approximations. In magnetostatic lenses the situation is more complicated. But there are cases which can be solved analytically. The simplest is a homogeneous field parallel with the optical axis and limited in length. The orbit equation can also be solved for a simple current loop and for a special kind of field, the so-called "clock field," which approximates very well the field distribution along the axis of a coil with iron and with a short, or very short, gap.

Of the coils in Fig. 4.20 the strongest and mostly used are types **c** and **e**. Type **e** has a very short gap with small aperture, and the field is reinforced by pole shoes. Such lenses are needed in the electron microscope, where the large magnification requires strong lenses. Type **d** is interesting because the rotation of the orbit is zero, if the lens is completely symmetric. Iron can introduce interference, so these lenses are sometimes constructed without iron.

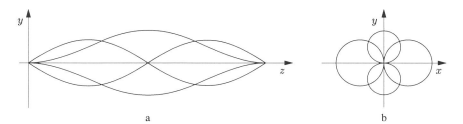

Fig. 4.21. Helicoidal orbits in a homogeneous magnetic field. (**a**) Side projection. (**b**) Projection plane perpendicular to the optical axis.

4.3.2.1. Homogeneous Magnetic Field

In a homogeneous magnetic field the projection of the particle orbits in a plane perpendicular to the optical axis are circles with constant radius. The paraxial differential equation (Eq. 4.35) becomes

$$r'' + \frac{r}{4\kappa^2} = 0,$$

with

$$\kappa = \sqrt{\frac{2V}{\eta B^2}}. \tag{4.67}$$

The general solution is

$$r = C_1 \sin \frac{z}{2\kappa} + C_2 \cos \frac{z}{2\kappa}. \tag{4.68}$$

For a bundle of particles with the same initial velocity, which start from a point on the axis, the initial conditions are $z = 0$, $r = 0$, and $dr/dz = r_0'$. Hence

$$C_1 = 2\kappa r_0', \qquad\qquad C_2 = 0, \qquad\qquad r = 2\kappa r_0' \sin \frac{z}{2\kappa}.$$

These particles move along helicoidal orbits and converge to a point on the optical axis, as seen in Fig. 4.21. The motion of each particle is a combination of a translation parallel with the optical axis and a rotation in a plane perpendicular to the optical axis. Such a lens can be used as an approximation for a short lens with iron, with two small holes in the yoke (Fig. 4.22).

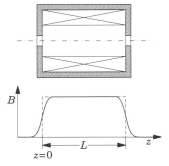

Fig. 4.22. Lens with homogeneous magnetic field. Magnetic field on the axis: - - - - approximation, ——— computed.

Observe the second principal orbit. At the entrance into the magnetic field $z = 0$, $r = r_0$, and $r_0' = 0$. With these initial conditions Eq. 4.68 becomes

$$r = r_0 \cos \frac{z}{2\kappa}.$$

After the passage through the lens, at $z = L$, the particle has a radial velocity

$$r_L' = -\frac{r_0}{2\kappa} \sin \frac{L}{2\kappa}.$$

The focal length of the lens is thus

$$f_2 = -\frac{r_0}{r_L'} = \frac{2\kappa}{\sin \frac{L}{2\kappa}}, \tag{4.69}$$

and the second principal plane is situated at

$$z_{H2} = z_{f_2} - f_2 = L - f_2(1 - \frac{r_L}{r_0}) = L - \frac{2\kappa}{\sin \frac{L}{2\kappa}}\left[1 - \cos \frac{L}{2\kappa}\right].$$

The orbits rotate according to Eq. 4.37:

$$\theta = \frac{L}{2\kappa}.$$

These simple results are quite useful. Besides being an approximation of a short solenoidal lens with iron, they show a characteristic feature of all magnetostatic lenses: The focal length varies periodically with the length of the lens. The strongest lens, with the shortest focal length, is obtained for $L/(2\kappa) = \pi/2$ — that is, when $L = \pi\kappa$. When $L = 2\pi\kappa$ the focal length becomes infinite, and the orbit of the particle turns 180°.

The focal length of a short lens, with $L \ll \kappa$, is

$$f_2 = 4\frac{\kappa^2}{L} = \frac{8V}{\eta B^2 L}. \tag{4.70}$$

That the focal length is inversely proportional to B^2 is valid generally for short magnetostatic lenses. Both principal planes coincide, and lie in the middle of the lens. At very short focal lengths the principal planes diverge, but are never further than $\pm 0.14L$ from the center.

A simple coil can be made with a single circular turn of wire. Integrating Biot-Savart's law the axial component of the magnetic field on the axis is

$$B = \frac{\mu_0 I R^2}{2\sqrt{(R^2 + z^2)^3}}, \tag{4.71}$$

where R is the radius of the turn and I is the current. Introducing Eq. 4.71 in the expression for the focal length of a magnetostatic lens, Eq. 4.53, the equation can be integrated:

$$f_2 = \frac{256 V R}{3\pi \eta \mu_0^2 I^2}. \tag{4.72}$$

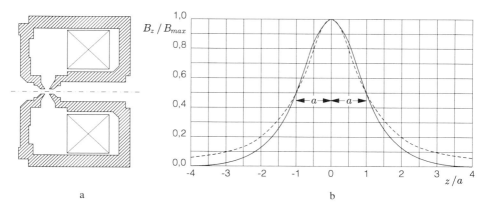

Fig. 4.23. (a) Objective lens of a transmission electron microscope. (b) Magnetic field distribution on the axis: - - - - Eq. 4.73, ——— computed by FEM.

A coil with air is in practice never made of a single turn. With n turns, factor n^2 must be inserted in the denominator of Eq. 4.72 and the whole expression multiplied by the coil form factor, which has a value between 1.0 and 1.25.

4.3.2.2. The Clock Field

In practically all electrostatic and magnetostatic lenses discussed up to now, we could obtain approximate expressions for the focal lengths and the positions of the principal planes. The weaker the lens, the better the approximations. The lenses used in electron optical devices are, however, usually strong. The approximations can thus only give a first hint how the electron optical system should be constructed, and numerical computations must give the answer.

A remarkable exception is the clock field, proposed by Glaser [88]:

$$B = \frac{B_{max}}{1 + (\frac{z}{a})^2}. \tag{4.73}$$

The equation approximates very well the field distribution on the axis of a coil with iron and with a short gap, shown in Fig. 4.23. B_{max} is the amplitude of the magnetic field in the middle of the gap at $z = 0$, and the parameter a is the half-width value. Compared with a real lens the approximation in Eq. 4.73 falls too slowly at high z/a values. It is possible to obtain a somewhat better agreement with the reality if an adjusted exponent, $\nu > 1$, is introduced,

$$B = \frac{B_{max}}{[1 + (\frac{z}{a})^2]^\nu},$$

but the paraxial differential equation (Eq. 4.35) cannot be solved analytically.

Introducing Eq. 4.73 into Eq. 4.35, the differential equation becomes

$$\frac{d^2 r}{dz^2} + \frac{\eta B_{max}^2}{8V} \frac{r}{[1 + (\frac{z}{a})^2]^2} = 0. \tag{4.74}$$

Glaser managed with two ingenious substitutions

$$x = \frac{z}{a} = \cot \varphi \qquad\qquad y = \frac{r}{a} = \frac{h(\varphi)}{\sin \varphi} = \sqrt{1 + k^2} h(\varphi) \qquad (4.75)$$

to transform Eq. 4.74 into

$$h''(\varphi) + (1 + k^2) h(\varphi) = 0, \qquad (4.76)$$

which can be solved analytically. Here

$$k^2 = \frac{\eta B_{max}^2 a^2}{8V} \qquad (4.77)$$

is a measure for the strength of the lens. The solution of Eq. 4.76, with $1 + k^2 = \omega^2$, is

$$h(\varphi) = c_1 \sin \omega (\varphi + c_2),$$

and thus

$$r(z) = c_1 \frac{\sin \omega (\varphi + c_2)}{\sin \varphi}, \qquad \varphi = \text{arccot}(z/a), \qquad (4.78)$$

where c_1 and c_2 are two constants. For the second principal orbit, $r(-\infty) = r_0$ and $r'(-\infty) = 0$, so φ becomes π for $z = -\infty$ in accordance with Eq. 4.75. Equation 4.78 then requires $c_2 = -\pi$ and $c_1 = -r_0/\omega$. The second principal orbit is

$$r(z) = -\frac{r_0}{\omega} \frac{\sin \omega (\varphi - \pi)}{\sin \varphi}.$$

Since $\varphi = \text{arccot}(z/a)$ the solution shows an oscillatory character. Depending on the value of ω the particles can cross the optical axis a few times inside the lens, as seen in Fig. 4.24.

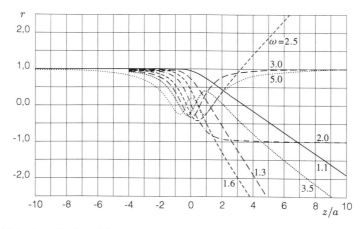

Fig. 4.24. Normalized particle orbits as a function of the strength of the lens. The particle starts parallel with the optical axis (second principal orbit).

This means that more than one focal length can be defined:

$$f_{2n} = \frac{a}{\sin n\frac{\pi}{\omega}}, \tag{4.79}$$

where n is an integer. Every change in sign of f_2 means a new crossing of the optical axis inside the field of the lens, while n increases by 1. This is used in the transmission electron microscope to make an immersion-like lens and to obtain a large magnification. In practice only one crossing is allowed, $n = 1$, so ω should be kept below 2.

4.4. ABERRATIONS IN ROTATIONALLY SYMMETRIC LENSES

4.4.1. Physical Description of Aberrations

Linearization of the paraxial equation makes the equation homogeneous. Design of an electron optical system should start with the paraxial approximation to compute the strength of the lenses in the system. In a later stage, lens errors must be considered, taking into account higher-order terms in Eq. 4.23. The serious problem in electron optics is the the lack of negative lenses, which in light optics are used, together with materials with different refraction index, to correct aberrations. For example, in a light microscope the opening angle is around $60°$, and in an electron microscope the electron beam is limited by apertures to about $0.5°$.

As long as the aberrations are weak, they do not limit the imaging to an appreciable degree. But when they become strong, it is important to know which possibilities exist to avoid them, or at least what can be made to change the construction and to reduce their influence. Already a simple discussion shows that different types of errors must exist. Assume, that we want to reproduce a small object in a magnified image. Linearizing Eq. 4.23 limits the orbits to small angles with the optical axis. The image can be improved by introducing an aperture somewhere between the object and the image. The aperture stops all particles which have a larger radial deviation than the aperture opening, and the particles which impair the image mostly are taken away. Another limiting case occurs if a big object should be imaged, but only the orbits which pass the lens near the optical axis can reach the image plane. Insertion of an aperture near the lens will not help in this case, and the aperture will not limit the orbits.

The simplest method to desribe the lens errors is to take into account the higher-order terms of the power series for the electric or magnetic field. If the terms are limited to third order in r, five different geometrical errors can be distinguished. It is important to have a clear picture of which types of errors are the most outstanding in different applications. It is often enough to correct — if possible — one or two types of errors which disturb the imaging mostly.

In calculating an electron optical system, numerical computations will be used at the end. It is then simpler to integrate directly the original differential equation (Eq. 4.23) numerically than to use the paraxial equation (Eq. 4.35). If the form of the electrodes and/or the shape of the coils is known, the field distribution can be computed, and thus we can determine where the particles hit the image plane. But the physical understanding of aberrations is necessary; otherwise there is no guidance how to make corrections.

Besides the geometrical aberrations, other sources of error deteriorate the image. The Maxwellian velocity distribution or time variation of the acceleration potential results in different orbits. Not one lens is geometrically perfect or perfectly centered and aligned to the optical axis. This also introduces different errors. Space-charge, if not negligible, influences the orbits and generates another error type.

Similarity with light optics stretches even to the analysis of the imaging errors. All five geometrical aberrations have an analogy in light optics. Anisotropic effects force electron optics to take into account azimuthal errors. In light optics there is a sixth type of lens error: All wavelengths are not refracted equally, because the index of refraction is a function of the wavelength. The error is the chromatic aberration. In electron optics it corresponds to errors caused by velocity distribution of the particles.

4.4.2. Calculation of Geometrical Aberration Coefficients

The lateral magnification is determined by the paraxial approximation (Eq. 4.38). Remembering that in the magnetic field the particle orbits rotate, it is suitable to write this equation as

$$\mathbf{r}_i = \mathcal{M}_l \mathbf{r}_o, \tag{4.80}$$

where $\mathbf{r}_o$ is the radial position of the object point, $\mathbf{r}_i$ is the radial position of the image point in the Gauss image plane, and $\mathcal{M}_l$ is the lateral magnification. $\mathbf{r}_o$ and $\mathbf{r}_i$ should in this context be seen as vectors in a plane perpendicular to the optical axis, as seen in Fig. 4.25. For example,

$$\mathbf{r}_o = r_o \, e^{j\theta_o}.$$

Geometric aberrations will cause a deviation in the Gauss image plane

$$\mathbf{r}_i = \mathbf{r}_{iG} + \Delta \mathbf{r}_i. \tag{4.81}$$

Subscript G refers to the paraxial solution. According to imaging laws the paraxial image point is unambiguously determined as a function of $\mathbf{r}_o$. This is not true for $\Delta \mathbf{r}_i$, because it depends also on $\mathbf{r}'_o$. This means that we must take into account both $\mathbf{r}_o$ and $\mathbf{r}'_o$ when calculating the aberration coefficients.

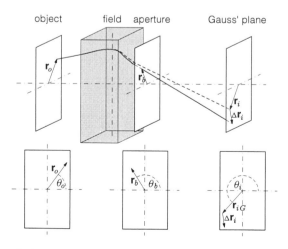

Fig. 4.25. Geometrical relations in calculation of aberrations. The dashed curve shows the orbit of a paraxial particle through the lens. The full line represents the real orbit.

Starting with the general orbit equation (Eq. 4.23) and using the same technique as in deriving the paraxial equation (Eq. 4.35), but this time with higher-order terms in the power series expansions for the electric and the magnetic field respectively, these terms can be seen as perturbations of the paraxial orbit equation. The method to solve the new equation was developed by Scherzer [89] and is the base for the aberration theory. Perturbation method has a drawback: The expressions for the five aberration coefficients are complicated and difficult to grasp. Scherzer succeeded, however, to show that with a suitable form of the field distribution in rotationally symmetric lenses, all geometrical errors can be eliminated, with the exception of the spherical aberration.

Scherzer's method for an electrostatic lens starts with Eq. 4.13. The power series for the electrostatic field, including the terms of the third order, gives

$$V = V_0 - \frac{r^2}{4}V_0'', \qquad -\frac{\partial V}{\partial r} = \frac{r}{2}V_0'' - \frac{r^3}{16}V_0^{IV}, \qquad -\frac{\partial V}{\partial z} = -V_0' + \frac{r^2}{4}V_0'''.$$

Hence

$$r'' + r'\frac{V_0'}{2V_0} + r\frac{V_0''}{4V_0} = \frac{r^3}{16}\left[\frac{V_0^{IV}}{2V_0} - \left(\frac{V_0''}{V_0}\right)^2\right] + \frac{r^2 r'}{8}\left[\frac{V_0'''}{V_0} - \frac{V_0'V_0''}{V_0^2}\right] - \frac{rr'^2}{4}\frac{V_0''}{V_0} - \frac{r'^3}{2}\frac{V_0'}{V_0}.$$
(4.82)

The three terms on the left side are the paraxial equation, and the right sides includes the perturbation terms. According to Eq. 4.81 we have

$$r = r_G + \Delta r,$$

where r_G is the solution of the paraxial equation, which after being introduced into Eq. 4.82 and after some algebraic manipulations, gives the perturbation equation

$$\Delta r'' + \Delta r'\frac{V_0'}{2V_0} + \Delta r\frac{V_0''}{4V_0} = \frac{r_G^3}{16}\left[\frac{V_0^{IV}}{2V_0} - \left(\frac{V_0''}{V_0}\right)^2\right] + \frac{r_G^2 r_G'}{8}\left[\frac{V_0'''}{V_0} - \frac{V_0'V_0''}{V_0^2}\right] - \frac{r_G r_G'^2}{4}\frac{V_0''}{V_0} - \frac{r_G'^3}{2}\frac{V_0'}{V_0}.$$
(4.83)

The equation is linear in Δr and can be solved analytically by expressing the right side with definite integrals with the potential distribution on the axis and its derivatives up to the fourth order. In this analytical solution the five geometric aberration coefficients can be distinguished.* Besides numerical computations of aberration coefficients the analytical expressions are used to optimize lens parameters in electron optical systems by variational methods.

We will use another approach to obtain a physical picture of the aberration coefficients by introducing an auxiliary plane perpendicular to the optical axis. This can be the aperture plane. Knowing the position of the particle orbit in two planes, object plane and aperture plane, the orbit of the particle is determined unambiguously. Choosing an orbit which passes the aperture plane through the same point as the paraxial orbit, the position in the aperture plane, r_b, can replace r_o'.

It is natural to introduce an aperture plane. In any rotationally symmetric electron lens there are electrodes or coils with or without iron, with a hole for the particle beam. Often apertures are introduced deliberately to limit the region with significant field strength,

---

* See, for example: Glaser, *Grundlagen der Elektronenoptik*, pp. 365-371; or Grivet, *Electron Optics*, pp. 126-127.

or to limit the beam in the radial direction. The position of the aperture plane is irrelevant; any z coordinate before, inside, or after the lens, or in the Gauss plane, is good enough. In the analysis of the imaging errors it is for practical reasons advisable to put the aperture plane in the image focus. The orbit position in the object plane, the aperture plane, and the Gauss image plane is given by r_o, r_b, and r_i. If an arbitrary direction, perpendicular to the optical axis, is chosen as the direction from which the angle θ_o, θ_b, and θ_i are counted, Eq. 4.81 becomes

$$\mathbf{r}_i = \mathcal{M}_l \mathbf{r}_o + \Delta \mathbf{r}_i.$$

The error in the Gauss image plane is a function of r_o and r_b:

$$\Delta \mathbf{r}_i = f(r_o, r_b). \tag{4.84}$$

The particle orbits rotate in a magnetic field. In the paraxial approximation the angle is the same for all orbits, according to Eq. 4.37. Assume that $\Delta \mathbf{r}_i$ contains a term const $\cdot r_o r_b$. If the angle θ_o of the object point vector $\mathbf{r}_o$ is changed by φ, both r_o and r_b must rotate by φ. The product const $\cdot r_o r_b$ will then rotate by 2φ. This would mean that a change in the angle of the object point in a rotationally symmetric lens would give an additional rotation of the orbit inside the lens. In a similar way we can see that products of type $r_o r_b^*$ (where r_b^* means conjugate complex of r_b) would not give any rotation at all of $\Delta \mathbf{r}_i$. We can conclude that $\Delta \mathbf{r}_i$ cannot include any even product of r_o and r_b and that triple products must be taken into consideration. Among them, all products which give a rotation by 3φ when the object point is turned by φ must be excluded. This limits the expression to third-order geometric terms which turn the orbit by φ, and

$$\Delta \mathbf{r}_i = \mathbf{A}\, r_b^2 r_b^* + \mathbf{B}\, r_o^* r_b^2 + \mathbf{C}\, r_o r_b r_b^* + \mathbf{D}\, r_o^2 r_b^* + \mathbf{E}\, r_o r_o^* r_b + \mathbf{F}\, r_o^2 r_o^*. \tag{4.85}$$

Equation 4.85 is valid for any position of the aperture plane. If the aperture plane lies outside the significant field, which means that the particle orbits between the aperture plane and the Gauss image plane are straight lines, the expression for $\Delta \mathbf{r}_i$ can be simplified. Because r_o and r_b are ambiguous — the only condition is that the paraxial approximation and the real orbit pass through the same point in the aperture plane — a change in r_b will not influence the rotation of $\Delta \mathbf{r}_i$, or

$$\Im\left(\frac{d\Delta \mathbf{r}_i}{dr_b}\right) = 0.$$

Assuming that $\mathbf{r}_o = 0$, we obtain $\Delta \mathbf{r}_i = \mathbf{A}\, r_b^2 r_b^*$, for example. Choosing a particle orbit with $\theta_b = 0$, we obtain $\Delta \mathbf{r}_i = \mathbf{A} r_b^3$. Put $\mathbf{A} = A\, e^{j\alpha}$. Derivation gives: $d\Delta \mathbf{r}_i / dr_b = 3A\, e^{j\alpha} r_b^2$, with $\Im(d\Delta \mathbf{r}_i / dr_b) = 0$, and that can be satisfied only if $\alpha = 0$, and therefore $\mathbf{A} = A$. In the same way it can be shown that $\mathbf{E} = E$ and $\mathbf{C} = 2\mathbf{B}^*$. Equation 4.85 becomes

$$\Delta \mathbf{r}_i = A\, r_b^2 r_b^* + \mathbf{B}\, r_o^* r_b^2 + 2\mathbf{B}^* r_o r_b r_b^* + \mathbf{D}\, r_o^2 r_b^* + E\, r_o r_o^* r_b + \mathbf{F}\, r_o^2 r_o^*. \tag{4.86}$$

Every term in Eq. 4.86 represents an aberration which resembles the corresponding aberration in light optics. In light optics all aberration coefficients are real, and also in electron optics for electrostatic lenses. The terms in Eq. 4.86 are:
1. Spherical aberration, coefficient A,
2. Coma, coefficient $\mathbf{B}$,

3. Astigmatism, coefficient **D**,
4. Curvature of field, coefficient E, and
5. Distortion, coefficient **F**.

These aberrations can be divided in two groups. In the first are spherical aberration and coma. These cause appreciable image errors even for small objects, and the important parameter is the dimension of the opening in the aperture. In the other group, astigmatism, curvature of field, and distortion, there are such errors that appear with large objects or large images. The dimension of the aperture is not so critical.

In a real lens all five terms in Eq. 4.86 are present. Depending on the application, different aberrations will influence the imaging in different ways. For example, in an electron microscope the biggest problem is the spherical aberration, because a very small surface — often smaller than 1 μm^2 — should be imaged with a magnification of a few hundred thousand times. In an oscilloscope tube the distortion will be the most important aberration. It is less important if the spot on the screen has a diameter of 0.1 or 0.3 mm. In an image converter for X-rays the photons cause photoemission from a cathode. The photocathode must in this case correspond in size to the organ it should image; that is, it must have a diameter of about 30 cm. The most important aberration will in this case be the curvature of the field, and therefore the photocathode is manufactured with a curved shape. In most cases one or two aberrations will dominate; these should be corrected firstly. In computation the light optics can be a guidance. From an object point, particle orbits are started which pass the aperture plane on a circle with radius r_b. In the Gauss plane an "aberration image" results. The form of this aberration image approximates the diffuse "image," which would be obtained in reality if all orbits which pass the aperture plane within an opening with diameter r_b would be taken into account. Exactly this is done in numerical computations. The form of the aberration image is a guidance which aberration is dominant.

4.4.2.1. Spherical Aberration

The coefficient of the spherical aberration is

$$\Delta \mathbf{r}_i = A r_b^3 \, e^{j\theta_b}, \qquad \mid \Delta \mathbf{r}_i \mid = A r_b^3, \qquad \theta_i = \theta_b. \qquad (4.87)$$

Spherical aberration is the only aberration which exists even for a point on the optical axis, and it sets the limit for the resolution which can be obtained by a lens even for very small objects — for example, in an electron microscope. The aberration image

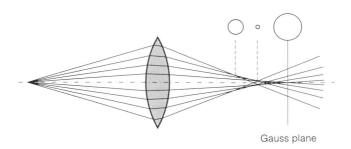

Gauss plane

Fig. 4.26. The origin of spherical aberration.

in the Gauss plane is a circle, the aberration disk, as seen in Fig. 4.26. A number of orbits which pass the aperture plane at radius r_b, lie in the Gauss plane on a circle. For different r_b the orbits will have different "focal points." With this aperture plane definition the numerical value of A is a function of its position. To avoid mixing this position into the definition of spherical aberration when different lenses are compared, it is convenient to use a definition based on the angle the orbits make with the optical axis on the image side, α_i:

$$\Delta r_i = C_s \alpha_i^3 \qquad (4.88)$$

In some special cases — for example, for the clock field — it is possible to compute the spherical aberration coefficient, C_s, analytically. C_s has the dimension of length.

Note that the diameter of the aberration disk varies with the position of the image plane if the image plane is moved from the Gauss plane. There is a position where the aberration disk has its smallest diameter, which is nearer to the lens, as seen in Fig. 4.26. The aberration image is in this case called "the disk of confusion."

4.4.2.2. Coma

The second and the third term in Eq. 4.86 give coma

$$\Delta \mathbf{r}_i = \mathbf{B}\, r_o^* r_b^2 + 2\mathbf{B}\, r_o r_b r_b^*. \qquad (4.89)$$

In the electrostatic case we have $\mathbf{B} = B$, and

$$\Delta r_i = B\, r_o r_b^2 [2 + e^{2j\theta_b}]. \qquad (4.90)$$

Particle orbits which pass the aperture plane on a circle with radius r_b lie in the Gauss plane on a circle, whose centrum is translated $2B r_o r_b^2$ from the Gauss image point and has a radius $B r_o r_b^2$, as shown in Fig. 4.27. While θ_b describes a circle in the aperture plane, $0 \le \theta_b \le 2\pi$, the corresponding circle in the Gauss plane is described twice.

All other orbits, which pass the aperture plane with different r_b, give aberration disks with different centra and radii. The centra lie on a straight line, and the common tangents to all circles have an opening angle of $60°$. In a magnetic field the aberration image is similar; the only difference is that coma is rotated.

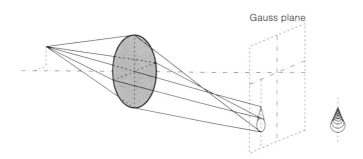

Fig. 4.27. Coma.

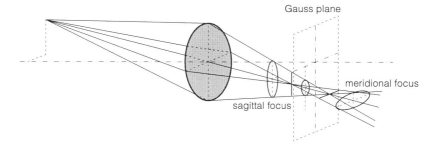

Fig. 4.28. Astigmatism.

Coma is an image error which cannot be neglected if the diameter of the aperture is large, even if the object is small. This can be the case when high current density is needed in focus — for example, in an X-ray tube, where the cathode must be imaged in a small spot. In electron optics it is difficult to demonstrate coma; the spherical aberration cannot be eliminated and dominates the image.

4.4.2.3. Astigmatism

The cause of the astigmatism is the fourth term in Eq. 4.86:

$$\Delta \mathbf{r}_i = \mathbf{D} r_o^2 \mathbf{r}_b^*. \tag{4.91}$$

In an electrostatic lens Eq. 4.91 is reduced to

$$\Delta r_i = D r_o^2 r_b \, e^{-j\theta_b}. \tag{4.92}$$

Particle orbits which describe a circle with radius r_b in the aperture plane give in the Gauss plane again a circle, with center in the Gauss image point, but the circle is described *counterclockwise* while the orbits in the aperture plane rotate *clockwise*, as seen in Fig. 4.28.

If the image plane is moved from its Gaussian position, the aberration image becomes an ellipse. In two positions, the one before and the other after the Gauss plane, the ellipses degenerate into straight lines, a *sagittal* (even called tangential) and a *meridional* focus. In these two positions all orbits starting from an object point assemble in a line focus independently on the position in the aperture plane. Different distances of the object point from the optical axis result, however, in different positions of the line foci. In a magnetic lens the same aberration image is obtained — a circle, ellipse, or a straight line — but is rotated as compared to the electrostatic case; the aberration is called the *anisotropic astigmatism*.

Astigmatism depends primarily on r_o^2. As long as the electron optical system is strictly rotationally symmetric, astigmatism causes problems only if the object or the image is large. But if the system is not centered or the field departs from cylindrical symmetry, a more serious type of astigmatism appears, which will be discussed in Section 4.4.4.

4.4.2.4. Curvature of the Field

The fifth term in Eq. 4.86 causes the curvature of the field

$$\Delta \mathbf{r}_i = E r_o^2 \mathbf{r}_b = E r_o^2 r_b \, e^{j\theta_b}. \tag{4.93}$$

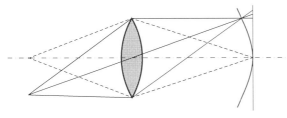

Gauss plane   **Fig. 4.29.** Curvature of the field.

The curvature of the field differs from the term causing astigmatism only in the sign in the exponent. Particle orbits, which describe a circle in the aperture plane, generate a circle in the Gauss plane. For each r_o there is a place between the aperture and the Gauss plane, where the image of an object point is a point. For different r_o all these points lie on a surface, which deviates from the Gauss plane proportional to r_o^2. The surface is a paraboloid, and in electron optics it is always concave toward the lens, as seen in Fig. 4.29. Because E is real, the error remains the same in magnetic lenses.

The error is pronounced only when a large object or a large image is encountered. The easiest way to eliminate the aberration is to give either to the object plane or to the image plane a form of a paraboloid, or as an approximation a spherical shape. This is used, for example, in image converters and large photomultipliers.

4.4.2.5. Distortion

The last term in Eq. 4.86 causes distortion:

$$\Delta \mathbf{r}_i = \mathbf{F} r_o^2 \mathbf{r}_o^*. \tag{4.94}$$

In an electrostatic lens this is reduced to

$$\Delta r_i = F r_o^3. \tag{4.95}$$

Distortion depends only on the radial position of the object point. A point is imaged as a point in the Gauss plane, but its position is displaced as compared to the Gauss image point. The displacement can be positive or negative depending on the sign

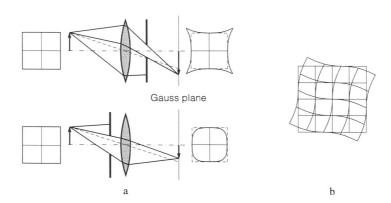

Gauss plane

a   b

Fig. 4.30. (**a**) Distortion. (**b**) Anisotropic distortion, image of a mesh.

of the aberration coefficient, as shown in Fig. 4.30. If F is negative the distortion is called *barrel*, but if F is positive the distorsion is called *pincushion distorsion*. In magnetic lenses the image is rotated, and the distortion is anisotropic (see Fig. 4.30b).

Distortion problems occur when the object or the image is large, and large image converters have the biggest problems.

4.4.3. Chromatic Aberration

The paraxial imaging and the theory for geometric aberrations assume a perfectly monoenergetic particle beam. In electron optics the analogy to index of refraction is the potential. In light optics the refraction index varies with the wavelength and causes chromatic aberration. An instability or a change of the potential will cause an image error which is similar to the chromatic aberration.

The potential is a measure for the kinetic energy of the particles. It is through the velocity/kinetic energy, via Eqs. 4.10 and 4.11, that potential enters the orbit equation (Eq. 4.23). There are two reasons which, from different origins, are the cause of energy spread. The first is the Maxwell's velocity distribution. In some applications — for example in the transmission electron microscope — the velocity spread is amplified by inelastic collisions. The second reason is the stability of the voltage or current sources. Even in the best stabilized power supplies there is noise on the output and the voltage or the current vary with time.

Generally, the effect of chromatic aberration is small in electron optical devices, except in accelerators and some spectrometers. The effects caused by electron emission or ion sources are often negligible. The thermionic emitted electrons have a mean kinetic energy kT, which is on the order of 0.1 to 0.2 eV. For photoemitted electrons the corresponding value is a few electronvolts. In most electron optical applications the acceleration voltage varies between 1 kV and a few hundred thousand volts, and the initial energy spread can therefore be neglected. This is different with the secondary emission. There are always reflected primary electrons whose energy is near the energy of the incident beam. If these electrons must be taken into account, the chromatic aberration can be large. The energy spread of the ions coming from an ion source is of the same order as the thermionic emitted electrons. In modern electron optical devices, power supplies with very high stability can be used, if necessary. The stability of 10^{-5} is common, with short time stability better than 10^{-6}. In most cases the chromatic aberrations caused by the stability of the power sources is not a problem.

One case must be treated separately — the transmission electron microscope. When passing through the specimen, some electrons lose energy by inelastic collisions. On the average an electron loses about 10 eV per inelastic collision. This is a quantum-mechanical effect, and the energy loss corresponds roughly to Fermi energy. By including an aperture with a small hole near the focus of the objective, most of these electrons can be eliminated and the contrast of the image improved.

The expressions for the chromatic aberration are derived by using perturbation methods. The change in the focal length in a magnetic lens, using Eq. 4.53, becomes

$$\frac{\Delta f}{f} = \frac{\Delta V}{V} + 2\frac{\Delta B}{B}, \qquad (4.96)$$

and for the electrostatic case, using Eq. 4.52,

$$\frac{\Delta f}{f} = 2\frac{\Delta V}{V^{1/4}V_m^{3/4}}, \tag{4.97}$$

where V is the arithmetic mean of the potential on both sides of the lens, and V_m is the potential minimum inside the lens. The lateral magnification is changed, as well as the rotation in a magnetic lens.

4.4.4. Mechanical Image Errors

In aberration discussions we assumed that the shape of all electrodes, coils or iron, is perfect. In the real life this is not so, because manufacturing is made with a limited precision. The installation is not perfect either, the surfaces are never perfectly parallel, and the axis of the system is never a perfect straight line. All of these factors cause mechanical aberrations.

Take as an example the electron microscope, which is very sensitive to mechanical and adjustment errors because of its extremely large magnification. The accuracy in manufacturing the lenses and aligning is of paramount importance for its optical properties. An objective in the electron microscope has an opening of a few millimeters, while an error in rotational symmetry of less than a micron can be seen. In light optics the manufacturing errors can be controlled easier. The lenses are grinded and polished, a manufacturing method which guarantees a high accuracy. They have either plane or spherical surfaces which easily can be checked by interference methods to a fraction of wavelength.

Mechanical aberrations can be divided into alignment aberrations and aberrations caused by errors in rotational symmetry. Calculations where it is assumed that the lenses in the system are positioned slightly eccentric of axis show that this kind of error does not greatly deteriorate the image quality if the displacement is small compared to the opening diameter. Using modern adjustment methods with laser beams, the system of lenses can be aligned good enough in most applications. It is also rather easy to make a mechanical construction where a fine adjustment by the particle beam itself is possible.

The errors which depend on the asymmetry in a lens are more difficult to calculate, measure, and correct. Generally, it is assumed that this type of error gives the holes in the electrodes or the iron in a magnetic lens a shape, which is best approximated by an ellipse. The electric or the magnetic field is no longer rotationally symmetric, because it has two symmetry planes which correspond to the axes of the ellipse.

The electric potential can then be approximated by

$$V(r, \theta, z) = V(z) - \frac{1}{4}V''(z)[1 - \epsilon(z)\cos 2\theta]r^2 + \cdots, \tag{4.98}$$

where the ellipticity, ϵ, is a function of z. Even if the object point lies on the optical axis, the imaging gives in the Gauss approximation two different focal points, which lie in two planes given by the axes of the ellipse. In these two foci the image of the object point are two short lines. The error is therefore called *elliptical astigmatism*. The disk of confusion lies midway between these two line foci. Its diameter is defined by

$$d = C_e\alpha, \tag{4.99}$$

where C_e is the coefficient of the elliptical astigmatism and α is the half opening angle of the beam as seen from the image plane. It is difficult to determine C_e, because normally neither ϵ nor its angle θ are known. Calculations with possible values show that the elliptical astigmatism gives an error proportional to an r^2 term, and not r^3, as in other geometrical aberrations. Therefore the elliptical aberration deteriorates the image to a greater degree than do the geometrical aberrations. The resolution only depends on the eccentricity — that is, on the accuracy in manufacturing — but not on the diameter of the opening in the lens. Thus it is still possible to choose as small an opening as necessary, to obtain a short focal length. Magnetic lenses can be treated in a similar way, but in lenses with iron the asymmetry in the field distribution in the pole shoes influences the elliptical astigmatism more than an error in manufacturing.

4.4.5. Numerical Methods for Lens and Aberration Computation

When designing an electron optical system we are confronted with two problems. The first one is general — a reasonable lens configuration. We must choose a suitable shape for the electrodes or pole shoes, object and image positions, focal lengths, aperture diameter, and voltage or current. With these values fixed the electron-optical system is defined. The second problem is to compute the paraxial approximation, the aberration image or the aberration coefficients, and the resolution.

The design starts with a decision concerning the type of the lens. This fixes the shape of the electrodes or of coils, maybe iron. The approximate value of the focal lengths can be found in diagrams, like those in Section 4.3, or from similar diagrams in the literature. The next step is the field problem solution, followed by computation of particle orbits. In numerical computations it is better to use Eqs. 4.13, 4.20, or 4.23 directly, instead of solving the paraxial equation (Eq. 4.35). It seems natural to start with the paraxial orbits and to compute the focal lengths and the positions of the principal planes. The numerical noise and the difficulty to determine the crossing point of two almost parallel lines — the paraxial solution is obtained only if the orbits make a very small angle with the optical axis — result in large errors. It is therefore advisable to start with the second principal orbit and to compute the position of the image focus, F_2, for more than one orbit, which start with different r_o.

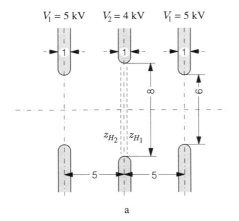 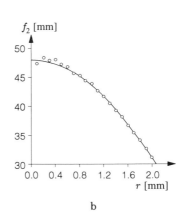

Fig. 4.31. (a) Einzellens. (b) Numerically computed focal length, $F_2 = f(r_o)$. The least square approximation gives for this lens: $f_1 = f_2 = 48.1$ mm, $z_{H1} = 0.27$ mm, $z_{H2} = -0.27$ mm.

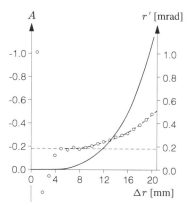

Fig. 4.32. The slope of the electron orbits (full line), r', and the spherical aberration coefficient (dashed line), A, as a function of the deviation from the Gaussian image point, Δr, for the einzellens in Fig. 4.31. The region where A can be assumed constant shows the limit for the extension of the third-order approximation. Observe the spread of A for small Δr.

An example shows Fig. 4.31; for small r_o the focal length exhibits a large spread. F_2 does not depend on the sign of r_o if the lens is rotationally symmetric. A series expansion of F_2 can only have even exponents of r_o:

$$F_2(r_o) = \sum_{i=0}^{n} a_i r_o^{2i}. \qquad (4.100)$$

If r_o is not far from the optical axis, it is enough to use three to four terms in the series, but two to three times as many orbits should be computed. The system is overdetermined, and with the least square method $F_{2_{r_o}} = a_0$ together with f_2 and z_{H2} is found. If the potential differs on both sides of the lens, the first principal orbit is used to compute F_1, f_1, and z_{H1}.

Aberration coefficients can be obtained by two different methods: either by using the expression based on the solution of Eq. 4.83, which by numeric integration directly gives the five aberration coefficients, or by direct numerical solution of the orbit equation. The latter method is preferred, and there are programs which automatically choose the orbits necessary to show the aberration images. Otherwise, by a suitable choice of the initial conditions, a number of orbits are computed, which pass through the lens in such a way that different aberration coefficients can be recognized. Figure 4.32 shows the result of such a computation for the same einzellens as in Fig. 4.31.

A number of attempts to automatically optimize an electron optical system, which is a multidimensional nonlinear optimization problem, was made. The result of each calculation are aberration errors, and a sequential search for an extremum is carried out. This is an iterative procedure, where the direction of the next movement and its distance govern the speed of the process. But the number of parameters which can be varied such as the lens position, size, and strength (possibly with angle and deflection in deflection systems), is so large that only programs to optimize certain parameters, mostly adapted to special equipment design, have been developed. These programs, developed primarily by those working with accelerator design, use matrix formalism. They take into account aberrations and can vary some parameters to optimize the system in order to obtain, for example, low chromatic aberration, a small spot, or a low beam divergence.

Finally, the design of all electrodes, coils, and pole shoes should be made as simple as possible — straight lines or spherical surfaces — to simplify the manufacture and to decrease mechanical errors.

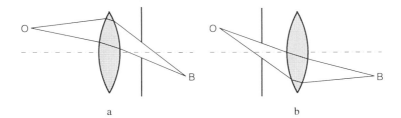

Fig. 4.33. The influence of the aperture position on the particle orbits.

4.4.6. Fighting the Aberrations

The aberrations *cannot be eliminated completely*. By ingenuous methods their effect can be eliminated in some parts of the image plane and the errors reduced. Light optics has gained spectacular success, for example with the fish-eye lenses with an opening angle of 180°, with lenses with aperture number less than 1.0, and with zoom objectives, where the focal length can be varied more than 1:10. The aberration in light optics can be corrected in two ways: by using glass with different refraction index and by using positive and negative lenses. In electron optics the possible variation of the refraction index is much greater than in light optics, but there are no negative lenses. The latter is such a big handicap that, from this point of view, the electron lenses are far behind glass lenses. Scherzer showed in his classical work [89] that it is impossible to compensate the spherical and the chromatic aberration with rotationally symmetric lenses. Other aberrations can be compensated, and this is done if necessary and applicable.

The frequently used method to fight aberrations is to install an aperture. Two things are achieved simultaneously: The beam size is limited, which decreases the spherical aberration, while other aberrations can be affected by the position of the aperture along the optical axis. The latter is illustrated by Fig. 4.33. Different orbits pass the aperture plane at different radii depending on the position of the aperture, between the object and the lens, or between the lens and the image plane. Therefore, different particles reach the image plane and all aberrations, except the spherical and chromatic, give different image errors. Distortion shows the biggest change. If the aperture is moved from the one side of the lens to the other, pincushion distortion changes into barrel distorsion, and vice versa.

The biggest problem in electron optics is the spherical aberration which limits the resolution, among other things, especially in the electron microscope. Great efforts were made to find methods to reduce the spherical aberration. Among other ideas, an interesting proposal was made by Gabor [90] to use positive space-charge with ions in the lens. Instead of Laplace's equation, Poisson's equation must be solved, and the field can be shaped differently. Later research has shown that even this proposal cannot help in the case of the electron microscope, because the spread of the electrons in the collisions with the ions reduces the correction and makes the aberration worse. The space-charge compensation was proposed, however, in another device: A lens to focus antiprotons in CERN will use positive space-charge to reduce both the spherical and the chromatic aberration.

It was shown during the 1940s that lenses without rotational symmetry can eliminate spherical aberration. Quadrupole and multipole lenses can be used for this purpose. Their equivalent in light optics are the saddle-shaped lenses. Scherzer computed a correction system [91], shown in Fig. 4.34, which eliminates all third-order aberrations in a microscope objective, but its size was to large. During the 1960s and 1970s

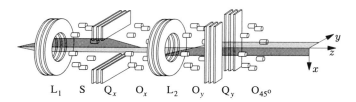

L_1 S Q_x O_x L_2 O_y Q_y $O_{45°}$

Fig. 4.34. Scherzer's correction system for a rotationally symmetric lens according to Seeliger's design. L_1, lens; S, stigmator; Q_x and Q_y, quadrupoles; L_2, rotationally symmetric intermediate lens; O_x, O_y, and $O_{45°}$ octupoles. (After *Handbuch der Physik, Band 33, Korpuskularoptik*, S. Flügge, Ed., © 1956, Springer-Verlag, with permission.)

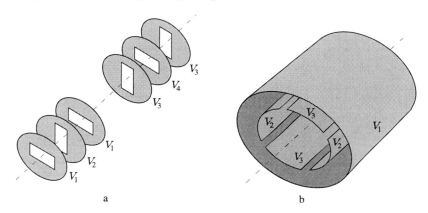

a   b

Fig. 4.35. (a) Correction with two einzellenses with rectangular openings. (b) Correction with five electrodes.

new magnetic materials were developed, with which very compact magnetic lenses can be constructed. More space becomes available for correction lenses, which use quadrupole-like electrodes (Fig. 4.35a), or lenses with five electrodes (Fig. 4.35b). Theoretical work has shown that these types of correction lenses can affect even fifth-order aberrations.

Design of systems with prescribed paraxial parameters and a minimum of aberrations is a goal of electron optics. A new approach is the use of dynamic programming, where a potential distribution or a magnetic field along the axis is searched for, which minimizes the aberrations and satisfies the paraxial equation. This field distribution is constrained to possible field amplitudes, while other constraints can be included, like vanishing field at object or image point. Approximating the field distribution or its derivative by straight lines in a limited number of regions, the dynamic programming problem becomes tractable. The reconstruction of the electrodes or pole shoes from the obtained field distribution is difficult, because the solution is numerical, and any error increases exponentially with the distance from the axis. Experiments were made with cubic spline functions, which mathematically is not sound but seems to give rather satisfactory results. The complicated electrode or pole shoe shapes are finally replaced by straight lines or circular segments and are checked by numerical orbit computations.

Mechanical aberrations usually have the character of elliptic astigmatism. The astigmatism induced by eccentricity can relatively simply be corrected by weak cylindrical lenses of the quadrupole or sextupole type. These lenses are called *stigmators*. The stigmator must be constructed so that it is possible to change both the strength and the direction of the field. Even if a stigmator is a much weaker lens than the objective lens, it must be manufactured with the same precision as other lenses in the system to avoid new aberrations.

4.5. QUADRUPOLE AND MULTIPOLE LENSES

The largest electric field between the electrodes in an electrostatic lens is limited because of breakdown to ~ 100 kV/cm, so the ratio between V_{max} and V_{min} at high particle energy decreases. The focal length becomes longer and the lens weaker, according to Eq. 4.52. Magnetostatic lenses have similar limitations, even if they can be made stronger than the electrostatic. Nature imposes a limit: The magnets with iron can generate a field of ~ 2.5 T before saturation. The field can be further increased by superconducting coils ~ 4 times. Translated into particle energy, with rotationally symmetric lenses the electrons can be focused up to a few hundred megaelectronvolts, while the protons can be focused up to about 1 MeV. Modern high-energy accelerators accelerate the electrons up to about 50 GeV, while the protons are accelerated up to 1 TeV. These accelerators have kilometer-long circumference. To keep the particle beam focused, stronger lenses are needed.

In rotationally symmetric lenses the electric and/or the magnetic field direction is primarily axial. These fields influence the radial particle motion only through their radial component, or because the particles have an azimuthal velocity component. An electric or a magnetic field which directly affects the particles perpendicularly to the axis would result in a much stronger focusing. Such field was proposed by Christofilos [92] and independently by Courant, Livingston, and Snyder [93]. These lenses have either four electrodes or four magnetic pole shoes symmetrically placed around the optical axis, as seen in Fig. 4.36 — hence the name *quadrupole lenses*.

Quadrupole lenses generate an electric or a magnetic field whose strength is proportional to the distance from the optical axis. In the radial direction the field gradient is approximately constant inside the useful region between the electrodes or between the pole shoes. The lens has four symmetry planes which cross each other perpendicularly to the optical axis with $\pi/4$ angles. Assume that the electrodes or the pole shoes are infinitely long. Denote the potential on the two diametrically opposite electrodes or pole shoes with V_1 — it can be electrostatic or magnetostatic. Thus the other two have the potential $-V_1$.

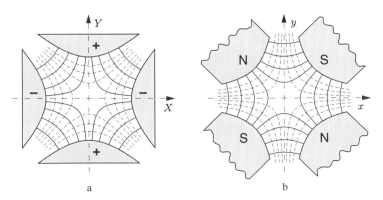

Fig. 4.36. (a) Electric quadrupole lens. (b) Magnetic quadrupole lens. —— field line, - - - - equipotential line.

The opening of the lens is $2a$. Choose two coordinate systems, OXY in the electrostatic and Oxy in the magnetostatic case. The electric force acts in, while the magnetostatic force acts perpendicular to, the direction of the field. The potential is a function of the coordinates x, y or X, Y:

$$W(s) = \Phi(x, y) + i\, V(x, y), \qquad s = x + i\, y.$$

Because of the fourfold symmetry we obtain

$$W(s) = \frac{1}{2}h_2 s^2 + \frac{1}{6}h_6 s^6 + \frac{1}{10}h_{10} s^{10} + \cdots.$$

The equipotential surfaces are in the electrostatic case,

$$V(X, Y) = \Re\{W(s)\} = \frac{1}{2}h_2(X^2 - Y^2) + \frac{1}{6}h_6\{X^6 - 15X^2Y^2(X^2 - Y^2) - Y^6\} + \cdots,$$

and in the magnetostatic case,

$$V(x, y) = \Im\{W(s)\} = h_2 xy + h_6 xy\left(x^4 - \frac{10}{3}x^2 y^2 + y^4\right) + \cdots.$$

At electrodes, $x = X = y = Y = \pm a$, the potential is $\pm V_1$. For an electrostatic lens we have

$$V(X, Y) = V_1\left\{\frac{H_2}{a^2}(X^2 - Y^2) + \frac{H_6}{a^6}[X^6 - 15X^2Y^2(X^2 - Y^2) - Y^6] + \cdots\right\}, \quad (4.101)$$

and for a magnetostatic case we obtain

$$V(x, y) = -V_1\left\{\frac{2H_2}{a^2}xy + \frac{H_6}{a^6}xy\left(x^4 - \frac{10}{3}x^2 y^2 + y^4\right) + \cdots\right\}. \quad (4.102)$$

The constants H_i depend only on the shape of the electrodes or pole shoes. If the field gradient should be constant, then $H_{2(2n-1)} = 0$ for $n \geq 1$. The electrodes or the pole shoes must be equipotential surfaces corresponding to $\pm V_1$, which with $H_2 = 1$ gives

$$\frac{X^2 - Y^2}{a^2} = 1 \qquad \text{or} \qquad xy = \frac{1}{2}a^2. \quad (4.103)$$

Both equations are hyperbolas, and the electrodes or pole shoes must be hyperbolic. They are not infinitely long, and $H_{2(2n-1)} \neq 0$ for $n > 1$, so higher-order terms enter the orbit equation and cause aberrations. Hyperbolic surfaces are difficult to manufacture; instead, lenses with circular arcs are used, as shown in Fig. 4.37.

A modification of the electrode shape or pole shoes does not greatly change the coefficients $H_{2(2n-1)}$. Therefore, practically all quadrupole lenses are made with circular cross section, mostly with $E = 2.5a$ and $R = 1.15a$, to obtain $H_6 = 0$.

Magnetostatic quadrupole lenses are used more often than electrostatic lenses. We will therefore describe only the magnetostatic quadrupole lenses, and we will limit the discussions to paraxial approximation, $H_6 = H_{10} = \cdots = 0$. The real quadrupole is not infinitely long, and as the electric or the magnetic field varies along the optical axis, assume that H_2 is a function of the axial coordinate z. For example,

$$H_2(z) = H_2(0)\, p(z), \qquad G(z) = 2H_2(z)V_1/a^2. \quad (4.104)$$

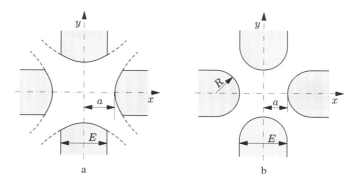

Fig. 4.37. Electrodes or pole shoes with **(a)** hyperbolic and **(b)** circular cross section.

Table 4.2

Comparison between hyperbolic and circular form

Form	H_2	H_6	H_{10}
Infinite hyperbola	1	0	0
Hyperbola, $E = 2a$	1	0.005	
Circular cross section, $E = 2.5a$, $R = 1.125a$	1	−0.005	−0.00125
$R = 1.147a$	1	0	−0.00124
$R = 1.25a$	1	0.02	−0.00124

According to Fig. 4.38 the function $p(z)$ equals 1 in the middle of the quadrupole, at $z = 0$, and falls to zero in infinity. Its value can, for example, be computed by a three-dimensional program. But for practical purposes, it is enough to replace the real quadrupole with an idealized one, which has an equivalent length:

$$L \approx \frac{1}{B_{z=0}} \int_{-\infty}^{+\infty} B(z)\,dz, \qquad (4.105)$$

where $B(z)$ is the magnetic field at constant distance from the axis. $L > l$, and measurements have shown that in most practical cases the equivalent length can be approximated by

$$L = l + 1.1a. \qquad (4.106)$$

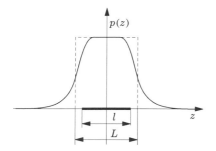

Fig. 4.38. Function $p(z)$ for a magnetic quadrupole. l is the mechanical length, and L is the equivalent length.

The pole surfaces can be seen as equipotential surfaces even at high magnetic field in the iron, because this region always is inside the yoke and never near the pole shoes. Assume therefore, that the whole magnetic field is concetrated in the region between the pole shoes; that is neglect the magnetization of the iron. The scalar magnetic potential as a function of the excitation is then

$$V_1 = \mu_0 N I, \tag{4.107}$$

where I is the current and N the number of turns in each of the four quadrupole coils, so that a constant

$$G = \frac{2\mu_0 N I}{a^2} \approx \frac{B_{tip}}{a} \tag{4.108}$$

can be defined.

Assume that the axial velocity, $v_z = v$, does not change appreciably inside the quadrupole. From the expression for the potential, Eq. 4.102, the field strength is computed:

$$B_x = -\frac{\partial V_1}{\partial x} = Gy, \qquad B_y = -\frac{\partial V_1}{\partial y} = Gx. \tag{4.109}$$

Introducing Eq. 4.109 into the equation of motion

$$\frac{d(m\mathbf{v})}{dt} = q\mathbf{v} \times \mathbf{B},$$

the components are — classically and relativistically if m is the particle mass —

$$m\frac{d^2x}{dt^2} = -qvGx, \qquad m\frac{d^2y}{dt^2} = qvGy,$$

or, in analogy with Eq. 4.12 where $\ddot{z} = 0$,

$$\frac{d^2x}{dz^2} = -\frac{qG}{mv}x, \qquad \frac{d^2y}{dz^2} = \frac{qG}{mv}y. \tag{4.110}$$

Define

$$k^2 = \frac{qG}{mv} = \frac{G}{B\rho}, \tag{4.111}$$

where $B\rho$ is the "magnetic rigidity" and ρ is the equivalent radius of the particle orbit.

The force acts on the particles differently in both directions. In the x direction the force on positive particles acts as a focusing force, whereas in the y direction it acts as a defocusing force. For negative particles the opposite is valid. A light optical analogy is a cylindrical lens; in a convex lens parallel light rays are brought into a line focus, whereas in a concave lens the rays diverge.

Equations 4.110 have two solutions:

$$x = a\cos kz + b\sin kz, \qquad y = c\cosh kz + d\sinh kz, \tag{4.112}$$

and the velocity components are

$$x' = -ak\sin kz + bk\cos kz, \qquad y' = ck\sinh kz + dk\cosh kz.$$

The constants a, b, c, and d depend on the initial conditions. At the entrance into the quadrupole, at $z = 0$, we have $x = x_0$, $y = y_0$, $x' = x_0'$, and $y' = y_0'$. At the exit, at $z = L$, where L is the equivalent length, we obtain

$$x = x_0 \cos kL + \frac{x_0'}{k} \sin kL, \qquad y = y_0 \cosh kL + \frac{y_0'}{k} \sinh kL,$$

and

$$x' = -x_0 k \sin kL + x_0' \cos kL, \qquad y' = y_0 k \sinh kL + y_0' \cosh kL.$$

The result can be written in matrix form (see also Appendix H). Introduce $kL = \theta$. In the focusing, x direction, the transformation matrix is

$$\mathbf{T}_c = \begin{pmatrix} \cos\theta & \frac{1}{k}\sin\theta \\ -k\sin\theta & \cos\theta \end{pmatrix}, \tag{4.113}$$

and in the defocusing, y direction we obtain

$$\mathbf{T}_d = \begin{pmatrix} \cosh\theta & \frac{1}{k}\sinh\theta \\ k\sinh\theta & \cosh\theta \end{pmatrix}. \tag{4.114}$$

The second principal orbit — which on the object side passes x_o, y_o — on the image side is given by

$$\begin{pmatrix} x_i \\ x_i' \end{pmatrix} = \mathbf{T}_c \begin{pmatrix} x_o \\ 0 \end{pmatrix}, \qquad \begin{pmatrix} y_i \\ y_i' \end{pmatrix} = \mathbf{T}_d \begin{pmatrix} y_o \\ 0 \end{pmatrix}.$$

In accordance with the definitions of the principal elements of a lens (Fig. 4.9), the focal distances and positions of the principal planes are

$$\begin{aligned}
f_{2x} &= \frac{1}{k\sin\theta}, & f_{2y} &= -\frac{1}{k\sinh\theta}, \\
z_{f2x} &= \frac{1}{k}\cot\theta, & z_{f2y} &= -\frac{1}{k}\coth\theta, \\
z_{H2x} &= \frac{\cos\theta - 1}{k\sin\theta}, & z_{H2y} &= \frac{1 - \cosh\theta}{k\sinh\theta},
\end{aligned} \tag{4.115}$$

where the distances are measured from the *exit side of the lens*. For weak quadrupole lenses or when the particle energy is high ($\theta < 0.2$), the quadrupole can be approximated by a thin lens

$$\sin\theta \approx \sinh\theta \approx \theta, \qquad \cos\theta \approx 1 - \frac{\theta^2}{2!}, \qquad \cosh\theta \approx 1 + \frac{\theta^2}{2!}.$$

The focal lengths are

$$f_{2x} = -f_{2y} = \frac{1}{k^2 L}. \tag{4.116}$$

The principal planes coincide and lie in the middle of the lens. If $\theta > 0.2$, the focal lengths cannot be approximated as equal and the principal planes move, as shown in Fig. 4.40.

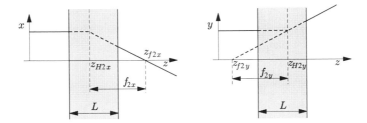

Fig. 4.39. Definition of the cardinal elements of the quadrupole. (**a**) In the focusing plane. (**b**) In the defocusing plane.

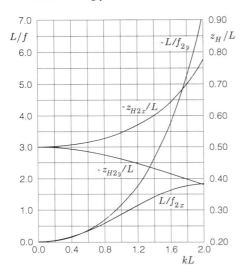

Fig. 4.40. Normalized focal length L/f_{2x} and L/f_{2y}, and principal plane distances H_{2x}/L and H_{2y}/L vs. kL.

Each quadrupole can be replaced by a thin lens and a drift space on each side, or by two thin lenses and a drift space between them. In the first case, in the focusing plane we have

$$\mathbf{T}_c = \begin{pmatrix} 1 & d_c \\ 0 & 1 \end{pmatrix} \times \begin{pmatrix} 1 & 0 \\ -1/f_{2x} & 1 \end{pmatrix} \times \begin{pmatrix} 1 & d_c \\ 0 & 1 \end{pmatrix}, \tag{4.117}$$

with

$$f_{2x} = \frac{1}{k \sin \theta}, \qquad d_c = \frac{1 - \cos \theta}{k \sin \theta}.$$

The length of the drift space, $d_c > L/2$. In the defocusing plane, f_{2x} is replaced by

$$f_{2y} = -\frac{1}{k \sinh \theta},$$

and d_c is replaced by

$$d_d = \frac{\cosh \theta - 1}{k \sinh \theta}.$$

The drift space, d_d, is less than $L/2$.

In the second case the transformation matrix is

$$\mathbf{T}_c = \begin{pmatrix} 1 & 0 \\ -1/f_{2x} & 1 \end{pmatrix} \times \begin{pmatrix} 1 & d_c \\ 0 & 1 \end{pmatrix} \times \begin{pmatrix} 1 & 0 \\ -1/f_{2x} & 1 \end{pmatrix}, \tag{4.118}$$

with

$$f_{2x} = \frac{\sin\theta}{k(1-\cos\theta)}, \qquad d_c = \frac{1}{k}\sin\theta.$$

In the defocusing plane we have

$$f_{2y} = -\frac{\sinh\theta}{k(\cosh\theta - 1)}, \qquad d_d = \frac{1}{k}\sinh\theta.$$

A single quadrupole cannot image an object. A combination of two quadrupoles, the first quadrupole focusing in x and the second in y direction, can focus in both directions. Such a combination, with a drift space in between, is called a *doublet*. In the general case the two lenses of a doublet, Q_1 and Q_2, have different lengths L_1 and L_2, have different excitations k_1 and k_2, and are separated by a drift space with the length D. The transformation matrix for the "focusing–defocusing" plane x–z is

$$\mathbf{M}_x = \mathbf{T}_{d2} \times \mathbf{T}_s \times \mathbf{T}_{c1}, \tag{4.119}$$

where $\mathbf{T}_s$ is the transformation matrix of the drift space,

$$\mathbf{T}_s = \begin{pmatrix} 1 & D \\ 0 & 1 \end{pmatrix}.$$

In the "defocusing–focusing" plane y–z a similar transformation matrix $\mathbf{M}_y$ is obtained where $\mathbf{T}_{d2}$ and $\mathbf{T}_{c1}$ must be replaced by $\mathbf{T}_{c2}$ and $\mathbf{T}_{d1}$. In the explicit form the expressions are complicated, so direct matrix multiplication of the numeric values is performed instead:

$$\begin{pmatrix} x_i \\ x_i' \end{pmatrix} = \mathbf{M}_x \begin{pmatrix} x_o \\ x_o' \end{pmatrix}, \qquad \begin{pmatrix} y_i \\ y_i' \end{pmatrix} = \mathbf{M}_y \begin{pmatrix} y_o \\ y_o' \end{pmatrix}. \tag{4.120}$$

Figure 4.41 shows the orbits in a doublet and the positions of its cardinal elements.

A doublet has two different "images" of an object which lies on the optical axis. These two "images" are the line foci, and the doublet is by nature strongly astigmatic. It is possible to bring together the line foci in only one focal point if the quadrupoles are excited separately or have different lengths. In such a case the doublet is called "stigmatic" or "pseudostigmatic" because the lateral magnifications are not equal, as seen in Fig. 4.42.

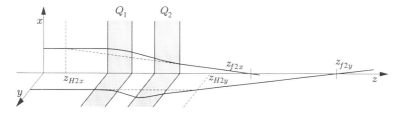

Fig. 4.41. Cardinal elements of a doublet. The quadrupoles Q_1 and Q_2 have the same excitation and length.

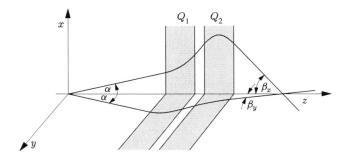

Fig. 4.42. Pseudostigmatic doublet. Quadrupoles Q_1 and Q_2 have different excitation, $k_1 L_1 \neq k_2 L_2$.

Besides doublets, triplets are frequently used. The first and the third quadrupole have the same excitation and are oriented so that their focusing planes coincide. The second quadrupole is turned 90° and has different excitation. The drift spaces are equal. The advantage of a triplet is its symmetry, which allows easier aberration control. In accelerator technology the quadrupoles are used to keep together particle beams in big accelerators or in storage rings. In a sequence of focusing and defocusing quadrupoles, mounted according a pattern around the rings with many kilometers in circumference, the particles can be kept circulating many days. Quadrupoles are also used to correct aberrations in rotationally symmetric lenses. For example, the lens combinations in Figs. 4.34 and 4.35b use quadrupoles, or quadrupole-like lenses, to correct the spherical aberration in electron microscopes.

A doublet is, as mentioned, by its nature astigmatic. A pseudostigmatic doublet cannot be used for imaging because the lateral magnification in both planes is not equal. Quadrupoles have 10 third-order geometric aberrations. The aberration image depends mostly on the the symmetry errors and on the diameter of the opening. The fourfold symmetry allows, however, a correction of the third-order aberrations by octupole lenses. The quadrupoles have also chromatic aberrations. There are two types of chromatic aberrations of the second order and two types of the third order.

The next steps are sextupole and octupole lenses. They are used to correct the aberrations in rotationally symmetric lenses, and frequently in accelerator technology. Dipole magnets, used to deflect particle beams, have chromatic aberrations, which can be corrected with sextupoles. This is, for example, the case in accelerators, where sextupoles are used to correct the errors in magnetic field distribution in dipoles during the acceleration of the particles and in polarized ion sources because of the interaction of the sextupole field components with the particles magnetic moment. Octupoles are used to correct aberrations in rotationally symmetric and in quadrupole lenses.

4.6. ELECTRON OPTICAL MIRRORS

In an electron optical system the particles turn if the potential falls below the potential of the electron emitting electrode, or becomes higher than the potential of the ion source. Depending on the field distribution, such a mirror can act as a concave mirror which generates a real image, or as a convex mirror from which a parallel particle beam diverges after the reflection, as seen in Fig. 4.43.

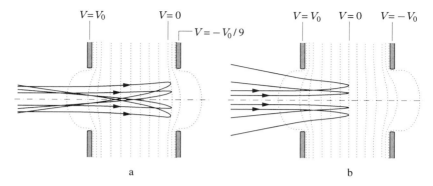

Fig. 4.43. Electron optical mirrors. (**a**) Concave mirror. (**b**) Convex mirror.

The mirrors are sparsely used in electron optics. The most important applications are solid-state measuring devices — spectrometers and electron microscopes. But mirrors must not always be used for imaging. Electrons can discharge a positively charged surface so that the potential falls to the cathode potential; thereafter the electrons are reflected. This is used in TV camera tubes to obtain the video signal.

In a light-optic planar mirror the incoming and the reflected ray make the same angle to the normal of the mirror surface, so the object and the virtual image are on the same distance from the mirror. In electron optics a homogeneous electric field corresponds to a planar mirror in light optics. Assume such a field between $z = 0$ and $z = d$, as shown in Fig. 4.44, generated by a mesh at potential V_0. Observe an electron beam which passes the plane $z = d$ at $-r_0$ with the velocity $v_0 = \sqrt{2\eta V_0}$ and an angle $\theta \ll 1$, so that $v_{0z} = -v_0 \cos \theta$ and $v_{0r} = v_0 \sin \theta$. Electrons are retarded in the field; at $z = 0$ they turn, and $v_z = 0$.

Again at $z = d$ we have

$$v_z^2 = v_0^2 \cos^2 \theta \approx 2\eta V_0 \qquad \text{and} \qquad d = \frac{\eta V_0}{2d}t^2,$$

giving the transit time in the field

$$t = \frac{4d}{v_0 \cos \theta}. \tag{4.121}$$

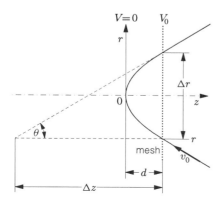

Fig. 4.44. Reflection of an electron orbit in a homogeneous electric field.

In the r direction the electron moves for

$$\Delta r = v_0 \sin \theta \ t = 4d \tan \theta,$$

so that

$$\Delta z = \frac{\Delta r}{\tan \theta} = 4d. \tag{4.122}$$

In a planar electron optical mirror the image lies behind the mirror at a distance which is three times longer than the object distance.

The spherical aberration has the opposite sign as in rotationally symmetric lenses (see Fig. 4.43a); however, a mirror cannot be used in combination with lenses to compensate the aberration. Because the incident and the reflected orbits must make an angle with the optical axis not to coincide, higher-order aberrations become too large.

4.7. DEFLECTION OF PARTICLE BEAMS

In many applications, electron beams are used as pointers. An example is the cathode ray tube which has a resolution of $>10^6$ pixels, and every pixel can be reached in ~ 1 ns. Other deflection examples are TV picture and camera tubes, sweep electron microscope, electron-beam technology and lithography, and so on. Compared with laser, the electron beam has the advantage in the deflection speed.

Electric or magnetic fields can deflect particles. Assume, as an approximation, that the field is two-dimensional; it varies along the z axis with y, but is independent of the transverse coordinate x. The linear deflection theory neglects the axial component of the deflecting field, caused by the fringing field, and corresponds to paraxial approximation in rotationally symmetric lenses. All deflection fields have the property that the deflection changes the beam focusing. If a beam is focused to a small spot without deflection, it will be defocused when deflected. The defocusing is a function of the deflection angle and is an expression for the deflection aberrations.

4.7.1. Electric Deflection

Assume that the deflection field, $E_d(z)$, only depends on the z coordinate. The electron beam comes from an electron gun, followed by one or more rotationally symmetric lenses. The gun accelerates the electrons to an axial velocity $v_0 = \sqrt{-2\eta V_0}$. Between the deflection electrodes the electrons are subject to a transverse electric field. The field, $E_d(z) = V_d/b(z)$, acts between $z = 0$ and $z = a$, and $b(z)$ is the distance between the deflection plates at z, as seen in Fig. 4.45. The deflection voltage is connected to the electrodes symmetrically, and the potential of the electrodes is $V_0 + V_d/2$ and $V_0 - V_d/2$.

Neglecting the fringing field at the entrance and at the exit of the deflection plates, and assuming that the distance between the plates changes slowly, $b'(z) \ll 1$, the equation of motion is

$$\frac{d^2 y}{dt^2} = \eta E(z) = \frac{\eta V_d}{b(z)}, \qquad \eta = -\frac{e}{m_0}. \tag{4.123}$$

The electron beam to be deflected has a finite diameter, which at the entrance almost fills the space between the deflection electrodes. Like in the paraxial approximation in rotationally symmetric lenses, assume that the axial velocity does not change during the passage between the deflection plates. Two effects are neglected by these approximations:

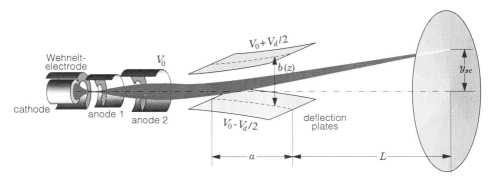

Fig. 4.45. Deflection in the electric field.

- The energy of the electron changes because the electron is accelerated in the deflection field and
- The axial velocity changes because the deflecting field always has an axial component (fringing field or deliberately introduced by the shape of the deflecting electrodes).

Conversion to the derivative of the space coordinates gives

$$y'' = \frac{\eta}{v_0^2} \frac{V_d}{b(z)} = \frac{V_d}{2V_0} \frac{1}{b(z)}. \tag{4.124}$$

The equation can be integrated twice; at $z = a$ we have

$$y = \frac{V_d}{2V_0} \int_0^a dz \int_0^z \frac{d\zeta}{b(\zeta)}.$$

When the electrons leave the deflection field the radial velocity y' remains constant, and if the distance from the deflection plates to the screen is L the deflection is

$$y_{se} = \frac{V_d}{2V_0} \left[\int_0^a dz \int_0^z \frac{d\zeta}{b(\zeta)} + L \int_0^a \frac{dz}{b(z)} \right]. \tag{4.125}$$

A special case is two parallel deflection plates at constant distance b, neglecting fringing field,

$$y_{se} = \frac{V_d}{2V_0} \frac{a}{b} [L + \frac{a}{2}]. \tag{4.126}$$

An important deflection parameter is the energy necessary to generate the field. The energy depends on the square of the field strength, and the deflection is linearly proportional to the field and to the length of the electrodes. It is therefore favorable to use long deflection plates. There are two reasons which limit the length. The beam must not hit the electrodes, so the electrodes are often made with increasing distance with z, as seen in Fig. 4.46, but this decreases the deflection sensitivity. The second reason is the transit time, which is of interest in high-frequency cathode ray tubes. The transit time between the deflection plates with the length a of an electron with the velocity v_0 is $t = a/v_0$. Introducing the transit angle instead of the transit time, $\alpha = \omega t$, gives

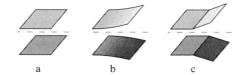

Fig. 4.46. Parallel, bend, and broken deflection plates.

$$\alpha = 2\pi \frac{a}{\sqrt{2\eta V_0}} f,$$

where f is the frequency of the signal. Acceleration voltage in cathode ray tubes varies between 3 and 20 kV. Assuming that the length of the deflection plates is 3 cm, the transit angle is about 35° at 100 MHz, if the acceleration voltage is 3 kV. Already at this frequency which is relatively low frequency for cathode ray tubes, the sensitivity decreases by 1.5 percent because the beam-coupling coefficient falls to 0.985 (Fig. 3.10).

To decrease the power necessary to obtain the deflection voltage, it is favorable to use the whole space between the electrodes. Every part of the space which is never transversed by electrons means unnecessary power, which the signal source must deliver. This is especially important at high frequency, where the displacement current which charges the capacitance of the deflection plates determines the current capabilities of the signal source.

One way to increase the deflection sensitivity is to decrease the acceleration voltage, but it must not be too small. In a high-frequency cathode ray tube the electrons must have at least an energy of 10 kV in order to obtain enough light from the spot when observing fast signals. A compromise is a low acceleration voltage in the electron gun, as well as after the deflection acceleration to full energy. This postacceleration and other methods to increase the upper frequency limit are discussed in Section 5.5.

4.7.2. Magnetic Deflection

Magnetic deflection is similar to electric deflection. Defining the magnetic field $B(0)$ at $z = 0$, the field as a function of the position z is $B(z) = B(0) \cdot b(0)/b(z)$. The deflection is

$$y_{sm} = B(0)b(0)\sqrt{\frac{\eta}{2V_0}} \left[\int_0^a dz \int_0^z \frac{d\zeta}{b(\zeta)} + L \int_0^a \frac{dz}{b(z)} \right]. \tag{4.127}$$

With parallel coils the approximation of a constant magnetic field is given by

$$y_{sm} = B\sqrt{\frac{\eta}{2V_0}} a[L + \frac{a}{2}]. \tag{4.128}$$

A comparison of Eqs. 4.126 and 4.128 shows that for the same deflection the ratio between the electric and the magnetic deflection is

$$\frac{E}{V_0} = \frac{B}{\sqrt{V_0}}.$$

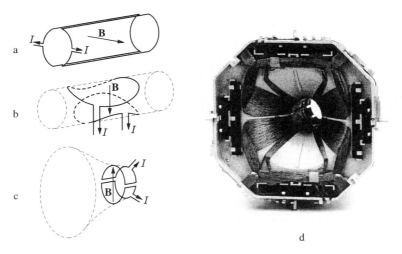

Fig. 4.47. Deflection coils for electron beam tubes. (**a**) Coils with parallel conductors. (**b**) Saddle coils. (**c**) Conical coils for TV picture tube. (**d**) Photograph of TV screen deflection coils (Philips).

It is thus easier to make a sensitive magnetic deflection, and the acceleration potential enters the equation in the square root instead of linearly. So with the exception of high frequency, magnetic field deflection is preferred.

4.7.3. Deflection Aberrations

A detailed analysis shows that there are 16 different errors caused by the second-order terms and 13 errors caused by the third-order terms. The second-order terms disappear if the electrodes or the coils are perfectly symmetric.

The aberrations can be divided into groups according to type; in each group there are similarities with the five geometrical aberrations in rotationally symmetric lenses, as seen in Fig. 4.48. The linear proportionality between the deflection and the deflection voltage or current exists only in paraxial approximation. The error manifests itself

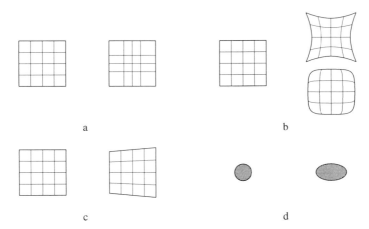

Fig. 4.48. Deflection aberrations. (**a**) Nonlinear distortion. (**b**) Cushion and barrel formed distortion. (**c**) Trapezoidal error. **d.** Astigmatism.

as a distortion; and the images of a quadratic mesh are rectangles of different size, depending on the distance from the optical axis. Even cushion or barrel formed distortion can appear, because the deflection in x and y direction can have different linearity errors. Asymmetry in the potential distribution between the deflection plates, or an oblique position of coils, results in a trapezoidal distortion. Astigmatism and the curvature of field can to an extent be cured by a curved screen. The deflection coma likes a circle with the paraxial image point at the periphery. Numerical computation of electron orbits shows directly which type of error dominates.

In contrast to an electron optical system with rotationally symmetric lenses, a system with deflection is always a combination of deflection and rotationally symmetric lenses. Even if the electron gun would be perfect, the deflection introduces errors. Observe an idealized system with parallel deflection plates and with negligible fringing field. Assume that an electron beam enters the deflection field. When the electrons come into the field, the deflection angle, θ, varies for different orbits because the potential is not the same for all orbits. According to Eq. 4.125 and taking into account the small change of the potential in the deflection field we obtain

$$y_{se} = \frac{V_d}{2(V_0 - \frac{y}{b}V_d)} \frac{a}{b}(L + \frac{a}{2}),$$

so the deflection angle is

$$\tan\theta = \frac{y_{se}}{L} \approx \theta \approx \frac{aV_d}{2(bV_0 - yV_d)}.$$

All orbits meet in a line focus at distance, f_v, where

$$\frac{1}{f_v} \approx \frac{d\theta}{dy} \approx \frac{a}{2b^2}\left(\frac{V_d}{V_0}\right)^2.$$

The deflection field focuses the orbits only in a plane and acts as a cylindrical lens, as seen in Fig. 4.49. The spot at the screen shows astigmatism.

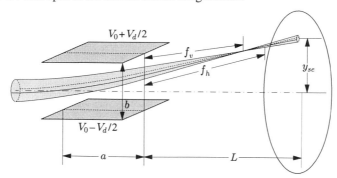

Fig. 4.49. Defocusing of a deflected beam.

4.7.4. Electron-Optical Prisms

Electron optical prisms also deflect particles, but unlike the deflection discussed in Sections 4.7.1 and 4.7.2, where the beam was supposed to be used as a pointer and already focused by a rotationally symmetric field, the prisms with constant or time-variable, homogeneous or inhomogeneous, electric or magnetic field can both deflect and focus. The prisms are used in spectrometers, spectrographs, particle separators, and similar devices, as well as in accelerator technology.

Most used are prisms with homogeneous and time-constant magnetic field between two plane and parallel pole faces. Assume that the magnet has finite dimensions and that pole faces are parallel with the x–y plane, so that $B_x = B_y = 0$ and $B_z = B_0 =$ const, and neglect the fringing field. In this field a charged particle moves along an orbit, whose projection in the x–y plane is a circle, with the radius of curvature

$$r = \frac{vm}{qB_0}. \tag{4.129}$$

An electron optical prism can thus be used to separate the particles according to their momentum or charge. It is therefore suitable to introduce a reference particle with mass m_c, velocity v_c, and charge q which moves on an orbit with the radius of curvature r_c. The reference particle enters the homogeneous field perpendicular to the edge of the magnet, and parallel with the pole faces at $z = 0$, as shown in Fig. 4.50. Inside the magnetic field the particle describes an arc of a circle with length $r_c\varphi_0$, which is defined as the optical axis. The motion of other particles in the beam can then be described by reference to this axis.

Observe a particle with the same magnetic rigidity (i.e., with the same mass, velocity, and charge as the reference particle) and with orbit in the $z = 0$ plane ($v_z = 0$), but which does not enter the magnetic field perpendicularly to the edge. If the distance between the optical axis and the entrance point is w_1 while the angle between the orbit and the optical axis is α_1, as shown in Fig. 4.51, the center of curvature of the orbit will lie in point A with coordinates

$$x_A = r_c \sin \alpha_1, \qquad y_A = w_1 + r_c(1 - \cos \alpha_1).$$

The particle orbit is an arc

$$[x - r_c \sin \alpha_1]^2 + [y - w_1 - r_c(1 - \cos \alpha_1)]^2 = r_c^2. \tag{4.130}$$

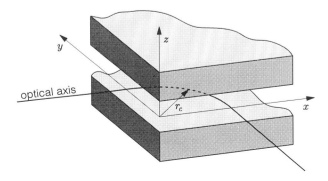

Fig. 4.50. Definition of the optical axis.

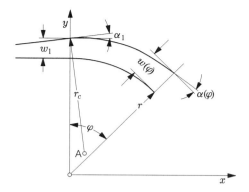

Fig. 4.51. Paraxial orbit.

Limiting the discussion to paraxial orbits (i.e., assuming that $w_1 \ll r_c$ and $\alpha_1 \ll 1$), all terms of higher order in Eq. 4.130 can be neglected, and

$$x^2 - 2r_c x \alpha_1 + y^2 - 2w_1 y \approx r_c^2 \tag{4.131}$$

remains.

The particle distance from the optical axis is denoted by $w(\varphi)$, where φ is the deflection angle, so the coordinates x and y are

$$\begin{aligned} x &= [r_c + w(\varphi)] \sin \varphi, \\ y &= [r_c + w(\varphi)] \cos \varphi. \end{aligned} \tag{4.132}$$

By geometrical considerations we have

$$\begin{aligned} w(\varphi) &= w_1 \cos \varphi + \alpha_1 r_c \sin \varphi, \\ \alpha(\varphi) &= -(w_1/r_c) \sin \varphi + \alpha_1 \cos \varphi, \end{aligned} \tag{4.133}$$

and the transformation matrix for an idealized magnetic prism with the deflection angle φ_0 is

$$\mathbf{T}_p = \begin{pmatrix} \cos \varphi_0 & r_c \sin \varphi_0 \\ -1/r_c \ \sin \varphi_0 & \cos \varphi_0 \end{pmatrix}. \tag{4.134}$$

For $\varphi_0 = 180°$ all orbits for particles with the same magnetic rigidity are focused to a point, if the particle starts from a point source with the velocity $v_z = 0$.

The situation is more complicated if the particles do not have the same magnetic rigidity. Observe a particle with rigidity mv/q and compare its orbit with the reference particle. If the reference particle has a radius of curvature $r_c = m_c v_c / q_c B$, a particle with another rigidity follows a circular orbit with radius

$$r = \frac{mv}{qB} = r_c(1 + \Delta) \tag{4.135}$$

along the circle

$$[x - r \sin \alpha_1]^2 + [y - w_1 - r(1 - \cos \alpha_1)]^2 = r^2.$$

Limiting the discussion to paraxial orbits, we obtain

$$x^2 - 2r_c x \alpha_1 (1 + \Delta) + y^2 - 2w_1 y \approx r_c^2 (1 + 2\Delta) \qquad (4.136)$$

and

$$w(\varphi) = w_1 \cos\varphi + \alpha_1 r_c \sin\varphi + r_c \Delta(1 - \cos\varphi) + \cdots,$$
$$\alpha(\varphi) = -(w_1/r_c) \sin\varphi + \alpha_1 \cos\varphi + \Delta \sin\varphi + \cdots. \qquad (4.137)$$

A time-constant magnetic field does not change the momentum of the particle; therefore Δ is constant and the transformation matrix is a 3×3 matrix:

$$\begin{pmatrix} w_2 \\ \alpha_2 \\ \Delta \end{pmatrix} = \begin{pmatrix} \cos\varphi & r_c \sin\varphi & r_c(1 - \cos\varphi) \\ -1/r_c \ \sin\varphi & \cos\varphi & \sin\varphi \\ 0 & 0 & 1 \end{pmatrix} \begin{pmatrix} w_1 \\ \alpha_1 \\ \Delta \end{pmatrix}. \qquad (4.138)$$

Observe now a magnet system with deflection angle φ_0 and a source which emits equal particles with the same energy. The source is situated at a distance l_1 from the magnet, as shown in Fig. 4.52. The particles enter the magnetic field with the deviation $w_1, \alpha_1, t_1, and \beta_1$ compared to the reference particle, where w_1 and α_1 refer to the deviation in the y direction while t_1 and β_1 refer to that in the z direction. After the particles passed the magnet, they can be observed in a plane which lies at a distance l_2 from the exit edge of the magnet. In this plane the position of the particles is given by w_2, α_2, t_2, and β_2. The transformation matrix in the x–y plane is

$$\mathbf{T}_{pxy} = \begin{pmatrix} 1 & l_2 & 0 \\ 0 & 1 & 0 \\ 0 & 0 & 1 \end{pmatrix} \begin{pmatrix} \cos\varphi_0 & r_c \sin\varphi_0 & r_c(1 - \cos\varphi_0) \\ -1/r_c \ \sin\varphi_0 & \cos\varphi_0 & \sin\varphi_0 \\ 0 & 0 & 1 \end{pmatrix} \begin{pmatrix} 1 & l_1 & 0 \\ 0 & 1 & 0 \\ 0 & 0 & 1 \end{pmatrix}, \qquad (4.139)$$

and that in the z direction is

$$\mathbf{T}_{pz} = \begin{pmatrix} 1 & l_2 \\ 0 & 1 \end{pmatrix} \begin{pmatrix} 1 & r_c\varphi_0 \\ 0 & 1 \end{pmatrix} \begin{pmatrix} 1 & l_1 \\ 0 & 1 \end{pmatrix}. \qquad (4.140)$$

Equation 4.140 describes the motion of the particle along a drift space; the homogeneous magnetic field does not influence the velocity component v_z. Equation 4.139 describes the focusing and dispersive properties of a magnetic prism.

A simple geometric construction shows the image point if the distance, l_1, from the source to the edge of the prism is known. The image point, B, lies on a straight line from the source, A, through the center of curvature of the reference particle, M. The construction

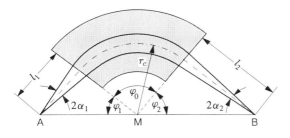

Fig. 4.52. Focusing in a magnetic prism. The points A, B, and M lie on a straight line (Barber's rule).

is known as Barber's rule [94]. The proof is simple: Assume particles with same magnetic rigidity; that is $\Delta = 0$. From Eq. 4.139 we obtain

$$\frac{l_1}{r_c} \cot \varphi_0 + 1 = \frac{l_2}{r_c} \left(\frac{l_1}{r_c} - \cot \varphi_0 \right). \tag{4.141}$$

The sum of the angles $\varphi_1 + \varphi_0 + \varphi_2$ is π, and thus we have

$$\cot \varphi_2 = -\cot(\varphi_0 + \varphi_1) = \frac{1 - \cot \varphi_0 \cot \varphi_1}{\cot \varphi_0 + \cot \varphi_1}. \tag{4.142}$$

According to Fig. 4.52, $\cot \varphi_1 = r_c/l_1$ and $\cot \varphi_2 = r_c/l_2$, so Eqs. 4.141 and 4.142 are identical.

The particles with different magnetic rigidity follow different orbits, a property used in spectrometers and particle separators. Note that particles with $v_z \neq 0$ at the entrance continue to move in the homogeneous magnetic field in the same direction. Therefore, a prism with homogeneous magnetic field focuses only in the x–y plane. The particles which start from a point source — for example, β particles from a radioactive source — come together in a line focus in a spectrometer with 180° deflection. However, it is not necessary to use a homogeneous field. If gradient fields are used, double focusing action is possible, and the particles converge to a point.

The focusing properties of the magnetic prisms can be drastically changed if the reference orbit makes an angle with the magnet edge. Figure 4.53 shows the consequence of such edge focusing in the median plane of a magnet. The path of an orbit which leaves the magnet at a distance w from the reference orbit, but parallel to it, makes a longer path inside the magnetic field. If α is the angle between the particle orbit inside and outside the magnet, we obtain

$$\alpha = \frac{d}{r} = \frac{w \tan \varepsilon}{r}.$$

As the chosen particle orbit corresponds to the second principal orbit (parallel with the optical axis) and lies in the median plane, the focal length is

$$f_m = \frac{w}{\alpha} = \frac{r}{\tan \varepsilon} > 0, \tag{4.143}$$

where ε is the edge angle, which is positive if the normal to edge surface is on the side of the center of curvature of the reference orbit.

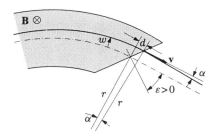

Fig. 4.53. Edge focusing in the median plane. Positive edge angle.

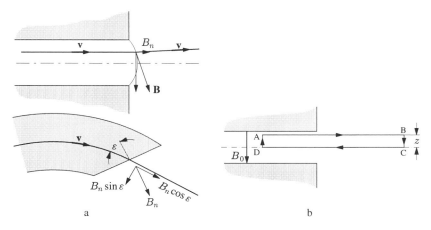

Fig. 4.54. (a) Edge focusing in the axial plane. Positive edge angle. (b) Integration path.

In the axial plane, the fringing field results in a magnetic field component in the horizontal direction, B_n, as soon as the particle orbit does not lie exactly in the median plane of the magnet (Fig. 4.54a). A component of B_n, $-B_n \sin \varepsilon$, is perpendicular to the velocity of the particle and acts with a force $-qvB_n \sin \varepsilon$. The corresponding change in the momentum is

$$\Delta p = \int qvB_n \sin \varepsilon \, dt = q \int B_n \sin \varepsilon \, ds = q \tan \varepsilon \int B_n \cos \varepsilon \, ds.$$

The integral can be computed by observing that the particle motion takes place in vacuum and therefore $\mu_0 \oint \mathbf{H} d\mathbf{s} = 0$. Figure 4.54b shows the closed integral path resulting in

$$\int_A^B B_n \cos \varepsilon ds + \int_B^C B_z(C) dz - \int_C^D B_n \cos \varepsilon ds - \int_D^A B_0 dz = 0.$$

Part C–D lies in the median plane, and this integral is zero. If path B–C is far away from the magnet, so that the fringing field can be neglected, even this integral is zero. What remains is

$$\int_A^B B_n \cos \varepsilon ds = \int_D^A B_0 dz = B_0 z$$

Assuming that the orbit inside the magnet corresponds to the second principal orbit with $v_z(A) = 0$, the particle velocity in the axial direction is

$$v_z(B) = \frac{\Delta p}{m},$$

and the angle of the orbit is thus

$$z' = \frac{\Delta p}{mv} = \frac{q \tan \varepsilon B_0 z}{mv} = \frac{\tan \varepsilon}{r} z.$$

The focal length in a thin lens approximation becomes

$$f_a = -\frac{z}{z'} = -\frac{r}{\tan \varepsilon} < 0. \qquad (4.144)$$

With a positive ε the edge focuses in the median and defocuses in the axial plane. Both focal lengths are equal in value but opposite in sign:

$$f_m = -f_a. \qquad (4.145)$$

The same result is obtained on the entrance side of a magnet with an edge angle ε.

This simple magnet focusing theory has limitations, and many corrections are necessary. More elaborate calculations take into account the fringing field. Different aberrations also exist; for example, the spherical aberration in a $180°$ prism results in a disk of confusion with the diameter $r_c \alpha_1^2$, where α_1 is the opening angle of the beam before the entrance into the prism. The spherical aberration can be eliminated by refraining from plane entrance and exit surfaces. If the fringing field and space-charge cannot be neglected, other aberrations manifest themselves.

Electrostatic systems work similarly. The simplest example is a system of electrodes, shaped as two concentric cylinders. It is possible to obtain double focusing action if the electrodes are given suitable shape — for example, toroidal — instead of cylindrical curvature.

4.8. THERMAL VELOCITY EFFECTS

In most electron tubes and devices which use electron optics for imaging or as a pointer, electrons start from a heated cathode. Their velocity distribution is Maxwellian, with the mean velocity ~ 0.1 eV. Because the lowest practically used acceleration potential is a few hundred volts, it seems that it is allowed to neglect the initial velocities. However, thermal velocity effects cannot always be neglected. Velocity distribution of the emitted electrons influences the beam diameter in front of the cathode, and it determines the maximum current density which can be obtained in focus. Experiments show that the cross section and current densities of electron beams depend on cathode temperature, probably because of changing emission from different cathode parts.

4.8.1. Cathode Imaging

Observe the space in the immediate neighborhood of a cathode. The cathode is an equipotential surface, so the field lines must be perpendicular to its surface, and the radial electric field directly in front of a planar cathode is zero. Figure 4.55 shows a planar cathode, with the surface perpendicular to the electron-optical system axis. In front of the cathode, at a distance z_1, is an equipotential surface with potential V_1. Assume a very small z_1, $\sim 0.05 - 0.1$ mm. In the region between the cathode and z_1 the electric field is homogeneous, $E_z = V_1/z_1$. The velocity distribution of the emitted electrons is Maxwellian. The mean radial velocity is then

$$<v_r> = \sqrt{2\eta V_T}, \qquad V_T = \frac{1}{2}\frac{kT}{e}, \qquad (4.146)$$

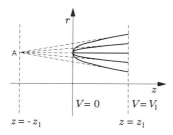

Fig. 4.55. Electron orbits in front of a cathode.

where T is the cathode temperature. The transit time between the cathode and z_1 is

$$t = \frac{2z_1}{\sqrt{2\eta V_1}}. \tag{4.147}$$

An electron which starts from the cathode with the radial velocity according to Eq. 4.146, reaches the equipotential surface at z_1 with a deviation

$$\Delta r = <v_r> t = \sqrt{2\eta V_T} \frac{2z_1}{\sqrt{2\eta V_1}} = 2z_1\sqrt{V_T/V_1}. \tag{4.148}$$

The orbit is a parabola. The tangent to the parabola has the slope

$$\tan \theta \approx \theta \approx \sqrt{V_T/V_1}. \tag{4.149}$$

Extrapolating the tangent behind the cathode, it looks as if all orbits which start from a point on the cathode are coming along straight lines from a point which is

$$z_{(\Delta r=0)} = z_1 - \frac{\Delta r}{\theta} = -z_1 \tag{4.150}$$

behind the cathode.

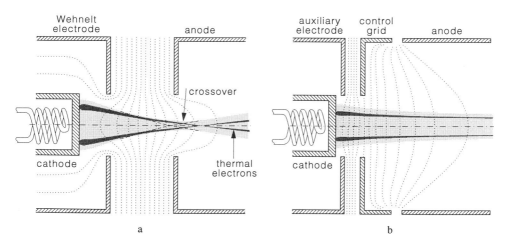

Fig. 4.56. Electron orbits in an electron gun of triode type. (**a**) Crossover gun, (**b**) collimated gun (auxiliary electrode is on cathode potential).

The cathode and the electrodes in the immediate neighborhood of the cathode form the electron beam. The construction is called an *electron gun.* There are two methods to form the electron beam: In electron guns with high perveance, only the cathode and the anode are used, and the construction should keep the beam laminar. When the emitted current and the perveance are low, a triode system with an auxiliary electrode, called the *Wehnelt electrode* [95], is usually used. This electrode controls the intensity of the emitted current. In an electron gun of triode type the Wehnelt electrode is connected to a negative potential. At high negative potential the current is cut off, as the potential approaches the cathode potential, the current increases. In a crossover-type gun, seen in Fig. 4.56a, the negative potential of the Wehnelt electrode makes together with the anode a lens with a short focal length. Without the Maxwellian velocity distribution the electron orbits would cross in the focal point (neglecting the aberrations). In reality, a crossover forms, where the electron beam reaches its smallest diameter. This crossover is imaged in a spot in most electron tubes and devices, which use the electron beam as a pointer. In a collimated-type gun, seen in Fig. 4.56b, the electrons are kept almost parallel after leaving the cathode. It has a higher luminance than the crossover gun.

4.8.2. Charge Density Effects

In many electron optical applications it is desirable to obtain such a big current as possible in a small spot in focus. The shortest sweep time — for example, in a cathode ray tube — depends on the acceleration voltage and the current density in the spot. Generally, all these systems have an electron gun and a number of lenses which focus the beam to a spot. There are four factors which limit the current density in the spot: Final wave length of the electrons, lens aberrations, space-charge, and velocity distribution at the cathode.

At an energy of 1.5 eV the wavelength of the electrons is ~ 1 nm. The wave properties of electrons can thus be neglected. The electron beam is focused to a spot by lenses. The spherical aberration plays a decisive role in how a small spot diameter can be reached. To obtain a spot of a few nanometers, apertures and small opening angles, θ, must be used because the spherical aberration is proportional to θ^3. The aperture limits the current which reaches the spot. We could be enticed to believe that by increasing the cathode emission any current density can be obtained in the spot, but the space-charge is a real problem. In the neighborhood of the spot the current density can be so high that the space-charge forces prevent the electrons from converging to a point and beam spreading determines the smallest diameter of the spot. In many applications — for example, in cathode ray tubes and other similar devices — the current density is so small that the beam spread caused by the space-charge can be neglected.

In most cases, aberrations and Maxwellian velocity distribution limit the current density which can be obtained in a spot. It is surprising that the velocity distribution of the emitted electrons, which at the surface of a cathode is ~ 0.1 eV, or in photocathodes about 1 eV, can influence the focusing in an electron optical system with acceleration voltages of a few tens or even hundreds of kilovolts. However, the acceleration occurs in the direction of the optical axis. In the transverse direction the electron velocity is changed only little in connection with focusing in the lenses. The motion from the cathode to the spot, shown in Fig. 4.57, must be seen in the phase space and is governed by Liouville's theorem [96] (see Appendix B): *Under the influence of conservative forces a volume element of the phase space remains constant.* If a beam in a scanning electron microscope

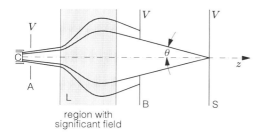

Fig. 4.57. Configuration of an electron optical system to calculate the Maxwellian velocity distribution influence. C, cathode; A, anode; L, lens; B, aperture; S, target.

which comes from a cathode with an area of 0.1 mm^2 is focused to a spot with an area of $1\,\mu$m^2, the lateral magnification is $M = 0.01$. To keep the volume element in the phase space constant, the transverse energy must increase 10^4 times. With $kT \sim 0.2$ eV, corresponding to a cathode temperature of 2400 K, the "equivalent electron temperature" in the spot is $\sim 10^7$ K, or ~ 1 keV.

Even if we assume that the lens in Fig. 4.57 is perfect and free from all aberrations, the aperture has a finite opening and limits, to an angle 2θ, the electrons which can reach the target.

In the field free space after the aperture, the number of particles is constant;

$$N = N_0.$$

The charge in a volume element is

$$dq = qN\,dx\,dy\,dz\,dv_x\,dv_y\,dv_z. \tag{4.151}$$

According to Eq. 2.13 the Maxwellian velocity distribution at the cathode is

$$N_0 = K\,e^{-mv_0^2/(2kT)}, \qquad v_0^2 = v_{x0}^2 + v_{y0}^2 + v_{z0}^2. \tag{4.152}$$

The electrons are accelerated in the z direction. After the acceleration and after passing the aperture the velocity is

$$v^2 = v_0^2 + 2\eta V, \tag{4.153}$$

where V is the acceleration voltage. The function $N(v, V)$, which gives the number of electrons in a volume element of the phase space, is then

$$N(v, V) = K\,e^{-m(v^2 - 2\eta V)/(2kT)}. \tag{4.154}$$

Note that the magnetic field does not enter Eq. 4.154. Even if a magnetic field influences the position of the electrons, it also influences the velocity components so that the density in the phase space remains unchanged.

The current density in the z direction is

$$dJ_z = v_z\,d\rho, \qquad d\rho = \frac{dq}{dx\,dy\,dz}.$$

Introducing Eqs. 4.151 and 4.154 we obtain

$$dJ_z = Kv_z\,e^{-m(v^2 - 2\eta V)/(2kT)}\,dv_x\,dv_y\,dv_z. \tag{4.155}$$

To integrate Eq. 4.155, the integration limits and the constant K must be known. At the cathode the potential V equals 0 and the current density J_z equals J_0. The velocity components have the limits:

$$0 \le v_z \le +\infty, \qquad -\infty \le v_x \le +\infty, \qquad -\infty \le v_y \le +\infty.$$

Hence

$$J_0 = K\, e^{eV/kT} \int_0^\infty e^{-mv_z^2/((2kT))} v_z dv_z \int_{-\infty}^\infty e^{-mv_x^2/(2kT)} dv_x \int_{-\infty}^\infty e^{-mv_y^2/(2kT)} dv_y. \quad (4.156)$$

The integrals over x and y are

$$\int_{-\infty}^\infty e^{-a^2 u^2} du = \frac{\sqrt{\pi}}{a},$$

leaving the integral over z:

$$J_0 = K\frac{2\pi kT}{m} \int_0^\infty e^{-mv_z^2/(2kT)} v_z dv_z. \quad (4.157)$$

Substituting

$$u = \frac{mv_z^2}{2kT}, \qquad du = \frac{m}{kT} v_z dv_z,$$

the integral becomes

$$J_0 = 2\pi K \left(\frac{kT}{m}\right)^2 \int_0^\infty e^{-u} du = 2\pi K \left(\frac{kT}{m}\right)^2$$

and the value of the constant is

$$K = \frac{J_0}{2\pi} \left(\frac{m}{kT}\right)^2. \quad (4.158)$$

The current density is thus

$$J_z = \frac{1}{2\pi} \left(\frac{m}{kT}\right)^2 J_0\, e^{eV/kT} \int e^{-mv^2/(2kT)} v_z dv_z dv_x dv_y. \quad (4.159)$$

Observed from the spot, the beam has spherical symmetry, so a conversion to spherical coordinates, as shown in Fig. 4.58, is suitable.

The volume element is

$$dv_x\, dv_y\, dv_z = v^2 \sin\alpha\, dv\, d\alpha\, d\varphi,$$

and with $v_z = v\cos\alpha$ we get

$$v_z\, dv_x\, dv_y\, dv_z = v^3 \cos\alpha \sin\alpha\, dv\, d\alpha\, d\varphi.$$

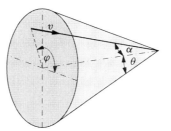

Fig. 4.58. Electron beam convergence.

The current density integral in spherical coordinates is thus

$$J_z = \frac{1}{2\pi}\left(\frac{m}{kT}\right)^2 J_0\, e^{eV/(kT)} \int\limits_{v_{min}}^{\infty} v^3\, e^{-mv^2/(2kT)} dv \int\limits_{0}^{\theta} \sin\alpha\cos\alpha\, d\alpha \int\limits_{0}^{2\pi} d\varphi, \qquad (4.160)$$

with

$$\int\limits_{0}^{\theta} \sin\alpha\cos\alpha\, d\alpha = \frac{1}{2}\sin^2\theta, \qquad \int\limits_{0}^{2\pi} d\varphi = 2\pi.$$

In the integral over v at the cathode, v_{min} is zero; after the acceleration we have $v_{min} = \sqrt{2\eta V}$. Introducing substitutions

$$u = \frac{mv^2}{2kT}, \qquad v^2 = \frac{2kT}{m}u, \qquad v\,dv = \frac{kT}{m}du,$$

v_{min} becomes

$$u_{min} = \frac{m}{2kT}2\eta V = \frac{eV}{kT}$$

and

$$\int\limits_{v_{min}}^{\infty} v^3\, e^{-mv^2/(2kT)} dv$$

$$= 2\left(\frac{kT}{m}\right)^2 \int\limits_{eV/kT}^{\infty} u\, e^{-u} du = -2\left(\frac{kT}{m}\right)^2 \{e^{-u}(u+1)\}\Big|_{eV/kT}^{\infty} = 2\left(\frac{kT}{m}\right)^2 e^{-eV/kT}\left(1+\frac{eV}{kT}\right).$$

The maximum current density in the spot is finally [97]

$$J_m = J_0(1 + \frac{eV}{kT})\sin^2\theta. \qquad (4.161)$$

The interesting point is that the maximum current density in the spot only depends on the potential at the spot, but not on the potential or magnetic field distribution between the cathode and the spot. This is valid under the assumption that a convergence angle, θ, can be defined with an aperture between the lens and the spot, which is usually true.

Take as an example the cathode ray tube. In such a tube the cathode has a diameter of ~ 1 mm; the opening in the last anode, which plays the role of the aperture, has a diameter of ~ 2 mm. The spot on the screen is ~ 0.3 mm in diameter, and the distance between the screen and the last anode is ~ 250 mm. The acceleration voltage is 10 kV. Introducing these values into Eq. 4.161 and taking into account the ratio between the area of the cathode and the spot, we can see that only about 24 percent of the cathode current reaches the screen.

Equation 4.161 is a consequence of Liouville's theorem, which means that it must be valid for photons, and it is an analogy to the conservation of the radiation strength in light optics.

The current density can be increased if the condition of field-free space is abandoned. If an axial magnetic field is used from the cathode to the screen, and the field can increase in strength with z, the limitation in Eq. 4.161 is circumvented. The magnetic field compresses the beam; assuming an infinitely large field in the spot, we obtain

$$J_m = J_0 \left(1 + \frac{eV}{kT} \right),$$

but in practice this cannot be reached because of the physical impossibility of designing such strong fields. Practically, this kind of beam compression is used in microwave tubes and will be discussed in Chapter 6.

There is another interesting phenomenon which is connected to the Maxwellian velocity distribution. The maximum current density in the spot is given by Eq. 4.161. The current is not equally distributed across the area of the spot; instead it can be assumed that the distribution is Gaussian-like. Assuming, nevertheless, that the current distribution is homogeneous, the diameter of the spot is

$$\frac{\pi}{4} d_m^2 J_m = I, \tag{4.162}$$

where I is the total current and d_m is an approximation to the spot diameter. Introducing J_m from Eq. 4.161 we obtain

$$d_m = \frac{2}{\sin \theta} \sqrt{\frac{I}{\pi J_0}} \frac{1}{\sqrt{1 + \frac{eV_0}{kT}}}. \tag{4.163}$$

Assume that the beam diameter at the entrance into the region of the deflection plates is d_0. In a cathode ray tube the distance, L, from the exit of the deflection plates to the screen is always much larger than the length of the deflection electrodes, a. Equation 4.126 can then be approximated by

$$y_{se} = \frac{V_d}{2V_0} \frac{a}{b} L.$$

The deflection voltage necessary to move the spot for the length of its own diameter can be computed by comparing d_m with y_{se}. Introducing in Eq. 4.163

$$\sin \theta \approx \theta \approx \frac{d_0}{2L},$$

and neglecting 1 in comparison with eV_0/kT, we obtain

$$V_d = 8\sqrt{\frac{kT}{\pi e}\frac{b}{d_0 a}}\sqrt{\frac{I}{J_0}}V_0. \tag{4.164}$$

A similar expression can be derived for the deflection in a magnetic field. The deflection voltage or the deflection current *does not depend* on L, which means that it is impossible to increase the information by making the distance longer, a conclusion which is valid quite generally for any kind of pointer.

5

Electron-Optical

Tubes and Devices

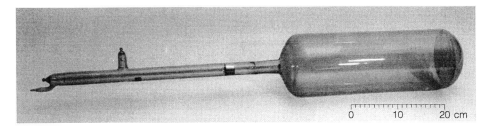

The first cathode ray tube was made in 1897 by K. F. Braun. It worked with secondary electron emission generated by ions in the residual gas. The ions produced a self-focusing effect, and the tube did not have any electron-optical components except a beam-limiting aperture. Inside, behind the front glass, was a mica sheet covered by fluorescent material, and the spot was deflected by a magnetic field (Telemuseum, Stockholm).

5.1. ELECTRON TUBES

5.1.1. Vacuum Microelectronics

HISTORICAL NOTES. When transistors, and later integrated circuits, enabled an unprecedented miniaturization of electronic components, it was generally believed that the era of the electron tube was over. Tubes have many drawbacks: They are large and fragile, the cathode must be heated to about 800°C to start emission, they generate much more heat than transistors for the same output power, and they need a vacuum to operate. Tubes also have some advantages: They can work at elevated temperature, they tolerate radiation, and, of most interest for the military, they are not destroyed by electromagnetic pulse from nuclear explosions (EMP) and high-power microwaves (HPM). Therefore, some effort was made to design small electron tubes, both in the USA and in the USSR, but the real breakthrough took place at the end of 1980s when a technique borrowed from integrated circuits construction allowed small cathodes using field emission (Section 2.2.5) to be an electron source. Cathodes with a packing density of more than a million cathodes per square centimeter and a current density of more than 100 A/cm² are now routinely constructed.

5.1.1.1. Vacuum Microelectronic Triodes

The structure is very similar to traditional triode tubes: The field-emitting "cathode" generates electrons, whose current is controlled by a "grid" voltage and is collected on an "anode," as seen in Fig. 5.1. First proposed in 1961 [98], these devices use very sharp tips, $\sim$ 20–50 nm, to enhance the local electric field and thus obtain field emission. In 1968 [99] the first working cathode had a tip of molybdenum. Many shapes and configurations, including wedges, have been investigated, but this "Spindt" cathode is the one which gives the most interesting results. Tip building is rather simple and follows similar processes used in constructing semiconductor electronic components. The microelectronic triode base is a silicon plate with an isolating SiO_2 layer, about 1–1.5 μm, covered by a metal film 0.5 μm thick. Electron-beam lithography produces small, 2-μm-diameter holes in the film trough to the top of the base-plate (Fig. 5.2a). A suitable evaporant, directed at an angle, deposits a thin release layer over each hole (Fig. 5.2b). A cone is formed by evaporation

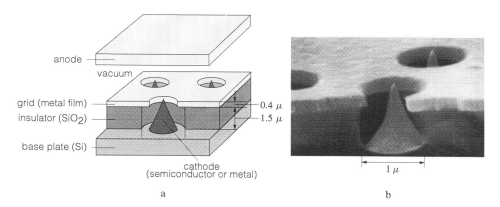

Fig. 5.1. Vacuum microelectronic triode. (**a**) Structure. (**b**) A scanning electron microscope micrograph of an SRI-developed Spindt-type gated field emission cathode (SRI International, Menlo Park, California).

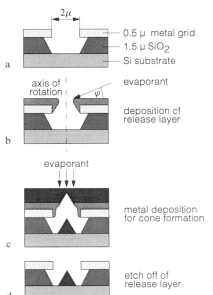

a   0.5 μ metal grid
     1.5 μ SiO$_2$
     Si substrate

b   axis of rotation   evaporant
     φ
     deposition of
     release layer

c   evaporant
     metal deposition
     for cone formation

d   etch off of
     release layer

Fig. 5.2. The construction process by Spindt. (After C. A. Spindt, *J. Appl. Phys.*, **47**, 3504, 1968, with permission.)

2μ

of a metal (Mo or W) or a semiconductor, which is automatically centered at the bottom of the hole upon the base plate (Fig. 5.2c). Finally, the release layer is etched away leaving the finished cathode and grid (Fig. 5.2d).

After evacuation the voltage is connected to grids and anodes. The cathodes emit current essentially independent of temperature (4 K–750 K). A current of between 10 and 100 μA can be drawn from a single tip, and thus current densities between 100 and 1000 A/cm^2 can be obtained, which exceed by several orders of magnitude those obtained by thermionic cathodes. Power dissipation at the anode limits the devices using arrays of field emitting cathodes. The anode heats up the cathode by radiation and sometimes also evaporates material (Fe, Cr, Ni) onto the cathode, leading to a "forming" effect whereby the emission can increase by several orders of magnitude, a process which is not yet completely understood or controlled.

Molybdenum is the most frequently used material for the cathodes, despite its high work function. Tungsten and Ti-coated or Ta-coated cones have also been produced. The material choice influences the performance not only through the work function, but also by the radius of the tip. Much work was done to achieve the sharpest tip, but often these efforts were based on known and easy fabrication processes rather than the choice of the best material.

An important vacuum triode parameter is its transconductance, defined by Eq. 3.21. In vacuum microelectronic triodes based on a Spindt-type cathode a transconductance of 5 μS per tip has been demonstrated, giving a total transconductance of about 50 S/cm^2, which is roughly 10^4 times larger than that for conventional thermionic triodes.

Other emitters have been tried, some using composite materials with metal or metal oxides bases, on which a dense crowd of W, Mo, TaC or NbC fibers were grown, randomly spaced very close to each other. The fiber tips were sharpened by etching. An Al$_2$O$_3$ isolating layer, obtained by electron-beam evaporation and covered by a molybdenum film, was made in the next step. The electron-beam evaporation results in circular holes,

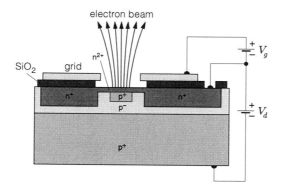

Fig. 5.3. Construction of an avalanche type cathode. (After E. M. E. Hoeberechts and G. G. van Gorkom, *Philips Techn. Rev.*, **43**, 49, 1987, with permission.)

which are self-aligned around the tip of the emitter, with a structure very similar to that shown in Fig. 5.1.

A silicon junction can be biased to produce an avalanche breakdown as used in EPROM and EEPROM devices. The electrons accelerated in the junction can obtain sufficient velocity so that they can escape into the vacuum, despite scattering in the top layer and overcoming the work function. Figure 5.3 shows the construction of such an emitter. A heavily doped p^+ substrate makes one contact of the diode, on which a p^- layer is epitaxially grown. An n^+ layer is built up by diffusion and serves as the electrical contact of the other diode terminal. Inside the p^- region a new p^+ region is produced by ion implantation, on top of which a very thin (about 10 nm) n^{2+} region is made, and also by ion implantation. The isolating SiO_2 layer and a thin metal film, the grid, are deposited in the last step.

The p^+–n^{2+} junction breaks into avalanche at lower reverse voltage than the p^-–n^+ region. In the depletion region a high electric field builds up and the electrons inside the region gain kinetic energy. A few percent of these electrons (3–4%) gain enough energy to escape into the vacuum. The useful current sets an upper limit on the efficiency of this type of cathodes. With avalanche voltage between 6 and 9 volts, the power losses in the diode are almost 10 watts for an emitted current of 50 mA. GaAs (with its higher bandgap) or other structures, like the PIN or the Schottky diode, may give improved performance.

Other possibilities were also investigated. One is the tunneling of electrons through a very thin (just a few atomic layers) isolator placed between two conducting surfaces, one of which is also extremely thin. By applying a voltage between the conducting surfaces, some electrons will have a chance to tunnel through the thin conductor on the vacuum side.

5.1.1.2. Applications and Future Possibilities

The advantages of using microelectronic tubes will be many when they overcome some children's diseases. Electrons move in vacuum at moderate voltages of a few tens of volts much faster than in conventional semiconductors (even if technology improvements in the submicrometer region could allow "ballistic" behavior of electrons in semiconductors, if the distances covered are less than the mean free path). Several researchers believe that vacuum microelectronic devices will perform usefully up to the gigahertz region.

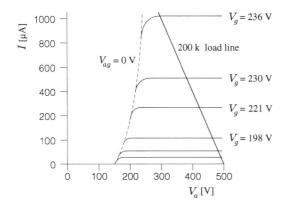

Fig. 5.4. Anode current curve families of a vacuum microelectronic triode. (After C. E. Holland et al., *IEEE Trans. Elect. Dev.*, **38**, 2368, © 1991 IEEE, with permission.)

In contrast to semiconductors, microelectronic tubes are insensitive to radiation and high temperature (up to about 500°C). Heat-resistant electronics could, for example, be directly inserted into jet engines or drilling crowns. Radiation-resistant electronics has applications in satellites or as sensors in reactor cores. Field-emitting cathodes are the most successful part of microtube development. However, working devices, except electron guns and displays for cathode ray and TV tubes, have yet to be made.

Microelectronic triode behavior differs in some respects from that of a conventional triode. In the latter, the grid controls the current by the potential in the star point (Eq. 3.18), which depends on both the grid and the anode potentials, the grid potential being biased negatively with respect to the cathode. In a microelectronic triode the current is determined by the electric field at the tip, so the grid potential, which must be positive in respect to the cathode, alone controls the emission. The high tip field accelerates the electrons to high velocity, where direction is little changed by the grid, so practically all electrons hit the anode.

Figure 5.4 (when compared to that for a conventional triode, Fig. 3.5) shows that the microelectronic triode anode voltage does not have much influence on the emitted current. Above a certain anode voltage the dynamic anode resistance (Eq. 3.22) becomes very high, and the behavior of the tube is more like a five-electrode tube, the pentode. The high-frequency response of a microelectronic triode is limited by the transconductance, g_m, and cathode–grid capacitance C_{gc}, not by transit time effects (Section 3.4). For conventional microwave triodes it will be shown (Eq. 7.7) that the product of the bandwidth and amplification is

$$\Delta\omega \cdot A_{f_0} = \frac{g_m}{C},$$

where $\Delta\omega$ is the bandwidth, and A_{f_0} is the amplification at the frequency f_0. Setting $A = 1$ the cutoff frequency is

$$f_c = \frac{g_m}{2\pi C_{gc}}.$$

If g_m is increased by increasing the number of cathodes (keeping the size of the device below a quarter of the wavelength), the cathode-grid capacitance will increase proportionally. To obtain high-frequency amplification the transconductance must be increased by using materials with lower work function, or a clean layout which will decrease the cathode-grid capacitance, as seen in Fig. 5.5. Until the end of 1992, triodes had rather poor

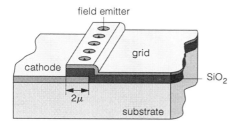

Fig. 5.5. Cathode structure to reduce grid-cathode capacitance. (After C. E. Holland et al., *IEEE Trans. Elect. Dev.*, **38**, 2368, © 1991 IEEE, with permission.)

high-frequency response, with cutoff frequency of a few megahertz. Much development work will be necessary before microelectronic triodes become routine devices, let alone having vacuum microelectronic integrated circuits become standard.

Field emission electron sources used in cathode ray and TV tubes, electron microscopes, electron-beam lithography, and flat display tubes have been more successful. An example is the avalanche diode of Fig. 5.3, which is used in an otherwise conventional small TV tube. The cathode has a much lower power dissipation (only 10 mW), the modulation can be obtained by a low-voltage signal, and it becomes operational instantly on turn-on.

Flat displays are based on **F**ield **E**mission **A**rrays (Fig. 5.6b). In one design the cathodes are molybdenum tips on a glass substrate, spaced 10 μm apart, resulting in 10,000 tips per square millimeter. SiO_2, 1 μm thick, makes the isolation between the cathode and the grid. The grid is 0.4 μm niobium. To support scanning, the cathodes and the grids are organized in orthogonal striplines. The anode is glass coated with indium-tin oxide and the three colored phosphors, as seen in Fig. 5.6a.

At V_a 300 V the current is almost zero with 40 V between the grid and the cathode. Modulating the grids between 40 and 80 V and the corresponding cathodes between 0 and 40 V, a current of about 0.1 μA per tip can be obtained, sufficient to activate the screen phosphor. The cathode current varies, and some cathodes are not emitting at all,

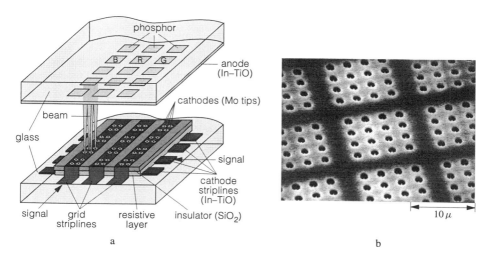

Fig. 5.6. Display screen with field emitting cathodes. (**a**) Structure (Adapted from *Cold Cathodes*, by Gary Stix. Copyright © October 1990 by Scientific American, Inc. All rights reserved). (**b**) A scanning electron microscope micrograph of the cathodes and grids (SRI International, Menlo Park, California).

so to achieve acceptable luminance variation a 2 μm layer of silicon (resistivity $\sim 10^5$ Ωcm) is deposited between the glass substrate and the cathodes. The emission of each cathode, given by the load line, will result in similar current. To avoid that accidental shorting of a grid and a cathode renders a complete pixel unadressable, each pixel is divided in 12 groups of cathodes, with 36 cathodes per group. If one cathode group becomes inoperable, the luminance changes by ~ 2 percent.

Many problems in vacuum microelectronics will be solved in the next few years. However, vacuum microelectronics will not supersede integrated circuits, but will provide additional components, which in many applications will complement semiconductor electronics.

5.1.2. High-Power Amplifying and Transmitting Tubes

5.1.2.1. Triodes and Tetrodes

The radiated power of commercial radio and television transmitters varies between 1 kW and a few megawatts, much more than semiconductor components produce at these operating frequencies. So even if transistors have replaced electron tubes at low power levels, no semiconductor component can replace classical amplifying and transmitting tubes at the highest power level.

The construction of these tubes depends on the output power required. Up to about 1 kW an oxide cathode, a molybdenum grid, and a copper or nickel anode are used. For higher output power a thoriated tungsten cathode is common, with a cathode-grid distance of a few tenths of a millimeter; the grid is made of molybdenum, maybe replaced by one of pyrolytic graphite (a material with good electric and thermal conductivity in only one direction). The anode is made of copper, tantalum, or molybdenum. Cathode and grid are mounted in a metal-ceramic envelope.

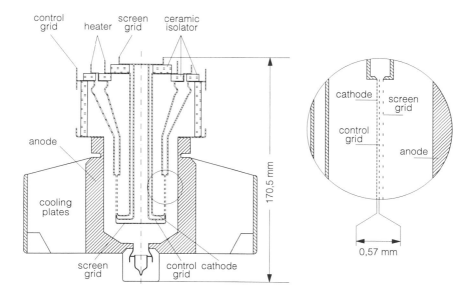

Fig. 5.7. Cross section of the transmitting tetrode RS2032CL (Siemens).

Fig. 5.8. Transmitting tetrode RS2032CL (Siemens). **(a)** The tube. **(b)** Cathode and control grid.

Modern transmitting tubes have one more electrode between the control grid and the anode. This screen grid is kept at a constant potential to the cathode. The capacitance between the anode and the control grid and also between the control grid and the cathode are strongly reduced and so is the influence of the change in anode potential on the anode current. This tube is called a *tetrode*.

The cross section of a modern transmitting tetrode is shown in Fig. 5.7, and a photograph of the tube is shown in Fig. 5.8. The tube is adapted to be mounted in coaxial circuits. The thoriated tungsten cathode is a directly heated mesh, and the control and the screen grid is a seamless molybdenum tube with punched holes, which guarantees stability. Both cold and warm the punched holes retain their relative positions, so that the control grid shadows the screen grid and the current to the screen grid remains small. When cold, the distance between the control grid and cathode is about 0.5 mm, while at the working temperature (the cathode at 2000 K and the control grid at 1300 K) it changes by less than 20 percent. Forced air cools the copper anode. All electrodes are isolated and restrained by ceramic isolators (Al_2O_3).

For output power up to 10 kW, forced air cooling is used. Higher power requires water cooling. At the highest power level (a single tube can deliver a peak power of 1 MW), the anode is sprayed with water which evaporates to cool its surface. The heat of vaporization of water, 2260 kJ per liter, allows for a much higher peak power while the tube dimensions can remain small. This is important at high-frequency, where the interelectrode capacitances and conductor inductances must be kept small.

Transmitting tubes used in radio and TV transmitters convert DC power into rf power with an efficiency greater than 90 percent. Usually two tubes are push-pull connected to work as a C-class amplifier. As oscillators these tubes are used in industrial inductive rf heating, as a power source for accelerators and as pulse generators and amplifiers when high voltage or high current is required.

High-power transmitters can be built with lower-power tubes connected in parallel with the advantage that if one or more tubes cease to work, the transmission continues, and defective tubes can be replaced while the transmitter is operating. For example,

Table 5.1

Some high-power triode and tetrode parameters (Siemens)

Tube	Tetrode	Tetrode	Triode	Triode	Tetrode
Type	YL1057	RS2032CL	RS3061CJ	RS3061CJ	RS2074HF
Use	TV transmitter	FM transmitter	Oscillator	Pulse amplifier	AM transmitter
Frequency [MHz]	600	110	130		50
Anode voltage [kV]	3.2	7.5	15	13.5	16
Screen grid voltage [V]	600	800			1500
Control grid bias [V]	−100	−100	−800	−300	−500
Anode current [A]	0.8	2.3	20	22.6	53
Output power [kW]	1.3	12	50	220	600
Heater voltage [V]	3.8	10	10	10	13.5
Heater current [A]	19.5	86	190	190	920
Cooling	Air	Air	Water	Water	Evaporation
Length [mm]	375	170.5	375	375	740

the protons and the antiprotons in the CERN's Super Proton Synchrotron are accelerated by sixteen 37.5 kW tetrodes, connected in parallel, so at 200 MHz the transmitter can deliver 600 kW. By contrast, for a two-tube push-pull C-class amplifier, a second pair of tubes with a heated cathode must always be on standby. In small slave broadcasting transmitters with output power of about 1 kW, the tubes are being replaced by transistors connected in parallel.

Table 5.1 gives examples of parameters for commercial high-power triodes and tetrodes.

5.1.2.2. EBS Tube

The **E**lectron **B**ombarded **S**ilicon tube is a hybrid. An electron beam, controlled by a grid, hits a biased silicon diode, called "active anode," as seen in Fig. 5.9. The electrons generate charge carriers in the p–n junction, which are separated because the diode is biased. The grid has a negative potential compared with the cathode, and it is modulated by pulses with amplitudes of a few tens of volts. The tube can work with pulse trains up to 1 GHz. With a 1 Ω load resistor, currents of a few hundred amperes can be obtained. The tube is used, for example, in accelerator technology to generate short current pulses to deflect particle beams.

In another type of EBS tube the electron beam is deflected by plates to one of two different active anodes, usually push-pull connected over the load resistor, but they can be used in two separate circuits. Deflections up to a few hundred megahertz are possible, and the tube can operate as an extremely linear wide-band amplifier.

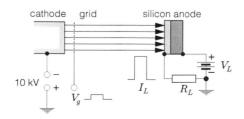

Fig. 5.9. EBS tube. Schematic.

5.1.3. Secondary Electron Multipliers

HISTORICAL NOTES. Among the earliest photomultipliers was a tube developed by Iams and Salzberg [100], with a single stage of secondary emission and an amplification factor of six. A tube constructed by Weiss [101] had 8 mesh dynodes and an amplification of a million. The mesh is transparent for many of the primary and secondary electrons, which give no secondaries, and was soon replaced by solid dynodes and combined magnetic and electrostatic [102] or pure electrostatic focusing [103].

Toward the end of the 1950s, photomultipliers with miniature dynodes in many columns were produced with the goal of making two-dimensional imaging with a large light amplification [104]. These tubes were the precursors of microchannel plates, and one of the channels became the channel multiplier.

In many physical measurements, individual particles must be detected. By hitting a surface with high secondary emission gain, δ, a single particle can trigger emission of secondary electrons (Section 2.2.4). These secondary electrons are accelerated and then strike other electrodes which also have high secondary emission yield. The charge is increased by δ at each electrode, and a single particle can finally produce over 10^8 electrons. Many types of secondary electron multipliers have been developed, and the most used are the photomultiplier and the channel multiplier.

5.1.3.1. Photomultiplier

A photomultiplier tube has a photocathode and a number of electrodes, called *dynodes*, the surface of which is covered by a material with a high δ. Photoelectrons from the cathode are accelerated by an electric field, hit the first dynode, and generate secondary electrons. The process continues on other dynodes. With dynodes of the appropriate shape, an electric field is generated by an external voltage connected to the dynodes, which accelerates the secondary electrons and directs their orbits into the next dynode. The amplified electron current is finally collected on the last electrode, the anode of the multiplier tube. For direct particle detection the photocathode is omitted and the tube is delivered ready for mounting in high vacuum systems and thus must not be exposed to air for a longer time.

Photomultipliers have many different structures, and the two most popular ones are shown in Fig. 5.10. In the linear model of Fig. 5.10a, the electron orbits are focused directly from dynode to dynode, while the avalanche continues along the tube. The transit time is short ($<$ 20 ns), all electrons follow similar orbits through the tube, and these multipliers are fast, with risetime sometimes less than a nanosecond. In the Venetian blind model of Fig. 5.10b, the first dynode covers almost the whole cross section of the tube and thus is very effective in collecting photoelectrons with simple electron optics around

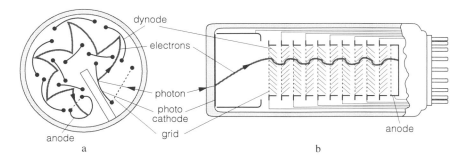

Fig. 5.10. Photomultipliers with (**a**) curved and (**b**) with "Venetian blind" dynodes.

Fig. 5.11. Photomultipliers (Hamamatsu).

the photocathode. The number of dynodes can be large, up to 20 in commercial models, and thus the amplification is larger as is the transit time (e.g., in a large tube with 20 dynodes it can exceed 50 ns).

All photomultipliers have high amplification. If the secondary emission gain for every dynode is δ, the current increases on the dynode n by δI_{n-1}. Often a material with a larger secondary emission gain, δ_1, is chosen for the first dynode, and thus for an n dynode tube the amplification is

$$A = \frac{I_a}{I_0} = \delta_1 \delta^{n-1}.$$

For a total voltage over the tube it is possible to optimize, or taper, the dynode potentials if δ is known. Resistive voltage dividers in the tube base or in the external circuit define the individual potentials. The amplification varies with the voltage between the photocathode and the anode as V^β, where β is a factor between 5 and 8, depending on the number of dynodes (e.g., for a tube with 10 dynodes, β is about 7). Thus the voltage source must be very stable since a 0.1 percent voltage change results in an amplification change of over 1 percent.

The limit for amplification is determined by the dark current, the anode current present when no photons hit the photocathode. It varies between 0.1 and 50 nA. The largest part of this dark current comes from the photocathode thermionic emission. Cooling the tube by 10°C will reduce the dark current by a factor of 10. Dark current increases if the tube is exposed to light because of the phosphorence in the glass. If the tube is returned to darkness for 30 minutes, the dark current is reduced by a factor of 10.

Photomultipliers are linear to about 2 percent. It is, however, impossible to simultaneously optimize linearity and amplification with a single voltage divider. To get the highest amplification for photometry or nuclear spectroscopy, all the resistors in the divider must be equal. To improve the capture of photoelectrons, the voltage drop between the photocathode and the first dynode is increased over that of the other dynodes. To obtain the highest possible linearity, the voltage drop is increased from dynode to dynode with a reduction in amplification. Tube manufacturers provide the optimal circuit parameters for different applications.

Thin layers of CuBeO, BeO, or alkaline antimonides cover the dynode surfaces, and photocathodes are made of alkaline materials (e.g., SbCs, AgOCs, SbRbCs) tailored

Table 5.2

Some photomultiplier tube parameters

Type	XP1922	XP2052	XP2050	R2083	R2490-05	R3600-02
	Philips	Philips	Philips	Hama-matsu	Hama-matsu	Hama-matsu
Application	General	General	General	Fast	General	General
Diameter [mm]	19	38	127	52	52	508
Spectral range [nm]	$300 - 650$	$300 - 650$	$280 - 620$	$300 - 650$	$300 - 650$	$300 - 650$
Maximum at [nm]	400	420	400	400	400	400
Number of dynodes	6	10	10	8	16	11
Construction	Linear	Linear	Linear	Linear	Fine mesh	Ven. blind
Working voltage [V]	1100	800	1270	3000	2500	2000
Amplification	3.5×10^4	6.5×10^5	1.4×10^5	2.5×10^6	5.0×10^6	1.0×10^7
Rise time [ns]	2.2	3.0	16.0	0.7	2.1	10.0
Transit time [ns]	17	36	90	16	8.5	90
Linear up to [mA]	30	30	10	100	500	20
Sensitivity [μA/lm]	8.5	11.5	11	10	8	8
Dark current [nA]	0.5	2	5	100	200	200

for a specific spectral region, as seen in Fig. 2.26. Table 5.2 gives examples of parameters for commercial photomultiplier tubes.

To detect electrically charged particles a scintillator is mounted on the face of the photomultiplier tube. Material in scintillators is NaI- or CsI-activated by thallium or organic compounds like anthracene. Incoming particles or high-energy photons lose energy by collisions with the scintillator atoms giving rise to light, which produces photoelectrons in the photomultiplier.

5.1.3.2. Channel Multiplier

Channel multiplier, a thin and curved glass tube, seen in Fig. 5.12, has an inner wall with high surface resistivity and a high secondary emission gain. With a voltage connected between its both ends the channel multiplier works like a continuous dynode photomultiplier. The resistance between the connections is about 10^8 Ω, and the voltage over the tube is a few kilovolts.

A charged particle or a high-energy photon which hits the channel multiplier somewhere near the entrance generates secondary electrons on collision with the wall. The electrons are accelerated along the tube by the voltage drop, hitting the tube wall many times and generating new secondary electrons. This avalanche leaves the tube on the positive side.

On the inner tube wall is a conducting layer which emits secondary electrons. Many electron orbits are possible, and it is impossible to say how often secondary emission takes place. However, if it happens sufficiently often the channel multiplier will practically have a constant amplification which depends on the length of the tube. Two channel multipliers with the same length-to-diameter ratios, usually between 50:1 and 100:1, will have the same amplification.

Channel multiplier tubes are always curved. When the electron avalanche nears the output, the electrons can ionize any residual gas molecules. If the tube were straight, in the electric field these ions could achieve a high enough energy (a few hundred electronvolts)

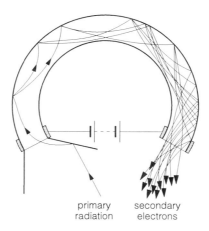

Fig. 5.12. Channel multiplier.

Fig. 5.13. Channel multipliers. Short models are current amplifiers, and the long ones are pulse amplifiers (Philips).

to generate new secondary electrons near the entrance of the tube. This would result in a new sequence of pulses until the tube electric field breaks down in about a microsecond. With a curved tube, the ions hit the wall before reaching a high enough energy to produce new electrons (see Section 9.3). The electrons, on the other hand, need only an energy of a few tens of electronvolts to create new secondary electrons and can achieve this easily in a curved tube. Channel multiplier amplification is therefore nearly constant up to pressures of 50 mPa.

Channel multiplier tubes saturate when approximately 10^9 electrons reach the end of the tube. The space-charge, which can be neutralized only slowly (in hundreds of nanoseconds) because of the high resistivity, breaks down the electric field, thereby keeping the electrons from reaching an energy sufficient to produce new secondary electrons. The amplification and the saturation current depend on tube voltage. In saturation the channel multiplier resembles a Geiger counter (see Section 10.2.2), in that it produces pulses with the same amplitude independent of the incoming particle energy. The channel multiplier transit time is roughly 10 ns, and only particles which hit the multiplier within an interval longer than the transit time will produce distinct pulses. If the space-charge along the tube walls cannot be reduced fast enough, the electric field breaks down and new pulses will not be registered. The channel multiplier is saturated. Commercial channel multipliers have limits between 10^3 and 10^5 pulses per second. At lower amplification

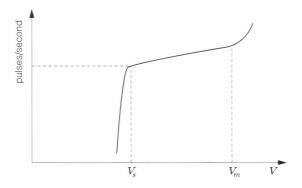

Fig. 5.14. Plateau (Philips).

the channel multiplier linearity is worse than that of the photomultiplier. The electrons take many different orbits through the tube, and the same energy of the primary particle does not result in the same output pulse amplitude. The amplitude distribution is Gaussian.

To serve as a multiplier of secondary electrons, the electric field along the tube must exceed a certain minimum value. If the field is to low, the kinetic energy of the electrons will not be sufficient to generate secondary electrons. When this minimum is exceeded, the number of pulses with primary excitation (number of particles per second) reaches a plateau, as seen in Fig. 5.14. At the end of this proportional plateau the number of pulses increases again because the high voltage produces spurious pulses.

The channel multiplier can also be used as a current amplifier which saturates at a current of about 5 μA. The response is linear up to about 20 percent of the saturation current. The tube shape for pulse and current amplification is shown in Fig. 5.15.

A frequent application of the channel multipliers is as a pulse amplifier for particle detection in, for example, spectrometers (see Section 5.2.2). The output side is closed and the signal is capacitively coupled to an amplifier, counter, or discriminator. When the tube is used as a current amplifier (i.e., as an analog component), the outcoming electrons are collected separately. Figure 5.15c shows a short model in which a prolongation, with a wall resistivity of about 5 percent that of the amplifying part, is added. This lower resistivity extension maintains the necessary electric field even for a large number of electrons. These channel multipliers, which have a saturation current of 10 μA, have often a built-in guard ring in the collector to reduce leakage current.

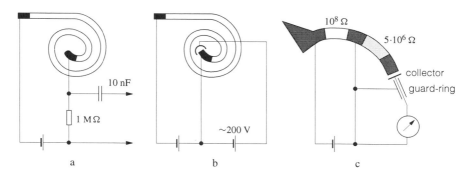

Fig. 5.15. Working principle (Philips). **(a)** Pulse amplifier, **(b)**, **(c)** current amplifier.

Table 5.3

Some channel multiplier parameters (Philips)

Type	X651	X636	X810	X710
Model	Short analog	Short analog	Spiral pulse	Spiral pulse
Amplification at [kV]	$5.0 \cdot 10^6$ 2.0	$5.0 \cdot 10^7$ 2.0	$1.2 \cdot 10^8$ 2.5	$3.0 \cdot 10^8$ 3.5
Resistance [Ω]	$8.0 \cdot 10^7$	$1.2 \cdot 10^8$	$6.0 \cdot 10^8$	$1.5 \cdot 10^8$
Maximum current [μA]	5	10		
Linear up to [μA]	2	2		
Starting voltage [kV]			1.6	1.6
Background [p/s]			0.05	0.05
Tube diameter [mm]	11.0	6.0	1.1	2.2
Highest pressure [mPa]	50	50	50	50

The tube diameter varies from 1 to 10 mm. Sometimes a cone-shaped surface up to 25 mm in diameter, covered by material with high secondary emission gain, is added in front.

The channel multiplier's lifetime depends on the cleanliness of the vacuum system, and most show an amplification decrease when the accumulated charge exceeds about 50–100 C, or after about 10^{12} pulses. Table 5.3 gives examples of parameters for commercial channel multiplier tubes.

5.1.3.2. Channel Plate

A **M**icro**C**hannel **P**late, or shorter channel plate, is constructed of thousands of short channel multipliers with a common cathode and anode, as seen in Fig. 5.16a. Two applications are especially important: image intensifier and particle detector. Individual tubes in the MCP have a diameter of 5–30 μm and a length between 0.25 and 1 mm. Independent of how a channel plate is used, an electronic image on the cathode is amplified between 10^2 and 10^4 times at the anode.

MCPs are made of two kinds of glass: The first has a high surface resistivity and secondary emission gain, and the second is easily chemically etched. Both glasses must have similar thermal expansion coefficients and viscosities. Starting with a tube

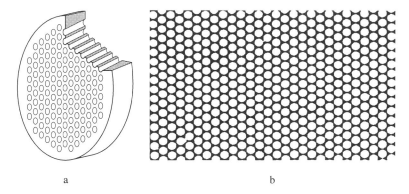

a   b

Fig. 5.16. Microchannel plate. (**a**) Sketch. (**b**) Photograph of the channels magnified 100 times (Philips).

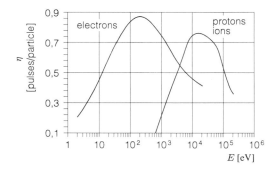

Fig. 5.17. Channel multiplier and channel plate. Detection efficiency for different particles as function of energy (Philips).

of 30 to 40 mm outer diameter and of 25 to 35 mm inner diameter, the dissolvable glass cylinder is introduced. The tube is then drawn slowly through a furnace where the tube diameter is decreased to about 1 mm. It is then cut into sections a few centimeters in length. These sections are fused into hexagons, again with diameters of 30–40 mm. These hexagonal blocks are once more drawn through the furnace, reducing their diameter to about 1 mm. They are then cut to appropriate lengths, fused in vacuum to the required diameter, cut into plates, ground, and polished to the desired thickness. Next the inner cylinders are etched away. The plate is then heated in a reducing atmosphere, creating in each channel a high-resistivity surface with high secondary emission gain. Finally, the polished surfaces are covered by a nickel–chromium alloy to make the electric contact. Channel plates are delivered vacuum-packed and have little exposure to air during mounting. They are very brittle.

The MCP terminal voltage is between 1 and 1.5 kV, and amplification can reach 10^5. Resistance between the cathode and anode is between 10^8 and 10^9 Ω, which, with a working voltage of 1 kV, gives a current between 1 and 10 μA. When the current in one channel (or in a few channels simultaneously) approaches this value the MCP saturates. The highest allowed pressure is roughly 10 mPa. Efficiency depends on the energy of the incoming particles and is different for electrons and ions, as seen in Fig. 5.17.

The MCP resolution depends on the channel diameter, wall thickness, and electron focusing between the plate and the readout screen. Commercial MCPs have channel diameters of 6–25 μm, and the distance between the centra is 8–31 μm, respectively,

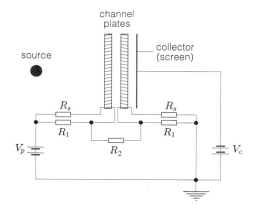

Fig. 5.18. Two channel plates connected in series.

and give a resolution of 20–100 lines per millimeter. The channels are usually made with a small slope (6° to 15°) with respect to the normal to the end surfaces. Two matched channel plates with opposite slope connected in series produce an amplification of up to 10^8 without ion induced feedback. The MCP schematic is shown in Fig. 5.18, where resistors R_s limit the current at saturation.

Two or three MCPs connected in series must be very near to each other to preserve the resolution. The orbits strongly diverge when the electrons leave the first plate, and some electrons enter another channel in the second plate. The same is true for readout, independent of whether one or two plates are used. Usually a screen is used for the readout, often with a TV camera behind it. The distance between the channel plate and the screen should not exceed 1 mm to preserve resolution. A strong electric field is applied, and the acceleration voltage is generally a few kilovolts. Magnetic field focusing can also be used.

Transit time in an MCP is short (less than 1 ns) and noise is low, without excitation less than 1 p/s cm^2. Amplification varies with the voltage across the plate as seen from Fig. 5.19. Table 5.4 gives examples of parameters for commercial channel plates.

Table 5.4

Some channel plate parameters

Type	F4142-07	G12-25SE	G12-46DT	G25-70	F2395-04
Manufacturer	Hama-matsu	Philips	Philips	Philips	Hama-matsu
Diameter [mm]	16.0	28.8	46.0	70.0	114
Useful diameter [mm]	15.4	24.8	42.0	68.0	105.0
Thickness [mm]	0.24	0.5	1.0	1.0	1.0
Channel diameter [μm]	6.0	12.5	12.5	25.0	25.0
Distance between channels [μm]	7.5	15.0	15.0	31.0	31.0
Open area [%]	60.0	60.0	60.0	60.0	57.0
Resistance [MΩ]	$100-500$	$200-700$	$30-100$	40	$5-500$
Amplification at 1 kV	10^3	10^3	10^3	10^3	10^4
Channel slope	13°	13°	13°	13°	8°
Maximum pressure [mPa]	1	13	13	13	1

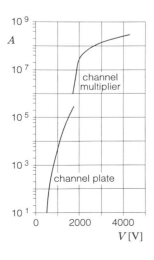

Fig. 5.19. Channel plate and channel multiplier, amplification as function of voltage (Philips).

5.1.4. Image Converters and Intensifiers

HISTORICAL NOTES. An evacuated glass envelope with a photocathode on one side and a fluorecent screen on the other, constructed in 1934, was a prototype of all proximity-type image intensifiers [85].   The light amplification was modest, and the intensifier could only be used for bright objects.   The main problem was spurious emission at high anode voltage causing screen shimmer. To circumvent background light, models with a longer distance between the photocathode and the screen were developed, and the electrons were guided by a longitudinal magnetic field [105].

Zworykin and Morton [106] used a lens in the first image inverter. The photocathode and the screen were both plane surfaces at the end of two tubes of the same diameter. In the same year Morton and Ramberg [107] introduced a curved photocathode, improving considerably the image quality.

The early image intensifiers and converters suffered from low light amplification, and later the models with curved photocathode gave a gain of 10 and could also be used for infrared detection. Further work was pioneered in Germany [108] during the second world war, where infrared converters were used instead of radar. Brightness enhancement was obtained by connecting three such tubes in series inside a common envelope.

Image converters and intensifiers make bright visible images from weak incoming images from wavelengths ranging from X-rays to infrared.

Two variants are common.   The first, the so-called proximity converter, seen in Fig. 5.20a, is constructed of two plane, parallel plates, about 1 mm apart. In some models a channel plate is interposed between the plates. One plate is a photocathode, and the other is a screen which can be observed directly, photographed, or connected optically to a TV camera tube. These converters are employed primarily as fast electronic shutters with exposition time as short as 1 ns. When the radiation should arrive, a positive pulse (a few kilovolts) is connected to the anode. The short distance between the photocathode and the anode allows all electrons to travel along parallel orbits so the imaging is almost aberration-free. The amplification, which depends on the construction, can be as high as 20,000 times for channel plate models, and about 20 times in others.

The so-called image inverter, seen in Fig. 5.20b, incorporates an electronic lens which turns the image on the photocathode upside-down before it is seen on the screen. The lens and anode voltages are high, between 10 and 20 kV, which reduces the chromatic and the geometric aberrations, especially the spherical. Large-diameter photocathodes are curved to reduce distortion and curvature of field. The light amplification is about 100.

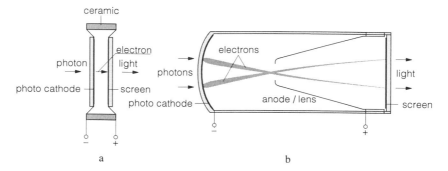

Fig. 5.20. Image converter of (**a**) proximity type and (**b**) inverter type.

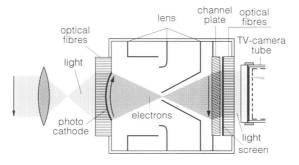

Fig. 5.21. Image inverter with optical fibers and channel plate.

Fig. 5.22. Image converters (Philips).

Modern high amplification (up to 100,000) image inverters are constructed with optical fibers in front of the photocathode and behind the screen, as seen in Fig. 5.21. These optical fibers restrict the incoming and outgoing light rays to be perpendicular to the photocathode and the readout equipment (e.g., a TV camera tube). The lens generates the photocathode image on the cathode side of the channel plate. By changing the voltage across the plate the MCP provides easily changed amplification and a higher contrast, because MCPs saturate at high currents.

Figure 5.22 shows a few image converters and intensifiers. Table 5.5 gives some typical parameters. The screen image can be observed visually, on a video camera, or be photographed as seen in Fig. 5.23. The dark current limits the faintest illumination that can be detected, around 10^{-7} lux.

A special image converter, seen in Fig. 5.24, has revolutionized medical X-ray diagnostic procedures. In X-ray examinations both the patient and physician are exposed to radiation; but by using an image converter almost no radiation reaches the physician, and the dose received by the patient is an order of magnitude less for similar- or better-quality images. X-rays are converted at the first screen into visible light, which hits the photocathode and causes photoelectrons to be emitted. These electrons are focused to the second screen where the visible light produces the diagnostic image. A TV camera tube (e.g., a vidicon or CCD) presents the image to the physician on a TV screen or it can be stored in analog or digital form. X-ray image converters have diameters of up to 40 cm so they can image large inner organs. X-ray image converters are also used in material analysis.

Table 5.5

Some image converter and intensifier parameters (Philips)

Type	XX1500HG	XX1380	XX1332	XX1390	XX1410	XX2020
	Inverter	Inverter	Inverter	Proximity	Proximity	Inverter
Diameter photocathode [mm]	18	20	50	18	18	230
Diameter screen [mm]	18	30	40	18	18	20
Amplification [$\times$1000]	85	22	45	15	10	
Resolution [l/mm]	33	50	45	29	30	45
Sensitivity (white) [μA/W]	350	350	320	400	420	
Sensitivity (800 nm) [mA/W]	35	35	28	35	40	
X-ray absorption						65 %

Fig. 5.23. Photograph taken with an image intensifier. This exposure was made on a moonless, overcast night, far from other light sources (Mullard).

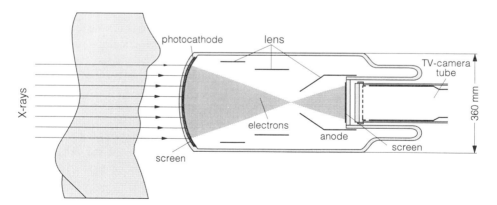

Fig. 5.24. Image converter for X-ray diagnostic.

We describe here the construction of a modern X-ray tube with accelerating voltages between 30 kV (for dental or soft tissue applications) and 200 kV (for skeletal imaging) and currents of up to a few tens of milliamperes. The cathode and the anode are made of tungsten. When the electrons hit the anode, they are slowed down, radiate X-rays,

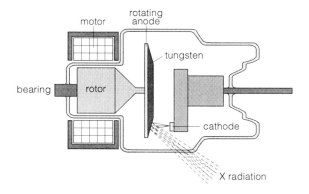

Fig. 5.25. X-ray tube with rotating anode.

and generate heat. At high power the heat must be distributed over the anode, which rotates. Rotation is accomplished by an induction motor whose stator is outside and the rotor inside the tube. Low-power X-ray tubes (dental X-ray) have stationary anodes.

A special image inverter is the streak camera tube, seen in Fig. 5.26a, which uses a pair of deflection plates to display the light intensity variation in time. The light from the source illuminates a slit, which is imaged on a photocathode, possibly through optical fibers. The deflection plates sweep the emitted photoelectrons.

This tube can work in different modes — a single signal can be swept over the screen, or a few optical fibers can be used to simultaneously observe different light sources (Fig. 2.27b). For sources that repeat in time, the deflection signal can be synchronized with the light signal. Some tubes have a long persistence screen which allows very faint light signals to be detected. A pulsed voltage can be connected to the acceleration electrode so that the electrons can then pass the tube in short intervals. Here the deflection voltage is increased in a staircase fashion, while the acceleration voltage is held constant during the pulse so that up to six two-dimensional images are seen.

Streak camera tubes have picosecond time resolution and a variety of photocathodes covering the wavelengths from infrared to X-rays. Table 5.6 gives some technical details of commercial tubes.

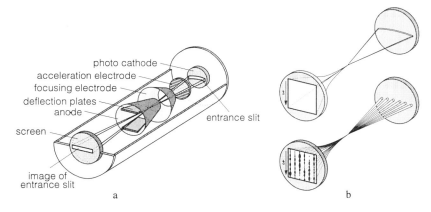

Fig. 5.26. Streak camera tube (Philips). (**a**) Construction. (**b**) Slit image with one and with several signals.

Table 5.6

Some streak camera tube parameters (Philips)

Type	P510	P510	P920	P500X
Wavelength	Visible	Visible	Visible	X-ray
Acceleration electrode	Slit	Grid	Grid	Slit
Photocathode [mm]	35×4	35×4	12	15×3
Enlargement	1.3	0.75	1.5	1.3
Diameter screen [mm]	64	64	18	64
Resolution [line/mm]	10	25	25	10
Time resolution [ps]	5	150	1.3	20
Deflection sensitivity [V/cm]	500	500	400	500
Anode voltage [kV]	15	15	15	15

Fig. 5.27. Two streak camera tubes (Philips).

5.1.5. Signal–Image and Image-Signal Converter Tubes

Tubes are made that convert electric signals into images or convert images into electrical signals. Among these the most important are cathode ray tubes, computer monitor tubes, radar screens, television screens, and television camera tubes.

HISTORICAL NOTES. In 1897 the first cathode ray tube was built by Braun [19] (see the photograph on the chapter title page). Wehnelt [21] replaced the discharge electron source by an oxide cathode and introduced between the cathode and the anode a control electrode [95], which today is called the Wehnelt electrode in the European literature. By adjusting the potentials, Wehnelt was able to focus the beam on the screen, and to deflect it by time-varying potentials connected to two pairs of deflection plates. However, his tube relied on self-focusing by ions, and lenses were introduced in 1930s. Modern cathode ray tubes, monitor tubes, and TV tubes are constructed according to the same principles, the important improvements concerned reducing the spot size, increasing the deflection linearity, and the high-frequency limit.

The first dynamic separation of an image was proposed by Nipkow [109] in 1884. He patented a rotating disk with small holes lined up in spiral rows, along with a selenium photocell, to obtain an electric signal. On the receiving side, rotation of the polarization plane by flint glass in a magnetic field should control the light passing through another disk, synchronously rotating with the first one. The system was to insensitive and was never built. The first public transmission using Nipkow's system, with a strong light source, a potassium photocell, and a neon lamp, was made in Great Britain in 1925, and regular broadcasting started in 1929. At the same time similar demonstrations were given in the USA.

The next step in the development of television utilized film as an intermediate medium. An electron beam in a cathode ray tube was used as the light source illuminating the film. The screen was equipped with a very short persistence phosphor. While the beam was scanned over the film a photomultiplier tube detected the light intensity [110]. A cathode ray tube generated the picture. The system was later used in commercial TV to send the movies and was called a "flying spot scanner."

The first completely electronic TV system was designed by Farnsworth [111] in 1934. The image of the object was generated on a photocathode. The emitted electrons were focused and deflected by two deflection coils, so that only electrons from a picture element could reach the photomultiplier and generate the video signal. The sensitivity of this image dissector tube was too low for regular TV broadcasting. The next step was made by Zworykin [112] in the same year. His tube, the iconoscope, was the forerunner of modern TV camera tubes. On an insulating layer a mosaic of photosensitive elements was deposited, on which the image was projected through a lens. A conducting coating, the signal plate, covered the other side of the layer. Photoelectrons were emitted from the photosensitive elements, resulting in a positive charge on the illuminated mosaic elements. When a focused electron beam hit the element the potential was returned to an equilibrium value. The charge change released from the signal plate indicated the light intensity which hit the element between the scans. Zworykin improved also the CRTs for use in TV receivers.

The first regular TV transmission using radio waves started in Great Britain in 1929, followed in 1931 by the first color TV experiment. The receiver was called "televiser."

5.1.5.1. Cathode Ray Tubes

In Cathode Ray Tubes, which display time-varying electric signals, the electron beam is generated by an electron gun (Section 4.8.1), focused by cylindrical lenses (Section 4.3.1), and deflected (Section 4.7.1), as seen in Fig. 5.28.

The tube construction depends on bandwidth, trace velocity (sweep-time), linearity, and screen luminosity, the latter two increase with accelerating voltage. But a high electron energy results in low deflection sensitivity (Eq. 4.125) and thus a limited trace velocity and bandwidth. Deflection sensitivity must be as high as possible, especially for high-frequency signals, because it is difficult to create high electric field between the deflection electrodes. At frequencies beginning at a few hundred megahertz the cables to these electrodes must be correctly terminated, usually into 50 Ω.

Postacceleration, often used to increase the deflection sensitivity, is achieved by three methods, depicted in Fig. 5.29. In all three, deflection occurs in a region of lower electron energy than that with which the electrons hit the screen. The simplest method, used up to a few tens of megahertz, employs a resistive helix (graphite) between the deflection plates and the screen. The electrons hit then the screen with sufficient energy to give a luminosity which is enough even for fast signals. Unfortunately, the deflection sensitivity is considerably decreased because the acceleration is axial. The second method

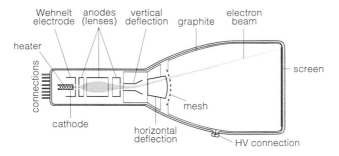

Fig. 5.28. Cathode ray tube.

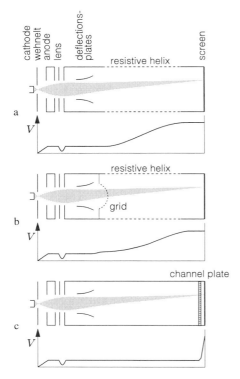

Fig. 5.29. Different methods to increase the deflection sensitivity. (**a**) Resistive graphite helix, (**b**) spherical mesh after deflection, (**c**) channel plate in front of the screen.

uses a mesh and a resistive helix after the deflection to emulate an electric field with spherical symmetry. After passing the mesh, electrons have almost straight orbits while near the screen the orbits curve, because the electric field looks like that between two parallel electrodes. Even with this method the deflection sensitivity is reduced. In the third method a channel plate amplifies the current while allowing a high acceleration voltage to be connected between the channel plate and screen, a distance of a millimeter, providing a high trace velocity and a high luminosity. The deflection, along with the electron motion up to the input of the channel plate, occurs at low electron energy, so these tubes operate up to a bandwidth of a few gigahertz.

The CRT screens use different phosphors. The blue and the white have the shortest persistence, while the yellow-green and orange have the longest persistence. High-frequency tubes use a blue or a white phosphor.

The CRT spot size, about 0.3 mm, is a compromise between voltages, alignment accuracy, and deflection sensitivity. The most important error comes from alignment, so elliptical astigmatism dominates. By changing the voltage between the last lens electrode and the symmetry potential of the vertical deflection plates (astigmatism knob on an oscilloscope), this aberration can be partially compensated. Modern CRTs often have a built-in magnetic correction for deflection astigmatism and orthogonality between the horizontal and vertical deflection together with an external coil, which allows these deflections to coincide with the viewing screen grid. Table 5.7 gives some parameters for a few cathode ray tubes.

Modern oscilloscopes employ sampling techniques to obtain larger bandwidth, up to 50 GHz. These oscilloscopes do not use conventional CRTs, but favor tubes more like those in computer monitors.

Table 5.7

Some cathode ray tube parameters

Type	D18-180GY	D18-180GH	T7100
	Philips	Philips	Tektronix
Screen [mm]	120×96	120×96	100×85
Acceleration voltage [kV]	2.5	16.0	2.5 + 12.5
Anode voltage [kV]	2.5	2.0	2.5
Lens voltage [V]	275 − 440	400 − 800	*
Line width [mm]	0.3	0.35	0.3
Vertical deflection [V/cm]	19.0	3.4	1.2
Horizontal deflection [V/cm]	26.0	6.4	1.4
Bandwidth [MHz]	25	100	2500
Screen phosphor	Green	Blue	White
Persistence	Medium	Short	Short

* T7100 uses channel plate.

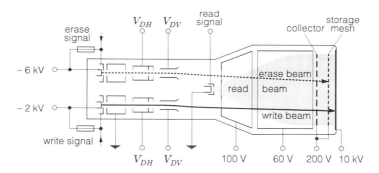

Fig. 5.30. Storage CRT.

A special storage CRT retains a signal picture for many minutes when the picture can either be erased or refreshed. Sampling technique opens the same possibilities, and storage CRTs are used in some special constructions.

Figure 5.30 shows an example of a storage CRT with three electron guns. An electric signal modulates the writing beam, which is then deflected in a conventional way. The electrons hit a fine metallic storage mesh covered on the gun side by an isolating layer (e.g., MgT_2). On the collector the resulting secondary electrons are collected, creating regions of positive charge on the storage mesh. The reading beam, a broad electron beam which covers the whole screen, is influenced by the charge on the storage mesh; its electrons are deflected and hit the screen with high kinetic energy gained between the storage mesh and the screen. The third gun erases the picture by sweeping over the whole storage mesh, discharging the places earlier charged by the writing beam.

5.1.5.2. TV Screens and Monitor Tubes

TV tubes and monitor tubes for computers are CRTs with magnetic deflection and a large rectangular screen, with deflection coils mounted on the outside of the glass neck. Modern tubes, seen in Fig. 5.31, use a 110° deflection along the screen diagonal.

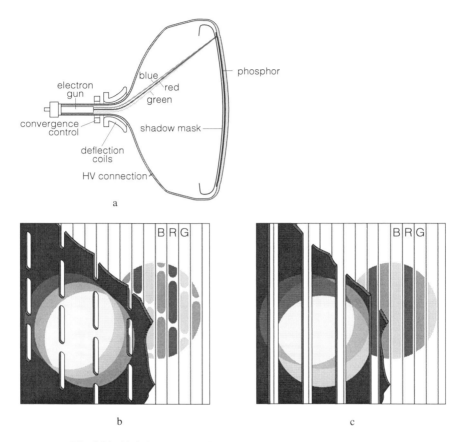

Fig. 5.31. (a) Color TV tube. **(b)** Shadow-mask and **(c)** Trinitron design.

A color TV tube has three electron guns, one each for blue, green, and red. Each electron beam is intensity-modulated by a corresponding TV or computer signal while all three are swept over every second line of the entire screen, first the even and then the odd. Two systems dominate the color CRT market. The screen has in both systems narrow vertical strips of three phosphors, one for each color. In the shadow-mask mask system a plate (frequently made of invar) with ~500,000 oblong holes, ~ 10 mm from the screen (Fig. 5.31b), allows the beams to pass through the holes, focused to the screen. The beam positions and the distance from the shadow-mask to the screen are well defined, so the mask will allow only that part of each beam to reach the phosphor that corresponds to the correct color. About 80 percent of the current is lost in the mask, so it is made black by a chemical surface treatment to increase its heat radiation. The mask holes are etched, often from both sides, to reduce the reflected and secondary electrons generated by the primary beam. In the new **H**igh **D**efinition **TV** screens the shadow-mask holes are round, and all three electrons beams pass through the same hole. The Trinitron (Fig. 5.31c), has a common lens for all three beams which decreases the spherical aberration, and a grid structure which does not affect the vertical resolution, giving a better time resolution and which is responsible for a subjectively sharper appearance.

Fig. 5.32. Modern electron gun (Philips).

Black-and-white TV screens use a simpler technology. A single electron gun, no shadow-mask, and the phosphor (Zn/CdS doped with Ag) give a gray scale. Monitor tubes can have a green or yellow phosphor.

In all TV tubes the phosphor is covered by a thin aluminum layer to increase the luminosity by reflection and to prevent ions (which originate in the residual gas) from hitting the phosphor and "burning" it when the TV set is switched off. The ions are focused along the tube axis and can burn in an "ion spot" in the middle of the screen. A graphite layer covers the remainder of the tube interior (except the neck) and provides a well-defined potential, absorbs the dispersed light, and gives a picture with more contrast. To increase the contrast and decrease the reflections, modern TVs use a screen of a dark glass (so-called Black Line screen). Improvements in phosphor strips production and higher acceleration voltage and current give better contrast, which is important when the picture is seen in daylight. HDTV with its increased resolution puts new demands on the spot size (~ 0.1 mm) and aberrations. New triode-type electron guns have been developed using three-dimensional computer programs. The electron orbits, which start at different radii at the cathode, cross so that the spherical aberrations are strongly reduced. The openings in the last electrode are elliptical with a well-defined eccentricity (Fig. 5.32), so that the electric field components are not rotationally symmetric, and the image errors can be reduced.

Color is defined by the choice of the screen phosphor. Blue phosphor is ZnS doped by Ag, green ZnS doped by Cu, and red YVO_4 (yttriumvanadat) doped by Eu (Europium). Figure 5.33 shows the spectral distribution of the radiation density for the three phosphors for a white screen. Both the color and the saturation depend on the current,

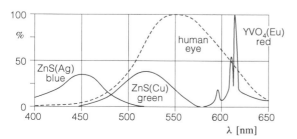

Fig. 5.33. Spectral distribution of the radiation density for the three phosphors on a TV screen for color TV when rendering white. (After *Moderne Vakuumelektronik*, J. Eichmeier, © 1983, Springer-Verlag, with permission.)

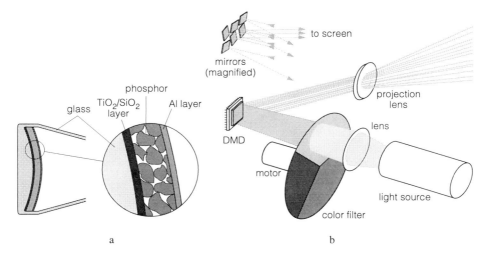

Fig. 5.34. (a) Screen construction in a projection TV tube (Philips). (b) TV screen with mirrors (Texas Instruments).

which hits the phosphor. Manufacturers specify the relation among the three gun currents, for example, for the above-mentioned phosphors: 25.9 percent for the blue, 35.8 percent for the green, and 38.3 percent for the red gun.

Besides deflection coils, there are several other magnetic adjustments for color, so tubes often have a built-in inner magnetic screen which should be demagnetized each time the tube is switched on. On the outside of the glass neck, surrounding the gun, are several rings with small ferrite magnets with which the three electron beams can be adjusted relative to each other and to the deflection coil in a process called convergence control. Table 5.8 gives examples of parameters of a few TV screen and monitor tubes.

Large displays are made of electronically divided images on a number of conventional TV screens, but special projection TV screens are constructed differently. To obtain the much higher luminosity, the tube works at higher voltage and current. In one construction, for example, a special layer, composed of TiO_2 and SiO_2, interposed between the phosphor and the glass, only allows light rays to pass, which make a very small angle with the layer normal (Fig. 5.34a). The reflected light returns to the phosphor, where it is reflected anew or is reflected on the thin aluminum layer backing the phosphor. Since both the layer and phosphor are thin, almost all the light produced by the electrons in the phosphor layer comes out perpendicularly to the glass surface at the place where the beam hits the phosphor. The TV screen is then optically reproduced on a projection screen. Another construction uses small mirrors, 16×16 μm, in a matrix; one mirror for each image pixel. The matrix, a **D**igital **M**irror **D**evice consists of up to 2.3 millions of mirrors and is made using sputtering and plasma etching processes. Each mirror can tip $\pm 10°$ from the neutral position (Fig. 5.34b). TV picture is stored in digital form in a CMOS–SRAM memory. This memory controls the inclination of each mirror, and the light from a light source is reflected either to the screen or to a light absorbing layer. The light intensity on the screen is controlled by pulse-width modulation; the mirror remains "on" during a time interval proportional to the amplitude of the TV signal. Colors can be obtained either by modulating the light with a motor driven filter, or by using three light sources and three DMD units.

Table 5.8

Some TV screen and monitor tube parameters (Philips)

Type	A36E	A59E	M78JU	M47E	M37-108X
Usage	TV	TV	TV	Monitor	Monitor
	Color	Color	Color	Black/white	Color
Screen diagonal [mm]	356	590	784	500	356
Acceleration voltage [kV]	23.0	25.0	27.5	20.0	25.0
Anode voltage [kV]	7.2	8.8	7.5	20.0	6.6
Lens voltage [V]	$310 - 650$	$575 - 825$	$425 - 885$	400	$300 - 800$
Max. current [μA]	1000	1500	1500	100	450
Deflection angle [°]	90	110	110	114	90
Distance between two red stripes [mm]	0.52	0.8	0.83		0.29
Glass transmission [%]	65	53	47.5	46	57

Related to monitor tubes is the **E**lectron-**B**eam-**A**ddressed **S**patial **L**ight **M**odulator. The tube has an electron gun producing a modulated electron beam. The beam is scanned across the surface of an electro-optical crystal, resulting in charge accumulation on the crystal surface. These charges change the refractive index of the crystal. When irradiated by a linear polarized laser light the crystal changes the polarization of the reflected light into elliptical. By passing through an analyzing plate the light reveals the image generated by the electron beam modulation. The response time of the crystal is ~ 100 μs, and the storage time is more than 24 hours. EBASLM converts serial signals into parallel output.

5.1.5.3. Radar Monitor Tubes

Radar monitor tubes are similar in construction to black/white monitor tubes, but their viewing area is circular. The difference is in the deflection system, where deflection coils generate a magnetic field which rotates synchronously with the radar antenna and a field which deflects the beam radially outward from the center. The radial deflection, fast compared to the rotation, shows the position of the radar echo. Orange phosphor with long persistence is used so that an echo can be seen on the screen until it is written over in the next rotation. Modern radar equipment use computer processing and display on a computer monitor tube. For air traffic surveillance, the computer also writes on the screen the aircraft coordinates, velocity and height, against a background of a stylized map of the terrain.

5.1.5.4. TV Camera Tubes

In a TV camera tube, seen in Fig. 5.35, a thin electron beam sweeps over a photoconductive layer. A system of optical lenses reproduces the object, the image of which is converted into an electric signal on a photoconductive layer. The front of the layer is connected to the positive DC source of 10–50 V. The layer has resistivity, which depends on the intensity of the incoming light, so depending on the local conductivity, the back of the layer will be charged more or less positive. An electron gun generates a beam with a very small diameter, which is focused by an axial magnetic field and deflected by horizontal and vertical deflection coils. The beam is swept over the photoconductive layer and neutralizes the positive charge (Fig. 5.36a). When the potential drops to zero compared with the gun cathode potential, the electrons turn because the back of the layer operates

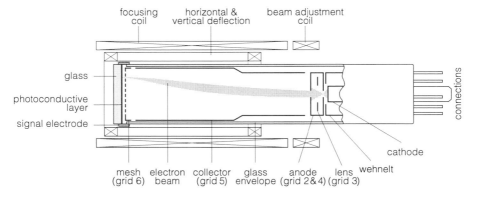

Fig. 5.35. Sketch of a TV camera tube (Philips).

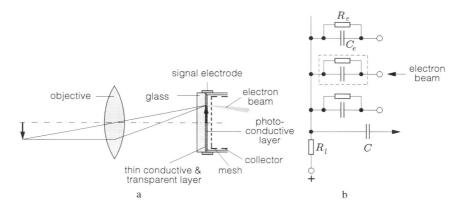

Fig. 5.36. TV camera tube (Philips). **(a)** Signal generation. **(b)** Equivalent circuit.

now as an electronic mirror. The electron current which hits the layer is thus a measure of the intensity of the incoming light.

The photoconductive layer has high resistance, roughly teraohms, which varies in different TV camera tubes. Small surface layer elements can thus be replaced with an equivalent circuit as in Fig. 5.36b — that is with the small capacitors, C_e, connected in parallel with light-dependent resistors, R_e. Using the electron beam the inner photoconductive layer potential is set to the cathode potential and the capacitors are charged to the DC voltage connected to the front of the layer. In the dark the layer acts almost as an isolator. Only a very small part of the C_e charge is lost between sweeps, and the electron current compensates the loss by "dark" current. Illuminated elements will lose a part of the charge through R_e. When the beam hits an element (about 10^{-7} s) the capacitor C_e is charged and the capacitive current to the signal electrode results in a voltage drop over the load resistor R_l connected via the capacitor C to an amplifier, providing the video signal.

Many TV camera tubes, which differ primarily in the composition of the photoconductive layer, are manufactured under various names: In the vidicon the photoconductive layer is antimonytrisulfide (Sb_2S_3); the newicon has a combination of two layers, zincselenite (ZnSe) and a mixture of zinc telluride (ZnTe) and cadmium telluride (CdTe); the silicon vidicon works with a monocrystal of silicon in which a few hundred thousand

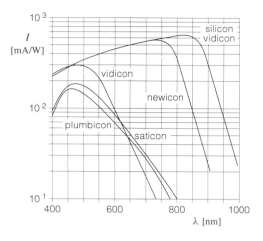

Fig. 5.37. Spectral sensitivity of a few TV camera tubes (Philips).

Fig. 5.38. Two Plumbicon tubes (Philips).

planar diodes are diffused; the plumbicon has a layer of lead oxide (PbO); and the saticon has a layer of selenium-tellurium-arsenide (SeAsTe). Even tubes sensitive to far infrared were developed using pyroelectrical layers (e.g., triglycinsulfate). Photoconductive layers are characterized by two parameters, "lag," which is reaction delay to a fast illumination change, and "blooming," which is a remnant charge on the inside of the photoconductive layer, which is not neutralized by the electron beam if the illumination is very strong. Table 5.9 gives some parameters of a few TV camera tubes. Figure 5.37 shows the spectral sensitivity.

"Lag" and "blooming" are not important for industrial and surveillance applications, but critical in TV broadcasting. For saticon and plumbicon tubes, which are used in broadcasting TV cameras, a construction change is made to reduce lag by reducing the photoconductive layer thickness to between 10 and 18 μm. The photoconductive layer in these tubes shows a small delay at normal or high illumination. However, when the illumination is faint, the large layer resistance cannot bring enough charge to the inner side. Thus the plumbicon solves this problem (Fig. 5.39), by placing behind the socket a small light bulb, which through a filter and an optical fiber feebly illuminates the inner side of the photoconductive layer. This results in weak current a few times larger than the dark current (4–8 nA), thus reducing the delay to below human perception.

The remnant charge depends on the beam current saturation. A local strong illumination of the photoconductive layer builds a large charge, which cannot be neutralized during one beam sweep. If the TV camera is moved, a comet-like tail appears on the TV receivers. To avoid such blooming, so-called anti-comet tail (ACT) guns were developed,

Table 5.9

Some TV camera tube parameters (Philips)

Type	XQ1240	XQ1276	XQ1410	XQ4187
Model	Vidicon	Newicon	Plumbicon	Plumbicon
Spectral region [nm]	300 − 800	300 − 1000	350 − 650	350 − 850
λ at max. sensitivity [nm]	550	750	450	430
Light sensitivity	Low	High	Medium	Medium
Diameter [mm]	16	11	30	18
Useful area [mm]	9.6×12.8	6.6×8.8	12.8×17.1	4.8×6.4
Resolution [TV-lines]	800	650	650	650
Voltage: grid 6 [V]	500	750	675	750
grid 4 [V]	300	750	600	250
grid 3 [V]	300	350	300	30
Signal [V]	45	20	45	20
Dark current [nA]	20	5	<2	<1
Normal illum. current [nA]	300	320	150 − 300*	25 − 100*
"Lag"	Medium	High	Low	Low
"Blooming"	Very low	High	Low	Low
Application	Surveillance, industry, X-rays, police, military	surveillance (infrared), industry, police, military	TV studio, fast processes, X-rays	portable TV camera (studio qual.)

* These tubes are manufactured for the three colors: red, green, and blue.

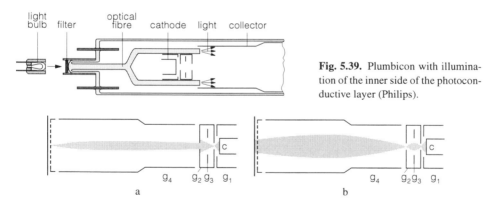

Fig. 5.39. Plumbicon with illumination of the inner side of the photoconductive layer (Philips).

Fig. 5.40. Electron beam in a plumbicon tube (**a**) during the sweep and (**b**) during the return of the beam (Philips).

like that of Fig. 5.40, which shows the ACT gun in a plumbicon. While the beam sweeps over the photoconductive layer it is focused by lenses (grid 2, 3, and 4) to a small spot on the layer (Fig. 5.40a). After a final sweep the beam returns to its initial position while grid 3, the central lens electrode, is pulsed negative, and grid 1, the current control electrode, is pulsed positive. The emitted current is increased to about 100 μA, and the beam is defocused to cover a much larger layer region (Fig. 5.40b). To keep all information stored on the layer from being erased, the cathode is simultaneously pulsed to +10 V. All electrons reaching regions where the potential is below 10 V are reflected, and the stored information is retained, except in highly illuminated regions.

Camera tubes are often now replaced by a semiconductor construction, the **C**harge **C**oupled **D**evice. For small video cameras, in industrial and in surveillance applications the CCD dominates, but for TV studio cameras the plumbicon and saticon are used. New camera tubes are developed for high-definition TV (HDTV). Plumbicon performances are improved: smaller beam diameter, better signal/noise, and improved modulation depth. A high-gain avalanche rushing amorphous photoconductor (HARP) saticon has a photo-electrode with four layers: a transparent signal electrode (SnO_2), a CeO_2 layer, a light-sensitive selenium layer, and an antimonytrisulfide (Sb_2S_3) layer which, during readout, prevents the electrons from entering the photoconductive layer. Both the signal/noise ratio and the sensitivity are improved.

Related to monitor tubes are **I**mage **D**issector **T**ubes and **M**icrochannel **S**patial **L**ight **M**odulators. In an IDT the electrons emitted from a photocathode are focused in the image section of the tube (constructed like an image inverter) to the aperture plate. The aperture plate has a small hole, $20-100\ \mu$m in diameter. Behind the aperture is a secondary electron multiplier with dynodes. The image is scanned across the aperture plate by deflection coils or electrodes. The process is fast and allows for random access of the desired pixel element of the image. IDTs are used in high-speed tracking, measurements of optical fibers, two-dimensional photon counting and random access cameras. MSLM is a channel plate image inverter with an electro-optical crystal as readout instead of a screen.

5.2. Electron-Optical Devices

5.2.1. Electron Microscopes

HISTORICAL NOTES. The first electron-optical image was made by Brüche [105] in 1932, and in 1933 the first transmission electron microscope was built by Ruska and Knoll [84]. Microscopes have been constructed with electrostatic and magnetostatic lenses. Initially most had electrostatic lenses, because magnetic lenses could not reach sufficiently short objective lens focal length with the then existing magnetic materials. These electrostatic lenses were, however, limited by the breakdown between the electrodes at high fields necessary to get short focal lengths. Today, commercial microscopes almost exclusively have magnetic lenses.

The first scanning electron microscope was built by von Ardenne [113] in 1938. In 1937, Müller [114] constructed the first field emission microscope; in a later model the positions of individual atoms on a tungsten tip were seen (Fig. 5.50).

The scanning tunneling microscope, constructed by Binning and Rohrer [115] in 1982, can image individual atoms on a surface, but it can also detect atomic binding between the atoms in a crystal lattice. Binning, Gerber, and Quate [116] in 1986 devised a similar device, which uses forces between the atoms instead of current tunneling to image individual atoms on the surface of a sample.

The resolution limit of a conventional optical microscope depends on the wavelength of the light and its diffraction. Two image elements (e.g., two parallel lines) have diffraction maxima separated by a distance, d, when

$$n\lambda = d \sin \varphi,$$

where λ is the wavelength, φ is the diffraction angle, and $n = 0, 1, 2, \ldots$. An image is formed when the zeroth- and first-order diffraction maxima reach the imaging plane and the diffraction angle must thus be larger than the microscope opening angle. Thus for $n = 1$, $60°$ opening angle, and the blue light ($\lambda \sim 400$ nm), the resolution of an optical microscope is $\sim 0.5\ \mu$m.

Table 5.10

Electron microscopes

Type	Property and application	Resolution
TEM Transmission microscope	Thermionic or field emission transilluminates the sample. Imaging with lenses for direct viewing, photograph of the screen, or amplification by MCP.	< 0.2 nm
STEM Scanning transmission microscope	Thermionic or field emission transilluminates the sample. The beam scans over the sample. Imaging with lenses. Direct viewing, photograph of the screen, or amplification by MCP.	0.2 nm
EEM Electron emission microscope	Emission of electrons with thermionic, secondary (electrons or ions), or field emission. Gives information about the structure of the surface, its chemical composition, or temperature distribution.	10–50 nm
IEM Ion emission microscope	Emission of positive or negative ions from the emitting surface with a primary ion beam of a few kilovolts in energy. Electrons are removed by magnetic field. Gives information on the surface structure or its chemical composition.	A few hundred nanometers
SEM Scanning electron microscope	Electron beam sweeps over the sample. Emitted secondary or reflected primary electrons are detected. The signal controls the current on a monitor tube. Deflection is synchronous with the deflection of the electron beam in the microscope. Depth of field a few millimeters.	~ 1 nm
Electron mirror microscope	Reflection of a slow (~ 10–50 eV) electron beam in front of the sample. Surface roughness influences potential distribution and the electron orbits of the reflected electrons.	A few hundred nanometers
FIM Field ion microscope	Acceleration of ionized gas atoms near the tip. Image of the atomic structure of the tip on a screen behinnd an MCP. Individual atoms can be "seen."	~ 0.1 nm
STM Scanning tunneling microscope	A fine tip of tungsten is swept over the sample. Tunnel current is measured. By piezoelectric rods the tunnel current is kept constant. Deformation of the rods is measured and the result is displayed on a computer screen.	~ 0.01 nm in height, ~ 0.2 nm lateral
AFM Atomic force microscope	Similar construction as in the scanning tunneling microscope. The force between the tip and the sample is measured.	~ 0.01 nm in height, ~ 10 nm lateral

Electrons accelerated to 1 MeV have a wavelength of 2.4×10^{-12} m. If an electron microscope has an opening angle of $1°$ (spherical aberration sets the limit), the resolution is 0.1 nm. Thus, much better resolution is obtained by an electron microscope.

The electron microscopes can be divided into two groups: those where the electrons pass through the sample, and those where they are reflected or generate secondary electrons on a small surface. A summary of electron microscopes and their properties and applications is given in Table 5.10.

Besides the electron microscopes of Table 5.10, there are additional variants like the scanning ion microscope, the environment scanning electron microscope, the Auger electron microscope, and others.

5.2.1.1. Transmission Microscope

The **T**ransmission **E**lectron **M**icroscope is the electron-optical analog to the conventional light microscope (Fig. 5.41). Here an electron beam with an energy between 20 keV

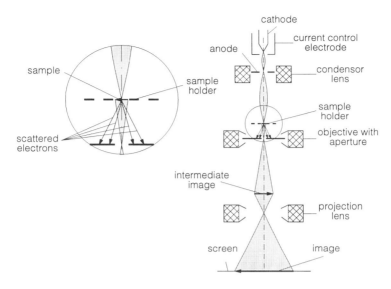

Fig. 5.41. Simplified diagram of a transmission electron microscope.

and 1 MeV passes through a sample whose thickness is between 10 and 100 nm. The illumination circle is between 0.1 μm and 3 mm in diameter.

The electrons are emitted from either a field-emitting cathode or a directly heated cathode made from either a thin V-shaped tungsten wire, possibly with a tip welded on the top of the "V", or a lanthanum-hexaboride crystal. A surrounding electrode (Wehnelt electrode) controls the cathode current. The cathode is connected to the negative side of the power supply while the acceleration electrode (anode) is connected to the positive side, often grounded. The condenser lens concentrates the electrons on the sample from which they scatter. The objective lens, which includes an aperture that limits the microscope opening angle, creates an intermediate image. This image is then magnified in a projection lens and displayed on the screen. Magnification depends on the partial magnification in each intervening lens whose individual magnification is (Eq. 4.44)

$$M = \frac{b}{a} \frac{f_1}{f_2}$$

Letting $a \approx f_1$, because the sample and the intermediate image both lie approximately in the focal plane of the lens, the total magnification is

$$M_T = \frac{b_1}{f_{2_1}} \frac{b_2}{f_{2_2}} \cdots \frac{b_n}{f_{2_n}}$$

Thus a large magnification indicates a short focal length (e.g., $f = 1.5$ to 2 mm), with all image distances roughly equal (~ 30 cm). Intermediate lenses can be added (see Fig. 5.43) and magnification of up to 1,000,000× can be obtained in commercial microscopes. The photographed image can then be magnified optically about 20 times, as seen in Fig. 5.42.

The microscope resolution is determined not only by the magnification but also by the ability of the human eye to resolve two lines (~ 0.2 mm) at the normal sight distance (~ 25 cm). The image is recorded on film with ~ 5 μm pixel size or TV camera

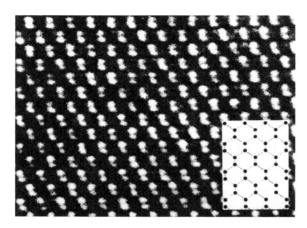

Fig. 5.42. High resolution image of silicon in [100] orientation showing dumbbells at the site of silicon atom pairs. The insert is the silicon structure. Magnification 24,500,000×. Distance between silicon atoms is 0.14 nm (Philips).

Table 5.11

Transmission electron microscope CM200 FEG (Philips)

High voltage	20, 40, 80, 120, 160, 200 kV		
Beam current	< 1 pA − 150 nA		
Diameter of the beam spot	0.4 − 2 mm		
Resolution: point	0.24 nm		
line	0.1 nm		
maximum	0.1 nm		
Magnification	25× until 1,100,000× in 39 steps		
Smallest imaging area	<0.2 μm		
Radial distortion	<1%		
Spiral distortion	<1.5%		
Curvature of field	< Resolution		
Chromatic error	No loss in sharpness for energy losses less than 1/1000 of acceleration voltage		
Stigmator for	Condenser, objective and diffraction lens		
	Stability		Drift
High voltage	1×10^{-6}/min		0.5×10^{-6}/min
Objective	1×10^{-6}/min		0.2×10^{-6}/min
Other lenses	10×10^{-6}/min		2.0×10^{-6}/min

with $\sim 20\times$ aftermagnification. At a magnification of 1,000,000× the TEM resolution is ~ 0.1 nm while the best optical microscope resolution is ~ 200 nm.

An optical microscope has an opening angle of $\sim 60°$ while electron microscope has an opening angle of $< 1°$, because it is impossible to correct aberrations, especially the spherical, in the latter. Consequently, the optical microscope depth of field limit is $\sim 0.2\ \mu$m, of the same order as its resolution, and thus it is easy to focus the image, because only a very thin slice of the sample is in view at a time. The rest is fuzzy and disappears from the image. The TEM depth of field is large (e.g., at a magnification of 20,000× a sample area of $\sim 5\ \mu$m^2 has a depth of field 0.2 mm); thus when part of the sample is correctly focused, the entire sample will also be correctly focused, creating clutter.

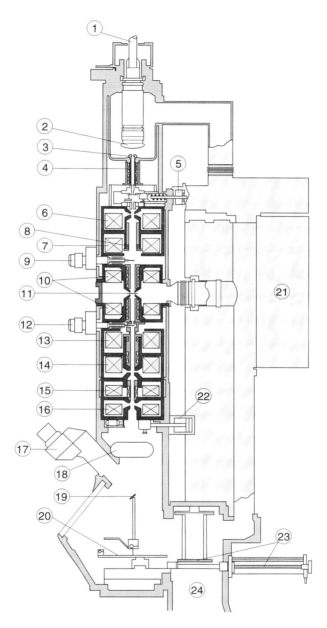

Fig. 5.43. Cross section of the Philips transmission microscope CM200. 1, High voltage cable; 2, cathode; 3, anode; 4, electron gun steering coils; 5, gun vacuum valve; 6, first condenser lens; 7, second condenser lens; 8, beam steering coils; 9, condenser aperture; 10, objective; 11, sample holder; 12, diffraction aperture; 13, diffraction lens; 14, intermediate lens; 15, first projection lens; 16, second projection lens; 17, binoculars (12×); 18, camera; 19, focusing screen; 20, main screen; 21, ion pump; 22, vacuum valve to magnetic components; 23, vacuum valve to pumps; 24, turbomolecular pump and roughing pump.

It is thus difficult to localize relative position in the direction of the beam. By applying a weak periodic deflection of the beam, or wobble, a change in the opening angle can be simulated, causing the image to oscillate. By adjusting the focusing parameters the image can be steadied, slightly out of the image plane, due to the spherical aberration of the image.

The interpretation of an electron microscope image is more difficult than that for an optical microscope, since the scattering processes in the sample must be understood. The most important scattering parameters are specimen density, thickness, and crystal structure (for crystalline materials).

Preparation of a sample depends on its material. Material science samples can be self-suspended in a sample holder or mounted on a fine metal mesh. Biological samples are either mounted on a metal mesh (made by electrolytic methods, or by laser or ion etching) or bonded with an agent (e.g., collodion) which allows them to be sectioned by microtome into thin layers as small as ~ 5 nm.

Parameters for a modern TEM are given in Table 5.11, and its cross section is shown in Fig. 5.43. These modern microscopes can work with the so-called dark field, by slanting the beam slightly so that the image is created only by electrons scattering at large angle. This radically changes the contrast, thereby easing the image interpretation. TEM used in the diffraction mode can examine the crystal structure giving a diffraction image (Laue diagram).

5.2.1.2. Electron Emission Microscope

In an **E**lectron **E**mission **M**icroscope the sample emits electrons either by heating or by photon or ion bombardment. If the sample itself emits, the temperature distribution over the surface or the emission from different parts of the surface can be studied. Ion beams are more often used, and the generated secondary electrons are focused by one or more projection lenses to form the image. The EEM is mostly used in metallography. Resolution increases with the electric field strength in front of the sample. It is easy to observe the crystal structure and foreign intrusions on the borders between the crystals. The image contrast results from the chemical composition of the sample, crystal orientation to the incoming ion beam, and the microscope optical axis (relief contrast). Models have been developed where the ion beam generates secondary ions that are imaged.

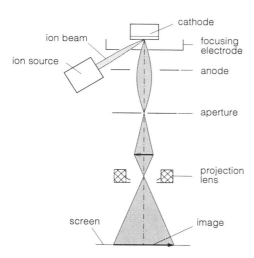

Fig. 5.44. Simplified diagram of a electron emission microscope.

Fig. 5.45. Photograph of a scanning electron microscope (Philips).

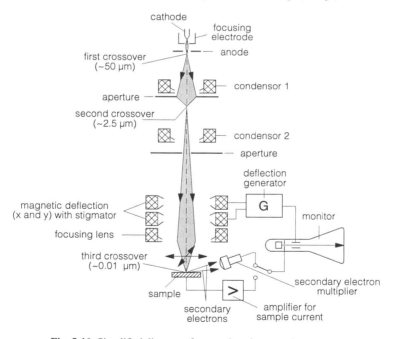

Fig. 5.46. Simplified diagram of a scanning electron microscope.

5.2.1.3. Scanning Electron Microscope

In a Scanning Electron Microscope a sharply focused electron beam is scanned over the sample by a pair of deflection coils. Where the beam hits the sample, secondary electrons are emitted, most with an energy of < 50 eV. Primary electrons with energies above 100 eV are also reflected, so it is possible to collect only secondary or primary electrons.

The slow secondary electrons provide topographic information about the structure of the sample, because the yield is proportional to the local tilt following the expression $y = 1/\cos\alpha$. A secondary electron collector is placed at a distance from the sample,

Fig. 5.47. Al_2O_3 spheres. Magnification $1650\times$ (Philips).

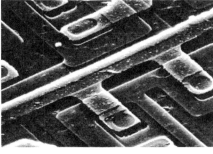

Fig. 5.48. Integrated circuit with (top) secondary electrons and (bottom) primary electrons. Magnification $1000\times$ (Philips).

and the electrons are collected by a voltage of ~ 200 V. The reflected primary electrons are less deflected, and they are detected typically by a solid-state detector mounted on the pole pieces of the last lens. Reflected electrons provide information on the local atomic number and are primarily two-dimensional maps of material density vs. position. The signals are amplified and modulate the beam current in a monitor tube. The electron-beam deflection is synchronous with the beam scanning in the microscope.

The SEM image comes from the interaction of the beam focus with the specimen. An extremely large depth of field results (Fig. 5.47) because the beam divergence, as defined by the focusing lens apperture, is on the order of milliradians. Samples some millimeters thick can be sharply depicted. The SEM image contrast originates in different secondary emission from different sample regions and from the sample surface contour and relief. Images show differences in chemical composition and the geometry of the sample. Some sample parts can, for example, be hidden for the collector and appear dark. Such samples can be imaged with a SEM, which are not convenient for a TEM, like very rough objects, biological samples (covered by a thin metal layer — often Pt or Au), or integrated circuits (Fig. 5.48). By measuring the secondary electron current, a transistor electrode can be seen either as high or low, as well as time variation of an analog signal detected. Advances in low-aberration SEM and higher-brightness electron sources opened the possibility for high-resolution **Low-Voltage** SEM, with beam energy of ~1–2 keV. For biological samples and integrated circuits the radiation damage, the beam-induced surface contamination, and the interaction volume are considerably reduced.

The fine-focused SEM beam can also generate X-ray emission from the sample surface, characteristic of the material being investigated (Section 5.2.2).

5.2.1.4. Field Emission Microscope

A fine tungsten tip is mounted in the center of a field emission microscope, which is sharpened by etching to a radius of ~ 100 nm. The tip surface is then leveled by brief strong heating. The tip is connected to the negative voltage of a few kilovolts, with the result that the electric field at the tip is so strong that electrons are emitted by field emission. The spherical screen around the tip produces a radial electric field at the tip. The field emission current depends on the orientation of the atoms on the surface of the tip, and it varies from point to point. A scattering image of the atom lattice is obtained on the screen with magnification up to 300,000×.

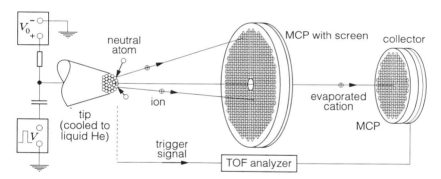

Fig. 5.49. Field ion microscope.

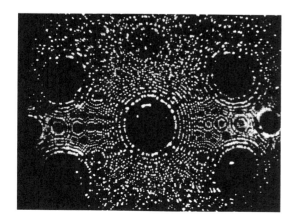

Fig. 5.50. FIM image of a tungsten tip. Pearl-like rings image the borders between crystal lattice of low order. [Photograph E. W. Müller, *Z. Phys.*, **131**, 136 (1955), © Springer-Verlag, with permission.]

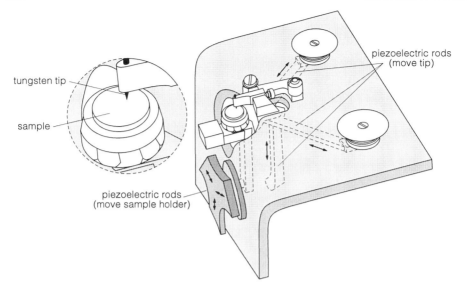

Fig. 5.51. Simplified diagram of a scanning tunneling microscope (Adapted from *The Scanning Tunneling Microscope*, by G. Binning and H. Rohrer. Copyright © August 1985 by Scientific American, Inc. All rights reserved).

It is possible to introduce low-pressure gas into the container and convert the instrument into a **F**ield **I**on **M**icroscope. The tip is connected to a positive voltage of a few kilovolts. The gas atoms near the tip then become polarized, are pulled toward the tip and are ionized near the crystal lattice atoms where the electric field is the strongest (Fig. 5.49). These ions then travel to an MCP with a screen and image the tip with a magnification of up to 1,000,000×. The center of the MCP can have a hole, to allow evaporated cations from the tip surface to reach a second MCP. In this device the tip voltage is additionally increased by nanoseconds pulses that allow mass analysis by a time-of-flight spectrometer. The first image of individual atoms was made with an FIM (Fig. 5.50).

5.2.1.5. Scanning Tunneling and Atomic Force Microscope

A very fine tungsten tip placed near (a few interatomic distances) a sample surface and connected to a voltage source will cause a tunnel current to flow between the tip and the surface. The tunnel current comes from the cloud of electrons, which envelop both the tip and the surface. The electron density decreases exponentially with the separation distance, and thus a slight change in separation distance will result in a dramatic change in tunnel current. The STM height resolution is ~0.01 nm, much less than the interatomic distance. Thus it is possible to "see" electron distribution of individual atoms and examine theirs bonds.

Figure 5.51 shows the construction of an STM, which is mounted on a rugged table, to which a plate is connected by three piezoelectric rods. These rods can be moved by connecting a suitable voltage. The sample is fixed on the plate extension. Three more piezoelectric rods control the position of another plate with a tungsten tip. Two of the rods scan electronically the tip over the sample. The third rod is moved to keep the tunnel current constant. All movements are controlled by the position change of the piezoelectric rods using electronics, which registers corresponding voltages. The scanning voltages, plotted on the screen, give the analog image of the tip movement (Fig. 5.52).

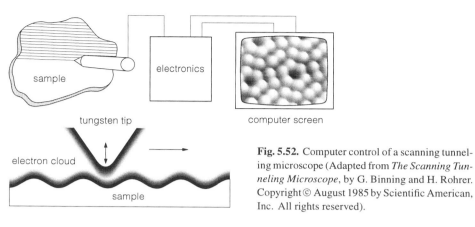

Fig. 5.52. Computer control of a scanning tunneling microscope (Adapted from *The Scanning Tunneling Microscope*, by G. Binning and H. Rohrer. Copyright© August 1985 by Scientific American, Inc. All rights reserved).

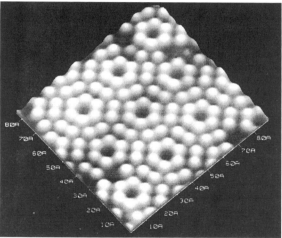

Fig. 5.53. STM image of a 7×7 reconstructed silicon (111) crystal surface. The bright protrusions correspond to atoms (12 per unit surface cell). Tunneling current 1 nA and sample bias voltage -2 V. (From M. Björqvist and M. Göthelid, Royal Institute of Technology, Stockholm, with permission.)

A very similar construction, the **A**tomic **F**orce **M**icroscope, uses its control system to monitor the force between a cantilever mounted tip and the atoms on the sample surface. This feeble force is measured by a mechanical spring. The vibrational frequency of atoms bound in molecules is $\sim 10^{13}$ Hz, while the mass is $\sim 10^{-25}$ kg, so the interatomic spring constant, $\omega^2 m$, is ~ 10 N/m. The cantilever, made of household aluminum foil 4 mm long and 1 mm wide, has a spring constant of ~ 1 N/m. Modern AFM cantilevers are made photolythographically from silicon, silicon oxide, or silicon nitride and have lateral dimensions of $\sim 100 \, \mu$m and a thickness of $\sim 1 \, \mu$m. Their spring constant is ~ 0.1–1 N/m. Small diamonds can be glued on the cantilever so that the position is detected optically, either by interferometry or by the deflection of the reflected light beam from a diode laser.

Because of the small force, 10^{-7} to 10^{-11} N, the tip can be in direct contact with the sample. AFM, in contrast to STM, can image isolators and organic substances. The latter is imaged by immersing the tip and the sample in water or ethanol to investigate, for example, proteins and DNA.

Coulomb or magnetic forces can also be used in microscopy. The Coulomb force microscope can image surface electric charge distribution, with sufficient sensitivity, in principle, to detect single electrons. The magnetic force microscope with a tip of a magnetized nickel, iron wire, or magnetic coated silicon cantilever can image magnetic devices (e.g., magneto-optical disks) even if they have a nonmagnetic coating. In both microscopes the resolution is ~ 20 nm.

5.2.2. Spectrometers

HISTORICAL NOTES. The first surface analysis was made in the 1920s by Davisson and Farnsworth [117], who studied secondary electron emission from different materials bombarded by electrons with energy of a few hundred volts. They observed electron diffraction, which was correctly interpreted by Germer and Davisson [80] as electron de Broglie waves [59]. Low-energy electron diffraction was the only way to investigate crystals and surface structures until the mid-1950s.

By 1925, Auger [118] had discovered, among the secondary electrons emitted from a bombarded surface, electrons whose energy was characteristic for individual elements. In 1952 Lander [119] asserted that Auger electrons could be used for surface analysis and in 1967 Harris [120] constructed the first Auger spectrometer.

Photoelectric emission, discovered by Hertz, caused Rutherford [121] to realize in the 1910s that photoelectrons give information about the chemical identity, but he lacked suitable detectors to measure the electron energy. In the 1950s a number of devices were constructed, which used X-ray electron sources for surface analysis [122], and in 1954 K. Siegbahn borrowed the technique from a double-focusing beta-spectrometer he had developed, to construct a high-resolution X-ray photoelectron spectrometer (XPS).

It was understood that even lower-energy photons can give interesting information, and devices were constructed which use an UV photoelectron source (UPS). Siegbahn and his collaborators realized also that with photoelectrons are Auger electrons and coined the phrase Electron Spectroscopy for Chemical Analysis for both UPS and XPS.

In 1910 J. J. Thomson [123] investigated ion emission from ion-beam-irradiated samples and could distinguish atoms of different masses in 1912. Dempster [124] and Aston [125] constructed, independently of each other, mass spectrometers in 1918, which helped in the discovery of the isotopes of elements. In the 1930s and 1940s, measurements of secondary ions were made, but the first useful instrument, based upon new and better ion sources, was built by Viehböck and Herzog [126] in 1949. Herzog and Leibl [127] then worked out the theory of ion optics, and in the mid-1960s the precursor of modern secondary ion mass spectrometer was constructed, and later the sputtering neutral mass spectrometer.

All these devices are in commercial production today, but there is a multitude of devices in scientific laboratories that use similar or equal photo- and secondary effects, or their combination, for surface analysis.

Interaction between a charged particle beam and matter can create photons or secondary electrons or ions. Similarly, electromagnetic radiation reacting with matter can create new photons or photoemission electrons. The energy of these radiations is often a characteristic of the element, and thus devices which use such interactions are called *spectrometers*.

Of the many spectrometers, we describe here only those that involve free charged particles, which interact with electric or magnetic fields. A summary of the most important commercially available spectrometers is given in Table 5.12.

Different spectrometers are suitable for different analyses. With some only the outermost layer of the sample can be investigated, others can give information at a larger depth, up to 1 μm, and still others can make an analysis of trace elements or measure isotope distribution in the sample. Spectrometry is one of the most important tools which gives qualitative and quantitative information about matter, independent of whether it is used in the laboratory, a technological process, or the environment.

Table 5.12

Spectrometers

Type	Property and application	Penetration	Resolution
EMA Microanalysis by electron beam	A fine-focused electron beam hits the sample and creates characteristic X-rays. Chemical analysis.	10 μm	100 nm
PIXE Particle induced X-ray emission	Primary high-energy ion beam from an accelerator creates X-rays which are analyzed. Analysis of trace elements.	10 μm	1 μm
ESCA Electron spectroscopy for chemical analysis . . . . UPS UV photoelectron spectrometer	Monochromatic UV radiation hits the sample and creates electron emission. Electron energy is analyzed. Surface analysis.	1 nm	5 μm
. . . . XPS X-ray photoelectron spectrometer	Monochromatic X-rays hit the sample and create electron emission. Electron energy is analyzed. Surface analysis.	1 nm	5 μm
AES Auger electron spectrometer	A small number of secondary electrons from a sample are Auger electrons. Spectrum gives information on the chemical composition in the surface layer of the sample. Surface analysis.	1 nm	5 nm
MS Mass spectrometer	Ions from an ion source are mass analyzed, often in connection with gas chromatography and liquid chromatography.		
SIMS Secondary ion mass spectrometer	Primary ion beam creates secondary ions whose mass is analyzed. Surface analysis.	3 nm	20 nm
SNMS Sputtering neutral mass spectrometer	Neutral particles by sputtering. Particles are ionized and pass through an electric and magnetic analyzing system where they are separated.	3 nm	50 μm
TOF Time-of-flight spectrometer	Monoenergetic ions pass a tube. The time of entrance and exit is measured. The flight time depends on ion mass.		
ITS Ion trap spectrometer	Electric and magnetic fields keep ions in a potential well. An RF modulation makes specific ions unstable depending on their mass.		
LEED Low-energy electron diffraction	Reflected primary low-energy electrons with wavelength adjusted to crystal lattice in the sample create an image on the screen.		
RMEED Reflection medium energy electron diffraction	Primary medium energy electrons tangentially hit the crystal. By an MCP/screen assembly, diffracted image is observed, highly sensitive to surface structure.		
RBS Rutherford backscattering spectroscopy	Primary ion beam hits the sample. Angular distribution of scattered ions is measured with respect to primary ions. Surface structure analysis.	30 nm	1 mm
CLS Cathode luminescence spectrometer	Low-energy electron beam hits the sample and creates characteristic luminescence spectrum.	10 μm	1 μm

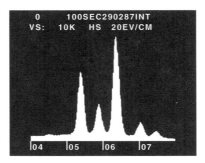

Fig. 5.54. Microprobe analysis of phosphor (Philips).

Many modern spectrometers can be used like an electron microscope, scanning over a sample surface and producing, after a computer analysis, a two-dimensional image of its chemical composition. Computers now make possible automatic analysis, storing a large reference set of spectral lines against which the measured spectra are compared.

All spectrometers can only obtain useful results from extremely clean samples. It is sometimes necessary to move the sample under vacuum from its place of manufacture to the instrument, which itself has ultra-high vacuum, 10^{-6} to 10^{-8} Pa.

5.2.1.1. Electron Microprobe Analysis

Electron Microprobe Analysis, often called the *microprobe,* is usually an accessory for a scanning electron microscope. In an SEM, when the primary electrons hit the sample, they release X-rays and luminescence (i.e., visible photons), which are detected by suitable detectors, and the energy of the X-rays or the wavelength of the light is measured. An SEM electron beam can be aimed at a specific place on a sample and the EMA brought in. Lower-magnification devices, especially built for EMA, are widely used in metallurgy and are aimed with a light beam and mirror.

EMA provides the chemical composition of the sample from the X-ray photon energy, determined by a crystal monochromator (see the description in the section on electron spectroscopy for chemical analysis). However, commercial devices usually use a proportional counter or a photomultiplier with output connected to a multichannel analyzer to obtain the spectrum directly. The wavelength and intensity of the emitted X-rays determine which chemical elements exist on the illuminated sample surface, with area $\sim 1~\mu m^2$ and depth $\sim 1~\mu m$. EMA is used mostly for elements with atomic number greater than magnesium, has a detection limit of ~ 50–100 ppm, and has an analysis accuracy of ~ 1 percent. Figure 5.54 shows an X-ray spectrogram from an EMA analysis.

In a similar process, the Proton-Induced X-ray Emission, protons from accelerators, with an energy between a few hundred kiloelectronvolts and a few megaelectronvolts illuminate the sample. PIXE is sensitive to the elements with low atomic number, has a detection limit of ~ 0.1 ppm, and an analysis accuracy of ~ 1 percent. Ion-Induced X-ray Spectroscopy, which uses α-particles or heavy ions, has the advantage that bremsstrahlung emission of photons decreases with increasing the bombarding particle mass, so the signal/noise ratio is improved.

5.2.1.2. Electron Spectroscopy for Chemical Analysis

Electron Spectroscopy for Chemical Analysis uses UV or X-ray beam to eject photoelectrons in a UV Photoelectron Spectrometer or X-ray Photoelectron Spectrometer.

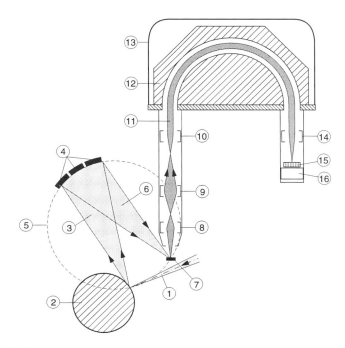

Fig. 5.55. X-ray photoelectron spectrometer (ESCA, type XPS). 1, Primary electron beam from an electron gun; 2, rotating water-cooled disk, coated with Al or Mg; source of 3, X-rays; 4, monochromator with quartz crystals; 5, Rowland's circle; 6, monochromatic X-rays; 7, sample; 8, first lens; 9, second lens; 10, third lens; 11, electron beam; 12, 180° hemispheric electrostatic analyzer; 13, mu-metal screen; 14, fourth lens; 15, channel plate; 16, detector. (After *Journal of Electron Spectroscopy,* P. W. Pamberg, **5**, 691, © 1974, with kind permission of Elsevier Science – NL, Sara Burgerhartstraat 25, 1055 KV Amsterdam, The Netherlands.)

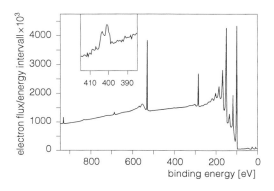

Fig. 5.56. X-ray photoelectron spectrum of silicon. The insert shows detail in the region around 400 eV energy. (Fysiska Institutionen, Uppsala University, with permission.)

In an XPS a variable-energy electron beam from an electron gun is focused on a rotating water-cooled disk, coated with aluminum or magnesium, from which a bundle of X-rays emerges and is then partially reflected by quartz crystals. The X-ray source, quartz crystals mounted on a spherical surface, and the sample lie all on the Rowland circle. Thus only incoming radiation with wavelength λ, which satisfies Bragg's law (e.g., $n\lambda = 2d \sin \varphi$, where n is an integer, d the crystal lattice constant), will be diffracted and reach the sample.

These monoenergetic photons release photoelectrons from the sample, among which are those from the upper layer that lose little energy in escaping. Such electrons, which will arrive at different detector elements depending on their momentum, are focused by a system of lenses and enter a 180° hemispherical electrostatic analyzer (Section 4.7.4). The analyzer channel energy is chosen by the electrodes voltage. After passing through a focusing lens, the electron current is amplified in an MCP before it hits the detector, a CCD camera if a sample image is sought, or a multichannel analyzer if an energy spectrum is needed. An XPS spectrum, shown in Fig. 5.56, has maxima which correspond to the electron binding energy of the surface layer atoms. Both XPS and UPS provide information about an ~ 1 nm layer on the sample surface, with resolution better than 0.3 and 1 percent respectively.

UPS photon source is a special gas-discharge lamp with emission in the UV region.

5.2.1.3. Auger Electron Spectrometer

In an **A**uger **E**lectron **S**pectrometer, electrons with a few kiloelectronvolts of energy bombard the sample, and among the emitted electrons a few are Auger electrons, seen in Fig. 5.57.

Here, an atom is ionized by collision ejecting an electron from an inner shell. This hole is filled by an electron from an outer shell by two processes. An electron from a higher shell falls into the shell with missing electron, and a photon with an energy $h\nu = E_K - E_{L_1}$ is emitted. Alternately, an electron from a higher shell can take the surplus energy and is emitted. For example, if an electron is missing in the K shell, an electron from the $L_{2,3}^{\star}$ shell can be emitted with a kinetic energy $E_K - E_{L_1} - E_{L_{2,3}}^{\star}$. Here $E_{L_{2,3}}^{\star}$ means that the energy in the $L_{2,3}$ shell does not correspond to the ground state but instead the one with a missing electron in the K shell. The end state is that the atom is missing two electrons.

The Auger emission probability decreases with increasing atomic number. For zirconium (Z = 40) it is 20 percent of that for lithium, so AES is used to investigate low-atomic-number elements except hydrogen and helium.

In an Auger spectrometer, shown in Fig. 5.58, an electron gun produces a beam with an energy of a few kilovolts that illuminates the sample. The electric field between the outer and the inner electrode only allows secondary electrons of a certain energy to pass.

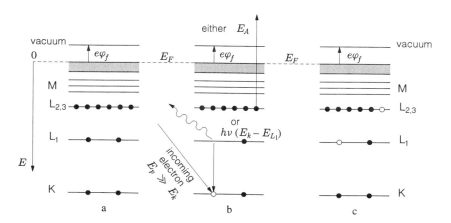

Fig. 5.57. Auger effect. (**a**) Initial state, (**b**) ionization and emission, (**c**) end state after Auger emission. (After *Practical Surface Analysis*, D. Briggs and M. P. Seah, © 1990, John Willey & Sons, with permission.)

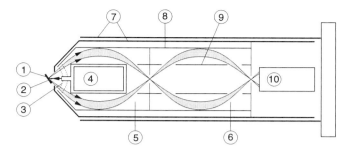

Fig. 5.58. Auger electron spectrometer. 1, Sample; 2, primary electron beam; 3, spherical retardation mesh (equipotential surface); 4, electron gun; 5, first electron-optical step; 6, second electron-optical step; 7, mu-metal screens; 8, outer electrode; 9, inner electrode; 10, detector. (After *Journal of Electron Spectroscopy and Related Phenomena,* P. Coxon et al., **52**, 821, © 1990, with kind permission of Elsevier Science – NL, Sara Burgerhartstraat 25, 1055 KV Amsterdam, The Netherlands.)

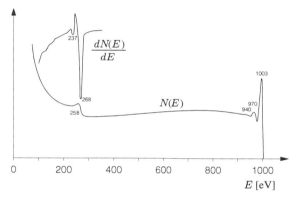

Fig. 5.59. Spectrum of carbon taken by an Auger electron spectrometer. (After *Practical Surface Analysis,* D. Briggs and M. P. Seah, © 1990, John Willey & Sons, with permission.)

To increase the electron-optical system sensitivity, double focusing is used. An MCP increases the detection sensitivity.

Changing the voltage between the inner and the outer electrode allows Auger electrons of different energy to reach the detector so that a complete spectrum can be obtained, as seen in the lower curve of Fig. 5.59. Secondary electrons produced by ~ 1 keV energy of the primary beam have a small peak near 250 eV from Auger electrons. The upper curve, the derivative of the lower, is used for interpreting Auger spectra. This detailed spectrum is obtained by modulating the electrode voltage by a weak AC voltage (a few volts at a few tens of kilohertz) and modulating the detected signal with the same frequency. The signal is amplified in a very narrow bandwidth lock-in amplifier. This improves the signal/noise ratio significantly and automatically gives the derivative. Resolution is about 0.3 percent.

5.2.1.4. Mass Spectrometer

In a Mass Spectrometer the ions from an ion source are analyzed in a mass filter to determine elemental composition of the sample. High-resolution mass filters can be of magnetic sector (often double focusing) type, quadrupoles, time-of-flight separators, and ion traps.

In a Secondary Ion Mass Spectrometer the sample is bombarded by monoenergetic ions (argon or cesium), with energy of a few kilovolts, so sputtering remains low and the sample surface changes only slowly. Secondary electrons and both positive and negative secondary ions are emitted. A double-focusing SIMS, shown in Fig. 5.60,

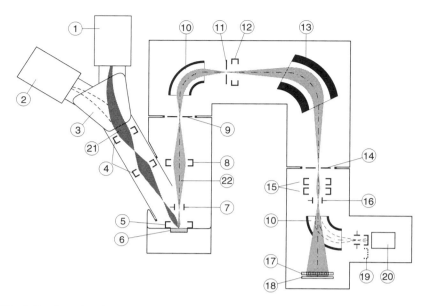

Fig. 5.60. Secondary ion mass spectrometer. Double-focusing model. 1, Duoplasmatron ion source; 2, cesium ion source; 3, mass filter for primary beam; 4, lenses; 5, immersion lens; 6, sample; 7, dynamic deflection system; 8, lens; 9, entrance slit; 10, electrostatic prism; 11, energy slit; 12, lens; 13, magnetic prism; 14, output slit; 15, projection lenses; 16, deflection system; 17, microchannel plate; 18, fluorescent screen or RAE detector; 19, movable Faraday cage; 20, electron multiplier; 21, primary ion beam; 22, secondary ion beam (CAMECA).

has a duoplasmatron ion source (see Section 10.6) for argon or cesium beam. The beam with a current of $\sim 5\ \mu A$ passes a mass filter and a lens system, which focus the beam on the sample. The sample is held at a potential of ~ 4.5 kV above ground. Secondary ions are extracted by a special immersion-type lens and are accelerated to ground potential, passing through the deflection system and a focusing lens. A 90° electrostatic prism accepts these ions, all of which do not have the same energy and therefore have different orbits. Ions with the same mass but different velocities are focused (Section 2.7.4) and pass through an energy slit, a lens, and a 90° magnetic prism. Ions with different masses pass the magnetic prism along different orbits and leave the prism separated by mass but focused on an output slit. When used as an ion microprobe, a small-diameter ion beam is scanned over the sample surface. Secondary ions are sequentially collected in a Faraday cup or electron multiplier so that the sample information comes from a defined position of the primary beam. As an ion microscope the large-diameter primary beam illuminates the sample surface. Secondary ions are emitted from the whole surface and displayed by stigmatic electron optics simultaneously, either on a fluorescent screen or by a **R**eactive **A**node **E**ncoder, a real-time pulse-counting, position-sensitive detector.

In 1958 Paul, Reinhard, and von Zahn [128] mass-separated ions using a long electric quadrupole, seen in Fig. 5.61, the opposite surfaces of which were connected, and a DC voltage with a superimposed AC voltage $2(U + V \cos \omega t)$ was applied. Here the potential on the symmetry axis is zero, and the two poles held at a potential of $+U$ create a well in which the positive ions reside. When these ions pass through the quadrupole, they are subjected to forces which change the sign with the phase of the AC voltage and the ion position.

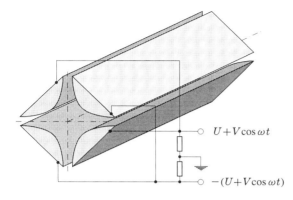

$U + V\cos\omega t$

$-(U + V\cos\omega t)$

Fig. 5.61. Sketch of quadrupole mass spectrometer.

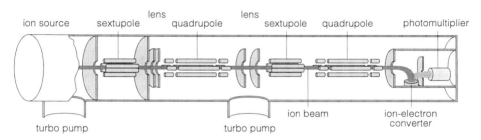

ion source   sextupole   lens   quadrupole   lens   sextupole   quadrupole   photomultiplier

turbo pump   turbo pump   ion beam   ion-electron converter

Fig. 5.62. Mass spectrometer. Quadrupole model. (Fisons Instruments.)

Thus during a part of the AC voltage period the ions will be on a potential hill — that is, in an unstable situation. Heavy ions are too inert to be influenced by the short instability. Light ions, however, will be lost after a few periods and can be collected on the electrodes. Thus this electrode pair acts as an "high-pass filter." The other electrode pair, connected to the potential $-U$, disperses positive ions and with the superposed AC potential light ions are conserved while heavy ions disperse after a few periods in the manner of a "low-pass filter."

Thus the electric quadrupole acts like a "band-pass filter," in which only ions with a certain mass range will experience stable periodic oscillations as they move down the quadrupole axis. If both U and V are varied but in a constant ratio, a complete spectrum over different masses will be the result of a voltage scan. The quadrupole-based SIMS is simpler than the double-focusing model and has an advantage in speed, since a complete spectrum can be obtained in milliseconds.

A modern quadrupole MS, shown in Fig. 5.62, uses a combination of three quadrupoles for mass separation. The short quadrupole sections mounted in front and behind the main units are connected only to the AC source, and act as strong focusing lens elements which reduce the defocusing effect of the DC fringing fields at the entrance and exit of the main quadrupoles. These instruments use different and interchangeable ion sources which in combination with gas or liquid chromatography can analyze organic compounds including proteins. The gas phase is ionized by electron beam ionization. The ionization of the liquid phase is performed by formation of charged droplets. These droplets desolve in smaller ones and emitt ions when they reach such a size that Coulomb repulsion between the ions in the droplet exceeds the surface tension. Liquid phase can also be ionized

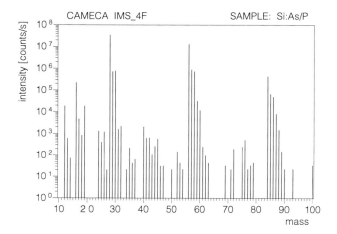

Fig. 5.63. Mass spectrum from a silicon phosphorus doped substrate obtained by 13.5 keV, Cs^+ sputtering ions. (Solid State Electronics, Royal Institute of Technology, Stockholm.)

by fast atom bombardment, like in a SIMS. The ions are extracted by electrostatic lenses and enter the MS. The sextupole element increases the fragmentation efficiency and maximizes the transfer of the ions from the source to the analyzer. After the quadrupoles separate, the ion beam enters the detector assembly, where it is deflected by a deflector plate and hits an ion–electron converter (a secondary electron emitter). The secondary electrons are focused by the deflector electric field and detected in a photomultiplier.

Double-focusing SIMS mass resolution is $M/\Delta M \geq 10^4$, or spectral lines separated by ~ 0.01 amu at $M = 100$ can be resolved. Isotopes below 1 ppm with respect to the dominating peak can be detected (Fig. 5.63). Quadrupole MS has a mass resolution of $M/\Delta M \sim M$. The mass range of the instrument extends to $m/z > 2000$. With multiple charged ions the effective mass range exceeds 100,000 amu.

The **T**ime-**O**f-**F**light spectrometer and the **I**on **T**rap **S**pectrometer also analyze particle masses. In a TOF spectrometer, short pulses of approximately monoenergetic ions enter a tube, and thus their velocity is mass-dependent. The time to pass through the tube will differ by ion mass, and thus a detector, which measures the time of arrival, will reveal the mass. The detector response ($\sim$ nanoseconds) determines the resolution. In an ion trap, combined electric and magnetic fields can keep ions in a potential well for times extending to months. ITS adds to the DC a variable RF voltage, which allows ions with higher and higher ratio of m/z to attain unstable oscillation amplitude and to leave the trap. These ions are counted, and by knowing the RF signal amplitude their mass can be deduced.

5.2.1.5. Electron Diffraction

Low-**E**nergy **E**lectron **D**iffraction was the first method used to analyze surfaces. In a LEED device, low-energy electrons from a small electron gun are sent toward the sample, as seen in Fig. 5.64a. The wavelength of these electrons must be approximately that of the sample crystal lattice constant. These electrons diffract near the sample surface and reach the viewing screen. Retarding grid stops electrons, which have lost part of their energy.

Reflection **M**edium-**E**nergy **E**lectron **D**iffraction uses higher-energy electrons, which hit the sample surface tangentially and are diffracted. On a screen, integrated with an MCP, the image is generated and viewed by a CCD on a monitor.

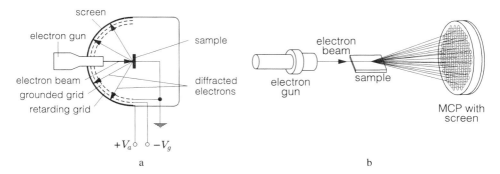

Fig. 5.64. (**a**) Low-energy electron diffraction. (**b**) Reflection medium-energy electron diffraction.

5.2.3. Electron Beam Technology

Electron beams can be focused into a small spot with high-power density. If such a beam hits a solid, it will quickly break all chemical bonds, thereby melting and evaporating the material.

There are many devices based on electron-beam technology, but all have the drawback that they must operate in a vacuum. Each uses some kind of electron gun, whose design depends on the appropriate perveance. Up to $p = 10^{-8}$ A/V$^{3/2}$ a triode gun similar to the ones in conventional CRTs is usual. For higher perveance it is necessary to employ Pierce-type guns (see Section 6.2) because of their higher current density. Tungsten or (for higher current density) LaB$_6$ is used as the cathode. These cathodes allow for air inlet when cold, without degrading the performance. The lenses are similar to those used in electron microscopy. A typical gun used in electron-beam technology, shown in Fig. 5.65, can give currents of over 10 A and an acceleration voltage of over 100 kV.

Electron beams are used for thermal material processing when other methods (e.g., gas discharges, electric arc, or heating by coal, oil, or laser) are not suitable — for example, in processing metals with high melting point, in thermal processing in vacuum, or when tolerances must be small. Laser and electron beams give the highest energy density

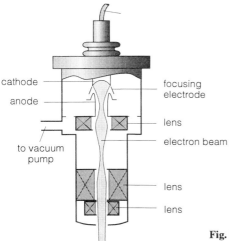

Fig. 5.65. Typical electron gun in an equipment for electron-beam technology.

Table 5.13

Electron-beam processing methods

Processing	P (W)	P/surface (W/m²)	V_a (kV)	d (mm)
Melting	10^5–10^7	10^3–10^4	20–50	10–50
Evaporation	10^3–10^6	10^3–10^4	10–40	2–30
Welding	10^2–10^5	10^5–10^7	15–180	0.1–5
Narrow beam	10–10^3	10^5–10^9	20–150	$5 \cdot 10^{-3}$–0.1

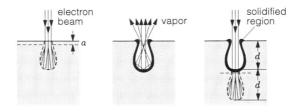

Fig. 5.66. Pulsed electron-beam processing. (After *Moderne Vakuumelektronik*, J. Eichmeier, © 1983, Springer-Verlag, with permission.)

of all processing methods. Down to 0.001 mm in diameter a power density of 10^9 W/cm² for electrons and 10^{10} W/cm² for laser beams can be attained. Table 5.13 reviews different electron beam processing methods.

An intensive local heating takes place when high-energy electrons hit a surface. If pulsed (Fig. 5.66), the electrons penetrate and heat the sample to some depth, d, leaving a thin uppermost layer, a, intact. Each successive pulse penetrates deeper into the solid, heats it so much that the material evaporates, and is thrown out. The heating occurs locally and the material solidifies between pulses. Continuous beams melt larger regions around the beam spot.

Electron-beam melting is used with high-melting-point materials (titanium, tantalum, niobium, molybdenum) and when a vacuum environment is required (e.g., alloyed steels). Figure 5.67a shows such equipment. Water cooling of the melt allows only a thin top layer to be liquefied, and many contaminations can be pumped away.

Electron-beam evaporation (Fig. 5.67b) is used to produce thin coatings on glass or plastic, in production of metal foils or thin film conductors in microelectronic for resistors and capacitors.

Electron-beam welding between two metals can be made to a depth of 15 cm at a speed of a few meters per minute. Thin foils can also be microwelded. Electron-beam welding is common in aerospace industries (safety) and when usual methods cannot be used (because of, for example, oxidation problems).

The high electron-beam energy can also be used to cut off thin layers and to polish, drill, mill, or engrave to predefine shapes. By the fast, simple, computerized deflection, tolerance of ~ 1 μm can be realized.

Electron beams are also used in zone melting (Fig. 5.68) to recrystallize material. A ring-shaped electron gun moves slowly along a rod. Near the gun the material melts, and it solidifies when the gun moves on.

Low-power electron beams, up to 100 W, are used in thin-film technology to produce resistance networks with tolerance of ~ 0.1 percent.

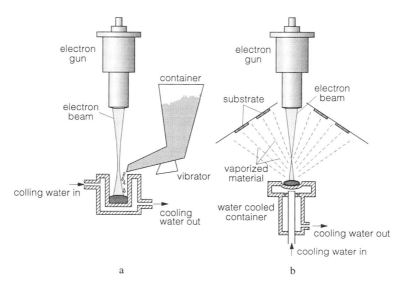

Fig. 5.67. Equipment for **(a)** melting and **(b)** evaporation with electron beam. (After *Electron Beam Technology*, R. Bakish, Ed., © 1962, John Wiley & Sons, with permission.)

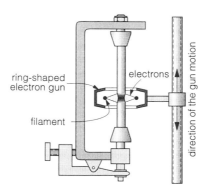

Fig. 5.68. Equipment for zone melting. (After *Electron Beam Technology*, R. Bakish, Ed., © 1962, John Wiley & Sons, with permission.)

5.2.4. Electron-Beam Lithography

Integrated circuit structure is transferred from a mask onto a photoresist layer by visible light, UV, or X-ray radiation. The radiation reaches the photoresist layer only in areas not covered by the mask. The exposed areas are then etched away (see Section 10.7.2) to produce the circuit. In contrast to photolithography, in electron beam lithography the electrons directly illuminate the photoresist layer.

A computer, equipped with a high-speed processor, generates the deflection signals and switches the beam on and off, by deflecting it into a knife-shaped aperture as seen in Fig. 5.69. The gun has an acceleration voltage of $\sim$ 20 kV, and a few computer controlled coils keep the beam centered. Two lens pairs focus the beam on the wafer, which is mounted on a movable table. A deflection generator controls two pairs of deflection coils for each coordinate axes. The table can also be moved under computer control,

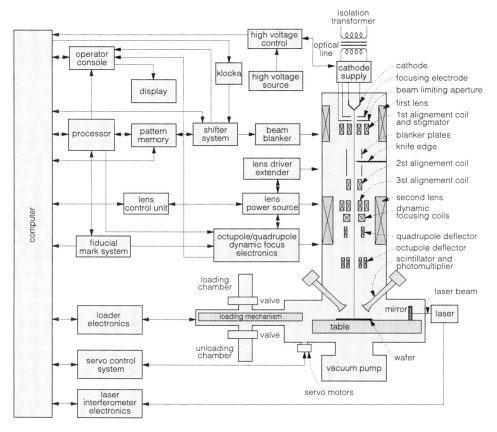

Fig. 5.69. Schematic picture of electron-beam lithography equipment (Hewlett-Packard).

the table position being read out by two laser beams, which by interference measure the distance to a mirror. The whole system is in high vacuum. At the wafer the electron beam diameter is ∼ 0.1 μm and current is ∼ 50 nA. To reduce aberration induced errors the electron beam scan is limited to a region of ∼ 2 × 2 mm while the wafer is a few centimeters in diameter. The table is therefore moved, and more than one circuit can be made on the same wafer in an "step-and-repeat" procedure.

Electron beam lithography is capable of better resolution and flexibility than photolithography. The resolution is about an order of magnitude better, with its narrowest line of 0.1–0.5 μm. CAD and programs to design the circuit, simplify production, so electron beam lithography is used in making small series of integrated circuits, in mask production, and in processes, which cannot be made by photolithography (e.g. production of vacuum micro-electronic devices). Disadvantage is an expensive machine, with low output compared to X-ray lithography, which has similar resolution.

The next miniaturization step in solid-state electronics will use quantum-mechanical behaviour of electrons, which needs nanometric structures. Using the SEM technology electron beam lithography can produce beams with < 5 nm diameter over a reduced surface ∼ 100 × 100 μm, small size mainly caused by deflection aberrations, Fig. 4.49.

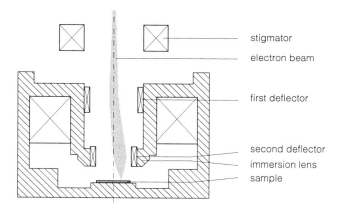

Fig. 5.70. Deflectors and immersion lens of a nanometer electron beam lithography equipment. [After *Adv. Electronics Electron Phys.*, Z. W. Chen, Nanometric scale electron beam lithography, **83**, (1992), with permission.]

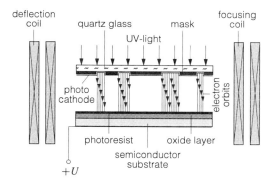

Fig. 5.71. Electron-beam copying. (After *Moderne Vakuumelektronik*, J. Eichmeier, © 1983, Springer-Verlag, with permission.)

These aberrations can be greatly reduced by using a lens with a second deflector placed inside the lens, which shifts, in the first approximation, the electrical centre of the lens. A short working distance brings about further improvement, thus the use of immersion lens. With a 4 nm beam diameter the working surface of the equipment shown in Fig. 5.70 is $\sim 250 \times 250 \ \mu m$.

Even direct copying by photoelectrons has better resolution ($\sim 0.8 \mu m$) than does photolithography with visible light. In electron-beam copying, chromium on a quartz glass mask is coated with palladium or CsI, to be a photocathode. When illuminated by UV light, the electrons are emitted only from places where it is transparent — that is, on the mask without chromium (Fig. 5.71). Photoelectrons are then guided by focusing and deflection (initial positioning) coils to the wafer, which is connected to a positive potential. A copy of the mask is made on the photoresist layer.

6

High-Perveance
Electron Beams

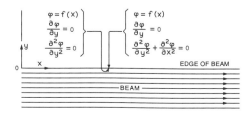

Boundary conditions of a parallel electron flow with plane boundaries. [From J. R. Pierce, *J. Appl. Phys.*, **11** (1940), with permission.]

6.1. INTRODUCTION

HISTORICAL NOTES. The first electron guns were designed for CRTs. They were all low-current guns with a perveance $\sim 10^{-10}$A/V$^{3/2}$. That an extrapolation of such designs to larger currents will not give the expected results became clear before the second world war. A high-current electron beam spreads because of the space-charge forces.

The breakthrough in the design of high-current beams came in 1940 in a work by Pierce [129]. He found the terms which establish correct boundary conditions for a two-dimensional beam. The electric field generated by these boundary conditions counteracts the space-charge and keeps a constant beam cross section between the cathode and the anode. However, the beam continues to spread after it passes the anode opening.

Brillouin [130] proposed the use of a longitudinal magnetic field. In this field any radial electron velocity results in an azimuthal motion, which gives a radial force directed opposite to the space-charge force. Appropriate strength of the magnetic field allows beam transport over any desired length. Brillouin proposal was later extended [131] to include different magnetic field distribution near the cathode, as well as periodic focusing with spatially alternating magnetic fields [132].

Electron beams are used in electron tubes and electron-optical devices. Those with low current density, where space-charge effects are neglible, can be designed using formalism of Chapter 4. However, when the current density is high, the repulsive forces between the charges distort the beam shape. In high-power microwave tubes, electron-beam currents up to a few hundred amperes must have small, unvarying diameter for decimeters to some meters along the length of the tube. The spread of the beam can be computed, but the lenses described earlier cannot prevent beam spreading caused by space-charge forces. Thus we assume a laminar beam (i.e., particle orbits never cross between the electron gun cathode and anode) and neglect the thermal velocity distribution and aberrations. This is justified from the mathematical formalism point of view when discussing the beam–field interaction.

Perveance, p, defined by Child–Langmuir's law (Eq. 2.31), is used to categorize electron guns:

$$p = \frac{I}{V^{3/2}}. \tag{6.1}$$

I is the emitted current and V is the gun acceleration, and thus an electron beam with a perveance of 1 μP can have a current of 1 mA at 100 V or 100 A at 215 kV. The perveance is defined by the electrode geometry, and in different electron guns it varies widely as seen in Fig. 6.1.

The beam spread increases with increasing space-charge. The line L/r_0 gives the distance along the beam where diameter will double because of its space-charge. For example, in a cathode ray tube with a perveance of $\sim 3 \times 10^{-10}$ P and a 0.2 mm beam diameter, the diameter will double after $\sim 1000\,r_0$ or ~ 200 mm. In a high-power microwave tube the perveance is $\sim 10^{-6}$ P and the beam diameter ~ 10 mm, so the beam travels ~ 120 mm before doubling its diameter. These tubes are ~ 500 mm long. Therefore some continuous focusing force is needed to overwhelm the beam spreading caused by space-charge forces.

The current density in modern high-power microwave tubes exceeds the maximum current density which can be obtained from a modern dispenser cathode by a factor between 10 and 100. This current density can be achieved by compressing the beam between the emitting surface and the region where it interacts with the high-frequency field.

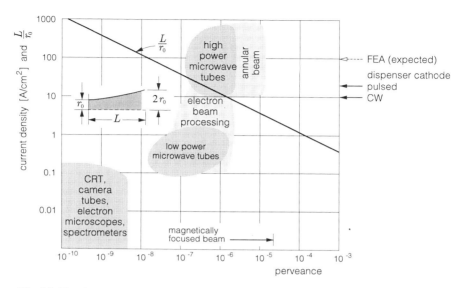

Fig. 6.1. The shaded areas are the perveance limits and the maximum current densities for various electron guns. The arrows show the upper current density limit for modern dispenser and field emission array cathodes.

The focusing is attained by shaping the electric field between the cathode and the anode so as to compensate for the space-charge forces in this region. An axial magnetic field in the later part of the tube provides the electrons with an azimuthal velocity, which, together with the axial magnetic field, generates a radial force, which compensates both the space-charge and centrifugal force.

6.2. HIGH-PERVEANCE ELECTRON GUNS

6.2.1. Pierce's Method

The first linear beam microwave tubes constructed in 1930s had electron guns, like those of cathode ray tubes, whose perveance, and even the current, was too low. Pierce [129] then developed the theory and the construction principles for high-perveance electron guns, which, with minor changes, is still used.

Assume a planar diode of Fig. 6.2a, infinite in the y direction, and compute the potential distribution along the x axis using Eq. 2.30:

$$V = \sqrt[3]{\frac{81}{32}\frac{1}{\eta}\left(\frac{J}{\epsilon_0}\right)^2}\, x^{4/3} = \text{const} \cdot x^{4/3}. \qquad (6.2)$$

If the cathode is made finite in the y direction the space-charge is reduced at the boundaries and the electron beam will expand when leaving the cathode (Fig. 6.2b). Pierce showed that by shaping the cathode and anode the same conditions inside the beam could be obtained as for an infinitely long diode (Fig. 6.2c). However, the electrons must travel in straight orbits normal to the cathode surface, so

- $dV/dy = 0$ at the edge of the beam and
- the potential inside and immediately outside the beam must be equal.

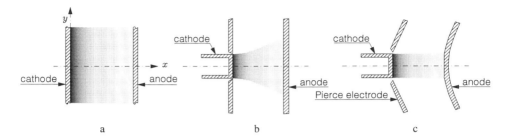

Fig. 6.2. (a) Space-charge-limited planar diode. The shading is proportional to the space-charge. (b) When the cathode is limited laterally the beam spreads because of the space-charge forces. (c) Pierce's construction of a planar diode with a focusing electrode which compensates the missing electron current.

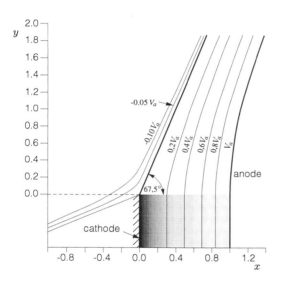

Fig. 6.3. Equipotential lines in a planar space-charge-limited electron beam according to Pierce's solution. One of the thick lines shows the shape of the focusing electrode, generally called *Pierce electrode,* the other the shape of the anode. The Pierce electrode is electrically connected to the cathode.

The potential inside the beam is given by Eq. 6.2, a solution of Poisson's equation, while immediately outside the beam it must be a solution of Laplace's equation, which Pierce showed to exist.

Using complex formalism, both the real and the imaginary part are solutions. Replacing in Eq. 6.2

$$x \longrightarrow \xi + j\zeta = r\,e^{j\varphi} \qquad \text{and} \qquad V \longrightarrow W = V_R + jV_I$$

we obtain

$$W = V_R + jV_I = \text{const} \cdot (\xi + j\zeta)^{4/3} = \text{const} \cdot r^{4/3}\,e^{j\,4\varphi/3}, \qquad (6.3)$$

where the real part is

$$V_R = \Re(W) = \text{const} \cdot (\xi^2 + \zeta^2)^{2/3} \cos\left[\frac{4}{3}\arctan\frac{\zeta}{\xi}\right]. \qquad (6.4)$$

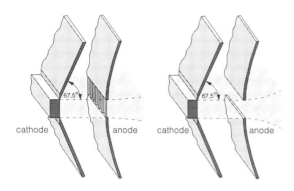

Fig. 6.4. Planar electron gun with Pierce's focusing electrode. (**a**) Construction with a grid. (**b**) Without the grid the lens effect increases beam spreading.

An equipotential line with $V_R = 0$ passes through $\xi = 0$ if

$$\cos\left[\frac{4}{3}\arctan\frac{\zeta}{\xi}\right] = 0,$$

where the slope is $(3/4)(\pi/2) = 67.5°$. Thus a planar electrode, connected electrically to the cathode, which makes the angle of $67.5°$ with the outermost orbit, creates near the cathode a potential distribution at the edge of the beam, which corresponds to the potential distribution given by Eq. 6.3. This focusing Pierce electrode in the planar case makes the electron orbits straight and forces them to be perpendicular when leaving the cathode. The anode determines the potential distribution of the cathode–anode region as a whole, and its shape must be such as to satisfy Eq. 6.4. A diode with these electrodes and some equipotential lines given by Eq. 6.4 is shown in Fig. 6.3.

Figure 6.4a shows a possible construction with Pierce's focusing electrode, with the anode having a grid to simulate an equipotential planar surface inside the beam. For high-perveance beams this solution cannot be used because the electrons will scatter from the grid, resulting in beam losses and heating. In another case, according to Fig. 6.4b the anode opening acts as a lens with properties of Fig. 4.14a. In both cases the beam will spread because of the space-charge forces after it has passed the anode. Planar cathodes with ribbon-like beams of Fig. 6.4 are therefore not used in practice. A rotationally symmetric cathode with a planar surface produces current density which is too low for many microwave tubes. By enlarging the cathode surface and constructing suitable electrodes a converging beam can be obtained. If a rotationally symmetric electron beam is desired, the cathode can be given a spherical shape. It is, however, impossible to use an anode satisfying Pierce's geometry because of the beam divergence caused by the lens effect. Figure 6.5 shows sketches of three high-perveance electron guns.

An annular beam can have higher perveance than a solid one, since the space-charge forces are appreciably reduced by the empty interior. Such electron guns achieve perveances which exceed 10^{-5} A/V$^{3/2}$. These guns use an axial magnetic field to launch the beam, described in Section 6.3.4.

6.2.2. Construction of High-Perveance Electron Guns

While in the planar case it is possible to analytically calculate the shape of the focusing electrode, it unfortunately cannot be done even in simple rotationally symmetric systems. Numerical attempts have also failed, because of instability near the cathode. But for all cathode shapes the focusing electrode should have the same $67.5°$ angle near the cathode,

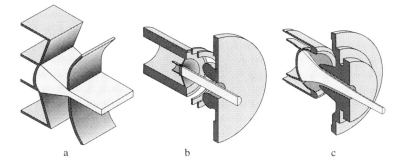

a   b   c

Fig. 6.5. Electron gun with **(a)** cylindrically convergent ribbon beam, **(b)** rotationally symmetric planar cathode giving confined flow, and **(c)** spherically convergent cathode for balanced flow.

because the cathode can be approximated by a plane surface in the immediate vicinity of the edge. This gives a more even current distribution over the whole cathode surface. By numerical computations, a suitable shape for the outer part of the Pierce electrode (Fig. 6.6), which depends also on the form of the anode, can then be found.

Approximations are used depending on the cathode shape: planar, cylindrical, or spherical. The current density is computed by solving Poisson's equation and finding an equipotential surface at a distance of 2–3 mesh units from the cathode. Langmuir and Blodgett [68,69] solved Poisson's equation in cylindrical and spherical coordinates for cylindrical or spherical cathodes, respectively. The current density in Cartesian, cylindrical, and spherical geometry is

$$J = -\frac{4\epsilon_0}{9}\sqrt{\frac{2e}{m}}\frac{V^{3/2}}{d^2} = -\frac{8\pi\epsilon_0}{9\gamma\beta^2}\sqrt{\frac{2e}{m}}V^{3/2} = -\frac{16\pi\epsilon_0}{9\alpha^2}\sqrt{\frac{2e}{m}}V^{3/2},$$

where

$$\gamma = \ln(r/r_k).$$

d is the distance between the cathode and the equipotential line, r_k is the radius of the cathode, and r is the radius of the equipotential line. β and α are given in Appendix M. In a high-perveance electron gun, the cathode does not cover the whole surface, so the angle subtended by the cathode as seen from the center of curvature is 2θ, and the result must be multiplied by θ/π for a cylindrical cathode and by $(1 - \cos\theta)/2$ for a spherical cathode.

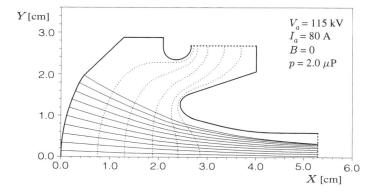

Fig. 6.6. Numerical computation of a high-perveance electron gun.

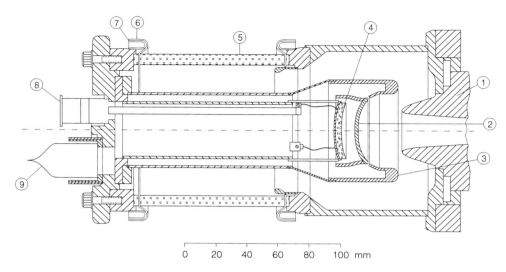

Fig. 6.7. Cross section of an electron gun for a high-power klystron (PK771, Royal Institute of Technology, Stockholm). 1, Anode; 2, cathode; 3, focusing electrode; 4, heater; 5, ceramic isolator; 6, metal–ceramic joint; 7, welding seam; 8, heater connection (the other connection is the cathode); 9, pumping tube.

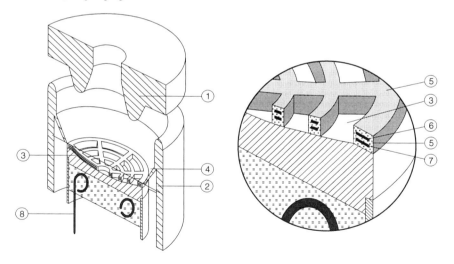

Fig. 6.8. Electron gun with control grid. Construction with a shadow and a control grid. 1, Anode; 2, focusing electrode; 3, cathode; 4, grid; 8, heater connection and heater wire baked in ceramic. Key for detailed sketch: 3, cathode; 5, layer of boron nitride; 6, control grid and 7, shadow grid; both grids are made of pyrolytic graphite. Graphite is coated by a thin layer of boron nitride for isolation and to reduce the emission from the grid.

Besides diode electron guns, with only cathode and anode, guns also can have a control grid to vary the emitted current as in a triode. For a control grid with a high negative potential with respect to the cathode, the current will be zero, independently of the anode voltage. If the grid potential is made less negative, or even positive, with respect to the cathode, the cathode begins to emit. Because the grid is often close to the cathode, high perveances can be realized while current is changed within wide limits at a constant anode voltage. In this construction, however, the grid is heated by electron bombardment, and the control requires relatively high power, because the grid current is finite, higher for larger cathode-to-grid distances. To avoid these disadvantages a shadow grid is included between the cathode and the control grid (Fig. 6.8), which limits emission from the cathode regions it covers and thus the control grid current remains low, even if its potential is positive.

Field emitter arrays, when available, will open a possibility to obtain much higher current densities as compared to conventional dispenser cathodes. The integral grid will allow pulse modulation at constant anode voltage, and while a high-current parallel beam can be designed without using spherical geometry, a much simpler magnetic confinement is obtained with the integral grid.

In high-perveance electron guns the anode must be constructed with a hole to allow the electron beam to pass. This changes the potential near the anode and the Pierce solution is no longer valid, even for a planar gun. The electrons will experience a radial force in the hole (e.g., in Figs. 4.14a and 6.4b). For a rotationally symmetric gun, the Eq. 4.62 approximation is satisfactory if the hole diameter is small compared to cathode–anode separation. In practice, the lensing is even stronger because of space-charge.

Thus when constructing high-perveance electron guns, one must place around the cathode a focusing electrode with the angle of 67.5°. This electrode must not be in direct contact with the cathode, because of thermal construction problems and losses through conduction. The extent of the focusing electrode must rather precisely hit the cathode edge and come as close as possible. However, thermal dilatation of the cathode holder, which may exceed 1 mm, will position the cathode differently at room and working temperature, which must be taken into account in the design. The cathode current density must be kept as constant as possible, which can be achieved by careful numerical computations. The anode hole influences the beam spreading and the current density.

6.3. MAGNETIC CONFINEMENT OF ELECTRON BEAMS

6.3.1. Beam Spreading

In a region free from external electric fields a *convergent laminar beam* attains a smallest diameter after which it begins to spread because of the repulsive forces between the charged particles. However, the orbits never cross each other. The beam modifies the potential in this region and thus the particle velocities are determined not only by the field distribution produced by the electrodes, but also by the beam shape and charge distribution. Among the important elements which determine the beam profile are the diameter of the beam as a function of the transport distance and current.

To estimate the influence of these factors we make some assumptions. We neglect the beam's nonlinear charge density over its cross section and its particle velocity distribution. We assume that the electrons have uniform axial velocity, that the beam is laminar, and that the divergence is small.

For a region limited by a conductive cylinder of length L, in accordance with Fig. 6.9a, we assume a convergent beam entering the cylinder, reaching its smallest diameter, $2b$, in the cylinder center. Because the beam divergence is small, the axial velocity, v_z,

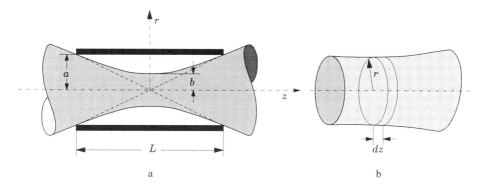

Fig. 6.9. (a) Spreading of a convergent high-current-density electron beam inside a conductive cylinder. (b) Integration region.

is aproximately constant. Applying Gauss' theorem to a small region inside the cylinder containing charge q and space-charge density ρ (Fig. 6.9b) we obtain

$$\oint \mathbf{E}_r \cdot d\mathbf{S} = \frac{q}{\epsilon_0}, \qquad \text{or} \qquad 2\pi r\, E_r\, dz = \pi r^2 \frac{\rho}{\epsilon_0} dz,$$

but

$$I_0 = -\pi \rho r_0^2 v_z,$$

so inside the beam (Fig. 6.10b) we have

$$E_{r_i} = -\frac{I_0}{2\pi \epsilon_0 r_0^2 v_z} r, \qquad (6.5)$$

and outside the beam obtain

$$E_{r_o} = -\frac{I_0}{2\pi \epsilon_0 v_z} \frac{1}{r}. \qquad (6.6)$$

The force on individual electrons is proportional to their distance from the beam axis, and we assume that no orbits will cross. Thus it is possible to determine the beam profile by following the orbits of the peripherial electron.

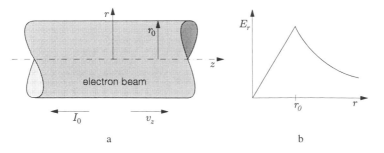

Fig. 6.10. (a) Definition of the coordinates. (b) Electric field distribution.

An electron in the beam experiences a force:

$$F_e = \frac{eI_0}{2\pi \epsilon_0 r_0^2 v_z} r.$$

The beam current generates a magnetic field. According to Ampères law we have

$$\oint \mathbf{B} \cdot d\mathbf{s} = \mu_0 I,$$

so

$$2\pi r B_\varphi = \mu_0 I = \mu_0 I_0 \left(\frac{r}{r_0}\right)^2, \qquad \text{or} \qquad B_\varphi = \frac{\mu_0 I_0}{2\pi r_0^2} r.$$

Electrons moving with the velocity v_z experience a Lorentz force

$$F_m = -\mu_0 \frac{eI_0 v_z}{2\pi r_0^2} r,$$

where the permeability, μ_0, is defined by

$$\mu_0 = \frac{1}{c^2 \epsilon_0},$$

so the Lorentz force can be rewritten as

$$F_m = -\frac{eI_0}{2\pi \epsilon_0 r_0^2 v_z} \frac{v_z^2}{c^2} r.$$

The total radial force acting on a beam electron is thus

$$F_r = F_e + F_m = F_e\left(1 - \frac{v_z^2}{c^2}\right). \tag{6.7}$$

In the nonrelativistic case, $v_z \ll c$ and the force generated by the own magnetic field is negligible. But as $v_z \to c$ the beam spreading ceases.

An electron at the beam periphery, $r = r_0$, feels in the classical approximation the radial field

$$E_r = -\frac{I_0}{2\pi \epsilon_0 r v_z}, \tag{6.8}$$

and its equation of motion is

$$m\frac{d^2 r}{dt^2} = -eE_r. \tag{6.9}$$

We make a coordinate transformation assuming that the axial velocity remains constant:

$$\frac{d^2}{dt^2} = v_z^2 \frac{d^2}{dz^2},$$

and

$$\frac{d^2 r}{dz^2} = -\frac{eE_r}{mv_z^2}.$$

If the electrons were accelerated by a voltage V before entering the region inside the cylinder, their velocity is given by Eq. 3.3:

$$v_z = \sqrt{-2\frac{e}{m}V} = \sqrt{-2\eta V}$$

and

$$\frac{d^2r}{dz^2} = \frac{I_0}{4\pi\epsilon_0 r \eta^{1/2} V^{3/2}}, \tag{6.10}$$

which describes beam spreading in a region free of external electric and magnetic fields. Multiplying both sides by $2dr/dz$ and integrating with $r = b$ and $dr/dz = 0$ for $z = 0$ (initial conditions for a completely parallel beam at $z = 0$), we get

$$\left(\frac{dr}{dz}\right)^2 = \frac{I_0}{2\pi\epsilon_0 \eta^{1/2} V^{3/2}} \ln\frac{r}{b}. \tag{6.11}$$

Define

$$\kappa = \frac{1}{2\pi\epsilon_0 \eta^{1/2}} = 3.034 \times 10^4 \left[\frac{V^{3/2}}{A}\right]$$

and

$$R = \frac{r}{b}, \qquad Z = \frac{z}{b}\sqrt{\kappa\frac{I_0}{V^{3/2}}} = \frac{z}{b}\sqrt{\kappa p},$$

where p is the beam perveance. After rewriting Eq. 6.11, we obtain

$$\frac{dR}{dZ} = \sqrt{\ln R}. \tag{6.12}$$

Further integration can only be achieved numerically and results in the *universal beam-spreading curve* of Fig. 6.11.

Knowing the beam perveance, beam diameter at a distance z can be found. The maximum current transmitted through a tube of diameter $2a$ is found by drawing a tangent to the curve through the origin. The quotient a/b is 2.35.

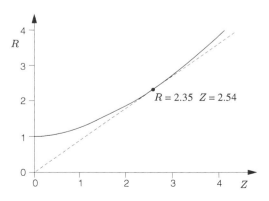

$R = 2.35 \quad Z = 2.54$

Fig. 6.11. Universal beam-spreading curve.

An approximate solution to Eq. 6.12 was given by Vibrans [133] as

$$Z = \int\limits_1^R \frac{dR}{\sqrt{\ln R}} \approx 2.09\sqrt{\frac{r}{b} - 1}. \tag{6.13}$$

In this idealized beam, spreading is caused only by space-charge. Real beams also have a divergence which depends on Maxwellian velocity distribution, nonuniform cathode current density, and aberrations of any lenses. Spreading will always be greater than that for an idealized beam.

EXAMPLE: In a high-power S-band klystron ($f \sim 3$ GHz, $U_a = 115$ kV, $L = 450$ mm) the beam current is transported through a 19 mm tube. For a convergent beam on entrance and the same diameter on exit, the beam current is limited to 1.5 A, but ~ 80 A is required in practice. So we need to find a way to confine the beam within the pipe.

6.3.2. Brillouin's Focusing

A high-perveance electron beam from an electron gun soon spreads quickly, so that many electrons hit the tube walls after exiting the gun. Therefore, some kind of focusing is necessary, usually using a constant or periodical axial magnetic field, with the choice depending on application. In stationary devices, solenoids are usually used; in low-power tubes and in satellites, permanent magnets with uniform or longitudinally periodic field are common.

In describing magnetic focusing of particle beams we assume that the beam cross section is circular, its flow laminar, and its current density at emission constant; it is focused by a rotationally symmetric axial magnetic field, $B = B(r, z)$, $B_\theta = 0$, and no external electric field exists, only the rotationally symmetric field from the own space-charge, $E_r = f(r)$, $E_z = E_\theta = 0$. The second assumption allows the computations to be limited to the peripherial beam particle, for which the equations of motion are Eqs. 1.74–1.76. Introducing

$$v_\theta = r\dot\theta, \qquad v_r = \dot r, \qquad v_z = \dot z$$

and rewriting, we have (for electrons)

$$\ddot r - r\dot\theta^2 = \eta(-E_r - r\dot\theta B_z), \tag{6.14}$$

$$r\ddot\theta + 2\dot r\dot\theta = \frac{1}{r}\frac{d}{dt}(r^2\dot\theta) = \eta(\dot r B_z - \dot z B_r), \tag{6.15}$$

$$\ddot z = \eta(-E_z + r\dot\theta B_r). \tag{6.16}$$

The vector potential of Eq. 1.19 can be introduced so that

$$B_r = -\frac{\partial A_\theta}{\partial z}, \qquad\qquad B_z = \frac{1}{r}\frac{\partial}{\partial r}(r A_\theta),$$

and

$$\frac{d}{dt}(r^2\dot\theta) = \eta\left[\dot r\frac{\partial}{\partial r}(r A_\theta) + r\dot z\frac{\partial A_\theta}{\partial z}\right]. \tag{6.17}$$

In the steady state the bracketed expression is a total differential (like Eq. 4.17),

$$\frac{dr}{dt}\frac{\partial}{\partial r}(r A_\theta) + \frac{dz}{dt}\frac{\partial}{\partial z}(r A_\theta) = \frac{d}{dt}(r A_\theta),$$

and Eq. 6.17 is

$$\frac{d}{dt}(mr^2\dot{\theta} - er A_\theta) = 0. \tag{6.18}$$

The parenthetical expression is the θ component of the generalized momentum of an electron in a magnetic field

$$\mathbf{p} = m\mathbf{v} - e\mathbf{A},$$

which is a constant of the motion.

Using the initial conditions at the cathode — r_k, $\dot{\theta}_k$, and $A_{\theta k}$ — we write

$$p_{\theta k} = mr_k^2\dot{\theta}_k - er_k A_{\theta k}.$$

With an absence of magnetic materials near the beam axis, Stoke's theorem can be used to compute the magnetic flux:

$$\Psi = \iint_S \mathbf{B}d\mathbf{S} = \iint_S \nabla \times \mathbf{A}d\mathbf{S} = \oint_C \mathbf{A}d\mathbf{s} = 2\pi r A_\theta. \tag{6.19}$$

Integrating Eq. 6.18 gives the angular velocity,

$$\dot{\theta} = \frac{\eta}{2\pi r^2}(\Psi - \Psi_k) + \frac{r_k^2}{r^2}\dot{\theta}_k. \tag{6.20}$$

For electrons starting at the cathode and neglecting Maxwellian velocity distribution, we put $\dot{\theta}_k = 0$, so that

$$\dot{\theta} = \frac{\eta}{2\pi r^2}(\Psi - \Psi_k). \tag{6.21}$$

The angular velocity an electron attains in a rotationally symmetric magnetic field only depends on its radial position and the magnetic flux which has passed through a flux tube defined by the orbit, and it is independent of the orbit shape between the cathode and the actual axial position. The expression of Eq. 6.21 is valid for all particle orbits under all circumstances between any two beam cross sections if the rotationally symmetric magnetic field is stationary and if the azimuthal electric field, E_θ, equals zero. The above is called *Busch's theorem* [134] and is illustrated in Fig. 6.12.

Busch's theorem is used to solve the motion of beams in rotationally symmetric magnetic fields. If the magnetic field at the cathode, B_k, is zero, the electron beam rotates after entering the magnetic field with the angular velocity,

$$\dot{\theta} = \frac{\eta\Psi}{2\pi r^2}.$$

If the magnetic field is homogeneous, $B(r, z) = B_0$, the magnetic flux is $\Psi = \pi r^2 B_0$ and Eq. 6.21 becomes

$$\dot{\theta} = \eta\frac{B_0}{2} = \frac{\omega_c}{2} = \omega_L, \tag{6.22}$$

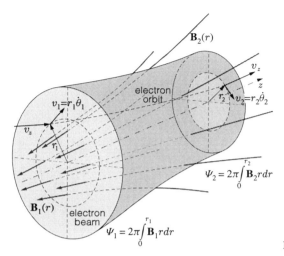

Fig. 6.12. Illustration of the Busch's theorem.

a result already obtained for the orbital rotation in a rotationally symmetric magnetic lens. All electrons rotate with the same angular velocity called *Larmor velocity*, ω_L, and ω_c is the cyclotron frequency, defined in Eq. 3.10.

A cathode of radius r_k will include the flux $\Psi_k = 2\pi r_k^2 B_k$. By defining two cyclotron frequencies

$$\omega_k = \eta B_k \qquad \text{and} \qquad \omega_s = \eta B_z,$$

where B_z is given by the magnetic field at z, Eq. 6.21 becomes

$$\dot{\theta} = \frac{1}{2}\left(\omega_s - \omega_k \frac{r_k^2}{r^2}\right). \tag{6.23}$$

Introducing this into the radial equation (Eq. 6.14) and observing that $\eta B_z = \omega_s$, we obtain

$$\ddot{r} = \frac{r}{4}\left(\omega_s - \omega_k \frac{r_k^2}{r^2}\right)^2 - \eta E_r - \eta \frac{r}{2}\left(\omega_s - \omega_k \frac{r_k^2}{r^2}\right)B_z.$$

By substituting E_r according to Eq. 6.8, we obtain

$$\ddot{r} = \frac{\eta I_0}{2\pi\epsilon_0 r v_z} + \frac{r}{4}\left(\omega_k^2 \frac{r_k^4}{r^4} - \omega_s^2\right). \tag{6.24}$$

The negative summand, if made large enough, will cancel the two positive terms and $\ddot{r} = 0$. In this case the electron beam maintains its diameter, because the force caused by the axial magnetic field and the beam rotation cancel the space-charge electrical force.

Brillouin [130] proposed this method to control high-perveance electron beams. For an electron gun with no magnetic field at the cathode, $B_k = 0$ and $\omega_k = 0$. So for the laminar flow of a peripherial electron $r = r_c$, we have

$$\ddot{r}_c = \frac{\eta I_0}{2\pi\epsilon_0 r_c v_z} - \frac{r_c}{4}\omega_s^2,$$

and the electron beam can be balanced if

$$\frac{\eta I_0}{2\pi\epsilon_0 r_c v_z} = \frac{r_c}{4}\omega_s^2 = \frac{r_c}{4}\eta^2 B_B^2, \tag{6.25}$$

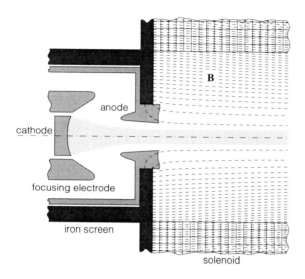

Fig. 6.13. Construction of an electron gun to start the Brillouin flow.

where B_B is called the *Brillouin field*. Brillouin proposed that the axial magnetic field abruptly starts after the beam passes the anode. Such an electron gun can be constructed (see Fig. 6.13) with the cathode and anode surrounded by an iron cylinder, and the solenoid is made with a plate as part of the yoke. This magnetic field will be very weak inside the electron gun, with the field lines in the solenoid starting mainly from the iron plate, so the field inside the solenoid is rather homogeneous.

The electron velocity inside the beam at radial position r can be expressed by an equivalent potential, V_r. To see the physical meaning of V_r and how the energy equation is balanced, assume that the beam electrons which pass through the anode hole have no radial velocity when they enter the magnetic field. In the transition from the field-free to the solenoidal field region there must be a radial magnetic field component, B_r. The magnetic force caused by v_z and B_r acts azimuthally by producing an azimuthal velocity, v_θ. The azimuthal velocity and the axial magnetic field, B_B, produce a force acting in the negative r direction. This Brillouin field can be adjusted so that the magnetic force exactly cancels the electric space-charge force and the centrifugal force; the beam will continue its motion through the solenoid, maintaining a constant radius.

This physical picture can be represented by mathematical equations if at every point the energy equation,

$$\frac{1}{2}m\dot{z}^2 + \frac{1}{2}mr^2\dot{\theta}^2 = eV_r, \tag{6.26}$$

is satisfied. Assuming that all electrons move on cylindrical surfaces parallel to the solenoid axis, we obtain

$$\ddot{r} = 0 = -\eta E_r - \frac{r}{4}\omega_s^2,$$

and the radial electric field must be

$$E_r = -\frac{r\omega_s^2}{4\eta} = -\frac{\partial V_r}{\partial r}.$$

Integrating we obtain

$$V_r = \frac{r^2\omega_s^2}{8\eta} + C,$$

which with $V_{r=0} = V_0$ for $r = 0$ becomes

$$V_r = V_0 + \frac{r^2\omega_s^2}{8\eta}. \tag{6.27}$$

Substituting Eq. 6.27 into Eq. 6.26, we obtain

$$\dot{z}^2 + r^2\dot{\theta}^2 = 2\eta V_r = 2\eta V_0 + \frac{r^2\omega_s^2}{4},$$

and comparing it with Eq. 6.23, where $\omega_k = 0$, it remains

$$\dot{z}^2 = 2\eta V_0. \tag{6.28}$$

So — all beam electrons, independently of their distance of the axis, have the same axial kinetic energy, despite their azimuthal velocity being proportional to the radius. They all have *the same axial velocity*. The electrons obtain the energy for their azimuthal velocity at the expense of the initial energy from their axial velocity. The *radial* electric force caused by the beam space-charge accelerates the electrons *azimuthally* with the help of **B** until the magnetic force at every radial distance r comes in equilibrium with the electric force.

The on-axis electric potential is V_0. For an electron at the beam periphery, $r = r_c$, the potential which corresponds to the kinetic energy of the electrons, V_{r_c}, is equal to the anode potential, V_a (assuming a beam in free space),

$$V_{r_c} = V_a = V_0 + \frac{r_c^2\omega_s^2}{8\eta},$$

and

$$V_0 = V_a - \frac{r_c^2}{8}\eta B_B^2. \tag{6.29}$$

The velocity with which the beam rotates always corresponds to the voltage difference between the electron radial position, r, and the axis.

At the beam periphery $V_{r_c} = V_a$ and so we express the Brillouin field by introducing Eq. 6.28 into Eq. 6.25, such that

$$B_B^2 = \frac{2I_0}{\pi\epsilon_0\eta r_c^2\sqrt{2\eta V_0}} = \frac{\sqrt{2}I_0}{\pi\epsilon_0\eta^{3/2}r_c^2\sqrt{V_0}},$$

or

$$B_B = 8.3 \times 10^{-4}\frac{1}{r_c}\sqrt{\frac{I_0}{\sqrt{V_0}}}. \tag{6.30}$$

The current which balances the Brillouin field is thus

$$I_0 = 1.45 \times 10^6 B_B^2 r_c^2\sqrt{V_0} \quad \text{[A]}. \tag{6.31}$$

Under the assumption that $V_0 \simeq V_a$ (i.e., the space-charge is not large, so that compensating does not require a big voltage drop), B_B or I_0 can be computed.

EXAMPLE: In the example in Section 6.3.1 it was mentioned that the required beam current is 80 A. If the beam diameter is 16 mm, according to Eq. 6.30 an axial magnetic field of 0.052 T is enough to compensate the space-charge forces and keep the beam diameter constant. Equation 6.29 shows that the potential on the axis is 3570 V lower than the anode voltage, and the approximation $V_0 = V_a$ is good enough.

Although the Brillouin flow can keep the beam size constant, even for high perveance, it is not used in microwave tubes. The assumption of uniform emission current density over the entire cathode cannot be realized in practice, so the space-charge in all beam cross sections, at all radii, will not be exactly compensated by the magnetic force: Electrons the same distance from the axis will not have the same exact angular velocity. Also, in any electron tube some residual gas atoms will be ionized and thus will partially neutralize the beam space-charge. This occurs randomly, and the beam will be noisy. Furthermore, in microwave tubes which use solid cylindrical electron beam, the current is velocity-modulated. The space-charge varies with axial position; it is not the assumed constant space-charge at the electron gun exit. While the Brillouin flow exactly compensates a constant space-charge, it cannot exactly do so for a variable space-charge of a velocity-modulated beam. In Sections 6.3.3 and 6.3.4 we discuss two other methods of beam balancing without introducing the problems associated with the Brillouin flow.

To compute the highest perveance obtainable by Brillouin flow we introduce in the expression for the current, Eq. 6.31, instead of V_0 its value according to Eq. 6.29,

$$I_0 = 1.45 \times 10^6 r_c^2 B_B^2 \sqrt{V_a - \frac{r_c^2}{8}\eta B_B^2}. \tag{6.32}$$

To find the maximum of Eq. 6.32 as a function of B_B we define

$$f = B_B^2 \sqrt{V_a - CB_B^2},$$

and the derivative is

$$\frac{df}{dB_B} = 2B_B\sqrt{V_a - CB_B^2} - \frac{1}{2}B_B^2 \frac{2CB_B}{\sqrt{V_a - CB_B^2}} = 0,$$

giving

$$CB_B^2 = \frac{2}{3}V_a.$$

Thus

$$B_B = \frac{4}{\sqrt{3\eta}r_c}\sqrt{V_a}$$

and

$$I_{0_{max}} = \frac{16}{3\sqrt{6}}\pi\epsilon_0\sqrt{\eta}V_a^{3/2} = 25.4 \times 10^{-6}V_a^{3/2}.$$

Under these assumptions the Brillouin flow admits for a very high perveance

$$p = 25.4 \times 10^{-6}[\text{A/V}^{3/2}],$$

but at a big price: The on-axis equivalent potential is only one-third the anode potential, and the axial velocity is only $\sim 60\%$ the velocity which corresponds to the acceleration voltage. In Fig. 6.1 the text "magnetically focused beam" refers to the Brillouin flow limit.

6.3.3. Confined Flow

To avoid Brillouin flow problems we can either choose *confined flow*, a constant magnetic flux along the whole beam, $\Psi_k(r_k) = \Psi_z(r_z)$, where r_k and r_z refer to the radius of any electron orbit, or allow for the field to vary from a minimum at the cathode to the maximum down the tube, $\Psi_k \neq 0 \neq \Psi_z$. In the simplest case the confined flow can be realized by a uniform magnetic field, which means that we replace ω_k with ω_s and require that at the beam periphery

$$\ddot{r} = \frac{\eta I_0}{2\pi \epsilon_0 r v_z} + \frac{r}{4}\omega_s^2\left(\frac{r_k^4}{r^4} - 1\right) = 0. \tag{6.33}$$

Physically, starting at the cathode the electrons experience a weak radial space-charge electric force which gives them a radial velocity, v_r. With the axial magnetic field, B_z, it gives an azimuthal force and imparts an azimuthal velocity, which, together with the axial magnetic field, results in an inward radial force, counteracting the beam spreading.

An equilibrium radius, r_m, is established at which

$$\ddot{r} = \frac{\eta I_0}{2\pi \epsilon_0 r_m v_z} + \frac{r_m}{4}\omega_s^2\left(\frac{r_k^4}{r_m^4} - 1\right) = 0. \tag{6.34}$$

Define

$$K = \frac{I_0}{\pi \epsilon_0 \eta r_k^2 B_z^2 v_z} = \frac{\eta J_0}{\epsilon_0 \omega_s^2 v_z}, \tag{6.35}$$

where J_0 is the current density and K is constant over the whole beam. Equation 6.34 can then be written as

$$\frac{r_m^4}{r_k^4} - 2K\frac{r_m^2}{r_k^2} - 1 = 0$$

and has solutions

$$\frac{r_m^2}{r_k^2} = K + \sqrt{1 + K^2}.$$

We choose the positive root since the beam diameter increases when the electrons leave the cathode. For a small increase $K \ll 1$ we obtain

$$r_m \approx r_k\left(1 + \frac{K}{2}\right). \tag{6.36}$$

In confined flow the electron beam does not keep a constant radius, the expanding beam (from space-charge) is bent around by the confining field and acquires an azimuthal velocity component, and this confinement fights the space-charge forces. It is convenient to define

$$r = r_m(1 + \delta), \qquad \delta \ll 1, \qquad \text{and} \qquad \delta = \delta(t).$$

Introducing this into Eq. 6.34 we obtain

$$r_m\ddot{\delta} = \frac{\eta I_0}{2\pi \epsilon_0 r_m v_z}(1 - \delta) + \frac{\omega_s^2}{4}\left[\frac{r_k^4}{r_m^3}(1 - 3\delta) - r_m(1 + \delta)\right]. \tag{6.37}$$

At r_m, Eq. 6.34 is

$$\frac{\eta I_0}{2\pi \epsilon_0 r_m v_z} = -\frac{r_m}{4}\omega_s^2\left(\frac{r_k^4}{r_m^4} - 1\right),$$

which, after insertion into Eq. 6.37 and simplification, results in

$$\ddot{\delta} + \frac{\omega_s^2}{2}\left(1 + \frac{r_k^4}{r_m^4}\right)\delta = 0, \tag{6.38}$$

whose solution is

$$\delta = C_1 \cos \kappa \omega_s t + C_2 \sin \kappa \omega_s t, \qquad \kappa = \frac{1}{\sqrt{2}}\sqrt{1 + \frac{r_k^4}{r_m^4}} \approx \sqrt{1 - K}.$$

At $t = 0$ the peripherial electrons start at the cathode with $r = r_k = r_m(1 + \delta)$, which, using Eq. 6.36, gives

$$\delta_{min} = \frac{r_k}{r_m} - 1 \approx -\frac{K}{2}.$$

The solution is thus

$$\delta = -\frac{K}{2}\cos\sqrt{1 - K}\omega_s t. \tag{6.39}$$

The largest and the smallest beam radius are

$$r_{max} = r_m(1 + \delta_{max}) \approx r_k(1 + K), \qquad r_{min} = r_m(1 + \delta_{min}) = r_k.$$

The radius of the beam evolves with time as

$$r = r_k(1 + \frac{K}{2})\left(1 - \frac{K}{2}\cos\sqrt{1 - K}\omega_s t\right), \tag{6.40}$$

and the angular velocity evolves as

$$\dot{\theta} = \frac{\omega_s}{2}\left(1 - \frac{r_k^2}{r^2}\right) = \frac{\omega_s}{2}K(1 - \cos\sqrt{1 - K}\omega_s t). \tag{6.41}$$

At $t = 0$ at the cathode, $\theta = 0$; thus we obtain

$$\theta \simeq \frac{K}{2}(\omega_s t - \sqrt{1 + K}\sin\sqrt{1 - K}\omega_s t). \tag{6.42}$$

r and θ describe the electron motion in the confined flow. The cycloidal electron orbits are composed of two motions: rotation around the axis with constant angular velocity $\omega_s K/2$ and a circular motion with the radius $r_k K/2$ (Fig. 6.14). The electron beam is perturbed in this "scalloping" phenomenon with a frequency $\sqrt{1 - K}\omega_s \approx \omega_s$ and wavelength (distance between two maxima) $2\pi v_z/\omega_s$.

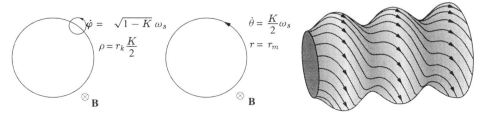

Fig. 6.14. "Scalloping" of electron orbits in confined flow.

The axial velocity of electrons is not constant in the confined flow. Starting with Eq. 6.34 we have

$$-\eta E_r + \frac{r_m}{4}\omega_s^2\left(\frac{r_k^4}{r_m^4} - 1\right) = 0$$

with $K \ll 1$

$$\frac{r_k}{r_m} = 1 - \frac{K}{2},$$

and the radial electric field becomes

$$E_r = -\frac{\partial V}{\partial r} = -\frac{r}{2\eta}K\omega_s^2.$$

Integrating with $V_{r=0} = V_0$ we obtain

$$V_r = V_0 + \frac{r^2}{4\eta}K\omega_s^2. \tag{6.43}$$

The energy equation

$$\dot{z}_r^2 + r^2\dot{\theta}^2 = 2\eta V_r,$$

with $\dot{\theta} = \omega_s K/2$, becomes

$$\dot{z}_r^2 = 2\eta V_0 + \frac{r^2}{4}K\omega_s^2(2 - K) \approx 2\eta V_0 + \frac{r^2}{2}\frac{\eta J_0}{\epsilon_0 v_z}. \tag{6.44}$$

At the beam periphery we have $r_m \approx r_k$ and

$$V_{r_m} \approx V_a = V_0 + \frac{J_0}{4\epsilon_0 v_z}r_k^2. \tag{6.45}$$

Comparing the potential distribution (Eq. 6.43) and the axial velocity (Eq. 6.44) we see that velocity and potential change with r follow each other, and both are parabolas (Fig. 6.15).

Thus in a confined flow the electron radial position determines its axial velocity. However, since we choose $K \ll 1$, the difference of the potential and velocity from the beam axis to edge is small. We define ΔV as the difference between the acceleration voltage and the axial potential by

$$\frac{\Delta V}{V_a} = \frac{J_0}{4\epsilon_0 V_a v_z}r_k^2, \tag{6.46}$$

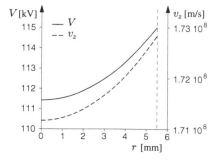

Fig. 6.15. Potential and velocity distribution with the radial position, r. $U_a = 115$ kV, $I_a = 80$ A, $r_k = 5.5$ mm. Relativistic value for the axial velocity is assumed.

which is a few percent. At the same perveance, however, $\Delta V/V_a$ is the same as that for Brillouin flow, and in both cases the potential drop is caused by the same space-charge forces which induce the axial energy loss. To satisfy $K \ll 1$ the magnetic field for the confined flow must be larger than that for the Brillouin, $B_z \gg B_B$.

For an electron beam in a conducting drift tube with radius r_D, the electric field is modified by mirror charges. Usually the drift tube potential equals the anode potential and

$$\frac{\Delta V}{V_a} = \frac{J_0}{4\epsilon_0 V_a v_z} r_k^2 (1 + 2 \ln \frac{r_D}{r_k}).$$

Thus to start a confined flow beam with a small scallop amplitude, we must keep $K \ll 1$. The current density is also limited by the cathode; ~ 5 A/cm^2 for an oxide, ~ 10 A/cm^2 for a dispenser in CW, and ~ 25 A/cm^2 in pulse regime. Therefore, confined flow is used in low-power microwave tubes, traveling wave tubes, and BWO tubes with power up to a few tens of watts. The magnetic field is 1.5 to 3 times larger than the Brillouin field. FEA cathodes, with current density ≥ 100 A/cm^2, might introduce a change in favor of confined flow even in high-power tubes.

6.3.4. Balanced Flow

In high-power microwave tubes the electron beam is compressed between the cathode and anode to achieve the required current density, so the magnetic field cannot be kept constant over the tube length. For example, in a 3.5 MW pulsed klystron, the cathode diameter is 45 mm, and the beam diameter in the drift tube is only 11 mm. The beam must be balanced so that it has an almost constant radius in the drift tube, with some possible scalloping. We must vary the magnetic field, $B = B(z)$, from a minimum at the cathode to $B = B_0 = $ const in the drift tube.

An electron gun which produces a balanced beam, shown in Fig. 6.16, has a spherical cathode. Its beam is compressed up to $z = z_0$, where the magnetic field becomes approximately constant. The solenoid, which generates the magnetic field, has pole plates on both sides of the yoke so that the magnetic field diverges outside the solenoid. The cathode, which encompasses the flux Ψ_k, has its field modified by a bucking coil.

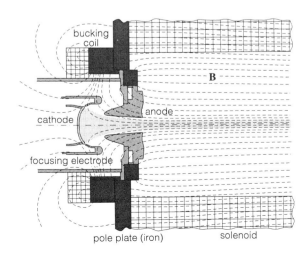

Fig. 6.16. Construction of an electron gun for balanced flow.

Note that the magnetic field lines cross the electron beam in the gun region (Busch's theorem), and so the electron beam gets the necessary rotation.

To determine the conditions to start a balanced beam in the drift tube, we write, using Eq. 6.21,

$$r^2 \dot{\theta} = \eta \frac{r^2}{2} B_0 + C,$$

with the initial conditions where the beam enters the homogeneous drift tube magnetic field

$$z = z_0, \qquad r = r_0, \qquad B = B_0, \qquad \omega_s = \eta B_0, \qquad \dot{\theta} = \omega_0,$$

and

$$C = r_0^2 \left(\omega_0 - \frac{\omega_s}{2} \right).$$

The resulting equation is

$$r^2 \dot{\theta} = \frac{\omega_s}{2} r^2 + r_0^2 \left(\omega_0 - \frac{\omega_s}{2} \right), \qquad \dot{\theta} = \frac{\omega_s}{2} \left[1 - \frac{r_0^2}{r^2} \left(1 - \frac{2\omega_0}{\omega_s} \right) \right]. \tag{6.47}$$

Introduce $\dot{\theta}$ in the radial differential equation (Eq. 6.14):

$$\ddot{r} = -\eta E_r - r \frac{\omega_s^2}{4} \left[1 - \frac{r_0^4}{r^4} \left(1 - \frac{2\omega_0}{\omega_s} \right)^2 \right]. \tag{6.48}$$

Like Eq. 6.35, use

$$K = \frac{I_0}{\pi \epsilon_0 \eta r_0^2 B_0^2 v_z},$$

and use Eq. 6.8, which gives

$$\eta E_r = -\frac{K}{2} \frac{r_0^2}{r} \omega_s^2.$$

The radial equation then becomes

$$\ddot{r} = \omega_s^2 \left\{ \frac{K}{2} \frac{r_0^2}{r} - \frac{r}{4} \left[1 - \frac{r_0^4}{r^4} \left(1 - \frac{2\omega_0}{\omega_s} \right)^2 \right] \right\}. \tag{6.49}$$

To have all forces cancel at $z = z_0$, use

$$\ddot{r} = \frac{K}{2} - \frac{1}{4} \left[1 - \left(1 - \frac{2\omega_0}{\omega_s} \right)^2 \right] = 0, \qquad \left(\frac{\omega_0}{\omega_s} \right)^2 - \frac{\omega_0}{\omega_s} + \frac{K}{2} = 0,$$

whose solution is

$$\omega_0 = \frac{\omega_s}{2} [1 \pm \sqrt{1 - 2K}], \tag{6.50}$$

where the solution with the negative sign must be used to obtain the correct magnetic flux. If also $\dot{r} = 0$, the beam would be completely stable. It is difficult to satisfy both $\ddot{r} = 0$ and $\dot{r} = 0$ at z_0. We must proceed numerically and some scalloping results. Figure 6.17 shows the result of such a computation made for the same klystron gun as in Fig. 6.6. The figure shows deliberately a much greater scalloping than the best result which can be obtained.

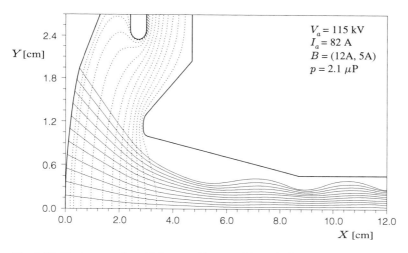

Fig. 6.17. Electron gun for a high-power klystron. Magnetic field balances the beam spreading but causes scalloping.

Busch's theorem can here be used to find the magnetic flux which the cathode must encompass. The difference between the flux at the cathode and the flux at $z = z_0$ must be

$$\Delta \Psi = \frac{2\pi}{\eta} r_0^2 \omega_0. \tag{6.51}$$

At the position where the homogeneous field starts, the flux is

$$\Psi_{z0} = \pi r_0^2 B_0. \tag{6.52}$$

Introducing Eq. 6.50 into 6.51 and using Eq. 6.52 the necessary flux at the cathode should be

$$\Psi_k = \Psi_{z0} \sqrt{1 - 2K}. \tag{6.53}$$

Even if it cannot be expected that the magnetic field over the whole cathode surface will be correct, its mean value should be

$$B_k = \frac{r_0^2}{r_k^2} B_0 \sqrt{1 - 2K}. \tag{6.54}$$

EXAMPLE: We continue with the example from Sections 6.3.1. and 6.3.2. With reasonable losses in the copper a magnetic field on the order of 0.1 to 0.2 T can be obtained. If we assume $B_0 = 0.15$ T, $I = 80$ A, $U_a = 115$ kV, $r_k = 22.5$ mm, $r_0 = 8$ mm, and relativistic v_z, then $K = 0.044$, and the mean magnetic field at the cathode should be ~ 0.018 T.

These computations only hint at how to design an electron gun and magnetic field to produce a balanced beam, because they describe the reality only in rough outline. The cathode current density varies over the surface, and so the space-charge electric field will also vary and a balance between all forces cannot be expected. It is uncertain if a laminar flow can be obtained. Maxwellian velocity distribution theoretically should not influence the beam; but practical measurements show that current density varies along the tube length

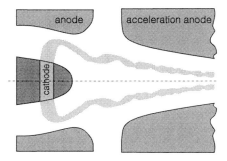

Fig. 6.18. Construction of a magnetron-type electron gun.

with varying cathode temperatures. In microwave tubes with velocity modulation the space-charge varies in space and in time, being much higher where the electrons are bunched. All these effects result in a local unbalance between the electric, magnetic, and mechanical forces which directly influence the beam and increase its scalloping.

With the balanced flow it is possible to achieve perveances of a few microperv. Modern microwave tubes (e.g., the gyrotron) need a beam with higher perveance; in these cases an electron gun with an annular beam is used, drastically reducing the space-charge forces.

The balance conditions for an annular and a solid beam are the same: The peripherial electrons must link a correct magnetic flux to generate the right azimuthal velocity to compensate the space-charge forces. The innermost electrons must link no flux at all, or else they would spiral toward the axis. It is difficult to make an annular beam with a spherical cathode which emits only along a circular segment.

With a magnetron-type gun, seen in Fig. 6.18, this problem can be avoided. The name comes from the magnetron, a microwave tube where a similar cathode construction is used. The gun has a cylindrical cathode, inclined a few degrees, surrounded by a conical anode, with an angle of $5-15°$. The whole gun is immersed in an axial magnetic field so that the electrons move outwards while a rotation of the beam is initiated. The anode voltage and the magnetic field are chosen so the radial deviation remains small and a thin beam is obtained. The cathode and anode slope produce an axial electric field which forces the electrons to move in the direction of the narrow cathode tip, where they glide away and thus form an annular beam. The electric field between the cathode and anode is shaped so that the emitted current density over the cathode surface is as constant as possible.

When the electron beam enters the drift tube the innermost electron must have no azimuthal velocity, because there is no space-charge interior to the beam. Busch's theorem states that the inner beam radius, r_0, must be equal to the radius of the cathode tip if the field is constant, or $\Psi_k(r_{0_k}) = \Psi_z(r_{0_z})$ in a $B = \mathrm{f}(z)$ field. The peripherial electron must, on the other hand, obtain a sufficiently large azimuthal velocity to cancel the space-charge and thus must link some magnetic flux. Therefore either the largest cathode radius must be smaller than the outer beam radius or the magnetic field must be inhomogeneous along the cathode surface.

Since an annular beam is more sensitive to perturbations than is a solid cylindrical beam, it is important to obtain as constant a current density along the cathode surface as possible. A local disturbance, like the one caused by an asymmetry in the cathode or anode construction, or an inhomogeneity in the space-charge distribution, can produce an instability which gradually builds up (Fig. 6.19).

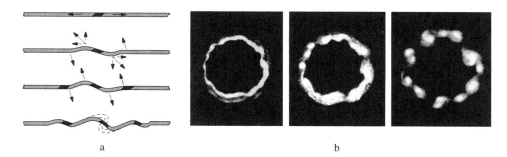

Fig. 6.19. Instability in an annular beam. **(a)** Forces acting on a thin annular beam with (in the axial direction) inhomogenous space-charge, causing beam instability. **(b)** Break up of an annular low-energy (80 eV, 0.058 T) beam because of space-charge waves. [From R. L. Kyhl and H. F. Webster, *IRE Trans El. Dev.*, **ED-3**, 172 (© 1956 IEEE), with permission.]

6.3.5. Periodic Focusing

High-power microwave tubes are large, so only solenoids are used to focus the beam; a 1 MW, 350 MHz CW klystron is about 4.75 m long. Solenoids require large excitation power and are not suitable for mobile units or satellites. Permanent magnets, which give homogeneous field, are heavy. But they can be made of magnets, coupled north-to-north pole and south-to-south pole, to get a periodic focusing, shown in Fig. 6.20, which can compensate the space-charge-induced beam spreading. The weight is decreased by $1/N^2$, where N is the number of magnets in the periodic structure.

The magnetic field varies along the axis in the first approximation as a cosine curve:

$$B_z = B_0 \cos \frac{2\pi z}{L}, \tag{6.55}$$

where B_0 is the field maximum and L the magnetic cell length. In the radial equation (Eq. 6.14) the angular velocity $\dot{\theta}$, expressed by Eq. 6.22, can be inserted, where B_z from

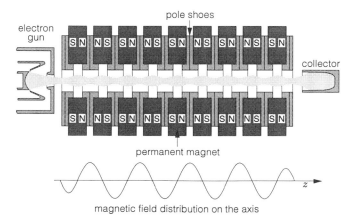

Fig. 6.20. Periodic magnetic focusing of an electron beam.

Eq. 6.55 replaces the constant magnetic field B_0. The resulting equation is

$$\ddot{r} = -\eta E_r - \frac{\eta^2 B_0^2 \cos^2 \frac{2\pi z}{L}}{4} r. \tag{6.56}$$

To compute the focusing properties, the time derivative $\ddot{r}$ is replaced by the space derivative r'' while the radial electric field is replaced by Eq. 6.8, giving

$$r'' + \left(\frac{\eta B_0}{2v_z} \cos \frac{2\pi z}{L}\right)^2 r - \frac{\eta I_0}{2\pi \epsilon_0 v_z^3} \frac{1}{r} = 0. \tag{6.57}$$

A suitable transformation of the variables can be made:

$$\sigma = \frac{r}{a}, \qquad T = \frac{2\pi z}{L},$$

$$\alpha = \frac{1}{2}\left(\frac{\eta B_0 L}{4\pi v_z}\right)^2 = 2.79 \times 10^8 \frac{B_0^2 L^2}{V_a},$$

$$\beta = \frac{\eta I_0 L^2}{8\epsilon_0 \pi^3 v_z^3 a^2} = \frac{385 I_0 L^2}{V_a^{3/2} a^2},$$

where a is the beam radius upon entering the periodic field and v_z is replaced by $\sqrt{2\eta V_a}$. Equation 6.57 can then be written as

$$\frac{d^2\sigma}{dT^2} + \alpha(1 + \cos 2T)\sigma - \frac{\beta}{\sigma} = 0. \tag{6.58}$$

Under the assumptions used in deriving the Brillouin flow, this equation can be solved numerically.

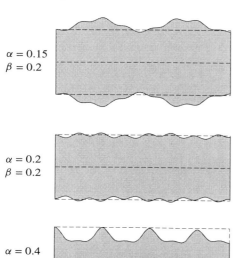

$\alpha = 0.15$
$\beta = 0.2$

$\alpha = 0.2$
$\beta = 0.2$

$\alpha = 0.4$
$\beta = 0.2$

Fig. 6.21. The shape of the electron beam in a periodic focusing structure as a function of the magnetic field parameter α. (**a**) Magnetic field too low. (**b**) Correct field. (**c**) Magnetic field too high.

Figure 6.21 shows the result of such a computation for different magnetic fields. Optimization is achieved if $\alpha = \beta$, so

$$B_{rms} = B_0/\sqrt{2} = 8.3 \times 10^{-4} \frac{1}{a} \sqrt{\frac{I_0}{\sqrt{V_a}}} = B_B,$$

the amount of the Brillouin field, since the radial force which compensates the space-charge spreading is proportional to B_z^2 and in a sinusoidal field with B_{rms} the mean value of the field is just the Brillouin field.

Numerical computations also show that periodic focusing exhibits stop bands. If the magnetic field strength is increased from zero, the electron beam becomes unstable for certain field values, becoming stable again at a still higher field strength.

6.4. COLLECTOR

After a magnetically focused high-perveance electron beam has done its work, the electrons must be accumulated on a collector, usually not a part of the microwave circuit. Such a collector is shown in Fig. 6.22.

As a special electrode, the collector must be large enough to allow for a low current density and local heating. For example, in a high-power klystron or traveling wave tube, the mean power can be many tens of kilowatts. By using a suitably shaped collector, the secondary and some of the primary electrons, which have lost energy in their interaction with the electromagnetic field, can be kept from entering the drift tube in the wrong direction. This is achieved by terminating the focusing solenoid yoke in an iron plate, so there is a large divergence of the magnetic field lines in front of and inside the collector.

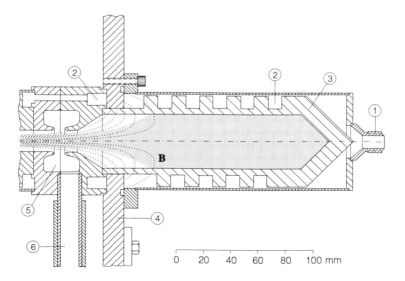

Fig. 6.22. Cross section of a collector of a high-power klystron (PK771, Royal Institute of Technology, Stockholm). 1, Water cooling entrance; 2, water channel; 3, collector; 4, iron plate terminates the yoke of the focusing solenoid; 5, last cavity; 6, waveguide.

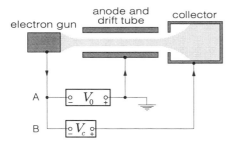

Fig. 6.23. Electrical connection of a microwave tube with depressed collector.

All devices which use high-perveance electron beams and high acceleration voltage generally have the last anode and the drift tube connected to ground and have the cathode connected to the negative supply voltage. A collector constructed according to Fig. 6.22, usually used in stationary units, is electrically connected ground — to the anode and drift tube potential — so many electrons hit the collector at high velocities. The electron kinetic energy is converted into heat on the surface of the collector and the power deposited can be ~ 50 percent, or more, so these collectors are normally water-cooled.

The collector can be connected to a separate power source, so it can have a lower potential than the beam potential inside the tube. The power losses can be decreased and the efficiency increased, which is especially important in mobile units and satellites, where the power must be conserved. If the collector is at a negative potential to the drift tube, a *depressed collector*, the electron velocities will be decreased when they hit the collector, and so the power losses are also decreased. Part of the power used to accelerate the electrons is recovered. Figure 6.23 shows a typical depressed collector circuit where the cathode–anode voltage, V_0, and cathode–collector voltage, V_c, are $|V_c| < |V_0|$.

Power supply A provides the acceleration voltage to only a small fraction, δ, of the beam current, I_0, namely the fraction intercepted by the drift tube because of scalloping and large radial displacement caused by modulation. Power supply B, with its lower voltage,

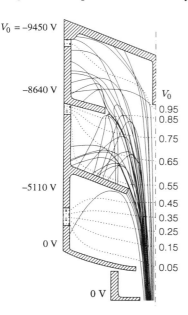

Fig. 6.24. Equipotential lines and electron orbits in a depressed collector with three steps [H. G. Kosmahl, *Proc. IEEE*, **70**, 1325 (© 1982 IEEE), with permission.]

supplies almost the whole current, $(1-\delta)\,I_0$. Power supply A must have good stability and relatively low power, while B must provide high power but with low stability. Together, both supplies deliver less power than would A alone, because energy is recovered, but the circuit becomes more complicated.

If electrons did not lose the energy in the tube, most of the power could be recovered, like in the electron cooling devices for accelerator storage rings. Here beams of a few amperes and a few hundred kiloelectronvolts are delivered by power supplies with currents of a few milliamperes, and collector efficiencies of more than 99.99 percent are obtained.

In microwave tubes, part of the electron-beam kinetic energy generates microwave power, so many electrons hitting the collector will have velocities lower than when they passed the anode, the reduction depending on their phase when passing the circuit where the microwave power is generated. With a depressed collector, the current to the collector and to the drift tube varies as collector voltage is changed. When $-V_c$ is decreased, at some voltage the electrons will start to be reflected, and the collector current decreases at the expense of the drift tube current, which overloads the power supply A. This limits the collector voltage and power saving. Thus a multistep collector can be constructed as in Fig. 6.24. Such collectors are used in microwave tubes, especially traveling wave tubes in satellites and in mobile military devices, where recovered power is important. Such tubes can have electrical efficiency exceeding 80 percent, compared to a conventional traveling wave tube with normal collector, which is ~ 20 percent.

7

Microwave Tubes

The first klystron (Varian).

7.1. INTRODUCTION

Microwave tubes generate or amplify high-frequency electric signals in a region from a few hundred megahertz to a few hundred gigahertz (i.e., wavelengths between 1 m and 1 mm). The division of the microwave region into frequency bands is given in Appendix N.

Transistors have replaced microwave tubes in low-power-level, low-frequency applications. But at high power and high-frequency the microwave tubes are now the only alternative because of the physical limitation in dissipating the generated heat in a semiconductor junction, which is frequency-dependent, at 1 GHz $\sim$ 1 kW and at 100 GHz less than 1 W. Current microwave tubes at 3 GHz produce a peak power of $>$ 20 GW, and at 100 GHz a few megawatts.

Microwaves themselves are used extensively in radar, communications, satellite control and communication, accelerators, medical equipment, microwave ovens, nuclear fusion, and military jamming and steering devices. Many different types of microwave tubes were developed: CW tubes with power up to 1 MW, pulsed tubes with power up to 20 GW,

Table 7.1

Properties and applications of microwave tubes

Tube type	Tube	Property and application
Classical electron tubes	Triode Tetrode	f < 4 GHz, low power, low price
Linear beam	Klystron	f < 50 GHz, narrow bandwidth. CW: medium amplification, medium power, TV transmitter Pulse: large amplification, large power, radar, accelerators
Linear beam	Reflex klystron	f < 250 GHz, low power, oscillator
Linear beam	Traveling wave tube	f < 100 GHz, medium amplification, low to high power, large bandwidth. Communication, satellite, signal amplifier, jamming transmitter
Linear beam	BWO tube	f < 500 GHz, low power, large bandwidth. Oscillator
Linear beam	Ubitron, peniotron	f < 100 GHz, high power and frequency. Radar, fusion
Crossed fields	Magnetron	f < 70 GHz, high power, narrow bandwidth, oscillator. Radar, accelerators
Crossed fields	CFA tube	f < 50 GHz, medium power and amplification, medium bandwidth. Radar, communication, driver to high power tubes, jamming transmitter
Relativistic electrons	Gyrotron Gyromonotron Gyroklystron Gyro traveling wave tube	f < 250 GHz, high power and frequency, small bandwidth. Radar, fusion
Relativistic electrons	Vircator	Very high power, many frequencies in output signal. HPM (High-power microwave)

wide-band tubes with the bandwidth greater than an octave, narrow-band tubes, oscillator tubes and amplifier tubes, and the process goes on. By construction, microwave tubes roughly fall into four groups: classical triodes and tetrodes, tubes with a linear beam (so-called O tubes), tubes with crossed electric and magnetic field (so-called M tubes), and tubes that use relativistic effects. There are, however, tubes that do not belong to these groups, as well as tubes that have elements of two groups. Among O-type tubes are the most important amplifiers, namely klystrons and traveling wave tubes, while the most important M-type tube is the magnetron. Gyrotrons, found in the fourth group, are tubes that simultaneously generate the highest frequencies and high power.

Table 7.1 summarizes properties and applications of different microwave tubes.

7.2. MICROWAVE TRIODES

Classical electron tubes have a frequency limit of $\sim$100 MHz, but triodes, and even tetrodes, can be used as microwave generators and amplifiers up to $\sim$4 GHz. The distance between the cathode and control grid in these tubes is very small, $\sim$80–100 μm. The control grid wire is $\sim$10 μm, and the cathode, grid, and anode therefore have a diameter of $\sim$1 cm, which limits the power to a few hundred watts CW and a few kilowatts pulsed. Figure 7.1 shows the cross section of a microwave triode made of metal and ceramic, the tube is suitable for direct insertion in coaxial circuit elements. The cathode is usually oxide, the grid gold, molybdenum, or tungsten mesh, the anode copper, and the heater wire tungsten.

Classical electron tubes at high frequencies are limited by transit time effects (Section 3.4) and interelectrode capacitance and wiring inductance. A high amplification

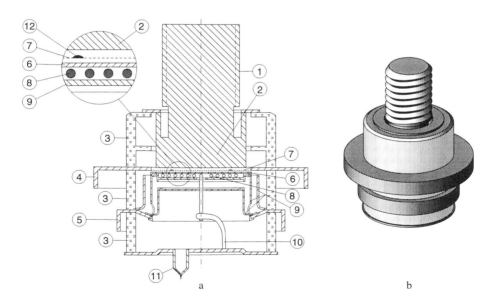

Fig. 7.1. Microwave triode 8940 (Eimac). In a pulsed oscillator circuit the triode can give a power of 10 kW at 2.5 GHz. The anode voltage is 5000 V, and the current is 6 A.
(**a**) Cross section. 1, Screw for the cooling flange; 2, anode; 3, ceramic isolator; 4, control grid connection; 5, cathode connection; 6, oxide cathode; 7, control grid; 8, heater wire; 9, thermal radiation shields; 10, heater connection; 11, pumping tube; 12, grid reinforcement. (**b**) Natural size.

cannot be obtained simultaneously with a large bandwidth. The transit time between the cathode and the control grid,

$$\tau = \frac{d}{v},$$

at high frequencies is comparable with the input signal period to be amplified. To obtain a significant amplification the beam coupling coefficient (Eq. 3.33),

$$M_B = \frac{\sin \omega\tau/2}{\omega\tau/2},$$

must be ≥ 0.9.

EXAMPLE: Assuming a potential between the triode 8940 control grid and cathode of 100 V, and an electrode separation of 80 μm, the transit time is 13.5 ns. The maximum frequency that gives a 0.9 beam coupling coefficient is ~ 2 GHz.

When the frequency exceeds 1 GHz the interelectrode capacitances and the wiring inductances present problems. The real part of the input impedance can be so small that it heavily loads the input circuit of the tube. Microwave triode and tetrode circuits usually include resonant elements, either parallel L–C circuits at lower frequencies or resonant cavities at higher frequencies. We analyze the behavior in such a circuit using Fig. 7.2.

The most important interelectrode capacitance is the cathode – control grid capacitance, C_{gc} (Section 3.3). Neglecting the feedback through C_{ga}, the input signal amplitude is

$$V_{in} = V_g + V_k = V_g + j\omega L_c(I_{in} + I_a). \tag{7.1}$$

The driving circuit must supply a current:

$$I_{in} = j\omega C_{gc} V_g. \tag{7.2}$$

Assuming $I_{in} \ll I_a = g_m V_g$, where g_m is the transconductance of the tube, we obtain

$$V_{in} = \frac{I_{in}(1 + j\omega L_c g_m)}{j\omega C_{gc}},$$

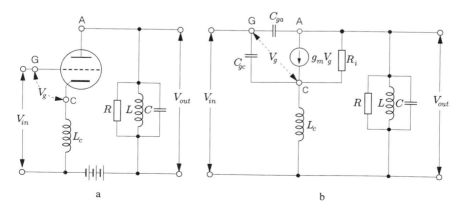

Fig. 7.2. (a) Triode circuit and **(b)** its equivalent scheme.

and the tube input admittance is

$$Y_{in} = \frac{I_{in}}{V_{in}} = \frac{j\omega C_{gc}}{1 + j\omega L_c g_m} \approx \omega^2 L_c C_{gc} g_m + j\omega C_{gc}, \tag{7.3}$$

where $\omega L_c g_m$ is negligible compared to 1. In practical cases the cathode wiring is short and thick (the circuit made as a resonant cavity), and the transconductance is on the order of millisiemens. At high frequencies, the real part of the input admittance is

$$G_{in} = \omega^2 L_c C_{gc} g_m, \tag{7.4}$$

and since it increases with the square of the frequency at high frequencies the input impedance looks like a short circuit. To preserve the output amplitude the power from the drive source must be increased. C_{gc} can be decreased if the electrode size is reduced, but this limits the power of the tube.

EXAMPLE: For triode 8940, C_{gc} is, with a heated cathode, 18 pF, the cathode wiring inductance is $\sim$ 60 pH, and tube transconductance at 0 V grid bias is 0.02 S. Then the real part of the input impedance at 3 GHz is 13 Ω.

By observing a simplified triode output circuit (Fig. 7.3), the output signal amplitude can be expressed as

$$V_{out} = \frac{g_m V_{in}}{G + j\left(\omega C - \frac{1}{\omega L}\right)}, \tag{7.5}$$

where

$$G = \frac{1}{R_i} + \frac{1}{R}.$$

The amplification at resonance is then

$$A_{f_0} = \frac{g_m}{G}.$$

The bandwidth is defined with the 3 dB points, so

$$G = \omega C - \frac{1}{\omega L}.$$

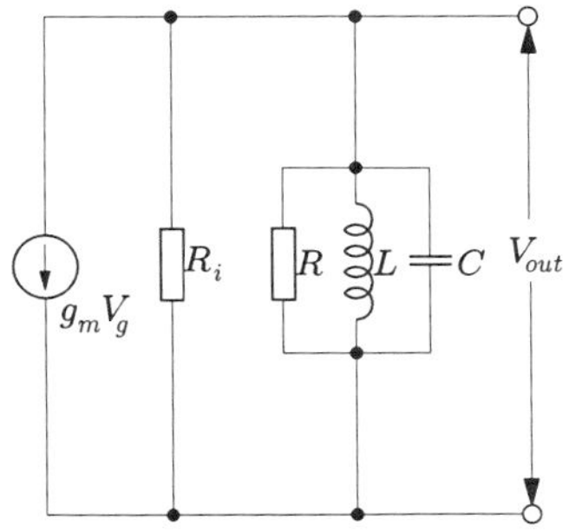

Fig. 7.3. The resonant output circuit of a triode.

Solving the equation

$$\omega_{1,2} = \frac{G}{2C} \pm \sqrt{\left(\frac{G}{2C}\right)^2 + \frac{1}{LC}},$$

the bandwidth becomes, under the assumption that $(G/2C)^2 \gg 1/LC$, which is always satisfied in practice,

$$\Delta\omega = \omega_2 - \omega_1 \approx \frac{G}{C}. \qquad (7.6)$$

The product of the bandwidth and amplification,

$$\Delta\omega \cdot A_{f_0} = \frac{g_m}{C}, \qquad (7.7)$$

is *independent of frequency* and large amplification can only be obtained with a small bandwidth. Narrow-bandwidth resonant circuits are therefore used in microwave and high-frequency amplifiers and oscillators.

7.3. LINEAR BEAM MICROWAVE TUBES

7.3.1. Introduction

Linear beam microwave tubes use an axial magnetic field to counteract beam spreading as the electrons move from the cathode to the collector. After emission, electrons are accelerated in an electron gun that they leave with nearly identical axial velocity. The electrons then enter a drift tube where they interact with a high-frequency electromagnetic field, either continuously over the whole tube length or locally in cavity resonators, and are accelerated or retarded by the field. Bunched together, they are moving along the tube to the output circuit where they induce current and generate output power and are finally collected.

Table 7.2

Microwave tubes with linear beam — O-type tubes

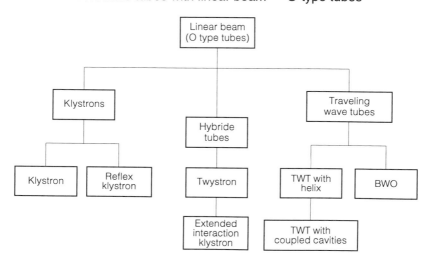

Table 7.2. surveys various linear beam microwave tubes, which are divided into three groups. In the first are klystrons and reflex klystrons, where interaction between electrons and the microwave field occurs discontinuously in cavities. In the second, which includes different kind of traveling wave tubes and the BWO tube, electrons continuously interact with the field along the entire tube. In the third, which contains elements of both groups (e.g., the twystron), the beam is modulated in cavities and interacts continuously with the field near the end of the tube. In the klystron with extended interaction, two or more consecutive cavities are coupled to increase the degree of modulation, to get broad band response, and to reduce peak fields in the output cavity improving efficiency. The name "O-type tubes" comes from the French — *tubes à propagation des ondes.*

7.3.2. Klystron

HISTORICAL NOTES: Research on microwave tubes began in the mid-1930 when in 1935 Arsenjewa-Heil and Heil [135] designed an oscillator for a linear beam. Based upon the mathematical formulation for velocity modulation of an electron beam between two grids, Brüche and Recknadel [136] discussed a process they called "phase focusing." The real breakthrough came when Varian brothers [137] at Stanford University in 1937 constructed the first working klystron, and in 1939 published the article "A high-frequency Oscillator and Amplifier," which explains so well the klystron working principle that it can even today be used to describe modern tubes. The article still talked about "cathode rays," the name for electron beams in those days. "Klystron" was from the Greek verb $\kappa\lambda\upsilon\zeta o$, which means breaking of waves against the shore. The electron gun of the first klystron was a scaled-up copy of an oscilloscope tube gun, the anode and resonators had grids, and the tube efficiency was only a few percent. Figure 7.5 is a copy of the original drawing.

A High Frequency Oscillator and Amplifier

RUSSELL H. VARIAN AND SIGURD F. VARIAN
Stanford University, California
(Received January 6, 1939)

A d.c. stream of cathode rays of constant current and speed is sent trough a pair of grids between which is an oscillating electric field, parallel to the stream and of such strenght as to change the speeds of the cathode rays by appreciable but not too large fractions of their initial speed. After passing these grids the electrons with increased speeds begin to overtake those with decreased speeds ahead of them. This motion groups the electrons into bunches separated by relatively empty spaces. At any point between the grids, therefore, the cathode-ray current can be resolved into the original d.c. plus a nonsinusoidal a.c. A considerable fraction of its power can then be converted into power of high frequency oscillations by running the stream trough a second pair of grids between which is an a.c. electric field such as to take energy away from the electrons in the bunches. These two a.c. fields are best obtained by making the grids form part of the surfaces of resonators of the type described in this Journal by Hansen.

Fig. 7.4. Title and summary of the original article about the first klystron.

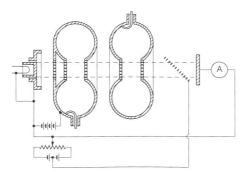

Fig. 7.5. Drawing of the first klystron. [R. H. Varian and S. F. Varian, *J. Appl. Phys.*, **10**, (1939), with permission.]

Modern tubes differ in important ways from the first klystron: The electron gun has a spherical cathode with a Pierce electrode (Fig. 6.7); an axial magnetic field counteracts beam spreading and balances the beam; the cavities are gridless; often there are more than two cavities; the collector is a long cylinder to decrease the current density and better distribute the thermal load (Fig. 6.22).

Large pulsed klystrons, which produce 150 MW of peak power, have currents of 500 A and acceleration voltages of 500 kV and are constructed with space-charge effects in the modulated beam allowed for. The space-charge influences the bunching of the velocity-modulated beam, so when the amplitude of the rf voltage across the cavities becomes high, nonlinear effects must also be considered. The theory of such klystrons is still under development, while many advancements have been made using numerical simulation, and so it is suitable to consider three different cases:

(a) *Ballistic theory.* For the small current, I_0, space-charge can be neglected.

(b) *Space-charge theory.* For I_0 large, space-charge cannot be neglected, but there is small signal amplitude.

(c) *Large signal theory.* Both the signal amplitude and the space-charge are large.

The Varian brothers considered their klystron as one-dimensional, because there were grids in the cavities, and they assumed that the field in the gaps was almost homogeneous. To follow their reasoning, we use the results of Eq. 3.37 for velocity modulation and rewrite as

$$v_z = v_0 + \frac{eM_BV_1}{mv_0} \sin \omega t_0 = v_0 \left[1 + \frac{M_BV_1}{2V_a} \sin \omega t_0 \right], \tag{7.8}$$

where $v_0 = \sqrt{-2eV_a/m}$ is the electron velocity before entering the gap, V_a the acceleration voltage, V_1 the modulating signal amplitude, ω the input signal angular frequency, and t_0 the time at which the electrons pass the gap center:

$$t_0 = t_1 + \frac{d}{2v_0}. \tag{7.9}$$

d is the cavity gap length, t_1 the time at which electrons enter the gap, and t_2 the time at which they exit. The beam coupling coefficient, M_B, is defined by Eq. 3.33 as

$$M_B = \frac{\sin \dfrac{\omega d}{2v_0}}{\dfrac{\omega d}{2v_0}}, \tag{7.10}$$

and the gap transit time is

$$\tau = t_2 - t_1 = \frac{d}{v_0}. \tag{7.11}$$

In modern large current density tubes, it is impossible to have grids, because they would melt. In a microwave cavity without grids, the electric field is position-dependent, $\mathbf{E} = \mathbf{E}(z, r)$, as seen in Fig. 7.6, computed for a high-power klystron input cavity, where the electron beam has a diameter of 11 mm, and the difference between the axial and edge ($r = 5.5$ mm) electric field, E_z, is not very large. If E_z in the gap is inhomogeneous, we introduce a time- and position-dependent beam coupling coefficient

$$M_B = M_B(r, t_0).$$

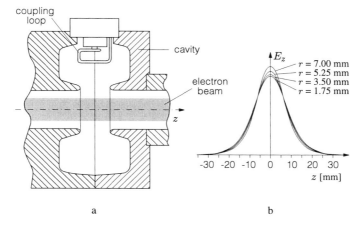

Fig. 7.6. Input cavity in a klystron. (**a**) cross section. (**b**) Electric field vs. axial position.

The transit time, τ, is almost always much less than $2\pi/\omega$ (for Fig. 7.6, $\omega\tau \sim 60°$), so the beam coupling coefficient is

$$M_B \simeq M_B(r),$$

assuming that the modulation amplitude V_1 is much less than V_a. The variation with r also is small, and we now find an equivalent gap length for which

$$\int E_{z\,(r=const)}\,dz = V_1 d_e(r),$$

where $d_e(r)$ is a slowly varying function of r. As

$$\frac{\omega d_e}{2v_0} \ll 1,$$

the beam coupling coefficient is

$$M_B = \frac{\sin\dfrac{\omega d_e}{2v_0}}{\dfrac{\omega d_e}{2v_0}} \approx 1 - \frac{1}{6}\left(\frac{\omega d_e}{2v_0}\right)^2. \tag{7.12}$$

The klystron can therefore be viewed as a one-dimensional device with the geometrical gap length, d, replaced by d_e. The effect of radial electric field inside the gap is neglected in the one-dimensional model, but with the assumption of $V_1 \ll V_a$, the radial field is rather small, except in the last two cavities of a multicavity klystron.

7.3.2.1. Ballistic Theory

Ballistic theory assumes a current density so small that space-charge effects can be neglected, which is never the case in modern high-power tubes. However, the theory is simple and can be a first approximation for the last part of a multicavity klystron, which allows for large signal amplitude, neglecting nonlinear effects.

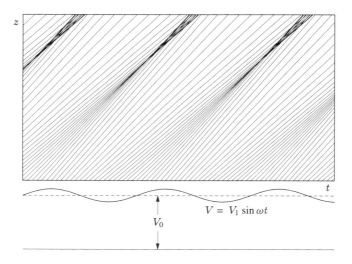

Fig. 7.7. Applegate diagram. V_0 is the acceleration voltage in the electron gun, and V_1 is the amplitude of the modulation signal over the first cavity.

In the first cavity of a klystron similar to Varian's shown in Fig. 7.5, the electron beam is velocity-modulated according to Eq. 7.8. The faster electrons catch up to the slower ones and bunch together, as seen in Fig. 7.7, the so-called Applegate diagram, where the electrons axial position with time is shown.

With the distance between the center of the first and the second cavity, L, the electrons reach the second cavity center at

$$t = t_0 + \frac{L}{v_z} = t_0 + \frac{L}{v_0\left(1 + \frac{M_B V_1}{2V_0}\sin\omega t_0\right)}. \tag{7.13}$$

Assuming that $M_B V_1 \ll 2V_0$ and expanding the denominator in a power series, we obtain

$$t \simeq t_0 + \frac{L}{v_0}\left(1 - \frac{M_B V_1}{2V_0}\sin\omega t_0\right).$$

Multiplying by the angular frequency, ω, we obtain

$$\omega t - \omega t_0 = \frac{\omega L}{v_0}\left(1 - \frac{M_B V_1}{2V_0}\sin\omega t_0\right) = \theta - X\sin\omega t_0. \tag{7.14}$$

$\theta = \omega L/v_0$ is called the *transit time angle* and

$$X = \frac{M_B}{2}\frac{V_1}{V_0}\theta, \tag{7.15}$$

the *bunching parameter.*

The phase angle when electrons arrive at the second cavity (Eq. 7.14) depends on their phase angle when they leave the first, as shown in Fig. 7.8. The charge passing the first gap during time dt_0 is

$$dq_0 = -I_0 dt_0,$$

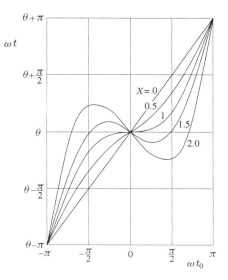

Fig. 7.8. Phase, ωt, when the electrons pass the center of the second cavity, vs. phase, ωt_0, in the first cavity.

where I_0 is the unmodulated beam current. Continuity requires that the same charge pass the second gap, $dq(t) = dq_0(t_0)$, where the current is

$$i(t) = \frac{dq(t)}{dt} = -I_0 \frac{dt_0}{dt}. \tag{7.16}$$

Observe that the arrival angle at the second cavity is a multiple function of the departure angle that depends on the bunching parameter. For $X > 1$ we have

$$dq = -I_0\{(dt_0)_1 + (dt_0)_2 + (dt_0)_3\}$$

and

$$i(t) = -I_0\left\{\left|\frac{dt_0}{dt}\right|_1 + \left|\frac{dt_0}{dt}\right|_2 + \left|\frac{dt_0}{dt}\right|_3\right\}, \tag{7.17}$$

where the absolute value is taken because the charge has the same sign independent of the order electrons arrive at the gap.

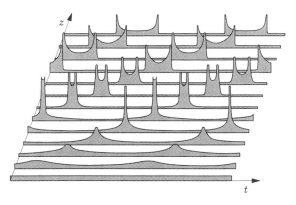

Fig. 7.9. Variation of the rf current along the tube according to ballistic theory. $X = 1$.

Figure 7.9 shows, for $X = 1$, the rf current variation along the tube and electrons bunching. We wish to amplify the current component that corresponds to the modulation frequency. The current in the second cavity is periodic with the period $\tau = 2\pi/\omega$, so it can be Fourier analyzed

$$i(t) = -I_0 + \sum [a_n \cos n(\omega t - \theta) + b_n \sin n(\omega t - \theta)], \qquad (7.18)$$

where all b_n are identically zero, while a_n can be written in terms of Bessel functions so that

$$i(t) = -I_0\{1 + 2\sum_{n=1}^{\infty} J_n(nX) \cos n(\omega t - \theta)\} \qquad (7.19)$$

The first rf current component is thus

$$i_1 = -2I_0 J_1(X)$$

with maximum at

$$X = 1.841,$$

the first Bessel function maximum. Using ballistic theory the optimal distance between the center of the first and second cavity is

$$L_{opt} = \frac{3.682\, v_0 V_0}{\omega M_B V_1}. \qquad (7.20)$$

Note that from Eq. 7.8 the position where the fastest electrons will catch up with the slowest is

$$L = \frac{\pi v_0 V_0}{\omega M_B V_1},$$

~ 15 percent less than the optimum distance, because the current density maximum does not coincide with the first harmonic current component maximum, since higher harmonics are present.

> EXAMPLE: In a 6-cavity CW klystron with acceleration voltage of 22 kV, the beam current of 2.8 A, and modulating frequency of 2450 MHz, the amplitude of the rf voltage across the gap of the fifth cavity is 14.5 kV. Using a beam coupling coefficient of 0.89, computed from the gap transit time, the optimum distance between the last two cavities is 34.8 mm. Large signal numerical computation gives 33 mm.
>
> In the first gap the signal amplitude is 1200 V, so according to Eq. 7.20 the optimum distance between the first and the second cavity is 415 mm.

7.3.2.2. Space-Charge Theory

In modern high-power klystrons the high current density results in space-charge influencing bunching. In the first cavity the modulation process can be described by Eq. 7.8, because the modulation voltage amplitude is so small that the electrons in the gap retain their same relative positions. Thus the finite gap length can be neglected compared to the distance between cavities. In the drift tube between the first and the second cavity, the faster electrons approach the slower ones and change the initially homogeneous beam space-charge. The Applegate diagram no longer looks like the one in Fig. 7.7 because the space-charge does not allow the fastest electrons to catch up to the slowest;

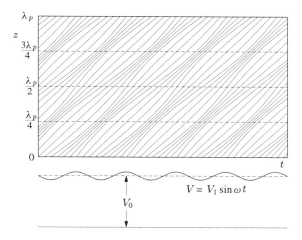

Fig. 7.10. Modified Applegate diagram.

the fastest will be retarded and the slowest accelerated. The space-charge increases, but can never reach the large values of ballistic theory, as seen in the modified Applegate diagram of Fig. 7.10.

The velocity modulation produces a small change in electron velocity with time. Assume an initial infinitely wide electron beam, so that only motion in the axial, z, direction will vary. Where electrons concentrate, space-charge forces produce a motion to lower space-charge regions. We observe the relative electron motion in a coordinate system moving to the right with the unmodulated beam velocity, v_0 (Fig. 7.11), and represent space-charge by disks of electrons, the higher the disks density, the greater the space-charge.

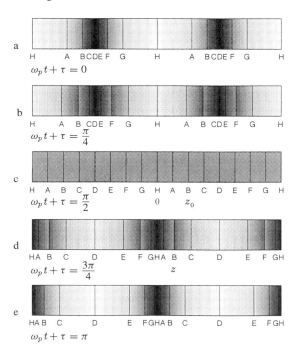

Fig. 7.11. Electron motion under the influence of the space-charge forces.

In Fig. 7.11a the highest space-charge is near disk D, the lowest near disk H, and both disks, D and H, will remain at rest in our coordinate system because of symmetry. The disks A, B, and C are forced to the left by space-charge forces while disks E, F, and G move to the right. For a half-period the disk motion is shown in Figs. 7.11a–7.11e.

Assume that the coordinate system origin lies in disk H, and that the velocity is low compared to v_0. The expression for the electron motion in disk B, for example, can be derived as follows: B is at z_0' in Fig. 7.11c, and a little later, in Fig. 7.11d, B has moved to z' where the space-charge is larger in the region $0 \leq z' \leq z_0'$.

$$\Delta q = -\rho_0(z_0' - z'),$$

and the electric field is

$$E_z = -\frac{\rho_0}{\epsilon_0}(z_0' - z'), \tag{7.21}$$

which acts on electrons such that

$$m\ddot{z}' = e\frac{\rho_0}{\epsilon_0}(z_0' - z'). \tag{7.22}$$

This elastic force acts in the direction of the equilibrium position of our coordinate system, so the solution of Eq. 7.22 is

$$z' = z_0' + A\cos(\omega_p t + \varphi), \tag{7.23}$$

where

$$\omega_p = \sqrt{\frac{e}{m}\frac{\rho_0}{\epsilon_0}} \tag{7.24}$$

is the *plasma frequency*.

A klystron beam is not infinitely wide but moves in a conductive drift tube, where it induces mirror charges. Referring to Fig. 7.11, representing the electron beam by disks with different space-charge, and the mean space-charge of the unmodulated beam defined by ρ_0, after modulation disks with larger and smaller space-charge will be found in the beam. An electric field according to Eq. 7.21 exists between these disks. The missing charge outside the beam causes the field lines to bend outwards with a resulting radial component and a decreased axial component, and the electric force of Eq. 7.22 is

$$E_z = -F^2\frac{\rho_0}{\epsilon_0}(z_0' - z'), \tag{7.25}$$

where the reduction factor for space-charge, which depends on the beam diameter, is $F^2 < 1$. The solution

$$z' = z_0' + A\cos(F\omega_p t + \varphi) \tag{7.26}$$

has the form of Eq. 7.24, and a *reduced plasma frequency* is defined as

$$\omega_q = F\omega_p. \tag{7.27}$$

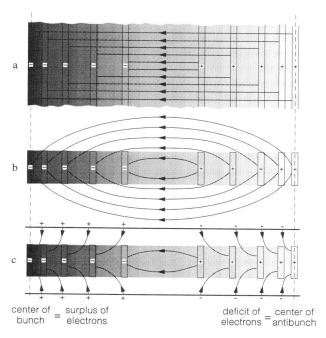

center of = surplus of   deficit of = center of
bunch   electrons   electrons   antibunch

Fig. 7.12. Electric field lines in electron beams with space-charge. The space-charge is symbolized by positive and negative disks. In disks with negative charge a surplus of electrons exists compared to the mean value of the space-charge in an unmodulated beam. In positive disks there is a deficit in electrons. (**a**) Infinitely wide beam. (**b**) Solid cylindrical beam with finite diameter. (**c**) Cylindrical beam in a drift tube.

In a klystron with a solid cylindrical beam enclosed by a conductive tube, moving mirror charges are created on which many field lines terminate. This further decreases the electric forces between bunch and antibunch (Fig. 7.12c), and so the plasma frequency reduction factor will be still lower, depending on the beam radius, b, and the tube radius, a. With the assumption that the scalloping is small, $b \sim$ constant, and Maxwell's equations can be solved so that

$$F^2 = \left(\frac{\omega_q}{\omega_p}\right)^2 = \frac{1}{1 + \left(\dfrac{S}{\beta_e b}\right)^2}, \tag{7.28}$$

where β_e is the propagation constant, S the solution of the transcendetal equation

$$\beta_e = \frac{\omega}{v_0}, \qquad S = \beta_e b \frac{J_o(S)}{J_1(S)} \frac{K_1(\beta_e b) I_0(\beta_e a) + K_0(\beta_e a) I_1(\beta_e b)}{K_0(\beta_e b) I_0(\beta_e a) - K_0(\beta_e a) I_0(\beta_e b)}, \tag{7.29}$$

and J_0, J_1, I_0, I_1, K_0, and K_1 are the Bessel functions.

Figure 7.13 shows graphically the plasma frequency reduction factor for a solid cylindrical beam with radius, b, transported in a perfectly conducting cylinder with the radius, a, and with propagation constant, β_e.

All the above is valid only under the assumption that the radial space-charge forces are compensated by an axial magnetic field, by some kind of beam focusing with confined or balanced flow, always the case for a klystron. Brillouin focusing is never used, because the rf bunching increases the local space-charge.

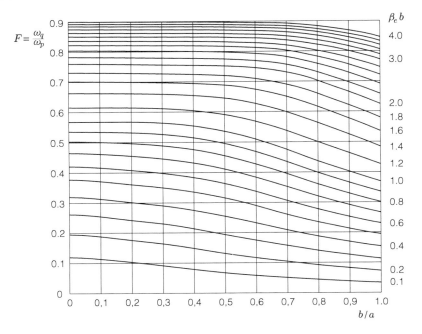

Fig. 7.13. Plasma frequency reduction factor.

EXAMPLE: For the klystron, as of the previous example we assume a beam diameter of 6 mm and a drift tube diameter of 9 mm. For $v_0 = 8.52 \times 10^7$ m/s (relativistically) we get $f_p = 748.6$ MHz, $\beta_e = 180.6$, and $\beta_e b = 0.54$. In Fig. 7.13 we read the plasma frequency reduction factor as $F = \omega_q/\omega_p = 0.29$ and the reduced plasma frequency is 217 MHz.

To define the integration constants (Eq. 7.26) we start in the input gap by choosing the reference electron position $z'_0 = 0$ for $t = t_0$, and thus

$$A \cos(\omega_q t_0 + \varphi) = 0$$

and

$$\omega_q t_0 + \varphi = \frac{\pi}{2}.$$

Thus

$$z' = A \cos\left(\omega_q t + \frac{\pi}{2} - \omega_q t_0\right) = -A \sin \omega_q (t - t_0). \qquad (7.30)$$

At $t = t_0$, taking into account the right side of Eq. 7.8, the rf component of the velocity is

$$v = \frac{dz'}{dt} = -\omega_q A \cos \omega_q (t - t_0) = v_0 \frac{M_B V_1}{2V_0} \sin \omega t_0, \qquad (7.31)$$

and

$$A = -\frac{v_0}{\omega_q} \frac{M_B V_1}{2V_0} \sin \omega t_0. \qquad (7.32)$$

Changing to the observer coordinate system by introducing

$$z = v_0(t - t_0),$$

that gives

$$t_0 = t - \frac{z}{v_0},$$

and defining two propagation constants

$$\beta_q = \frac{\omega_q}{v_0} \qquad \text{and} \qquad \beta_e = \frac{\omega}{v_0}, \tag{7.33}$$

Eqs. 7.30 and 7.31 become

$$z' = \frac{v_0}{\omega_q} \frac{M_B V_1}{2V_0} \sin \beta_q z \sin(\omega t - \beta_e z), \tag{7.34}$$

and

$$v = v_0 \frac{M_B V_1}{2V_0} \cos \beta_q z \sin(\omega t - \beta_e z). \tag{7.35}$$

The rf space-charge density is for negatively charged electrons

$$\rho = -\frac{dq}{dz},$$

and continuity requires

$$dq = -\rho_0 dz'.$$

Thus

$$\rho = \rho_0 \frac{dz'}{dz} = \frac{\rho_0 v_0}{\omega_q} \frac{M_B V_1}{2V_0} \{\beta_q \cos \beta_q z \sin(\omega t - \beta_e z) - \beta_e \sin \beta_q z \cos(\omega t - \beta_e z)\}.$$

We introduce, according to Eq. 7.33,

$$\frac{v_0}{\omega_q} = \frac{1}{\beta_q} \qquad \text{and} \qquad \frac{\beta_e}{\beta_q} = \frac{\omega}{\omega_q}$$

and the space-charge becomes

$$\rho = \rho_0 \frac{M_B V_1}{2V_0} \{\cos \beta_q z \sin(\omega t - \beta_e z) - \frac{\omega}{\omega_q} \sin \beta_q z \cos(\omega t - \beta_e z)\}. \tag{7.36}$$

For our numeric example we have $\omega/\omega_q \sim 11.3$, consistent with a value of the order of 10 or more for all klystrons, so the first term in Eq. 7.36 can be neglected:

$$\rho \simeq -\rho_0 \frac{M_B V_1}{2V_0} \frac{\omega}{\omega_q} \sin \beta_q z \cos(\omega t - \beta_e z). \tag{7.37}$$

The total space-charge density and the velocity can be written as a sum of the DC and rf components:

$$\rho_{tot} = -\rho_0 + \rho \qquad \text{and} \qquad v_{tot} = v_0 + v$$

Similarly we define the total current density,

$$J_{tot} = -J_0 + J,$$

where J_0 is the current density before the modulation, and J is the rf component. Introducing the expressions for ρ_{tot} and v_{tot}, the current can be expressed as

$$J_{tot} = \rho_{tot} v_{tot} = -\rho_0 v_0 + \rho v_0 - \rho_0 v + \rho v. \tag{7.38}$$

The space-charge theory described so far is valid only for small modulation amplitude, and so the product of the high-frequency terms can be neglected. If space-charge forces are large, nonlinear effects result that are not accounted for in this simple theory. With

$$J_0 = \rho_0 v_0$$

the rf component of the current becomes

$$J = \rho v_0 - \rho_0 v. \tag{7.39}$$

Using Eqs. 7.35 and 7.37, the current can be written as

$$J = -\rho_0 v_0 \frac{M_B V_1}{2V_0} \frac{\omega}{\omega_q} \sin \beta_q z \cos(\omega t - \beta_e z) - \rho_0 v_0 \frac{M_B V_1}{2V_0} \cos \beta_q z \sin(\omega t - \beta_e z).$$

Even here the summand without ω/ω_q can be neglected and the rf current density is

$$J \simeq -J_0 \frac{\omega}{\omega_q} \frac{M_B V_1}{2V_0} \sin \beta_q z \cos(\omega t - \beta_e z). \tag{7.40}$$

Here the reduced plasma frequency is much smaller than the modulation frequency; therefore $\beta_q \ll \beta_e$, and Eq. 7.40 represents a wave whose phase velocity equals the unmodulated beam velocity, v_0, but whose amplitude slowly varies with the distance from the modulating gap (the $\sin \beta_q z$ term). The same is true for the rf velocity component and the space-charge.

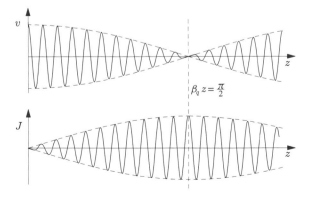

Fig. 7.14. High-frequency velocity and current density components vs. distance from the modulation cavity ($z = 0$) at $\omega t = \pi/2$. The shape of the envelope of the waves remains constant in time, and the waves move to the right with the velocity v_0.

Figure 7.14 shows how the velocity and the current density vary with the axial coordinate, z, at $\omega t = \pi/2$. Note that the rf velocity component has its maximum in the input gap. The amplitude decreases then as $|\cos \beta_q z|$. The rf current is zero in the input gap and its amplitude increases as $|\sin \beta_q z|$. The current density has its maximum at

$$\beta_q z_{opt} = \frac{\pi}{2}. \tag{7.41}$$

The distance corresponds to a quarter plasma wavelength where *the next cavity should be placed*. Note that the amplitude of the rf velocity component is zero on the same place. In a multicavity klystron the beam can again be modulated in the second cavity as it was in the first. The induced current, caused by the passing of the already modulated beam, is enough to excite the cavity, so no external signal is necessary.

EXAMPLE: We continue with our example klystron. With $f_p = 217$ MHz we have $\beta_q = 16$ m^{-1}, $z_{opt} = 98$ mm, compared to 415 mm for ballistic theory. The beam current is 2.8 A and corresponding perveance is 0.86 μP, so the space-charge theory gives the correct position for the second cavity.

Equations 7.35 and 7.40 describe the change of velocity and of current density along the drift tube. Figure 7.14 shows that the amplitudes of these slowly changing functions have the form of a standing wave and thus can be described as a sum of two advancing waves with the same frequency. Introducing

$$\beta_f = \beta_e - \beta_q \qquad \text{and} \qquad \beta_s = \beta_e + \beta_q \tag{7.42}$$

Eq. 7.35 can be written as

$$v = v_0 \frac{M_B V_1}{4 V_0}[\sin(\omega t - \beta_f z) + \sin(\omega t - \beta_s z)], \tag{7.43}$$

while Eq. 7.40 becomes

$$J = -J_0 \frac{\omega}{\omega_q} \frac{M_B V_1}{4 V_0}[\sin(\omega t - \beta_f z) - \sin(\omega t - \beta_s z)]. \tag{7.44}$$

One of the waves, with the phase velocity $v_f = \omega/\beta_f > v_0$, is called the *fast wave*, while the second one, with the phase velocity $v_s = \omega/\beta_s < v_0$, is called the *slow wave*. Description of a klystron in the space-charge theory is based on the interference between the fast and the slow wave. These two space-charge waves represent the two modes that are generated by the modulation of an unmodulated electron beam in the input gap of an excited cavity. In the input gap both phase velocity components have equal amplitudes and are in phase, while both current density components also have equal amplitudes, but are out of phase. A quarter plasma wavelength farther the current densities are in phase and the phase velocities are out of phase.

The power flow in the klystron is governed by the electron kinetic energy, W_k, through a surface S:

$$P = \frac{dW_k}{dt} = \frac{W_k}{V} S v_0,$$

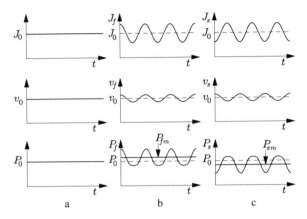

Fig. 7.15. Current density and power flow vs. time in a velocity modulated electron beam: **(a)** Without modulation, **(b)** fast wave and **(c)** slow wave. (After *Physical Electronics*, C. L. Hemenway, R. W. Henry, and M. Caulton, 1967, with permission.)

where W_k/V is the total kinetic energy per unit volume, or number of electrons per unit volume, $-\rho/e$, times the kinetic energy of the electrons, $mv^2/2$. Thus,

$$\frac{W_k}{V} = -\frac{1}{2}\frac{\rho}{e}mv^2 = -\frac{1}{2}\frac{Jv}{\eta} \tag{7.45}$$

and

$$P = -\frac{v_0 SJv}{2\eta}. \tag{7.46}$$

Figure 7.15 shows the current density, velocity, and power flow with time at a position z. In the case without modulation, $J = J_0$, $v = v_0$ and the power flow is constant:

$$P = \frac{I_0 v_0^2}{2\eta}.$$

For the fast wave we have

$$-J = J_0 + J_m e^{j\omega t} \qquad \text{and} \qquad v = v_0 + v_m e^{j\omega t},$$

and for the slow one we obtain

$$-J = J_0 - J_m e^{j\omega t} \qquad \text{and} \qquad v = v_0 + v_m e^{j\omega t}.$$

According to Fig. 7.15b the rf velocity and current density components are in phase in the fast wave. The mean of the $\sin^2$ term, which originates in the product $J \cdot v$, is positive over the whole period and the mean of the power flow $P_{fm} > P_0$. In the slow wave the components are out of phase, and $P_{sm} < P_0$ (Fig. 7.15c).

We can define the rf power flow as the difference between the case with and without modulation. The positive value, $P_{fm} - P_0 > 0$, in the fast wave means that it transports more energy than does the beam without modulation. In the slow wave we have $P_{sm} - P_0 < 0$. This power flow does not occur in the negative direction as the power is transported in both cases in the direction of electron motion. A positive flow means that the electrons are bunched in the region of the highest velocity.

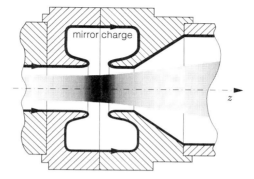

Fig. 7.16. Output cavity.

In the antibunch, the velocity of the electrons is the smallest. Since the slow wave transports a negative rf power flow, the beam energy can be extracted by increasing the amplitude of this wave. Similarly, by increasing the fast wave amplitude we add energy to the beam.

7.3.2.3. Interaction with Cavities

The klystron electron beam interacts with the drift tube and with the cavities. In the drift tube the interaction occurs through the mirror charges, which travel along the tube, change the plasma frequency, and introduce higher harmonic components at large space-charge amplitudes toward the end of a multicavity tube. When a bunch comes into a cavity, mirror charges travel along its walls. However, in principle, the situation is equivalent to the situation described by Fig. 3.7. In a klystron cavity there are no grids and the outer resistance in the figure must be replaced by the cavity and its load, but the conclusion can be taken directly. In the cavity and in the load, induced charges will flow, which depend only on the beam current in the klystron.

Figure 7.16 shows an output cavity where first harmonic component of the rf beam current is i_1. Since the gap in the cavity has a finite length, the beam coupling coefficient, M_{B_2}, must be allowed for. The induced current is then

$$i = i_1 M_{B_2} e^{-j\beta_e z_0}, \tag{7.47}$$

where z_0 is the z coordinate of the center of the gap. The induced current, in phase with the beam current, is *independent* of the load. The modulated electron beam acts on a cavity as a current source.

The voltage phase over the gap is determined by the induced current and the resonant frequency of the cavity. When the bunch is near the left side of the gap, the largest amount of mirror charge passes the gap tip. When the bunch has passed the gap, the largest mirror charge is at the right gap tip. In the meantime, the mirror charges have moved along the cavity walls and have induced a current. The cavity has a Q-value ~ 30, so the induced current is thus only a fraction of the total current on the cavity walls. But it is the induced current which supplies the losses in the cavity walls and drives the load.

If the output cavity is in resonance, the voltage over the gap will reach maximum when the bunch passes the gap center because at resonance the load is purely resistive, and the induced current and the voltage over the gap must be in phase. The electric field then has the same direction as the induced current, and as a consequence

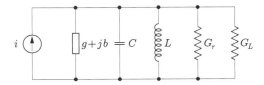

Fig. 7.17. Output cavity equivalent circuit.

the beam electrons feel a strongly retarding electric force. This explains the transfer of energy from the beam to the cavity and the load.

Besides the circuit components of the cavity (L, C, and G_r) and the load (G_L), the interaction of the electron beam with the electric field of the cavity must also be taken into account. During the transit through the gap the electric field changes, and so will the interaction between the electrons and the field. The interplay of the energy can be allowed for by defining beam loading admittance, $Y = g + jb$, also called *beam loading,* which loads the cavity and depends on the transit time (a more detailed description of this interaction is given in Appendix C). In the input and output gaps, like in all gaps of the klystron, the transit time angle should be shorter than $\pi/2$ because such a choice guarantees that the beam coupling coefficient has a value ~ 1. We see from Fig. C.1 in Appendix C that the beam loading admittance has the character of a susceptance, it is capacitive, and for small signal amplitudes it varies linearly with frequency. Therefore

$$b = \omega C_s.$$

The highest power a klystron can give to the load depends on the beam, the load, the cavity's Q-value, and its frequency. Like in all resonant circuits, attention must be paid to the Q of the circuit. A klystron cavity has two types of Q-values: when cold *without* the beam and when hot *with* the beam. The unloaded cold Q-value is

$$Q_0 = \frac{\omega_0 C}{G_r}, \tag{7.48}$$

and its hot equivalence is

$$Q_w = \frac{\omega_0(C + C_s)}{G_r + g}. \tag{7.49}$$

The cold unloaded Q-value is defined as for a normal resonant circuit. The hot one accounts for the beam loading admittance. The hot external Q-value is

$$Q_e = \frac{\omega_0(C + C_s)}{G_L}, \tag{7.50}$$

and the hot loaded Q-value is

$$Q_L = \frac{\omega_0(C + C_s)}{G_r + g + G_L}. \tag{7.51}$$

For the different Q-values the following expression is valid

$$\frac{1}{Q_L} = \frac{1}{Q_e} + \frac{1}{Q_w}. \tag{7.52}$$

The maximum power in the load is obtained if

$$G_L = G_r + g \quad \text{and} \quad \omega^2 L(C + C_s) = 1. \tag{7.53}$$

The admittance of the load can be changed by the coupling between the cavity and the waveguide, and by adjusting the load to the waveguide impedance. The later can be made, for example, by using a ferrite isolator between the klystron and the load. The resonance condition is easy to achieve in the output cavity, because its cold Q-value always is low, ~ 30. Brute force adjustment of the resonant frequency can be made by mechanical deformation of the cavity walls or by squeezing the gap.

The input cavity can be viewed in a similar way. The difference is how the cavity is excited. If, in Fig. 7.17, i is replaced by the current from the signal source, and the load conductance, G_L, is replaced by the conductance of the source, the equivalent circuit remains the same, and all results are unchanged. However, the unloaded Q of the input cavity is usually ~ 200 to ~ 800 — that is more than 10 times larger than the Q-value of the output cavity.

In a klystron with more than two cavities, only the first and the last are loaded. Sometimes the second cavity is artificially loaded to increase the klystron bandwidth. The unloaded cavities are excited by the first harmonic component of the beam current and their Q is usually high, a few thousands. The load conductance, the dominating term in Eq. 7.51, is missing in these cavities, and high voltages and strong electric fields are obtained, resulting in a strong modulation.

7.3.2.4. Large Signal Theory

Klystrons, used as high-power microwave amplifiers, have up to eight cavities. According to Eqs. 7.35 and 7.40, and their graphical presentation in Fig. 7.14, the maximum of the rf current component lies at a distance of a quarter plasma wavelength from the modulating gap, at the same place where the rf velocity component has its minimum. For small modulation signal amplitudes the rf velocity is theoretically zero. If the next cavity is placed there, a current will be induced with the consequence that an electric field will build up across the gap. If a frequency for this cavity that gives the right phase is chosen, the beam will be modulated in a similar way as in the first cavity, and the bunching will be increased. This can continue with the third and even later cavities.

Unfortunately, after each new cavity which the beam encounters, the situation becomes more complicated to describe. In the first two cavities, the simple linear space-charge theory gives satisfactory results. But with a higher degree of modulation, the higher harmonic space-charge, velocity, and beam current components will be more and more pronounced and thus cannot be neglected. Inside the drift tube, the increased nonlinearity of the bunching must be considered, which results in an increasing number of reduced plasma frequencies. To solve Maxwell's equation under these circumstances is a formidable problem. There exist numerical methods that are in good agreement with measurements for medium-power klystrons, where the relativistic effects can be neglected. In very-high-power pulsed tubes, with acceleration voltages of a few hundred kilovolts and relativistic effects, the radial electric field in the last cavities is so large that the one-dimensional theory does not represent physical reality.

The problem is tackled in different ways. The first part of a high-power klystron can be numerically simulated by an extension of the one-dimensional space-charge theory,

with higher-order harmonic components included. There are reasonable physical grounds
for this, in that the radial displacement of the electrons is small and negligible. In last
cavities this is not true for two reasons. The penultimate cavity is unloaded and the
induced gap voltage is somewhat smaller than the accelerating voltage. In the last cavity
the voltage is similarly high. In both cavities, some electrons will be retarded so much that
the influence of radial forces can no longer be neglected. Therefore, all one-dimensional
theories are unsuccessful in describing the behavior of the klystrons with the highest power.
There were attempts to use disks of electrons with different thickness and shape, and to
use the so-called displacement theory. The modern computers have changed the situation
now allowing two-dimensional simulations.

To begin with, equidistant disks of electrons with the corresponding charge, say between 20 and 30 per rf period, allowed the electric forces between the disks to be computed.
Taking into account the modulation in a cavity was more difficult. Since nothing is known
initially about the voltage across the gap, except in the first cavity where the circuit parameters and the drive power are known, the computations were iterative. In the first iteration
for every cavity an approximate value for the voltage is assumed, the induced current is
computed, and a better approximation for the voltage is obtained. However, the rf current
has many higher harmonic components, so the interaction between the disks themselves,
as well as between the disks and the cavities, is nonlinear. Attempts to divide the disks
radially were also made and pointed out the necessity of taking into account radial forces.

An alternate attempt to simulate a klystron was with displacement theory. A coordinate
system, which moves along the tube with the DC beam velocity, v_0, was used, and the
displaced variables, which denote the rf velocity, u, in the displaced position, s, were
computed (Fig. 7.18). Maxwell's equations give in this system of coordinates linear
integral equations which describe the space-charge effects. The coupling to the electric
field in the cavities is, however, still nonlinear. Displacement theory was used to compute
the interaction of the beam with the drift tube and with the cavities up to the last two
cavities where this theory gives satisfying results. The theory is one-dimensional and does
not solve successfully the high-power tube problems.

The plasma and accelerator physicist's PIC and CIC codes (Section 1.4), are now used
to simulate high-power klystrons, only with rotationally symmetric geometry, since the
computations are two-dimensional. These programs allow the electron orbits to cross and
even to turn back, which can happen in the last cavities under specific load conditions. The
simulation begins with a number of electrons with the initial velocity v_0 or with the position
and with the velocity computed in a computer simulation of the electron gun. The result
of the simulation of klystrons with a peak power of 50 MW agrees with the measurements
within a few percent. Next will be a three-dimensional simulation, which is necessary

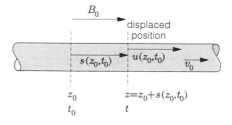

Fig. 7.18. Definition of displaced coordinates.

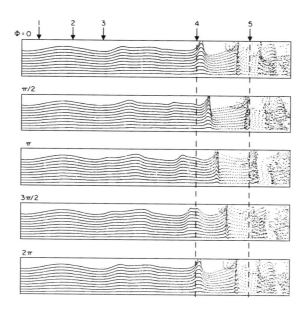

Fig. 7.19. Electron orbits in the SLAC klystron XK-5 computed by PIC code. The arrows show the positions of the five cavities. The figures show momentary pictures at five equidistant time intervals over a period. (After S. Yu, *Particle-in-Cell Simulation of High Power Klystrons*, SLAC/AP-34, 1984, with permission.)

since the last cavity is very asymmetrical because of its coupling to the waveguide. Figure 7.19 shows an example of the computed electron orbits in a klystron developed at SLAC (Stanford Linear Accelerator Center).

7.3.2.5. Construction

Figure 7.20 shows the cross section of the pulsed klystron PK771, made at the Royal Institute of Technology in Stockholm. With an amplification factor of 54 dB the tube can give 3.5 MW of power (see Table 7.3). We use this figure to describe some technical details that are applicable to all high-power klystrons.

The most klystron parts are made of the OFHC, electrolytically clean and free from oxygen copper, because of its high conductivity and because different details in manufacturing are brazed together in vacuum or hydrogen furnaces, so contamination from each process can cause fissures with the consequence that the tube will not be vacuum tight. Other important materials are ceramic (isolators and vacuum window), stainless steel (flanges and screws;), tungsten (heater wire), and molybdenum (cathode holder). The cathode is either of the oxide type or the dispenser type.

The electron gun is constructed according to the principles described in Section 6.2.2, while Fig. 6.7 is an enlarged drawing of the PK771's gun. Figure 6.22 shows its collector, described in more details in Section 6.4. The klystron has five cavities. The three intermediate cavities, although similar, have different resonant frequencies. The output cavity is considerably smaller because of the large coupling hole to the waveguide. Table 7.4 gives the relevant parameters of the cavities.

PK771 is, like other high-power klystrons, water-cooled — the collector, the drift tube, and the cavities. The power loss in the last cavities is large, and the gap tips

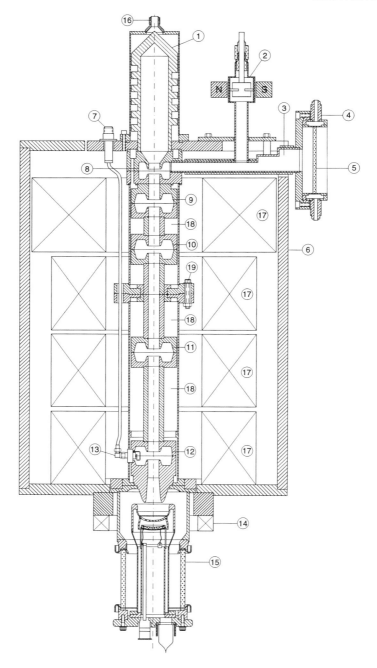

Fig. 7.20. Cross section of the pulsed klystron PK771 (Royal Institute of Technology, Stockholm). 1, Collector (see Fig. 6.22); 2, ion pump; 3, waveguide; 4, cooling water connection; 5, ceramic waveguide window; 6, magnet yoke; 7, input signal connection; 8, output cavity (see Fig. 7.16); 9–11, intermediate cavities; 12, first cavity (see Fig. 7.10); 13, coupling loop; 14, bucking coil for magnetic field correction at the cathode; 15, electron gun (see Fig. 6.7); 16, cooling water connection; 17, coils; 18, cooling water channels; 19, joint between both tube halves sealed by gold wire.

Table 7.3

Pulsed klystron PK771 — parameters

Acceleration voltage [kV]	120
Perveance [μp]	1.9
Beam current [A]	79
Frequency [MHz]	2998
Output power [MW]	3.5
Efficiency	0.37
Amplification [dB]	54
Pulse width [μs]	8
Beam diameter [mm]	11
Drift tube diameter [mm]	19
Length [mm]	980
Weight [kg]	28
Magnetic field [T]	0.12
Magnet weight [kg]	250

Table 7.4

Pulsed klystron PK771 — cavities

Cavity number	1	2	3	4	5
Cold frequency [MHz]	2993	2997	3020	3051	2998
Cold unloaded Q	520	430	8400	8400	33
Computed voltage across the gap [kV]	0.12	0.94	9.7	50.4	102
Gap length [mm]	12	12	10	10	7
Distance to the next gap [mm]	160	160	71	51	

and drift tube can be hit by the beam, because electrons at the beam edge can have large radial displacement in the last two cavities caused by scalloping and radial forces.

High-power klystrons are usually equipped with an ion pump, which is permanently connected to its power source and continually pumps the tube. Klystrons need high vacuum, $\sim 10^{-7}$ Pa, for three reasons. The cathode, either oxide or dispenser type, needs a vacuum better than 10^{-5} Pa so as not to be poisoned. Flashover across the gap of the last two cavities can occur if the pressure is high (e.g., in the SLAC's 5045 klystron, Table 7.5, the voltage in the last gap is ~ 400 kV). Third, ions generated by residual gas in the drift tube cannot leave the beam region because of the axial magnetic field, and thus are accelerated between the anode and the cathode, which they hit with large kinetic energy damaging its central part.

The power is extracted from the last cavity by a waveguide, the first part of which must have a smaller height than the corresponding standard waveguide, because this height is generally too large in relation to the cavity height. This problem can be solved by a $\lambda/4$ transformer. The waveguide must finish in a vacuum window, made of ceramic (pure Al_2O_3) so that the power loss in the window is not large compared to the tube output power [e.g., in the earlier mentioned 50 MW SLAC klystron the window glows red ($\sim 900°$C) at full power]. Therefore, the window is water-cooled.

To compensate for the space-charge forces, an axial magnetic field is used: permanent magnets in smaller klystrons, electromagnets in big, high-power tubes. Balanced beams are used in most tubes, since the current density obtained from the cathode is not sufficient to use confined flow. The beam is compressed between the cathode and the entrance to the drift tube, and so the magnetic field distribution is critical. Most tubes use special bucking coils around the electron gun to adjust the field distribution in this region.

The cavity frequencies in a multicavity klystron are adjusted in a different manner, depending on the tube application: different for high amplification, the highest possible power, or the largest bandwidth. If all cavities are trimmed to the design frequency of the tube (synchronous trimming), a large amplification is obtained — in a six-cavity klystron, ~ 110 dB. In practice this is not desirable since the noise, inherent because of Maxwell's velocity distribution and uneven cathode emission, is also amplified, and the signal/noise ratio is decreased. Reflected electrons from the collector and the last cavity can set the tube into oscillations. Thus synchronous trimming is seldom used, and the amplification of most klystrons is adjusted to ~ 40–60 dB.

When trimming for the highest power and efficiency, the frequency of the penultimate cavity is considerably higher than the construction frequency, to which the last cavity is trimmed, and the penultimate cavity becomes inductive. The voltage phase across the gap lags the beam-induced current. For a bunch nearing the cavity, the gap voltage retards the electrons in front of the bunch because the voltage has not yet passed through zero. The bunch itself is not greatly influenced by the electric field, since the electrons pass the gap center when the field is low. The electrons that straggle the bunch are accelerated. In this way the bunching is amplified in the last cavity, and the output power increased, as in the klystron PK771 (Table 7.4).

In klystrons, with power up to ~ 50 kW in the S-band, the efficiency can be further increased by introducing an intermediate cavity trimmed to twice the tube frequency. Numerical simulations show that such a second harmonic cavity amplifies the bunching and the current first harmonic component. Efficiencies more than 70 percent are obtained with such tubes. Regrettably, this method works only for medium-power tubes but not for high-power or higher-frequency tubes. In high-power S-band tubes the current density is so high that the beam diameter must be between 12 and 20 mm, the diameter of the second harmonic cavity is ~ 30 mm, and the drift tube must be between 20 and 25 mm, so there is no space left for such a cavity. In a medium-power klystron in the S-band, the beam diameter is usually less than 10 mm and a cavity trimmed to the second harmonic frequency can be built in.

The bandwidth of a klystron trimmed synchronously or for high efficiency is narrow, usually only a few tenths of a percent. If the intermediate cavities are "stagger tuned" (i.e., one cavity below and the next above the tube frequency), a bandwidth between 5 and 10 percent can be obtained at the price of a reduced amplification. If the bandwidth is only increased slightly and the efficiency is kept high, the Q of the intermediate cavities can be decreased, especially that of the second cavity. The reduced Q-value can be achieved by sputtering a lower conductivity material on the cavity walls, as in the klystron PK771. Note that even if the first three klystron cavities seem trimmed higher than the construction frequency of 2998 MHz, these frequencies are "cold," without the electron beam. Accounting for the beam loading admittance reduces those resonant frequencies to the construction frequency.

Table 7.5 surveys different commercial klystrons. The first, VA-517, is an oscillator with outer feedback. Such tubes are manufactured with frequency up to 50 GHz, have

Table 7.5

Klystrons

	Varian	Varian	Thomson	Thomson	Varian	Varian	SLAC
Type	VA-517	VA-928A	TH2054A	TH2089 B	VKX-7752	VKS-8262B	5045
Application	Oscillator	CW	CW	CW	Pulse	Pulse	Pulse
Frequency [GHz]	9.5–10.6	26–36	2.45	0.352	9–11	2.999	2.856
Bandwidth [MHz]	250	250	5	1 (-1 dB)	200	5	
Acceleration voltage [kV]	11	12	25	100	7	120	400
Perveance [μp]	0.09	0.76	0.81	0.63	1.0	2.1	2.0
Current [A]	0.1	1.0	3.2	20	0.6	86	506
Output power [kW]	0.05	1	50	1300	120	5000	65000
Efficiency [percent]		8	62	65	3	45	45
Amplification [dB]		50	48	42	50	50	50
Pulse width [μs]					4	5	3.5
Number of cavities	2	6	5	5	6	6	6
Length [mm]	230	280	1100	4770	330	889	
Weight [kg]	1.6	36	65	1600	4.5	45	
Magnet weight [kg]			230	650		114	
Magnet type	Permanent	Electro	Electro	Electro	Permanent	Electro	Electro

Fig. 7.21. Klystron TH2089. Specially developed tube for the LEP storage ring at CERN. The length of the tube is 4.77 m (Thomson Tubes Electroniques).

very low noise and good frequency stability, and find applications in Doppler radar and microwave spectrometry. The next three are CW tubes. The first, manufactured with frequency between 1 GHz and 50 GHz, is used by the military in airborne-, marine-, and earth-based radar and in the troposphere reflection links. The second and third L- and S-band CW tube with output power up to 1 MW are used in 400 to 800 MHz TV transmitters, accelerators, plasma heating in fusion research, and material heating. The fourth was developed for a CERN's synchrotron (Fig. 7.21). The fifth and sixth are pulsed klystrons, the first with low power and the latter with high power, whose most important applications are in scattering radars and accelerators. The seventh is a pulsed tube developed for the Stanford linear accelerator.

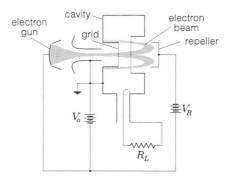

Fig. 7.22. Reflex klystron. V_a is the voltage source for the gun, V_R the source for the repeller, and R_L is the load.

An approach to obtain a high output power is a cluster of 42 klystrons inside a common vacuum chamber and coil with a magnetic field of 0.4 T. Every klystron has six cavities, with an acceleration voltage of ~ 400 kV and current of ~ 100 A. At the frequency of 11.4 GHz, ~ 1.2 GW with an efficiency of 70 percent should be obtained.

7.3.2.6. Reflex Klystron

Klystrons are narrow-band amplifier tubes, which in practice do not exhibit any tendency oscillate. If some output power feeds with the correct phase the input, the tube will oscillate. If an oscillator is wanted with a broader frequency region, the cavities must be tuned up for each frequency, which is extremely impractical.

Many microwave applications require rapid frequency changes of an oscillator at low power. Up to a few gigahertz, semiconductors can be used; but for millimeter waves, tubes are needed. A solution is the reflex klystron [138] (Fig. 7.22). Here electrons start from a Pierce-type electron gun and enter a gridded cavity, after which they come in a retarding electric field from a repeller electrode. The repeller power source potential is lower than the cathode potential, so electric mirror is thus created (Section 4.6). The electrons are turned back without ever hitting the repeller and reenter the cavity. The reflex klystron has no magnetic field so the electron beam spreads from space-charge forces even with low beam current of a few tens of milliamperes. The electrons eventually hit an electrode at earth potential, most of them the inner side of the cavity.

The reflex klystron is self-starting since Maxwell's velocity distribution and nonuniform cathode emission produce space-charge density variations. These density variations will, in turn, generate very weak plasma oscillations with all possible frequencies (e.g., white noise). From these frequencies the cavity resonant frequency will be selected and produce week modulation signal in the gap. The electrons are now velocity-modulated and the faster electrons catch up the slower ones and bunch the beam. An Applegate-like diagram for a reflex klystron is seen in Fig. 7.23.

Electrons passing the gap center at a time between A and C are accelerated, while those that pass between C and E are retarded. Those at B have their velocity increased the most, while for those at D the decrease is the largest. Electrons near B therefore need more time to return to the gap, whereas electrons around D require a shorter time, and so the electrons are bunched. If the repeller voltage is such that the electrons return

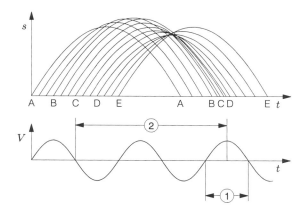

Fig. 7.23. Applegate diagram. 1, Bunched electrons with reversed velocity; 2, transit time in $1\frac{3}{4}$ mode.

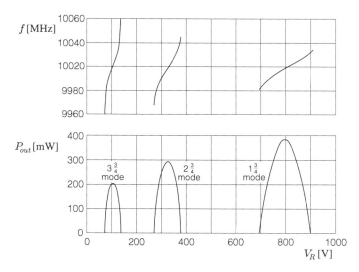

Fig. 7.24. Output power and oscillation frequency vs. repeller voltage (After *Principles of Electron Tubes*, J. W. Gewartowski and A. H. Watson, © 1965, Van Nostrand, with permission.)

to the gap coincidentally, the klystron will oscillate. The time, which depends on the repeller voltage, is $N\frac{3}{4}$ of the period of the cavity frequency, $N = 1, 2, \ldots$.

The output power with oscillation frequency for a reflex klystron working in the X-band is shown in Fig. 7.24. Modern tubes are manufactured with frequencies between 3 and 250 GHz and have a very low noise and very high-frequency stability; the latter is achieved by an external stabilizing cavity (e.g., tubes working at 100 GHz have a temperature coefficient of 1 MHz/°C). The reflex klystron frequency can be changed by changing the repeller voltage, and therefore these tubes are sources for low-output-power microwave devices where the oscillation frequency must be changed rapidly. Important applications are: local oscillator in AFC circuits, communications, military (missiles and their control), microwave spectrometers, parametric amplifiers, and maser pumps. Output power is from 1 W in the S- and X-band to a few milliwatts at 250 GHz.

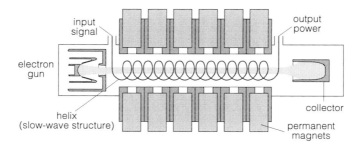

input
signal

output
power

electron
gun

collector

helix
(slow-wave structure)

permanent
magnets

Fig. 7.25. Sketch of a traveling wave tube.

7.3.3. Traveling Wave Tube

HISTORICAL NOTES: In 1942 Rudolf Kompfner [139], an Austrian architect interested in physics, drew in his notebook a sketch of a helix surrounded by a cylindrical metal tube through which both ends of the helix passed perpendicular to the helix axis. On one side he wrote "signal" and on the other he wrote "output" while inside the helix electrons were indicated moving from the signal side to the output. The text: "A completely untuned system! Would it work?" appeared followed by electric field distribution sketches which were wrong, but a friend of his, R. R. Nimo, drew a correct field picture. Both realized that Kompfner's idea was correct, and it was possible to build a wide-band microwave amplifier, a traveling wave tube.

The traveling wave tube is a microwave amplifier, characterized by a large bandwidth. Low-power tubes can cover more than an octave. The circuit in a TWT is not resonant and is always a slow-wave structure, and the electromagnetic wave phase velocity is fashioned to the velocity of the electrons. A survey of the slow-wave structure theory is given in Appendix D.

In a TWT, seen in Fig. 7.25, the microwave signal to be amplified is connected to one side of a slow-wave structure. The weak electric field of the wave which travels along the slow-wave structure provides the electron beam with a small velocity modulation. Like in a klystron, the electrons moving along the tube transform the velocity modulation into a space-charge modulation, which induces a current in the circuit, amplifying the signal. In contrast with the klystron, the interaction between the beam and the electromagnetic field is continuous. The field is strongly amplified at the end of the tube and is extracted by coupling to the outer circuit.

The modulation process principle is seen in Fig. 7.26, where disks represent space-charge. The microwave input signal generates a low-strength electromagnetic wave which spreads along the slow-wave structure. The equidistant disks are influenced by forces, F, from the electromagnetic field, so that disks A to E are accelerated, disks F to K are retarded, and a bunch starts forming around L (Fig. 7.26a). The initial velocity of the electrons, v_0, is chosen to be a little larger than the electromagnetic wave phase velocity. A little farther along the tube, disks L, I, J, and K, which were accelerated, are now in a retarding field with disks A to C. More disks are in the retarding field than in the accelerating, so this is a net energy transfer from the electron beam to the electromagnetic field, and the field amplitude increases. In Fig. 7.26c more disks come into the retarding field and the bunch space-charge increases until finally, at the end of the tube (Fig. 7.26d), the most disks have been driven into the retarding field region, and most of the electrons kinetic energy has been transferred to the electromagnetic field.

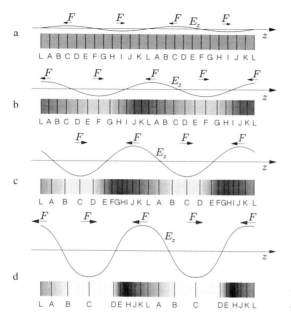

Fig. 7.26. Space-charge distribution in the electron beam.

The interaction between the electromagnetic wave and the electron beam can be divided into two parts: the motion of the electron beam under the influence of electric forces and its own space-charge, and then the electron beam generated induced current buildup. The first is described by the electronic equation, and the second is described by the circuit equation. Both equations can be solved simultaneously and give the relationship between the circuit parameters in the slow-wave structure and the rf component of the electron beam current. Solution is given in Appendix E, which can be summarized as follows: Input signal generates four traveling waves in the tube, a wave whose amplitude increases exponentially along the slow-wave structure, a wave whose amplitude decreases, a wave with constant amplitude, and a wave with a small amplitude, traveling in the opposite direction. The amplitude of the growing wave increases as

$$ e^{\frac{\sqrt{3}}{2}\beta_e C z}, $$

where C is the amplification parameter. For small input signal we have

$$ C = \sqrt[3]{\frac{Z_n I_0}{4V_0}}, $$

where Z_n is the beam coupling impedance, I_0 is the DC beam current, and V_0 is the electron gun acceleration voltage. If the number of electronic wavelengths along the slow-wave structure is N, the small signal amplification is

$$ A = -9.54 + 47.3CN \ [dB]. $$

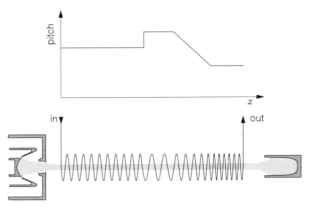

Fig. 7.27. A sketch of a TWT with a helix with variable slope angle to modify the space harmonic wave phase velocity.

7.3.3.1. Large Signal Problems

The traveling wave tube is an amplifier whose signal amplitude can vary over wide limits. For small signal amplitudes linear theory can be used but for higher amplitudes, there are two nonlinear phenomena which become important.

The mean velocity of the beam electrons being slightly larger than the phase velocity of the almost synchronous nth space harmonic wave is an easily fulfilled condition as long as the electrons do not transfer too much of their kinetic energy to the circuit. For high signal amplitudes, the electron velocity decreases toward the end of the tube because of this energy transfer, with the electron bunch coming into an accelerating field, which takes the energy from the microwave field in the circuit and the amplitude of the electromagnetic wave, starts to decrease instead of to increase. Output power and the efficiency of the tube decrease too and saturation is reached.

The other nonlinear effect is caused by the local space-charge in the bunch which increases at high signal amplitudes. The electrons obtain a velocity distribution from the nonlinear space-charge forces, which can no longer be neglected. In addition, the interaction of the beam with the electromagnetic field of the circuit generates higher harmonic components, which must be considered in the space-charge and circuit equation.

In the manner of large-signal klystron theory, disks of electrons were used in numerical simulations of TWT's behavior, and more recently PIC code simulations, which has led to some practical consequences.

In a TWT with a helix-type slow-wave structure the slope angle can be changed near the end of the tube to decrease the space harmonic wave phase velocity and to compensate for the decreased electron velocity, with an increase in output power and efficiency. But this introduces a phase error, which is counterbalanced by making the helix slope angle larger over a certain length of the helix (Fig. 7.27). In the first part of the tube, the slope angle according to linear theory is used. For a TWT built according to these principles an efficiency of ~ 60 percent is reached in the Ku-band, while a constant slope angle TWT has an efficiency of ~ 20 percent.

In the description of the linear theory the fourth wave with negative velocity was mentioned, a wave traveling from the output to the input. If the load is not perfectly matched to the TWT output impedance, part of the electromagnetic wave will be reflected by the load and will then travel along the tube to the input where it can be reflected again if the input impedance is not perfectly matched to the circuit impedance. This reflected wave will be amplified, and if the amplification in this loop is big enough, it can cause large variations in amplification as a function of the signal frequency. The worst-case scenario would result in oscillations. If A is the amplification of the tube, L the cold losses in the circuit (i.e., losses without the electron beam), ρ_o the reflection coefficient at the output, and ρ_i the reflection coefficient at the input, the oscillation condition is

$$A > L + \rho_i + \rho_o. \tag{7.54}$$

Matching the input and output to the waveguide or coaxial line without reflections is difficult. Generally, both reflection coefficients will have a value of ~ 0.1, which corresponds to a VSWR of 1.2 or ~ 10 dB attenuation. The cold losses in a TWT are usually ~ 5 dB. To suppress the oscillations, the amplification must be below 25 dB (usually around 20 dB for safety), much less than a TWT is capable of. Several methods are used to prevent the oscillations at high amplification: attenuation along the slow-wave structure, division of the slow-wave structure in two or more sections terminated by severs, or by varying the slope angle along the helix.

Attenuation is achieved with a thin variable conductivity carbon film, most conductive in the middle of the tube. The film is sputtered or evaporated on the ceramic support rods, which hold the helix in place (Fig. 7.28). Both the forward and the backward wave are attenuated by the carbon film. To adjust the impedance, the carbon film thickness is changed gradually. The space harmonic wave amplitude is still rather low in the region where the attenuation is large. Unfortunately, the bunching process is disturbed by the attenuation increasing the beam velocity spread. Therefore this method is used in low-power tubes and when the efficiency is important.

The backward wave can be attenuated also by matching a multisection tube (Fig. 7.29), and building in terminations (called *severs*) where the slow-wave structure is divided.

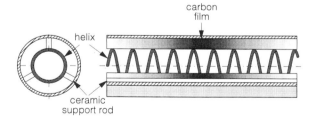

Fig. 7.28. Carbon film attenuation along the helix support rods. The film thickness is simulated by shading.

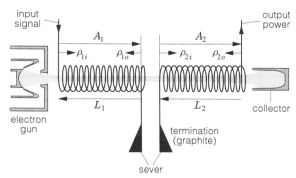

Fig. 7.29. TWT with a helix divided in two sections.

When the forward wave reaches the first sever, the total power is delivered to the matched load in the saver and there is no electromagnetic wave on the other side of the sever. The current and velocity modulation of the beam is, however, preserved and retains the information about input signal frequency and amplitude. In the next part of the tube, the current and velocity modulation generate a new space harmonic wave, which is amplified, reaches the end of the tube, and produces output power. Since the bunching is negatively influenced by the transition at the sever, the transition length must be as short as possible. A smaller amplification is chosen in the first part of the tube, and one as big as possible is choosen in the second for the best efficiency. Equation 7.54 is also valid for multisection TWTs, so in every section the amplification must be limited to ∼ 25 dB. The amplification can be slightly larger, because the severs can be matched better than matching between the slow-wave structure to the input and output.

Even a variable slope angle, according to Fig. 7.27, decreases the oscillation sensitivity, because the backward-going wave sees a wrong propagation constant and is attenuated by energy transfer to the electron beam. The slope angle must be specially adapted to be used in this manner.

Another problem with a helix TWT is the phase velocity, which influences and determines the bandwidth, but changes with frequency as a consequence of the change in the number of turns per wavelength with frequency. At lower frequency the number of turns per wavelength increases, which affects the electric and magnetic field coupling between the turns. The fewer turns per wavelength, the smaller the circuit inductance, and the larger the phase velocity. This can be partially compensated by using ceramic support rods, which fix the helix in place, since their capacitance decreases the phase velocity. Another reason for the phase velocity change is that the helix must be surrounded by a metal cylinder to which the support rods are attached. The electric field distribution depends on the ratio between the cylinder and helix diameters and varies with frequency. At high-frequency the field is concentrated between the turns, while at lower frequency the metal cylinder affects the field distribution more, because of the increase in the number of turns per wavelength. A big influence is the distance between the helix and cylinder: The smaller the distance, the lower the phase velocity, which suggests that this distance be made small to level off the frequency dependence and increase the bandwidth. However, if the distance is reduced, the beam coupling impedance (Eq. E.13, Appendix E) is decreased, and so are the amplification and efficiency. Three compromises are shown in Fig. 7.30 using capacitive copper rods that are brazed to the cylinder, thereby simulating a smaller-diameter cylinder and compensate for the influence of the ceramic support rods.

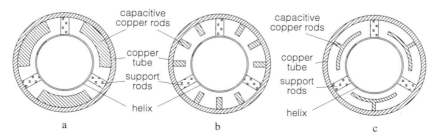

Fig. 7.30. Methods to control the phase velocity over the bandwidth.

7.3.3.2. Construction

Since Kompfner's idea became known, many slow-wave structures have been tried in traveling wave tubes, but only two types dominate: Low- or medium-power tubes use the helix — as a simple helicoidal wire, a rectangular cross-section helix, or the so-called ring-bundle helix. High-power tubes, over 10 kW, use coupled cavities.

A Pierce-type electron gun often with a control grid (or a shadow grid), according to Fig. 6.8, is employed. Some TWTs intended for short pulse operation have a special modulation anode between the cathode and anode. The main anode is kept at the potential of the slow-wave structure, usually that of the chassis. The collector can be simple in stationary units, but is often of the depressed type for mobile units and in satellites, sometimes with more than one stage. Low-power tubes are cooled by air, and the heat is radiated directly into space in the satellites. Forced water cooling for the collector and the slow-wave structure is used in high-power tubes, since ohmic losses in the structure generate heat, and toward the end of the tube the structure can be hit by electrons with large radial displacement.

The choice of materials is similar to that for klystrons: copper, ceramic, and stainless steel. In low-power TWTs the helix is made of molybdenum or tungsten wire, but in medium-power tubes, copper or coppered molybdenum with rectangular cross section is used. The helix is supported by ceramic support rods which must conduct away the ohmic heat generated in the helix. Because of its low heat conductivity, aluminum oxide ceramic (Al_2O_3) cannot be used so the rods in medium-power tubes are made of beryllium oxide or boron nitride, both with similar dielectric constant as Al_2O_3, but much better heat conduction.

To insert the helix in the metal tube, the tube is triangularly deformed by external pressure, the helix and the support rods are then slid in where the tube is the widest, and when the outer pressure is released the tube presses on the rods and makes a good thermal contact. Alternately, the helix and support rods are pressed into a cold or hot tube. In medium-power tubes, and for tubes used in satellites, which must have a long reliable life, the rectangular helix is brazed to the ceramic support rods, and these are brazed to the tube (Fig. 7.31).

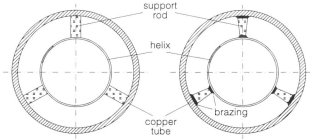

Fig. 7.31. Mounting of the support rods.

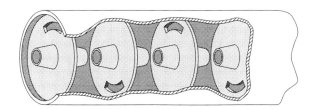

Fig. 7.32. Coupled cavities in a high-power TWT.

The TWT's linear dimensions can be magnified or reduced to obtain the desired frequency region. Experimental tubes with ~ 90 GHz frequency were made with ceramic 0.25×0.5 mm^2 support rods cross section, so heat conduction limited the power to only a few tens of watts. Two methods may overcome this difficulty. In the first, synthetic diamond rods, whose heat conductivity is much larger than copper, could be used; in the second the whole tube interior is coated with beryllium oxide by plasma sputtering.

High-power TWTs use coupled cavities as the slow-wave structure (Fig. 7.32). The cavities are inductively coupled, with a break at the places where severs are built in to attenuate the reflected wave. The first cavity is coupled to the input where the beam modulation starts as in a klystron, and the beam induces current in the next and all following cavities. A cavity electric field is a sum of the induced field and the field that comes through the coupling from the preceding and succeeding cavities. All three components modulate the beam in each cavity, and from the last cavity the power is extracted in a manner similar to that used in the klystron. Figure 7.33 shows a photograph of a high-power traveling wave tube, while Fig. 7.34 is a cross section of the same tube.

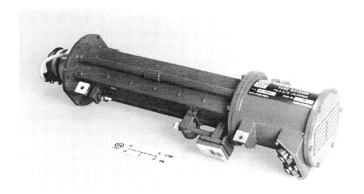

Fig. 7.33. Traveling wave tube VTX-5786 (Varian).

Table 7.6

Traveling wave tubes

	Varian	Varian	Thomson	Varian	Thomson	Varian	Thomson
Model	VTS-6252	VTM-6195	TH3640	VTX-6383	TH3619	VTA-6630	TH1304B
Application	CW	CW	CW	CW	CW	CW	P
Frequency [GHz]	$2-4$	$6-18$	$5.8-6.4$	$10.2-10.7$	$11.7-12.5$	$30-31$	$2.85-3.15$
Acceleration voltage [kV]	3.8	3.9	13.5	20.0	7.1	22.0	47.0
Current [A]	0.42	0.115	1.4	3.0	0.16	1.0	16.5
Output power [kW]	0.2	0.025	3	10	0.23	2	240
Efficiency [percent]	12	6	16	17	50	9	31
Amplification [dB]	37	50	44	50	50	40	32
Pulse width [μs]							25
Magnet type	Permanent	Permanent	Electro	Permanent	Permanent	Electro	Electro
Collector type	Depressed			Depressed	Depressed		Depressed

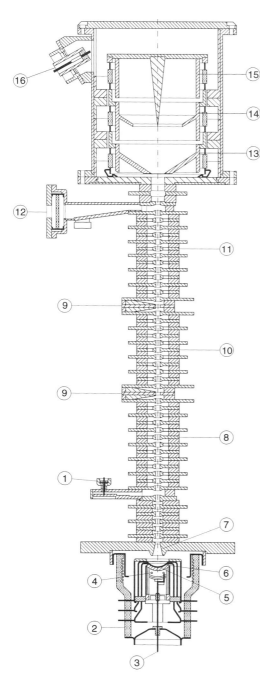

Fig. 7.34. Cross section of the traveling wave tube VTX-5786 with coupled cavities (Varian). Data: X-band, 10 kW pulse power, 1 kW mean power, anode voltage 23 kV, 3.2 A beam current, collector voltage 14 kV. The tube is grid-pulsed and water-cooled, has periodic magnetic focusing, weights 9 kg, and is 500 mm long. 1, Input signal connection; 2, high voltage isolator; 3, heater connection; 4, heater wire; 5, dispenser cathode; 6, control grid; 7, anode; 8, slow-wave structure, first part; 9, sever; 10, slow-wave structure, second part; 11, slow-wave structure, third part; 12, waveguide output with ceramic window; 13, first collector electrode; 14, second collector electrode; 15, third collector electrode; 16, connection to the second and third collector electrode.

Table 7.6 surveys some commercial traveling wave tubes. The first tube is used as a jamming transmitter in military applications, or as a wide-band amplifier in civil applications. The other tube is a "mini" TWT, to be used in airplane radar (civil and military) and as a jamming transmitter. A tube with similar performance is used in communication links. The third and fourth tubes are ground station-to-satellite transmitter tubes.

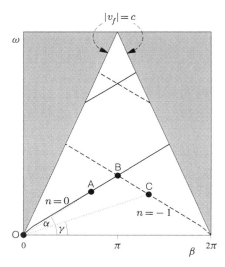

Fig. 7.35. Brillouin diagram of a helix. In the shaded area we have $v_f > c$.

The next is a direct TV signal satellite transmitter tube; similar lower-power (~ 20 W) tubes send signals from satellite to special receiving stations, or from space probes back to earth. The sixth tube is a millimeter-wavelength tube built with coupled cavities, which shows the commercial limits of high power and frequency. The last tube is a high-power TWT to be used in military radar.

7.3.4. BWO Tubes

Space harmonic waves in a slow-wave structure can satisfy the boundary conditions only if all harmonic components simultaneously exist in the structure (Appendix D). The Brillouin diagram, $\omega = f(\beta)$, of the actual structure allows the phase velocity for each space harmonic component to be found. Since the traveling wave tube works in the ground mode, $n = 0$, the phase and group velocities have the same direction, but in the $n = -1$ mode, the phase velocity and group velocity are in opposite directions. An analogy with the motion of water waves in Appendix F gives a simple physical explanation.

In the $n = -1$ mode a microwave tube with the input signal feed near the collector and the output power extracted near the cathode is possible, and these tubes can be both amplifiers and oscillators. Mathematically, the electronic equation (Eq. E.12) in Appendix E, derived for the traveling wave tube, is valid even for a tube with backward wave, because only the electron beam and its space-charge entered the derivation. The circuit equation (Eq. E.26) can be similarly derived, by taking into account the space harmonic wave with $n = -1$, which interacts with the electron beam, with the final result differing only in the sign of the right side. The solution gives again four waves, with the only important one being the wave which travels in the negative z direction and has a maximum at $z = 0$.

The Brillouin diagram for a helix-type slow-wave structure (Fig. 7.35), shows the striking difference between a TWT tube and a BWO tube. TWT uses the space harmonic wave with $n = 0$, and the line OA determines the phase velocity, $v_f = \tan \alpha$. Moving A along the $\omega = f(\beta)$ curve, it can be seen that ω can be varied in wide limits without changing the phase velocity much, which explains the large bandwidth of a TWT tube. A BWO tube works with $n = -1$, so by choosing the phase velocity $v_f = \tan \gamma$, represented by the line OC, the tangent at C to the curve $\omega = f(\beta)$ is negative and so is the group velocity.

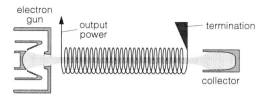

Fig. 7.36. Principle of a BWO tube.

For the field to strongly interact with the electron beam, the beam velocity must be approximately equal to the phase velocity. The beam velocity is proportional to the square root of the acceleration voltage, so by changing the voltage a strong interaction can be obtained over a wide frequency region, which corresponds in Fig. 7.35 to moving point C along the $n = -1$ curve. The angle between the line OC and the $\omega = f(\beta)$ curve is rather large and so the bandwidth is very narrow, when the interaction with the $n = -1$ wave is employed, which characterizes an oscillator. From here comes the name, backward wave oscillator, even if the tube can in principle work as an amplifier. Figure 7.35 also shows the frequency limits, since one must not come near point B, where interaction with $n = 0$ space harmonic wave is possible, and the reflected wave from the output can become significant and generate oscillations with undesired frequency.

Figure 7.36 shows a sketch of a BWO tube where the slow-wave structure is terminated on the collector side with a matched termination. The signal is taken out near the cathode. By varying the acceleration voltage, the tube can oscillate over a wide frequency region, an octave in the L-, S-, or X-band, less at higher frequencies. The BWO tube is used as a microwave signal generator and in applications where low microwave power is needed (e.g., as local oscillator). Commercial tubes cover the frequency region from the L-band to 600 GHz with modest power, a few watts in the L-band to a few milliwatts at the highest frequencies.

7.4. CROSSED-FIELD MICROWAVE TUBES

7.4.1. Introduction

Tubes with crossed fields *use a stationary magnetic field to achieve the interaction* between the electrons and the rf field. The electrons move in a region with mutually perpendicular constant electric and magnetic fields. The electron orbits in such a field combination, discussed in Section 3.2, are cycloidal and a drift velocity, $v_D = E/B$, can be defined, with which the electrons move perpendicularly to both the electric and magnetic field. Most of the crossed field tubes are coaxial, with the cathode cylinder usually near the tube axis.

Table 7.7 surveys different crossed field tubes which can be divided into two main groups. In the first are tubes with a large cathode that covers most part of the cylinder, and the electron emission is distributed. In the other group, injected beam tubes, electrons are emitted only from a small region near the central electrode, which must not have the same potential as the rest of the central electrode (normally called the cathode).

The most important among the crossed-field tubes is the magnetron, an oscillator that belongs to the first group. In both groups there are **Crossed-Field A**mplifier tubes — built with a slow-wave structure which allows forward waves. A **Magnetic BWO** tube works

Table 7.7

Microwave tubes with crossed fields

with backward waves; in this group there is also an oscillator, the voltage-tuned magnetron. There are many other combinations, among them linear tubes with crossed fields (more an exception from the rule), but most were manufactured only as research examples and are not manufactured commercially. The name M-type tubes comes also from French — *tubes à propagation des ondes dans le champ magnetique.*

7.4.2. Magnetron

HISTORICAL NOTES: An electron tube with a coaxial anode and cathode — the latter a tungsten wire — in a magnetic field perpendicular to the electric field was described by Hull [140] in 1921. With this tube he could show very weak rf oscillations. Over the next 15 years the power was pushed up to ~ 100 W in the UHF region, and the anode was divided into two half electrodes and later four quadrant electrodes, two and two connected together. These simple magnetrons did not oscillate with a constant frequency, and the output power was rather low. Posthumus [141] showed in 1935 that magnetron could achieve considerable efficiencies if the conditions of traveling wave operation were fulfilled. The consequence was a US patent [142] of a multicavity structure, which seems not to be observed.

In 1940, during the Battle of Britain, when the German bombers were destroying England, English aircraft used a radar constructed with such tubes with a wavelength of ~ 5 m. This radar had a very wide beam and a low resolution. Because of its shorter wavelength and higher power, the klystron could increase the range of the radar and make detection more difficult, and the British decided to manufacture such tubes.

In the same year, Boot and Randall [23] combined the British radar tube, with a copper cylinder anode with cavities, like the ones used in the klystron. These cavities where coupled by narrow slits to the region around a cylindrical cathode. The first Boot and Randall tube gave directly an output power of 10 kW at 10 cm wavelength. An industrial production started immediately, and after only three months the first English radar was installed in airplanes, on the ground, and in the navy. In a book describing the achievements of electronics during the second world war, J. J. Coupling (a pseudonym for J. R. Pierce) wrote in 1948:

The two most important weapons of the recent war were the atomic bomb and radar ... and it was Maggie, the magnetron, that made Allied radar incomparably superior to the Nazi's boat.

It is interesting to note that Boot and Randall did not know anything about the theoretical work, nor about the Russians, Alekseiev and Malairov [143], who in 1938 had constructed a magnetron with hole and slit cavities, with 10 cm wavelength and output power of 100 W.

Figure 7.37 shows a sketch of a magnetron. A cylindrical cathode emits the electrons over the whole surface. The anode is a copper cylinder with cavities, the inner sides of which are coaxial with the cathode. The electrons emitted from the cathode

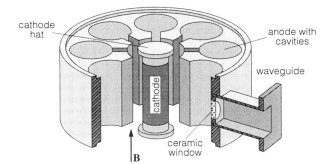

Fig. 7.37. Sketch of a magnetron.

form a cloud around it, and depending on the magnetic field strength, they either hit the anode or fly by and return to the cathode. The anode cavities form a slow-wave structure, and if the electron drift velocity is roughly equal to the space harmonic wave phase velocity traveling around the structure, an interaction between the electrons and the rf field occurs. Electrons retarded by the field transfer energy to the rf field, a current is induced in the cavities, and the power can be extracted by a loop or by coupling to a waveguide.

7.4.2.1. Linear Magnetron Model

Although the magnetron is always constructed as a coaxial device, it is acceptable to use a linear model to describe its functioning. Imagine a plane, infinitely long cathode, at a distance d and parallel to a plane, infinitely long anode (Fig. 7.38). Pretend initially that the anode does not have any slow-wave structure.

The electric field between the cathode and the anode is $|\mathbf{E}| = -E_y = E$, and the whole system is immersed in a homogeneous magnetic field, $|\mathbf{B}| = -B_z = B$, perpendicular to the x–y plane. The electron equation of motion in such a field, derived in Section 3.2, shows that if the initial velocity at the cathode is zero, Eqs. 3.14 and 3.15 can be rewritten as

$$
v_D = \frac{E}{B},
$$
$$
v_x = v_D(1 - \cos \omega_c t), \qquad v_y = v_D \sin \omega_c t, \qquad (7.55)
$$
$$
x = v_D\left(t - \frac{1}{\omega_c} \sin \omega_c t\right), \qquad y = \frac{1}{\omega_c} v_D(1 - \cos \omega_c t),
$$

where v_D is the drift velocity and ω_c the cyclotron frequency (Eq. 3.10). All electrons move to the right in the figure and the drift velocity is constant, parallel to the cathode. Even if Maxwell's velocity distribution is taken into account, the drift velocity remains constant and unchanged for all electrons.

The emitted electrons form a cloud, called the *Brillouin layer*, in front of the cathode. As in a diode, a space-charge is generated and the slowest electrons are retarded by the space-charge field and return to the cathode. The space-charge invalidates

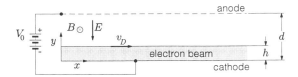

Fig. 7.38. Linear magnetron model.

the simple derivation of Eqs. 7.55, and a potential gradient is generated in front of the cathode, so that the drift velocity must be expressed by a variable E field,

$$v_D = \frac{E}{B} = -\frac{1}{B}\frac{dV}{dy}. \tag{7.56}$$

From the energy equation

$$\frac{1}{2}mv_D^2 = eV, \tag{7.57}$$

we obtain

$$\left(\frac{dV}{dy}\right)^2 = 2\eta B^2 V, \qquad \eta = \frac{e}{m},$$

which, after integration, is

$$\sqrt{\frac{2}{\eta}}\frac{\sqrt{V}}{B} = y + C.$$

The boundary conditions $y = 0$, $V = 0$ give $C = 0$. Potential inside the cloud is thus

$$V = \frac{\eta}{2}B^2 y^2. \tag{7.58}$$

At the top of the cloud, $y = h$, the space-charge ends, but the electric field

$$E(h) = \left(-\frac{dV}{dy}\right)_{y=h} = -\eta B^2 h \tag{7.59}$$

must be continuous at the transition and constant above it in the region without charge that continues to the anode. The anode voltage, V_0, is given by

$$V_0 = -\int_0^h E\,dy - \int_h^d E\,dy = V(h) + \eta B^2 h(d - h) = \eta B^2 h\left(d - \frac{h}{2}\right). \tag{7.60}$$

The Brillouin layer thickness varies with both the anode voltage and the magnetic field. If $h < d$ no electrons can reach the anode, and the anode current is zero. The electrons can reach the anode ($h = d$) if

$$V_0 \geq \frac{\eta}{2}B^2 d^2, \tag{7.61}$$

a relation called *Hull's parabola* [140] (Fig. 7.39), which determines the cutoff voltage of a linear magnetron. If the anode voltage is less than the corresponding value on Hull's parabola, theoretically no current reaches the anode.

In the coaxial magnetron, Hull's cutoff condition remains a parabola but the cutoff voltage is

$$V_0 = \frac{\eta}{8}B^2 r_a^2\left(1 - \frac{r_c^2}{r_a^2}\right)^2, \tag{7.62}$$

where r_a is the radius of the anode and r_c is that of the cathode.

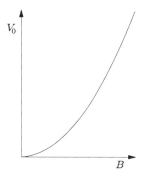

Fig. 7.39. Hull's parabola.

Even in a linear magnetron model the anode should be a slow-wave structure and the space harmonic wave phase velocity along the structure must be adjusted to the drift velocity at the top of the electron cloud, so that the electrons will interact with the rf field; some electrons will be retarded, thereby increasing the energy of the wave, while others will be accelerated. The rf field is superimposed on the DC field between the cathode and the anode.

The upper part of Fig. 7.40a shows a picture of the total electric field when the rf components reach peak value. The lower part shows the forces acting at that moment on the electrons. The field will, to some modest degree, change while the wave travels along the structure. Figure 7.40b shows the approximate motion of the electrons starting from B, D, F, and H. An electron which starts at B is influenced by a transverse force directed toward the anode and by a longitudinal accelerating force, so that the axis of its cycloidal motion, the dash-dotted curve, will bend toward the cathode, which, after a short orbit, the electron will hit. An electron which starts at D is accelerated in the same direction, but the axis of its cycloid is bent toward the anode, and the amplitude of its cycloidal motion

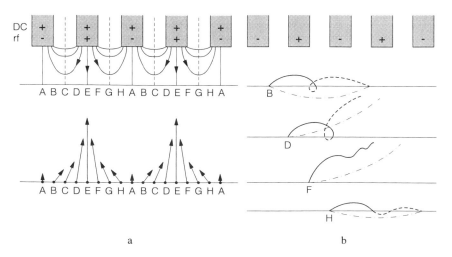

Fig. 7.40. (a) Forces which affect the electrons and (b) motion of electrons in the superimposed rf and DC fields. (After *Vacuum Tubes*, K. R. Spangenberg, 1948, McGraw-Hill.)

increases also. So even this electron is accelerated in the longitudinal direction and hits the cathode. An electron which starts near F is accelerated in the transverse direction, but is retarded in the longitudinal. The axis of the cycloid is also bent toward the anode, but as the electrons are retarded in the longitudinal direction, they give up a part of the kinetic energy obtained in the DC field to the rf field. These electrons remain in a retarding longitudinal rf field until they finally hit the anode. The lost DC energy is larger than that gained in the transverse direction of the rf field, and the difference drives, through induced currents, the slow-wave structure. In this way the energy transfer from the DC power source to the rf wave takes place. An electron starting at H feels a retarding force in the longitudinal direction, along with a weak force toward the anode in the transverse direction. The latter is not sufficient for the electron to make a long orbit, and so it also returns to the cathode. A bunching of electrons occurs between E and G, because the electrons which start near G are subject to a stronger retarding component of the longitudinal force and thus come closer to the electrons which starts at F. Electrons which start at E are less retarded, so even they come closer to electrons which start at F. This bunching, and the fact that most of the electrons which return to the cathode only take a little part of their kinetic energy from the rf field, contributes to the high efficiency of the magnetron.

At the top of the Brillouin layer the drift velocity according to Eqs. 7.56 and 7.59 is

$$v_D(h) = \frac{E(h)}{B} = \eta B h \tag{7.63}$$

and should be approximately equal to the phase velocity,

$$v_f = \frac{\omega}{\beta},$$

of the space harmonic wave to which it is desired to couple the electron beam:

$$v_D(h) \approx v_f.$$

Hence

$$\frac{\omega}{\beta} = \eta B h, \qquad \text{or} \qquad h = \frac{\omega}{\eta \beta B}. \tag{7.64}$$

If this expression for h is introduced into Eq. 7.60, we obtain

$$V_0 = \frac{\omega}{\beta}\left(Bd - \frac{1}{2\eta}\frac{\omega}{\beta}\right), \tag{7.65}$$

which is Hartree's condition [144] and shows the linear relationship between the anode voltage and magnetic field. For the magnetron to start, oscillating electrons at the top of the Brillouin layer must have the correct drift velocity, which happens only at a V_0 which, at the corresponding B, lies above the Hartree's line. On the other hand, Hull's parabola must not be exceeded, for then almost all electrons reach the anode. Thus the magnetron working point is chosen between Hartree's line and Hull's parabola, the shaded area in Fig. 7.41. For the coaxial magnetron Hartree's condition is

$$V_0 = \frac{\omega}{\beta}\left[B\frac{(r_a - r_c)^2}{2} - \frac{1}{2\eta}r_a^2\frac{\omega}{\beta}\right]. \tag{7.66}$$

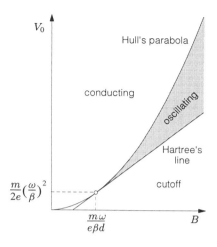

Fig. 7.41. Oscillation region between Hull's parabola and Hartree's line.

7.4.2.2. The Oscillation Modes

The magnetron is difficult to grasp theoretically; the basis are Hull's parabola and Hartree's line, but even these equations are based on the use of a laminar model. The electron beam interacts with DC and rf fields concurrently, which complicates the comprehension. Figure 7.42 shows anode current with anode voltage for a coaxial magnetron without slow-wave structure. Without the magnetic field, we expect that the current follows $V^{3/2}$. With the magnetic field and with an anode voltage higher than the corresponding Hull's parabola value (at 3 in figure), the current varies as $V^{3/2}$. However, for voltages below Hull's parabola the current should be zero. This is so at low anode voltages, but at 1 in the figure there is a small anode current. A further increase in voltage corresponding to the value on Hull's parabola causes a sharp current increase (at 2 in figure). Qualitatively, this sharp increase can be explained by electrons in the tail of the Fermi–Dirac distribution having nonzero initial velocity. Furthermore, the whole cathode surface does not emit evenly, which results in statistical deviations in the space-charge distribution; some electrons feel a strong force directed toward the cathode, whereas others feel it toward the anode. Finally,

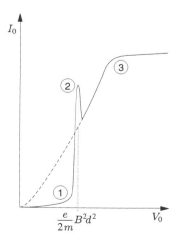

Fig. 7.42. Anode current vs. anode voltage in a coaxial magnetron without slow-wave structure. (After *Principles of Electron Tubes*, J. E. Gewartowski and A. H. Watson, © 1965, Van Nostrand, with permission.)

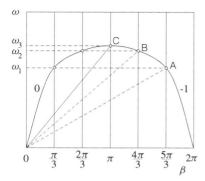

Fig. 7.43. Brillouin diagram for the slow-wave structure of a magnetron with 3 periods.

electrons hitting the cathode cause secondary emission and local cathode heating. All these effects produce irregularities and instabilities in the space-charge, generating an appreciable noise most noticed near the V_0 and B_0 corresponding to Hull's parabola.

The high inherent noise, which in first approximation is white, initiates the oscillations with the frequency corresponding to the design frequency of the slow-wave structure. However, the slow-wave structure allows infinitely many space harmonic waves and the noise has all frequencies, so without special precautions the magnetron will not start to oscillate at the design frequency. The Brillouin diagram for the magnetron slow-wave structure is like Fig. D.3 (see Appendix D), with the main difference that a magnetron structure is closed on itself. Therefore, only oscillations with such space harmonic waves are possible for which the total phase shift equals $2n\pi$ radians. Thus an N period slow-wave structure along the anode can only start the oscillations if

$$\beta = \frac{2n\pi}{N}. \tag{7.67}$$

Figure 7.43 shows oscillation modes if $N = 3$.

To start the oscillations, the beam drift velocity at the top of the Brillouin layer must approximately equal the phase velocity of the space harmonic wave. The magnetrons work mainly with pulsed anode voltage, so during the pulse risetime the anode voltage increases from zero to the peak value. In the Brillouin diagram, the slope of a line through the origin, which shows the phase velocity, must be proportional to the anode voltage for the tube to oscillate. So when the anode voltage starts to increase, the beam drift velocity in the Brillouin diagram will first pass point A, then B, and finish at C. Magnetrons are constructed to work in the π mode, corresponding to the frequency ω_3 in the figure, where the electron-beam interaction is the strongest, and the tube has the highest power and efficiency. However, the magnetron will start to oscillate with frequency ω_1, since it is the first opportunity when the condition for synchronism between the electrons and the wave is satisfied. This is avoided by making the beam coupling admittance for the undesired mode low, or resonant losses must be introduced in the slow-wave structure.

For oscillations in the π mode the beam drift velocity is synchronous with two space harmonic waves, $n = 0$ and $n = -1$. The group velocity for the first is in the direction of the beam drift velocity, while that for the second is in the opposite direction. The power flow coupled to these waves is equally large, and both are phase-shifted π per period of the slow-wave structure. The field distribution is similar to that of of any microwave cavity

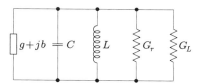

Fig. 7.44. Equivalent circuit of a cavity in the slow-wave structure.

at resonant frequency. The interaction with the load is described similarly to the klystron by defining the unloaded, loaded, and external Q-value.

7.4.2.3. Power, Efficiency, and Load

The magnetron can thus be described by an equivalent resonant circuit, shown in Fig. 7.44. Each cavity in the anode slow-wave structure can be seen as a resonant circuit with its L and C. The beam coupling impedance, $g + jb$, has a negative conductance, corresponding to the power generated in the circuit, and a small inductive susceptance that varies with the oscillation amplitude. The cavity C is concentrated near the cathode, and L is concentrated near the outer walls. Neglecting the small beam coupling susceptance, the unloaded Q-value is

$$Q_0 = \frac{\omega_0 C}{G_r}, \tag{7.68}$$

the external Q-value is

$$Q_e = \frac{\omega_0 C}{G_L}, \tag{7.69}$$

and the loaded Q-value is

$$Q_L = \frac{\omega_0 C}{G_r + G_L}. \tag{7.70}$$

These Q-values are for the whole structure, since they are defined as the ratio between the energy and the losses. Efficiency can be expressed by

$$\eta = \frac{Q_L}{Q_e} = \frac{G_L}{G_L + G_r}. \tag{7.71}$$

Maximum efficiency is thus obtained when the magnetron is as heavily loaded as possible, $G_L \gg G_r$; however, under such circumstances, a small change in the load susceptance strongly affects the oscillation frequency, an effect called *frequency pulling*. When the frequency changes, the beam coupling impedance is also changed, as are the resonant conditions. The magnetron working conditions are usually given in a Rieke diagram [145] (Fig. 7.45), a Smith diagram [146] with curves of constant power and frequency. The concentric, dotted circles give the **V**oltage **S**tanding **W**ave **R**atio of the load at the output coupling to the waveguide or to the coaxial line. For a certain load mismatch the VSWR depends on the electrical distance between the tube and the load. The VSWR also influences the frequency and the power output of the tube. The magnetron can see the load as resistive, inductive, or capacitive. According to Fig. 7.45, at a VSWR of 1.5, the power can vary $\sim \pm 25$ percent, and the frequency can vary between -6 and $+5$ MHz. In the early days it was only possible to couple the output to a magic T and send half the power to an artificial resistive load. Today, ferrite isolators are used, and a VSWR of less than 1.5 can be obtained even if the load is short-circuited.

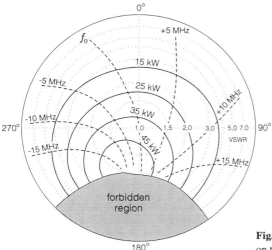

Fig. 7.45. Rieke diagram shows the load influence on the power and frequency of a magnetron.

The frequency of a magnetron also changes if the current in the tube changes. For example, if a pulse-to-pulse or intrapulse variation in the anode voltage exists, the frequency will vary in what is called *frequency pushing*.

7.4.2.4. Numerical Simulations

Magnetron construction has relied on Hull's parabola (Eq. 7.62), Hartree's line (Eq. 7.66), and the empirical experience collected over the 50 years since Boot's and Randall's first magnetron. Modern numerical simulations have not succeeded in producing sufficiently reliable models to explain all the experimental results. The complicated geometry, as well as the interaction between the electrons and simultaneously DC and rf fields, contributes to the difficulties. Furthermore, it is not clear if a simple laminar model can be used. As one magnetron expert says*:

"It is possible that such assumptions, made for convenience of analysis, may have already eliminated a significant amount of physics."

Computer simulation to get a grasp of how the magnetron electron orbits look is shown in Fig. 7.46a, where the electron cloud around the cathode builds a Brillouin layer of varying height, where "wheel spokes" extend to the anode. A modern simulation is seen in Fig. 7.46b, where only a linear model was used. It can be seen how the Brillouin layer takes shape in front of the cathode and how "wheel spokes" form on top of the layer. The interest in such simulation continues to be large because it is believed that there is much to be understood as to how a normal magnetron works. In addition, there are experiments with relativistic magnetrons, in which power output records were broken [e.g., in the S-band these magnetrons give > 10 GW of pulsed output power (Section 7.6.3)].

7.4.2.5. Construction

All magnetrons are constructed as coaxial devices. In the center of most is a cylindrical cathode, surrounded by a copper anode in the form of a slow-wave structure. But this order can be inverted to obtain higher current. The cathode ends on both sides with "hats" (Fig. 7.37), which limit the emission in the axial direction, shape the electric field,

---

* Y. Y. Lau, in Chapter 9 of *High-Power Microwave Sources*

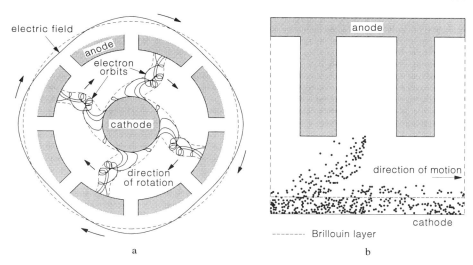

Fig. 7.46. (a) Electron cloud in form of wheel spokes in a magnetron with 8 cavities. [After H. A. Boot, and J. T. Randall, *IEEE Trans. Electron Devices*, (© 1976 IEEE), with permission.] (b) PIC code simulation of electron orbits in a linear model. [After *High-Power Microwave Sources,* edited by Victor L. Granatstein and Igor Alexeff, © 1987, Artech House, Inc., Norwood, MA, with permission.]

and help avoid transport of the emitting material thus reducing the risk of flashovers. A fourth benefit is that the emitted electrons cannot hit the iron pole pieces, which make a part of the magnetic circuit and lie on both sides of the cathode. Such current, if sufficiently high, can heat the pole pieces above their Curie temperature and cause them to lose their magnetic properties.

In the π mode each second cavity will be at the same rf potential, so it is possible to connect the lips of each second cavity galvanically with wire (straps) without changing the rf properties of the tube [147]. For the parasitic modes these connections produce a big change of the resonant conditions; the frequency is moved far enough and the slow-wave structure is loaded heavily, so that parasitic mode oscillations never start. Figures 7.47a and 7.47b show two methods used in practice.

To connect the cavities with wire works for the S-, C-, and X-band tubes, but at higher frequencies (in the millimeter region), it is difficult, if not impossible, to braze the short-circuiting straps. Another model was therefore developed, the so-called "rising sun magnetron" (Fig. 7.47c). The larger cavities determine the oscillation frequencies

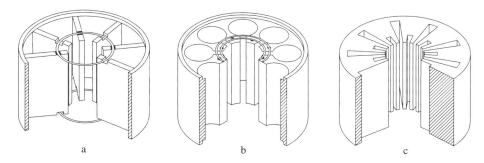

Fig. 7.47. Methods to avoid undesired mode oscillations.

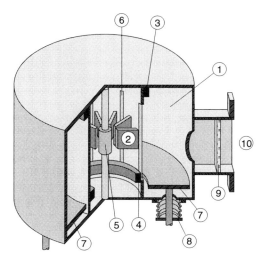

Fig. 7.48. Coaxial magnetron with frequency adjusting piston. 1, Coaxial cavity; 2, slow-wave structure; 3, attenuation ring for cavity; 4, attenuation ring for the slow-wave structure; 5, cathode; 6, coupling slit; 7, frequency adjusting piston; 8, vacuum feed-through for the piston; 9, ceramic window; 10, rf output. (After *Introduction to Coaxial Magnetrons*, Varian-Beverly.)

for all modes with phase velocity substantially lower than for the π mode. The phase velocity which corresponds to the oscillation frequencies of small cavities is substantially higher than that for the π mode. The π-mode frequency, and thus also its phase velocity, is approximately the mean between the resonant frequencies in the ground mode of both cavity types.

The magnetron works at a fixed frequency, which, if changed over a wide region, must be accomplished mechanically requiring bellows. The capacitance of the slow-wave structure cavities can be changed by inserting dielectric rods, metal plates, or rods into what corresponds to a gap in an ordinary cavity. The inductance can be changed by inserting cylindrical metal rods in the cavity peripheries. Another method is to build the cavities open on one side in the axial direction. A metal plate can be moved externally through a bellow in the axial direction to vary the frequency, and at the same time suppress the parasitic modes.

The coaxial magnetron of Fig. 7.48 has a slow-wave structure similar to a normal magnetron but surrounded by a coaxial cavity which works in TE_{011} mode. The coupling between the coaxial cavity and the slow-wave structure is made by slits in each second cavity of the structure. The coaxial cavity has a high Q-value and stabilizes the magnetron frequency. In the coaxial cavity and in the magnetron slow-wave structure, parasitic mode attenuation rings are built in. The movable piston allows the coaxial cavity frequency to be changed over reasonably wide limits compared to standard tubes. The coaxial cavity stabilizes the frequency; thus the tube rf field can be smaller, which prolongs the tube life as much as 10 times.

In radar a very clear echo which can be distinguished in the noise and among possible jamming signals is required. If the emitted signal frequency varies pulse-to-pulse, the probability of detecting the signal in strong interference increases. Magnetrons were developed which send each pulse with a different frequency, one example of which is the so-called spin-tuned magnetron, which has a rotor in vacuum, driven from the air side by induction. In the extension it has a copper cylinder with a number of holes (Fig. 7.49) that lie at the same height as a notch in the slow-wave structure. The rotor spins at a few thousand turns per minute, so every transmitted pulse has another randomly distributed frequency.

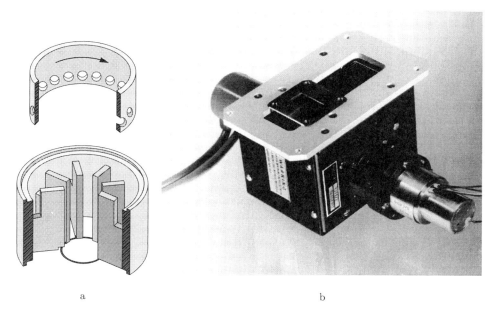

Fig. 7.49. (a) Sketch of the slow-wave structure and rotating anode in a spin-tuned magnetron. (b) Spin-tuned X-band magnetron YJ1172 with 1200 pulses/second, 100 kW pulsed power, and 5 percent bandwidth (CelsiusTech Electronics AB).

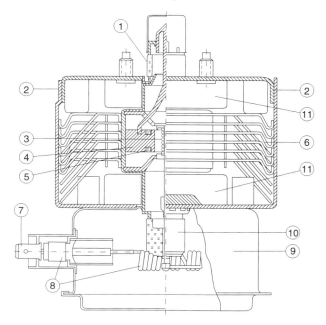

Fig. 7.50. Cross section of a microwave oven magnetron (Toshiba). 1, Antenna — rf output; 2, magnet yoke; 3, thoriated tungsten wire, cathode; 4, anode (resonance cavity); 5, strap rings; 6, air-cooling fins; 7, filament and cathode connection, anode grounded; 8, rf filter; 9, filter case, shields for fringing field; 10, high-voltage isolator; 11, permanent magnet.

Table 7.8

Magnetrons

	Toshiba	Thomson	Thomson	Varian	Celsius Tech	Varian
Type	2M172A	TH3094	TH3051	VMS1143B	YJ1180	SFD-332
Application	CW	CW	P	P	P	P
Frequency [GHz]	2.46	2.45	8.5 − 9.6	2.7 − 2.9	8.8 − 9.4	32.9 − 33.5
Tuning [MHz]		25	1100	200	450*	600
Acceleration voltage [kV]	4.0	5.6	22	64	22	18
Current [A]	0.3	1.6	27.5	97	26	16
Output power [kW]	0.7	6	200	3500	210	60
Efficiency [percent]	60	67	33	48	37	21
Pulse width [μs]			2.8	4	1.5	
Cooling	Air	Water	Air	Water	Air	Air
Model	Standard	Standard	Coaxial	Coaxial	Spin-tuned	Coaxial

* pulse-to-pulse frequency variation interval

For example, in a Ku-band tube with 65 kW output power, the frequency can vary over an interval of 670 MHz.

In all electron tubes that work with continuous cathodes and crossed fields, many emitted electrons are turned back and hit the cathode, adding to the cathode heating. In some high-power magnetrons the cathode heating must be completely turned off once the magnetron starts to oscillate. Manufacturers' instructions must be observed to avoid tube damage.

Table 7.8 surveys some commercial magnetrons. The first, used in microwave ovens (Fig. 7.50), is made in quantities 10 million per year and costs ∼ $15. These tubes are made by brazing copper plates into a slow-wave structure instead of machining the structure from solid copper. The cathode is thoriated tungsten wire. The second tube is used in industrial rf heating. All other tubes are pulsed, and their most important use is in radar. Three are of coaxial cavity type, and one is spin-tuned. The fourth tube has a pulsed power of 3.5 MW and is used in scattering radar and particle accelerators.

7.4.3. CFA Tube

A cross-field tube can be used also as an amplifier, a **C**ross-**F**ield **A**mplifier of which there are eight different types. CFAs work with a forward or a backward wave, with distributed emission or injected beam, with or without feedback, but today, only tubes with distributed emission are built commercially.

An example of the CFA tube (Fig. 7.51), is a sketch of a **F**orward-**W**ave **C**rossed-**F**ield **A**mplifier tube. Here the electron cloud rotates around the cathode like in a magnetron, but between the input and output is a drift region where the electron bunch disintegrates. An attenuating material is used to extinguish the electromagnetic wave, so the feedback between the output and input is broken. The input signal modulates the electrons and the bunching generated by the interaction with the slow-wave structure, and it amplifies the signal on its way around the tube.

Different types of slow-wave structures are used — some that support forward waves, and others that support backward waves. CFA tubes replace the more expensive TWTs with smaller, cheaper, and more robust tubes, which have higher efficiency.

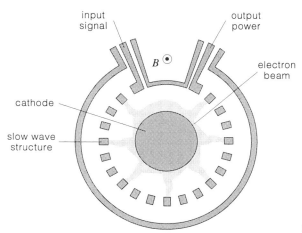

Fig. 7.51. Principle of a FWCFA tube.

Most of what has been said of magnetrons is valid for CFA tubes, with some differences. Many CFA tubes work with a cold cathode, there are always electrons between the cathode and anode, and some field emission exists at high electric field even at room temperature. The field-emitted electrons together with secondary emission can, with the correct anode voltage and input signal, build a large current in ~ 10 ns. In X-band a pulsed CFA tube can give an output power of 1 MW at 50 percent efficiency and a bandwidth of 500 MHz.

7.5. THE GYROTRON

HISTORICAL NOTES: O- and M-type tubes are together often called *tubes with slow-waves*. In these tubes the phase velocity of the electromagnetic wave is adjusted to the velocity of the electrons in the beam and is thus less than the velocity of light. At the end of the 1950s, Gaponov [148] in the Soviet Union, Twiss [149] in Australia, and Schneider [150] in United States independently discovered that microwave tubes can be constructed where the phase velocity is larger than the electron velocity, even larger than the velocity of light. The interaction mechanism between the electrons and the electromagnetic wave is based on the so-called **E**lectron **C**yclotron **M**aser or **C**yclotron **R**esonance **M**aser principle.

In the middle of the 1970s the Russians were first to obtain high power with tubes based on the ECM effect, working in millimeter band, and gave them the name gyrotrons, after small circles in which the electrons move. Today commercial tubes are available with 1 MW peak power at 100 GHz.

The gyrotron electron gun is of the magnetron type, generating an annular beam (Fig. 6.18). The axial magnetic field is weak at the cathode and increases in the interaction region (Fig. 7.52), and so a large rotation of the beam electrons is achieved. It is customary that the ratio between the electron azimuthal and axial velocity, $v_\perp/v_z$, be between 1 and 2. The beam interacts with the field in an open waveguide-type cavity with cylindrical cross section, constructed to work in different modes, like for example in the TE_{01} mode of Fig. 7.53a. In a simple cylindrical waveguide the limitations in size are not serious when made to work at 100 GHz, and even the heat conduction problems can be solved.

The ECM is based on a relativistic effect. When the electrons enter the cavity, they are at random position in their cyclotron orbits, so the transverse electric field, $E_\perp$, accelerates some and retards others. The retarded electrons lose mass as they give up energy to the rf field, and thus they rotate faster ($\omega_c = eB/m$) and advance in phase. For accelerated electrons the opposite is true; thus electrons bunch but most come into a retarding electric field,

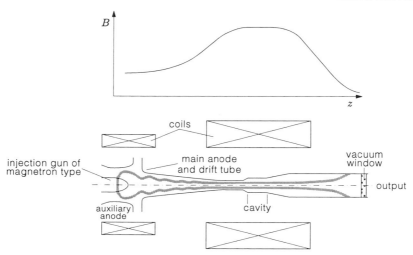

Fig. 7.52. Sketch of a gyrotron. The upper part of the figure shows the magnetic field distribution along the axis. (After *High-Power Microwave Sources,* edited by Victor L. Granatstein and Igor Alexeff, © 1987, Artech House, Inc., Norwood, MA, with permission.)

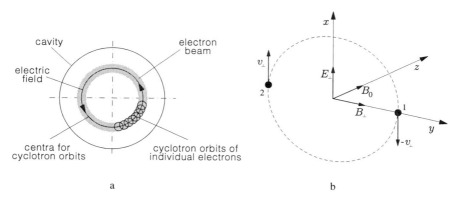

Fig. 7.53. (a) Waveguide cavity cross section. The electron-orbit centra lie on the same circle as the electric field maximum. **(b)** Two electrons in combined DC and rf field.

where they give up their kinetic energy to the rf field. If the cyclotron frequency is slightly lower than the rf field frequency, the electrons will remain bunched in the retarding field over half of the period of their cyclotron orbit

Observe two electrons in their cyclotron orbit in the cavity of Fig. 7.53b. At $t = 0$ their positions are chosen such that their velocities are parallel and antiparallel, respectively, to the rf field. In a cylindrical cavity we have

$$\mathbf{B}_\perp = \frac{\beta_z}{\omega_c}\mathbf{z}_1 \times \mathbf{E}_\perp, \tag{7.72}$$

where $\mathbf{z}_1$ is a unit vector in the z direction and $\mathbf{B}_\perp$ and $\mathbf{E}_\perp$ are rf field components. In the axial direction lies a DC magnetic field, $\mathbf{B}_0$. These three vectors are mutually perpendicular. Without the rf field, both electrons would move along the dashed orbit with a constant Larmor radius,

$$r_L = \frac{v_\perp}{\omega_c},$$

and their cyclotron frequency is

$$\omega_c = \frac{\omega_{c0}}{\gamma} = \frac{e|\mathbf{B}_0|}{m_0 \gamma}, \tag{7.73}$$

where γ

$$\gamma = \frac{1}{\sqrt{1 - v^2/c^2}} = 1 + \frac{e}{m_0 c^2} V_0, \tag{7.74}$$

and V_0 is the acceleration voltage. If, for example, with an acceleration voltage of 100 kV the gyrotron frequency is 35 GHz, the axial magnetic field must be 1.2 T; if the frequency is 100 GHz, a magnetic field of 3.5 T is needed.

The electrons move in the z direction with axial velocity $\mathbf{v}_z$. Thus their movement in respect to the rf field causes a Doppler-shifted frequency

$$\omega_D = \omega_c + \beta_z v_z = \frac{\omega_{c0}}{\gamma} + \beta_z v_z. \tag{7.75}$$

The change of ω_D per unit time bunches the electrons. The time derivative is

$$\frac{d\omega_D}{dt} = \frac{-\omega_{c0}}{\gamma^2}\frac{d\gamma}{dt} + \beta_z \frac{dv_z}{dt}, \tag{7.76}$$

where the right side is obtained from the energy equation

$$m_0 c^2 \frac{d\gamma}{dt} = -e\mathbf{v}\mathbf{E},$$

which, taking into account the directions, gives

$$\frac{d\gamma}{dt} = -\frac{e}{m_0 c^2}\mathbf{v}_\perp \cdot \mathbf{E}_\perp. \tag{7.77}$$

Newton's second law and the Lorentz force are

$$m_0 \frac{d(\gamma \mathbf{v})}{dt} = -e[\mathbf{E} + \mathbf{v} \times (\mathbf{B}_0 + \mathbf{B})]. \tag{7.78}$$

Since the change of the z component is of the main interest, we have

$$\frac{dv_z}{dt} = -\frac{e}{\gamma m_0}\frac{\beta_z}{\omega_c}|\mathbf{v}_\perp \times \mathbf{B}_\perp|.$$

Here, Eq. 7.72 and a substitution of $\mathbf{B}_\perp$ can be used, which gives a triple product

$$\mathbf{v}_\perp \times (\mathbf{z}_1 \times \mathbf{E}_\perp) = (\mathbf{v}_\perp \cdot \mathbf{E}_\perp)\mathbf{z}_1 - (\mathbf{v}_\perp \cdot \mathbf{z}_1)\mathbf{E}_\perp,$$

the second term of which equals zero, because $\mathbf{v}_\perp$ is perpendicular to $\mathbf{z}_1$ and so

$$\frac{dv_z}{dt} = -\frac{e}{\gamma m_0}\frac{\beta_z}{\omega_c} v_\perp E_\perp. \tag{7.79}$$

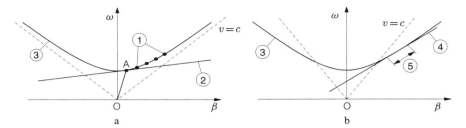

Fig. 7.54. Brillouin diagram for the gyrotron. (**a**) ECM mode. $v_f > c$, line OA. (**b**) Slow cyclotron wave mode, $v_f < c$. 1, Modes in the waveguide cavity; 2, fast cyclotron wave frequency line; 3, waveguide cavity ω-β curve; 4, slow cyclotron wave frequency line; 5, possible region for wide-band interaction (like in a TWT).

Introducing Eqs. 7.79 and 7.77 into 7.76, the result is

$$\frac{d\omega_D}{dt} = -\frac{e\beta_z^2}{m_0\gamma\omega_c}\left[1 - \frac{\omega_c^2}{\beta_z^2 c^2}\right] v_\perp E_\perp, \qquad (7.80)$$

where on the right side the first term comes from the Lorentz force and defines the motion of the electron in the z direction, while the second determines the change in energy because of acceleration or retardation and represents the relativistic effect. Which of the two terms dominates the Doppler shift depends on if

$$\frac{\omega_c^2}{\beta_z^2 c^2} = \frac{v_f^2}{c^2} > 1 \qquad \text{or} \qquad \frac{v_f^2}{c^2} < 1. \qquad (7.81)$$

If $v_f^2/c^2 > 1$ the ECM mechanism is responsible for the interaction. From here comes the name *fast cyclotron wave mode*. If $v_f^2/c^2 < 1$ the interaction is called *slow cyclotron wave, eigen-, whistle* or *Weibel mode*, of which we choose the first.

Let's return to Fig. 7.53b, where both electrons are where $\mathbf{v}_\perp \cdot \mathbf{E}_\perp$ is a maximum, but has different sign for each electron; if one is accelerated and the other is retarded, they must approach each other, or bunch in phase with a phase angle depending on the ratio v_f/c. After bunching the electrons remain in phase with the electromagnetic field in the waveguide cavity and transfer their kinetic energy to the rf field.

A difference between the two possible interactions can be seen in the Brillouin diagram of Fig. 7.54. In the left figure the small slope line shows the fast cyclotron wave mode frequency. The cyclotron wave line is tangent to the curve at point A, the gyrotron working point. The phase velocity, greater than the speed of light, is represented by the line OA, which crosses the curve at large angle. Thus the gyrotron can be used in the ECM mode as an oscillator (gyromonotron), or as a narrow-band klystron-type amplifier with two cavities (gyroklystron).

Figure 7.54.b is valid for the slow cyclotron wave mode. The line of the cyclotron wave and the cavity modes curve are tangent to each other over a wide region, indicating that a TWT-type interaction is possible.

Gyrotrons can work with higher cyclotron frequency harmonics, and thus lower magnetic fields can be used. A gyrotron working at the second harmonic can do so without superconductive coils up to 150 GHz.

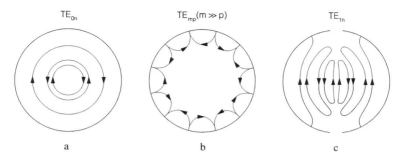

Fig. 7.55. Waveguide cavity modes in gyrotrons. (**a**) TE_{0n} mode, (**b**) whispering gallery mode, (**c**) TE_{1n} mode. (After *High-Power Microwave Sources,* edited by Victor L. Granatstein and Igor Alexeff, © 1987, Artech House, Inc., Norwood, MA, with permission.)

The first gyrotrons worked with TE_{01} mode cavities, with the advantage that the electric field is purely azimuthal and so the losses are small. The disadvantage is that the gyrotron easily can start oscillating in some of the TE_{2n} modes. The opposite extreme is the TE_{1n} modes where cavities can have a split wall without influencing the electric field distribution, retaining mode stability, but experience substantial losses. A third option is the so-called whispering gallery modes,[*] $TE_{m,p}$ (where $m \gg p$), have relatively low losses and good mode separation. Figure 7.55 shows the three mode types used in gyrotrons. If high efficiency and low losses are important as in CW operation, TE_{0n} modes are usually chosen. In pulsed operation, losses are less important, and a TE_{1n} mode is the usual choice.

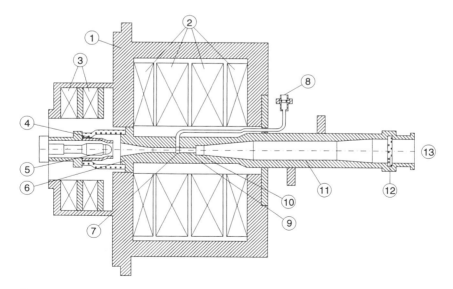

Fig. 7.56. Cross section of a gyroklystron (Varian). 1, Magnet yoke; 2, coils; 3, bucking coil; 4, auxiliary anode; 5, cathode; 6, main anode; 7, input cavity; 8, input signal; 9, drift tube; waveguide below cutoff; 10, output cavity; 11, collector; 12, vacuum window; 13, output power.

---

[*] The name comes from the gallery on the dome of St. Paul's Cathedral in London, where a very low whispering on one side can be heard on the opposite side.

Magnetron-type injection guns are used in gyrotrons where the electron beam has exactly the right shape to interact with an azimuthal electric field. These guns, however, have a relatively large beam velocity spread, $\Delta v_z/v_z$. In a gyromonotron a 15–20 percent energy spread can be accepted, because the cavity interaction region is short. In a gyroklystron and a gyro-TWT, the energy spread must be < 5 percent. In the slow cyclotron wave modes the wavelength in the cavity is shorter because of a low phase velocity, and a much lower velocity spread in the z direction is required, considerably smaller than the 5 percent which is the limit of the best magnetron-type injection guns.

Gyrotrons are like klystrons from the material point of view. The characteristics include the use of magnetron-type injection gun, cylindrical cavities without gaps, power extracted axially, and electrons collected on the output waveguide walls, as seen in Fig. 7.56. At high frequencies the use of superconductive coils is mandatory.

7.6. MICROWAVE TUBES IN DEVELOPMENT

Fifty years have passed since the beginning of microwave tube development with the first klystron, magnetron, and traveling wave tube. These tubes, with their successors O- and M-type tubes, cover a broad region of frequencies, output power, amplification, bandwidth, and efficiency. Their working principle is the interaction between the electrons and the electromagnetic field with phase velocity lower than light speed. However, at high-frequency and millimeter wavelengths, their manufacture becomes complicated and heat generation in the structure at high power is large, so that it cannot be conducted away.

Microwave tubes, which work in the fast wave mode, have widened the limits. The gyrotron is not the only tube, and the ECM is not the only possibility of interaction at high frequencies. New microwave tubes are being developed which use relativistic effects to produce tens of gigawatts of pulsed power.

Advances in computer simulations, and in material and vacuum technology played an important role in modified construction of conventional tubes and in the development of new tubes. A promising improvement in the electron gun design are the field emission arrays with their high current densities. Conventional O-type tube solid beam guns with these cathodes will use confined flow with all its advantages, but even magnetron-type guns will benefit in a simpler cathode–anode geometry.

7.6.1. Ubitron

In 1960 Philips [151] described a new type of microwave tube, the Unmodulated Beam Interaction elecTRON device, which uses an unmodulated beam entering a waveguide. An input signal generates a TE mode electromagnetic wave. The electron beam is focused by a strong periodic magnetic field which has both axial and transverse components, so the electron beam will therefore have a wavelike motion in the transverse direction, called *undulatoric*. If the electron-beam velocity is adjusted to the velocity of the fast electromagnetic wave in the waveguide, so that the beam changes the azimuthal direction in synchrony with the wave, a strong interaction will occur. The electrons will give up a part of their kinetic energy to the electromagnetic wave, and the ubitron will work as an amplifier.

The Free Electron Laser is based on a similar mechanism. A high-energy electron beam, usually from an accelerator, passes a periodically varying magnetic field in an undulator or in a wiggler, where the electrons move periodically in the radial direction,

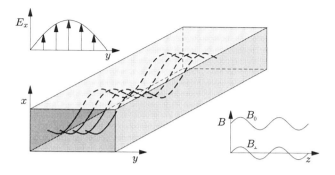

Fig. 7.57. Wavelike motion of the electrons in a waveguide in TE_{01} mode.

and if an electromagnetic wave is present in the region, they interact with the wave. In an FEL the electromagnetic wave is generated by spontaneous emission in the undulator or in the wiggler, while in the ubitron the field is from the input signal.

To effectively amplify the input signal, the electrons must be bunched, and the bunch must be situated in a retarding field. Figure 7.57 shows the electron motion in the waveguide together with the field distribution of the stationary magnetic field and the transverse rf electric field in the TE_{01} mode. The periodic transverse magnetic field is

$$\mathbf{B}_\perp = B_t(\mathbf{x}_1 \cos \beta_t z + \mathbf{y}_1 \sin \beta_t z), \tag{7.82}$$

where B_t is the periodic magnetic field amplitude, $\beta_t = 2\pi/\lambda_t$ is the propagation constant, and λ_t is the magnetic field spatial period. The electron velocity in this field is

$$\mathbf{v} = v_t(\mathbf{x}_1 \cos \beta_t z + \mathbf{y}_1 \sin \beta_t z) + v_\parallel \mathbf{z}_1, \tag{7.83}$$

with

$$v_t = \frac{\omega_t v_\parallel}{\omega_0 - \beta_t v_\parallel}. \tag{7.84}$$

Here,

$$\omega_0 = \frac{|eB_0|}{\gamma m_0}, \qquad \omega_t = \frac{|eB_t|}{\gamma m_0}, \tag{7.85}$$

where B_0 is the stationary axial focusing magnetic field, γ is the relativistic factor, and the denominator in Eq. 7.84 comes from the Doppler effect. The energy equation gives

$$v_\parallel^2 + v_t^2 = (1 - \gamma^{-2})c^2, \tag{7.86}$$

which has four solutions. The positive solution gives a strong interaction between the beam and the wave when the cyclotron orbit period and the periodic magnetic field period are synchronous. In analogy with the FEL this periodic magnetic field is called the *undulator field*.

Figure 7.58 illustrates the interaction. Electrons moving in the z direction can be divided into three groups in relation to the phase of the electromagnetic wave. One group has a favorable phase, the second unfavorable, and the third neutral. The figure shows an electron in the favorable phase at three different times, t_1, t_2, and t_3, phase-shifted by $90°$. At t_1 the electrons move obliquely, $v_x(t_1)$ is negative as is the direction

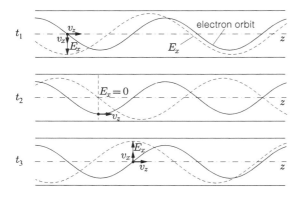

Fig. 7.58. Position of an electron at three different times in relation to the electromagnetic wave. (After *Microwave Tubes*, A. S. Gilmour, Jr., © 1986, Artech House, Inc., Norwood, MA, with permission.)

of the transverse electric field, and the electrons are retarded and give a part of their energy to the rf field. At t_2 the electric field is zero, and the electrons have only an axial velocity. At t_3 the situation is like the one at the time t_1, with the difference that both the direction of $v_x(t_3)$ and that of the transverse field are positive. Also these electrons are retarded. Electrons with the unfavorable phase are accelerated in the same fashion. The neutral ones are not influenced by the field and keep their position in relation to the field. The transverse velocity modulation, which is a consequence of the undulator field, causes bunching in this way.

When the electrons are bunched, the period of the magnetic field can be changed slightly as can the cyclotron frequency. The bunch then moves in a retarding field, remains there, and gives up a part of its kinetic energy to the electromagnetic wave.

In practice, not the rectangular waveguide of Fig. 7.57 but a cylindrical one is used. The electron gun is of the magnetron type with annular beam. The cylindrical waveguide is copper and steel disks, piled up one after another; this simultaneously produces focusing and the undulator field. The magnetic field is ~ 1 T. In pulsed operation, experimental ubitrons gave an output power of 1.6 MW at 16 GHz and 150 kW at 54 GHz.

7.6.2. Peniotron

Peniotron [152], from the Greek "πηνιο," to spool, is a relative of the gyrotron and uses ECM mechanism to generate microwaves; it does not use relativistic effects, but produces bunching in phase. A annular beam is injected into a TE_{21} mode cylindrical waveguide (Fig. 7.59), where the beam is focused by an axial magnetic field. The electrons move in the z direction, about which they rotate with the cyclotron frequency, ω_c. The magnetic field, constructed similarly to the gyrotron, is low in the gun region but increases sharply toward the interaction region. Usually the ratio between the azimuthal and axial velocities, $v_\perp/v_z$, is between 1.5 and 2.

The tube parameters are chosen so that

$$n = s + 1,$$

where s is the harmonic number of the cyclotron oscillations, and n is the harmonic number of the electromagnetic wave. Taking into account the Doppler effect, the tube works as an amplifier with frequency

$$\omega = (n - 1)\omega_c + \beta_z v_z,$$

where β_z is the propagation constant and v_z is the *constant* velocity in the axial direction.

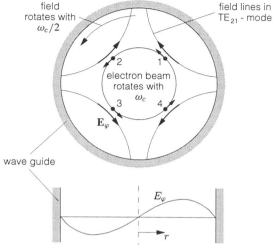

Fig. 7.59. Cross section of a cylindrical wave-guide in the TE$_{21}$ mode with four electrons, representing an annular beam. (After *High-Power Microwave Sources,* edited by Victor L. Granatstein and Igor Alexeff, © 1987, Artech House, Inc., Norwood, MA, with permission.)

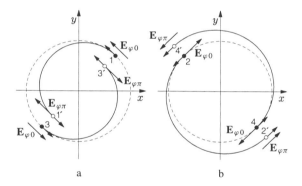

Fig. 7.60. Half turn for the four electrons from Fig. 7.59. (After *High-Power Microwave Sources,* edited by Victor L. Granatstein and Igor Alexeff, © 1987, Artech House, Inc., Norwood, MA, with permission.)

The electric field rotates so that it falls in phase by 2π per Larmor rotation, and each electron therefore sees a complete period of the rf field along its orbit.

From the side, in the $x–z$ plane, the electron orbits look like a helix as in an ubitron (Fig. 7.58). In the $x–y$ plane (Fig. 7.60a), the electrons 1 and 3, are in the favorable phase. These electrons will be retarded by the rf field; their Larmor radius, $r_L = v_\perp/\omega_c$, becomes smaller. Half a period later, these electrons are accelerated, but are now in a region with a lower field, as seen in the field distribution of Fig. 7.59, and the acceleration will not be as strong as the earlier retardation. Electrons 1 and 3, which give up a part of their azimuthal kinetic energy to the electromagnetic field, are retarded as electrons 2 and 4 are accelerated. Half a period later they will be in a stronger rf field and retarded at the same time. So even these electrons transfer more energy to the rf field than they take. The retarding field at the orbit outer side is always stronger than the accelerating field on the inside. The centers of the Larmor orbits move continuously outwards, and the electrons transfer a part of their azimuthal kinetic energy to the rf field.

The peniotron rf field only changes the electron azimuthal velocities, while the axial velocity remains constant once the electrons enter the drift tube with its constant axial magnetic field. The interaction of the electrons with the rf field can transfer all the energy of their azimuthal motion to the microwave field. In the peniotron the $v_\perp/v_z$ ratio must be as high as possible to obtain high efficiency, theoretically up to 70 percent.

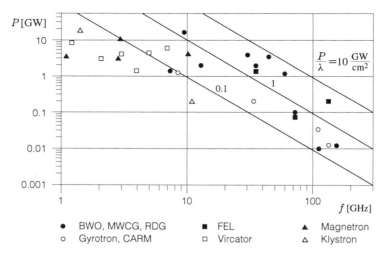

Fig. 7.61. Peak power of HPM devices. (After *High-Power Microwaves*, J. Benford and J. Swegle, © 1992, Artech House, Inc., Norwood, MA, with permission.)

7.6.3. High-Power Microwave Tubes

The introduction of pulsed power in the late 1960s, with electron beams over 10 kA at 1 MV or more, developed for simulation of nuclear weapons, inertial confinement fusion, and high-energy physics, opened the possibility of using these beams in generation of microwave power. A number of classical microwave tubes were extrapolated to relativistic energies. New devices, based on other types of electron-beam–electromagnetic-field interaction, were developed. Figure 7.61 shows the peak power of microwave tubes and devices using pulsed power at the beginning of 1990s, the P/λ^2 lines show the power density on target of a signal transmitted from a fixed-size antenna.

The development of HPM tubes has depended on advances in electronics and physics. Without high-voltage and high-vacuum techniques and without the modern material technology, megawatt and gigawatt power generation over the whole electromagnetic spectrum to the visible light would not have been possible. New cathodes, like FEA and explosive emission cathodes, allow current densities up to tens of kiloamperes per square centimeter. In the latter, high field, ~ 100 kV/cm in front of the aluminum, graphite, or stainless steel cathode extracts large currents from micropoints which rapidly heat and explode, forming plasma within nanoseconds, covering the cathode and forming the emitter. The process is not destructive for the surface as a whole, and the micropoints regenerate from one pulse to the next. These cathodes tolerate poor vacuum, and the limit is given by arcing between the cathode and surrounding electrodes.

A scaling law predicts how the power of a known tube can be scaled up to a new tube with a higher frequency:

$$P_0 = \alpha P_r \frac{f_r}{f_0},$$

where P_r is the power at the frequency f_r, α is a factor between 2 and 4, and P_0 is the expected power at the frequency f_0. This form of scaling law holds also for efficiency.

The construction of the HPM tubes was driven by the interest in Directed-Energy Weapons, fusion research, and accelerators. Among the DEW, HPM has the advantage of free propagation in the atmosphere as compared to laser and particle beams. HPM weapons can be designed for hard-kill on short distance where the solid-state components are destroyed by the electric field, or for soft-kill where critical components are disabled. For example, an energy of 10^{-7} J/cm^2 is enough to cause bit errors in computers, which can be achieved by a 100 ns, 1 GW pulse emitted from a 5 m aperture antenna at ~ 10 km distance. Among the different possibilities to heat plasma in a fusion research device the **E**lectron **C**yclotron **R**esonance **H**eating of plasma is most efficient, but needs CW power of ~ 1 MW at 100–200 GHz. The possibility to generate HPM opened a new sphere of interest: impulse radar, laser pumping and power beaming. An impulse radar sends an ~ 1 ns pulse, so the bandwidth is in gigahertz. Such a radar has a high resolution (~ 10 cm) and better target identification and probably could detect "Stealth" aircraft. HPM is used in pumping nitrogen and excimer lasers, with the advantage of an efficient and clean process. Energy beaming from earth to satellites, from satellites to earth, and between satellites has been investigated since the 1960s. Conventional CFA tubes can be used (for 10 GW 10^5 tubes), but it seems that HPM could decrease the investments. Another region of interest is to use HPM beams for ionospheric modification — that is to generate local artificial ionization in the ionosphere up to ~ 50 km, to facilitate wave propagation for behind the horizon radar and long-range communications during solar storms.

7.6.3.1. Relativistic Klystron

Conventional klystrons were extrapolated to relativistic energies and X-band frequencies. One of the devices uses a structure similar to the twystron, the power is extracted both from the last cavity and from the slow-wave structure. At 1.3 MV and 0.6 kA, 330 MW was extracted.

A markedly different design is the high-current relativistic klystron shown in Fig. 7.62. Bunching is obtained by the retarding action of the space-charge field caused by the intense annular beam from a magnetron-type gun. The beam is velocity-modulated in the first cavity. In the second cavity the peak current approaches the diode space-charge limiting current, decreasing the beam potential. The rf voltage across the gap is such that a periodic "gate" occurs, which generates a high current modulation after the cavity. The power is extracted by use of a converter operating as a coaxial transmission line, and it collects the electron beam in a carbon collector. Experimental tubes have reached ~ 20 GW, at ~ 50 percent efficiency and phase stability $< 1°$.

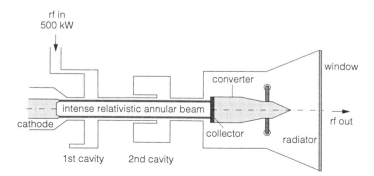

Fig. 7.62. High-current relativistic klystron amplifier. [After Y. Lau, M. Friedman, J. Krall and V. Serlin, *IEEE Trans. Plasma Sci.*, **18**, 553 (© 1990 IEEE), with permission.)]

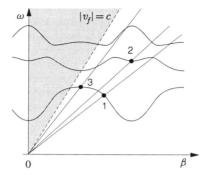

Fig. 7.63. Brillouin diagram of a sinusoidally rippled slow-wave structure. 1, BWO-type interaction; 2, RDG; 3, SWO. (After *High-Power Microwaves*, J. Benford and J. Swegle, © 1992, Artech House, Inc., Norwood, MA, with permission.)

7.6.3.2. Relativistic BWO and Related Devices

A BWO was the first true HPM tube [153] with an output power of 10 MW and efficiency of 0.05 percent. The basic configuration of this and other tubes that followed is an extrapolation of conventional BWO to relativistic energies. The slow-wave structure is either an iris-loaded waveguide, as in conventional high-power TWTs (Fig. 7.33), or a sinusoidally and helically rippled wall. Such BWO tubes have reached a power output > 1 GW.

BWO- and TWT-type are not the only possible interactions between a space-harmonic wave and the electron beam. Figure 7.63 shows three possible interactions in a sinusoidally rippled slow-wave structure. The **R**elativistic **D**iffraction **G**enerator intersects the second space-harmonic wave at a point which has group velocity ~ 0, and the wave shifts from one period to the next by 2π. The **S**urface **W**ave **O**scillator also has group velocity ~ 0, but the axial electric field likes a "surface wave," which has the peak amplitude near the structure wall. Both the RDG- and SWO-type interaction are suitable for thin annular beams, launched from magnetron-type guns.

The highest-power Čerenkov devices, the **M**ulti**W**ave Čerenkov **G**enerator and **M**ulti-**W**ave **D**iffraction **G**enerator, use a large cross section, d, which is a few times larger than the free-space wavelength, λ_0, and a slow-wave structure divided in two sections. The latter are necessary because of the difficulty in choosing the operating mode in a device where many space-harmonic waves open the interaction possibility, as $d \gg \lambda_0$. In MWCG the two sections have the same period, although they can have different lengths. Each section works as an SWO, and for itself it cannot start the oscillations. Together, they produce the output signal with a narrow spectral width. In MWDG and RDG the period differs in both sections, the first acting as a BWO and the second as a diffraction generator. The diameter of the drift tube is equal to the diameter of the slow-wave structure, and the electromagnetic feedback stabilizes the oscillating mode. MWCG has the highest output power: 15 GW at 9.4 GHz, 3.5 GW at 46 GHz, and 1 GW at 60 GHz, all with high efficiency reaching 50 percent at 9.4 GHz.

A phase velocity lower than the speed of light can also be obtained if the electromagnetic wave propagates in a dielectric. The advantage of the design is its simplicity: The dielectric is made by lining a copper waveguide, for example, with Lucite or polyethylene. The problem is the damage by breakdown and beam striking. Experimental devices have given 580 MW at ~ 10 GHz.

7.6.3.3. Relativistic Magnetron

Bekefi and Orzechowski [154] extrapolated in 1975 a conventional magnetron geometry to relativistic anode voltages of ~ 1 MV and could generate output power of 900 MW in the S-band. Relativistic magnetrons work with anode voltages 1–2 MV, currents up to a few tens of kiloamperes, magnetic field ≤ 1 T, the output power of a few gigawatts, pulse length ≤ 1 μs, repetition frequency from single shots to 250 Hz, and with field emitting cathodes. The frequency covers L-, S-, and X-bands.

Relativistic magnetrons are mostly extrapolations of conventional models, but some phenomena which depend on the extremely high tube currents were detected. First, the resonance conditions change when the current supersedes ~ 10 kA since the magnetic field from the electron current cannot be neglected. It is unclear how the electron orbits are changed by the influence of their own magnetic field on the axial motion. Second, irregularities on the cathode surface from which most field emitted electrons originate emit also ions. The ions and the electrons make a plasma that can directly short-circuit the cathode–anode gap and quench the oscillations. Third, electrons hit the anode with large kinetic energy, penetrate into the surface, and there cause local melting and disintegration, which limits the duty cycle and tube life. Power extraction uses conventional radial extraction, but from more than one cavity, and the waveguides are coupled outside the tube structure. The structure Q is reduced and an optimum number of output waveguides exist for a slow-wave structure. Axial extraction with a transmission mode conversion was also tried with similar extraction efficiency.

A new type of crossed-field device, the **M**agnetically **I**nsulated **L**ine **O**scillator, uses the self-magnetic field of the cathode current to cut off the electron beam and to isolate the anode. The electron drift velocity is in the axial direction, but otherwise a MILO works similarly to a magnetron and can be designed as a linear or coaxial device. Extraction is the main problem, having ~ 1 percent efficiency, because the radial extraction destroys the oscillation π-mode and the axial needs a special end diode to absorb the electron beam.

7.6.3.4. Fast Wave Tubes

Construction of high-power ECM tubes has a few goals: gyrotrons with 1 MW CW at ≤ 200 MHz for cyclotron resonance heating of plasma in thermonuclear fusion research, as well as HPM gyrotrons and development of new types of ECM tubes. The design of long pulse and CW gyrotrons has shown a steady upswing; using whispering gallery mode and large diameter cavity it has reached > 1 MW in 0.5 ms and ~ 400 kW in 500 ms operation at 140 GHz, with efficiency 28–38 percent. Experimental HPM gyrotrons have given ~ 250 MW at 35 GHz in ~ 40 ns pulses and up to ~ 1 GW in a mixed regime with three frequencies — 8.35, 8.5, and 9 GHz — simultaneously.

The limit of the fast cyclotron wave mode is defined, according to Eq. 7.80, with $v_f/c = 1$, when $d\omega_D/dt$ vanishes. As the electrons transfer energy to microwaves, the decrease in βv_z is compensated by an increase in frequency, and the electrons remain in resonance. The process is called **C**yclotron **A**uto**R**esonance **M**aser and implies a large axial velocity, v_z. It promises a high efficiency in HPM operation. A CARM tube, shown in Fig. 7.64, uses a pulsed magnetic field of ~ 2 T and a wiggler, made of slotted copper rings with spacing equal to the Larmor wavelength, to provide the transverse beam velocity. The large-diameter cylindrical high-mode cavity has Bragg mirrors on both ends of the interaction region.

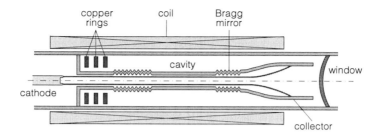

Fig. 7.64. Sketch of a CARM tube. (After *High-Power Microwaves*, J. Benford and J. Swegle, © 1992, Artech House, Inc., Norwood, MA, with permission.)

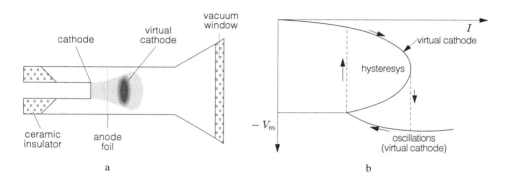

Fig. 7.65. (a) Sketch of a vircator. **(b)** Electrostatic potential minimum vs. current.

7.6.3.5. Vircator

In the 1920s it was already known that when the current in a diode exceeds some limit, a virtual cathode is formed. With a hole in the anode, electrons that enter the field-free region can cause the current to exceed the limit of the Child–Langmuir's law if the space-charge is high enough.

Vircator [155] (Fig. 7.64a), is a diode in which the anode has a thin foil or a very transparent grid. The cathode is pulsed negative from a high-voltage source. This pulse causes a current increase according to Child–Langmuir's law. The electrons are accelerated and come into the field-free region behind the anode. The space-charge generates an electric field with a potential minimum which decreases when the current increases (Fig. 7.65b), forming a virtual cathode from which the electrons are reflected. At a certain current the virtual cathode becomes unstable, the potential minimum decreases abruptly, and the electron cloud starts to oscillate in location and potential with approximately the plasma frequency of the beam. The curve, which shows the virtual cathode potential, makes a jump to the oscillating branch. When the anode voltage decreases, the oscillations persist until the current falls below a limit when another jump back to the normal curve takes place, and the tube ceases to oscillate. Vircator shows hysteresis.

A triode-like model uses an electron beam oriented perpendicularly to the region where the virtual cathode forms and where the power is taken out. Vircator uses anode voltages between 100 kV and 8 MV, with currents between a few kiloamperes and a few hundred kiloamperes, and works in short pulses up to ~ 1 μs. The efficiency is ~ 5 percent;

the triode models have somewhat higher efficiency. Vircator is a tunable oscillator; the frequency can cover two octaves and depends on the space-charge density. The radiation is broadband, and the output pulse consists of a series of spikes. Constructions with open cavity increase the frequency stability and efficiency. The generated power in a vircator reaches 20 GW in a tube with $V_a = 8$ MV and $I_a = 250$ kA.

8

Accelerators

Two cyclotrons were built in Scandinavia in the 1930s. The first to obtain a proton beam was the cyclotron at Nobel Institute for Physics (now Manne Siegbahn Laboratory), Stockholm, in August 1939.

8.1. INTRODUCTION

Particle accelerators have profoundly deepened our understanding of the fundamental nature of matter. Originally using radiation from natural radioactive sources, the nucleus and more recently elementary particles have been ellucidated by accelerator beams. Also there are many applications which make use of accelerators. For example:

- Archeology dating
- Tumor therapy
- Tracer isotope production
- Radiography
- Food and fodder preservation
- Medical sterilization
- Waste processing
- Ozone production
- Plastic polymerization
- Implantation of monomers in composite materials
- Ion implantation in semiconductors
- Coloring of glass and crystals
- SO_2 and NO_x elimination from industrial smoke
- Synchrotron light production of VLSI circuits

Accelerators use electric charge of particles to increase their energy in vacuum, sometimes ultrahigh vacuum, since some particles routinely travel a distance longer than that from the earth to the moon. In circular machines the beam is kept in orbit by magnetic and electric lenses and deflection prisms. Modern accelerator technology requires the most modern production processes, and thus it drives many technologies like high-voltage techniques, superconductivity, magnet design, electron guns, vacuum technology, microwaves, and, in its early stage, computer technology.

HISTORICAL NOTES. Rutherford [156] produced the first nuclear reaction in 1919 when he split nitrogen nuclei with α-particles from a radioactive source. It was soon understood that further progress would require particles with higher energy. In 1922 Slepian [157] suggested an accelerator, a betatron, and 2 years later Ising [158] proposed a linear accelerator powered by a high-frequency source. In 1928 Wideröe [159] constructed a primitive linear accelerator demonstrating Ising's ideas. The lack of powerful rf and microwave sources limited the energy.

Encouraged by Wideröe's article, Lawrence [160] invented the cyclotron and in 1931 his student Livingston [161] constructed the first simple machine which confirmed the principle and in 1932 a 1 MeV proton accelerator was operating. In the same year the first artificial nuclear reaction was made by Cockcroft and Walton [162] using a cascade generator to accelerate protons to 400 keV with which they split lithium nuclei.

The principle for electrostatic generators, discovered by Van de Graaff [163] in 1931, was applied in 1936, and the first betatron was built by Kerst [164] in 1940.

During the 1930s new machines with ever higher energy were constructed while the transverse stability and oscillations of the particle beams were investigated theoretically. Just before the second world war it became clear [165] that the relativistic increase of mass would prevent the acceleration of protons above ~ 15 MeV in a cyclotron, while electrostatic and cascade generators had much lower limits because of the high-voltage problems. A new idea, found independently by Oliphant (1943), Veksler [166] (1944), and McMillan [167] (1945), that, as the relativistic mass increases, the phase stability could be achieved either by changing the rf source frequency or by ramping up the magnetic field during the acceleration. Oliphant never published his proposal because of wartime censorship, while Veksler's proposal, published in a Soviet journal,

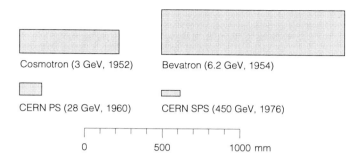

Fig. 8.1. Cross section of accelerating tubes in large proton synchrotrons.

went unnoticed until after McMillan's article was published. Veksler understood that the phase stability principle was also valid for electrons, and so he proposed the microtron.

In Berkeley a 184 inch-diameter magnet was quickly rebuilt in 1946 to produce a synchro-cyclotron, which accelerated protons to 190 MeV and α-particles to 384 MeV. The first microtron was built in 1948 in Canada, and the first electron synchrotron with an energy of 300 MeV appeared in the United States in 1949. By 1952 the first large proton synchrotron, the Cosmotron in the United States, was finished and attained an energy of 2.3 GeV, which was later increased to 3 GeV. Two years later the new Berkeley Bevatron produced a 6.2 GeV proton, an energy sufficient to produce antiprotons. The Cosmotron and Bevatron vacuum chamber and magnetic size caused each machine to weigh more than an aircraft carrier (Fig. 8.1). A new limit had been reached.

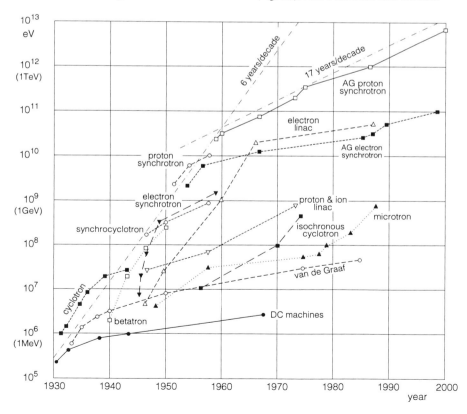

Fig. 8.2. Livingston diagram for accelerators with stationary target.

In 1950 Christofilos [92], a Greek elevator engineer, with an amateurish interest in accelerators, applied in 1950 for an US patent about a new alternating gradient (AG) focusing principle, but this was not observed by the accelerator community. Two years later Courant, Livingston, and Snyder [93] announced the same principle. Applying this new principle resulted in a drastic reduction in the vacuum chamber size and the magnet weight. Proton synchrotrons using AG-focusing were constructed at CERN and Brookhaven in 1960 and 1961 which achieved energies of 28 and 33 GeV, respectively.

The high-power microwave sources developed for radar during the second world war (Historical notes in Chapter 7) made it possible to realize the old idea of **LIN**ear **AC**celerators. At Stanford University a series of electron linacs was built, the largest being 2 miles long, with energies up to 50 GeV. Similar machines for protons were also built, first at Berkeley and the largest at Los Alamos, which produces 800 MeV protons in currents of 1 mA.

Larger machines continued to be constructed. At Fermilab a proton synchrotron, the Tevatron, accelerates protons up to 1 TeV. The diagram displaying the evolution of different types of accelerators is seen in Fig. 8.2. The initial tenfold energy increase every six years broke in the early 1960s.

The available energy in the center of mass system is drastically reduced when the target particle is at rest (see Appendix G, Eq. G.9) as seen in Fig. 8.3. A solution was to let two equal mass high-energy particles collide so that their total kinetic energy is available. These machines, where particles move in closed counterrotating orbits, are called *colliders*. This idea from O'Neill [168] was first realized at Stanford. The first high-energy collider, the **I**ntersecting **S**torage **R**ing at CERN, was built in 1964 and used protons from the 28 GeV proton synchrotron. The CERN's **L**arge **E**lectron-**P**ositron collider accelerates electrons and positrons up to 50 TeV, in a 27 km circumference ring. If the available energy in the center of mass for the colliders is expressed on a Livingston diagram, the tenfold increase in energy every six years is regained, as seen in Fig. 8.4.

With the **L**arge **H**adron **C**ollider in the LEP tunnel at CERN, conventional technology may be at its limit. New ideas are now again necessary if still higher energies are to be realized. Some of these are described at the end of this chapter.

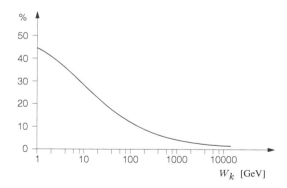

Fig. 8.3. Available energy in collision of a high-energy proton with a stationary proton, expressed in percent of the kinetic energy W_k.

Accelerators can be classified by their accelerating principle in groups, as seen in Fig. 8.5, according to the accelerating field, the space or time variation of the magnetic field, and the accelerated particles.

In accelerators with straight orbits the accelerating field is produced by a DC, low-frequency, high-frequency, or microwave source. The Cockcroft–Walton and Van de Graaff accelerators are DC machines. Low-frequency power is used in induction machines. Linacs for protons and heavy ions use rf sources, while for electrons the source is microwave power. In all linear accelerators the particles can be injected continuously.

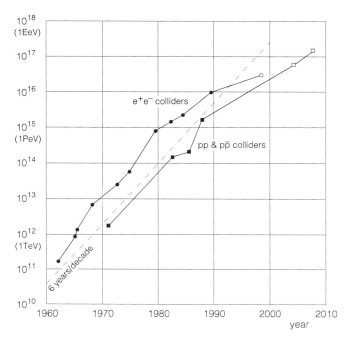

Fig. 8.4. Livingston diagram for colliders. The ordinate shows the equivalent fixed target energy. •, Working; ∘, in construction or upgrading.

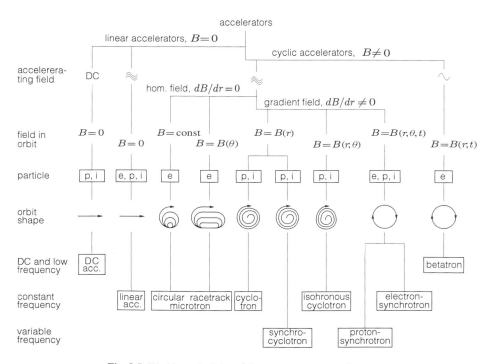

Fig. 8.5. Working principles of the most common accelerators.

The introduction of a magnetic field causes particle orbits to become curved, and so particles can be made to pass in each revolution through one or more accelerating gaps, where their energy is increased. The rf or microwave frequency is adjusted so that the particles remain synchronized with the electric field in the accelerating gap(s). These orbits are circular, roughly circular or spiral, depending on their distribution and the form of the magnetic field. If the magnetic field is spatially constant, the period of revolution is (Chapter 3.2)

$$T = \frac{1}{f} = 2\pi\frac{r}{v} = 2\pi\frac{m}{qB}, \tag{8.1}$$

where f is the rf power source frequency, r the radius of curvature, v the particle velocity, m its relativistic mass, q its charge, and B the magnetic field. The radius of curvature is

$$r = \frac{mv}{qB}. \tag{8.2}$$

Depending on the constant parameter, *cyclic accelerators* are of two kinds: those with constant revolution time or m/B constant and those whose orbit radius is constant, and so the magnetic field is increased proportionate to mv.

When relativistic mass increase is small m/B can be kept almost constant in a homogeneous magnetic field and the radius of the particle orbits increases. This accelerator, the cyclotron, is limited by the magnet size and the relativistic mass increase, so protons and ions are accelerated here. The microtron uses homogeneous magnetic field and accelerates electrons since their velocities differ little from the velocity of light; thus in Eq. 8.2, v can be changed to c. $r/T = 2\pi c$ is constant for all orbits and again magnet dimension determines the energy of the machine.

For machines where r is kept $\sim$ constant the ratio mv/B remains constant only if the magnetic field increases proportional to the particle momentum. In such constructions the magnetic fields generated by prism-like dipole magnets, focusing elements and straight sections around the periphery, close the orbit. This ordering of the elements, called the *lattice*, is used in electron and proton synchrotrons.

The accelerating electric field can also be created by induction while the magnetic field changes. Only electrons can be accelerated in these betatrons. The particle injection can only be made when the magnetic field is weak; hence the duty cycle is low and determined by the time required to cycle the magnetic field.

There are other intermediary forms. For example, in the synchrocyclotron the magnetic field is constant, but the power source frequency is varied to keep the ratio of fm/B constant. In isochronous cyclotrons, which combine features of the cyclotron (continuous beam current) and high-energy, the magnetic field increases with the distance from the center to keep m/B constant. The particle orbits are unstable in the magnetic field direction (Section 8.3.1) but a complicated magnetic field construction with hills and valleys, allowing the field to alternately be high and low, guarantees the stability.

8.2. DC ACCELERATORS

DC accelerators, mainly used to accelerate protons and ions, have common features indicated in Fig. 8.6. Particles from a source are accelerated by a large number of isolated electrodes in an evacuated tube. A resistive voltage divider or a corona discharge defines the electrode potentials so that an approximately uniform electric field is created

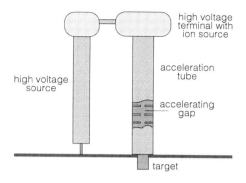

high voltage
terminal with
ion source

acceleration
tube

high voltage
source

accelerating
gap

target

Fig. 8.6. Sketch of a DC accelerator.

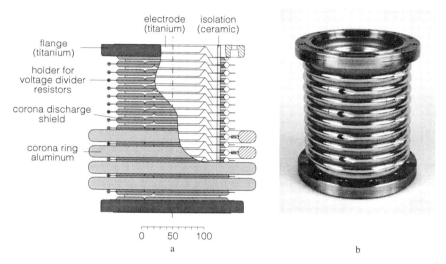

electrode   isolation
(titanium)   (ceramic)

flange
(titanium)

holder for
voltage divider
resistors

corona discharge
shield

corona ring
aluminum

0   50   100

a   b

Fig. 8.7. Example of an acceleration tube. (**a**) Sketch and cross section. (**b**) Photograph. (National Electrostatic Corp.)

and the gaps between the electrodes act as electrostatic lenses which keep the particles focused near the tube axis. Many kinds of high-voltage sources feed these accelerators; the most common are cascade and electrostatic generators. The high voltage of these machines, up to ~ 25 MV, need breakdown precaution in the form of polished electrode support rings, so-called corona rings, and both the high-voltage generator and acceleration tube are mounted in SF_6 filled tanks. This electronegative gas counteracts discharges (Chapter 9).

A typical acceleration tube, shown in Fig. 8.7, is made of ceramic rings brazed to thin titanium rings chosen for their similar thermal expansion coefficients. The inner electrodes screen the ceramic from the beam particles and photons to avoid charging.

A cascade generator is the high-voltage source in Cockcroft–Walton-type DC accelerators where it acts as a voltage multiplier. Two capacitor banks are interconnected by diodes (Fig. 8.8), and a transformer is connected to one bank. The capacitor voltage in this bank follows the transformer voltage where the AC voltage swings between $+V$ and $-V$, if V is the peak secondary winding transformer voltage, usually between 50 and 100 kV. The capacitors in the other bank are charged through the diodes and

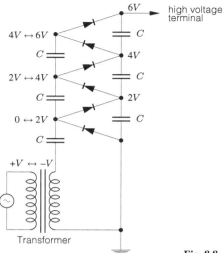

Fig. 8.8. Cascade generator for a Cockcroft–Walton accelerator.

the DC voltage is increased by $2V$ in each capacitor, in the absence of load current. For load current I and the number of capacitors $2N$, the maximum voltage is

$$V_{max} = 2NV - \frac{I}{fC}(\frac{2}{3}N^3 + \frac{1}{4}N^2 + \frac{1}{12}N),$$

with AC voltage driving frequency, f. For this reason most cascade generators use a frequency higher than the main frequency, often ~ 500 Hz.

Van de Graaff accelerators use electrostatic generators to obtain high DC voltage (Fig. 8.9). A motor-driven belt of insulating material transports the charge from ground to the high-voltage terminal. The charge is sprayed on the belt by a corona discharge.

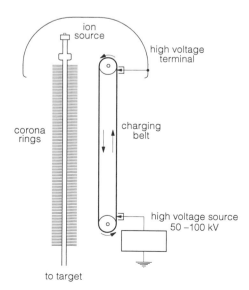

Fig. 8.9. Principle of a Van de Graaff accelerator.

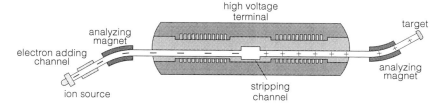

Fig. 8.10. Principle of a tandem accelerator. (After *Particle Accelerators*, S. M. Livingston and J. P. Blewett, 1962, McGraw-Hill.)

The high-voltage terminal is an equipotential surface, and so there is no electric field inside. A comblike electrode is placed near the belt to accumulate the charge and convey it to the terminal. The terminal voltage is determined by the transported charge and by the load current up to $\sim 4 \times 10^{-5}$ As/m^2 for an air isolated machine. For a 0.5 m wide belt and with a 25 m/s velocity the charging current is ~ 0.5 mA, but in SF$_6$ atmosphere it is 10 times greater.

The insulator belt can be replaced by spheres, cylinders, or metal plates connected with insulating interrupts giving longer life time and higher charge-carrying capability because the transfer is made by direct contact. This type of Van de Graaff accelerator is called a *Pelletron* or *Laddertron*.

A negative ion can be produced by a source, accelerated and then "stripped" by traversing a foil, or by interaction with a gas, leaving a positive ion. In the tandem Van de Graaff accelerator (Fig. 8.10), negative ions are formed by adding electrons immediately after the ion source, which is at ground potential. These negative ions accelerate toward the high-voltage terminal, inside of which terminal the stripper is located. The now positive ions are then accelerated to the ground potential where a target is located. Analyzing magnets before and after the accelerator remove ions with the wrong charge-to-mass ratio.

8.3. ORBITAL STABILITY

8.3.1. Betatron Oscillations

In large synchrotrons the particles can travel a distance longer than that from the earth to moon. Focusing is needed to keep these particles close to the ideal or equilibrium orbit over such distance. "Equilibrium" here is used in the sense of oscillations around the ideal orbit, which for the rest of the chapter we will call the *reference orbit*. Although a series of lenses and prisms could be used to ensure stability, a simpler method exists when there is an almost homogeneous magnetic field — let the magnetic field slowly decrease with the magnet radius.

Figure 8.11 defines a coordinate system which we will use for cyclic accelerators. The particle position is defined by its radial, x, and axial, y, deviation from the reference orbit and by its orbit location, s, or alternatively by its corresponding azimuthal angle, θ. We assume that radial and axial deviations are small compared to the orbit circumference, $2\pi r_0$.

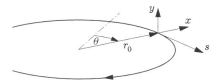

Fig. 8.11. Coordinate system for cyclic accelerators.

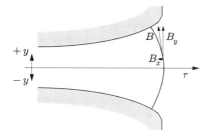

Fig. 8.12. Field components in a magnet with radially increasing distance between the poles.

To obtain a magnetic field which decreases with radius, the distance between the magnet poles must increase with radius (Fig. 8.12). Hence the magnetic field has a strong axial ($B_y \approx B$) and a weak radial ($B_x \ll B$), component. The axial component forces particles to follow a circular orbit while the radial component gives an axial force qvB_r. If $dB/dx < 0$, the force is directed toward the magnet median plane and is focusing (Kerst and Serber [169]).

The field index,

$$n = -\frac{dB/B_0}{dx/r_0},\tag{8.3}$$

indicates the relative decrease of the magnetic field with distance from the reference orbit center of curvature, and axial stability exists if $n > 0$.

Here the reference orbit is defined as that orbit where the centripetal force $F_p = -qvB$ is equal to $F_f = mv^2/r$ (Fig. 8.13). A particle not on the reference orbit will experience a force proportional to $1/r$ toward the reference orbit by choosing

$$0 < n < 1.\tag{8.4}$$

For small excursions from the reference orbit in the radial direction, we have

$$dF_x = \left(-\frac{mv^2}{r_0^2} + qvn\frac{B_0}{r_0}\right)dx,\tag{8.5}$$

and so the particle will oscillate around the reference orbit. Assuming a time-independent magnetic field, $\nabla \times \mathbf{B} = 0$, we have

$$\frac{dB_x}{dy} = \frac{dB_y}{dx},$$

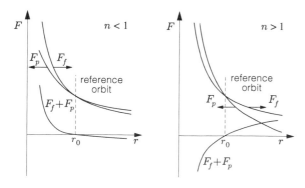

Fig. 8.13. Forces in a radially decreasing magnetic field.

and for small displacements we obtain

$$dF_y = qv\, dB_x = -nqv\frac{B_0}{r_0}dy. \tag{8.6}$$

So also in the axial direction the particle will oscillate around the reference orbit.

If the increase of energy or a change of the magnetic field is slow these results also hold. The oscillations can be described by

$$m\frac{d^2x}{dt^2} + qvB_0(1 - \frac{x}{r_0}n) - \frac{mv^2}{r_0}(1 - \frac{x}{r_0}) = 0$$

and

$$m\frac{d^2y}{dt^2} + qvB_0n\frac{y}{r_0} = 0, \tag{8.7}$$

where $1/r = 1/(r_0+x) \approx (1-x/r_0)/r_0$ and $B \approx B_0(1-xn/r_0)$ was used. Introducing (using Eqs. 3.9 and 3.10)

$$\omega^2 = \frac{qv}{m}\frac{B_0}{r_0}, \tag{8.8}$$

where ω is the revolution frequency, and eliminating time by $dt = ds/v$ and $v = \omega r_0$, Eqs. 8.7 are

$$m\frac{d^2x}{ds^2} + \frac{1-n}{r_0^2}y = 0$$

and

$$m\frac{d^2y}{ds^2} + \frac{n}{r_0^2}x = 0. \tag{8.9}$$

Since $s = \theta r_0$ the solutions are

$$x = X\sin(Q_x\theta + \varphi)$$

and

$$y = Y\sin(Q_y\theta + \psi) \tag{8.10}$$

The constants Q_x and Q_y are the number of oscillations per turn, equal to $\sqrt{1-n}$ and $\sqrt{n}$ respectively. This is called the *betatron tune* and the oscillations are called *betatron oscillations*.

This linearized theory assumes small oscillation amplitudes and no coupling between the axial and radial forces. The coupling is usually small, but if Q_x/Q_y is a fraction of the type of small integers (e.g., $n = 0.2$ and $Q_x = 2Q_y$) a resonance occurs where a beat results and the energy swings between the axial and the radial plane. If $1/Q_x$ or $1/Q_y$ are small integers or small fractions, any error in the magnetic field distribution or magnet position causes another type of resonance which is more disruptive. After a few betatron oscillations, particles come with the same phase to the same place on the orbit, and their oscillation amplitude is increased until they are lost.

8.3.2. Strong Focusing

Betatron oscillation (Eqs. 8.9) is like the motion of charged particles in a magnetic quadrupole (Eq. 4.110). By defining

$$K_x = \frac{\sqrt{1-n}}{r_0} = \frac{Q_x}{r_0} \quad \text{and} \quad K_y = \frac{\sqrt{n}}{r_0} = \frac{Q_y}{r_0} \tag{8.11}$$

the radial and axial transformation matrices are

$$\mathbf{T}_x = \begin{pmatrix} \cos K_x s & \frac{1}{K_x}\sin K_x s \\ -K_x \sin K_x s & \cos K_x s \end{pmatrix} \tag{8.12}$$

and

$$\mathbf{T}_y = \begin{pmatrix} \cos K_y s & \frac{1}{K_y}\sin K_y s \\ -K_y \sin K_y s & \cos K_y s \end{pmatrix}. \tag{8.13}$$

Assuming initially $y = 0$, the amplitude varies as

$$y = \frac{\sin K_y s}{K_y} y'(0).$$

If n could be increased, then the betatron oscillation amplitude would be appreciably reduced. However, $n \gg 1$ gives an imaginary $K_x = \sqrt{1-n}/r_0$ and the trigonometric functions in the radial matrix are changed into hyperbolic, which results in defocusing. The transformation matrix in the defocusing plane is

$$\mathbf{T}_{xd} = \begin{pmatrix} \cosh K_x s & \frac{1}{K_x}\sinh K_x s \\ K_x \sinh K_x s & \cosh K_x s \end{pmatrix} \tag{8.14}$$

A series of magnets along the orbit whose field index is alternated, as in Fig. 8.14, allow for a large n, with focusing. These magnets act as positive and negative lenses, and the focusing is strong in both directions. Figure 8.15 shows an light-optical analog. The method is called **A**lternating **G**radient focusing [92, 93].

Large n changes the betatron tune, and the number of betatron oscillations around the circumference of the accelerator becomes much larger as illustrated in Fig 8.16.

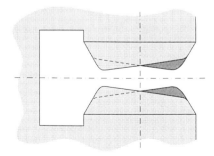

Fig. 8.14. Shape of the pole pieces in an AG machine.

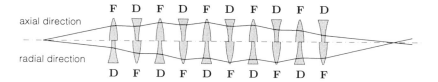

Fig. 8.15. Optical analogy to strong focusing.

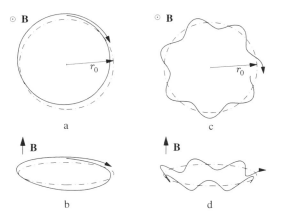

Fig. 8.16. Betatron oscillations. (**a**) Radial and (**b**) axial oscillations in a weak focusing machine ($n < 1$). (**c**) Radial and **d.** axial oscillations in a strong focusing machine.

The focusing (F) and defocusing (D) magnets and straight sections between them can be gathered in groups, each of which is called a *period*. Such a group may be symmetrical (1/2 F, D, 1/2 F, as in Fig. 8.17), and along the circumference of the accelerator there are N such periods. If $|n_F| = |n_D| \gg 1$, then all $K_{Fx} \approx K_{Fy} = K = K_{Dx} \approx K_{Dy}$. Assuming no straight sections between magnets and that all magnets have the same length $s = l$, so that $\theta = Kl$, the radial and axial transformation matrices for such a period are

$$\mathbf{R} = \begin{pmatrix} \cos\frac{\theta}{2} & \frac{1}{K}\sin\frac{\theta}{2} \\ -K\sin\frac{\theta}{2} & \cos\frac{\theta}{2} \end{pmatrix} \begin{pmatrix} \cosh\theta & \frac{1}{K}\sinh\theta \\ K\sinh\theta & \cosh\theta \end{pmatrix} \begin{pmatrix} \cos\frac{\theta}{2} & \frac{1}{K}\sin\frac{\theta}{2} \\ -K\sin\frac{\theta}{2} & \cos\frac{\theta}{2} \end{pmatrix} \quad (8.15)$$

and

$$\mathbf{A} = \begin{pmatrix} \cosh\frac{\theta}{2} & \frac{1}{K}\sinh\frac{\theta}{2} \\ K\sinh\frac{\theta}{2} & \cosh\frac{\theta}{2} \end{pmatrix} \begin{pmatrix} \cos\theta & \frac{1}{K}\sin\theta \\ -K\sin\theta & \cos\theta \end{pmatrix} \begin{pmatrix} \cosh\frac{\theta}{2} & \frac{1}{K}\sinh\frac{\theta}{2} \\ K\sinh\frac{\theta}{2} & \cosh\frac{\theta}{2} \end{pmatrix}. \quad (8.16)$$

For a complete accelerator cycle the transformation matrices are

$$\mathbf{M}_r = \mathbf{R}^N \quad \text{and} \quad \mathbf{M}_a = \mathbf{A}^N. \quad (8.17)$$

Using the matrix formalism it is possible to find the stability criteria even if N is large. Matrices in Eqs. 8.15 and 8.16 can be written

$$\mathbf{M} = \begin{pmatrix} \cos\mu + \alpha\sin\mu & \beta\sin\mu \\ -\gamma\sin\mu & \cos\mu - \alpha\sin\mu \end{pmatrix}, \quad (8.18)$$

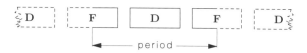

Fig. 8.17. A period of alternating gradient magnets.

where the argument of the matrix, μ, shows the betatron oscillations phase advancement in a magnet period. The betatron oscillations are stable only if μ is real, and since the trace of $\mathbf{M}$ is $2\cos\mu$, the stability criterion is

$$-2 \leq \text{Tr } \mathbf{M} \leq 2. \tag{8.19}$$

Since Eqs. 8.9 are valid for any n, for large n both for the radial and the axial direction they are

$$\frac{d^2w}{ds^2} + K^2w = 0, \tag{8.20}$$

where K varies periodically with the period L. α, β, and γ are three position-dependent parameters, related by

$$\beta\gamma - \alpha^2 = 1, \tag{8.21}$$

and it can be shown that

$$\gamma w^2 + 2\alpha ww' + \beta w'^2 \equiv \varepsilon = \text{const}, \tag{8.22}$$

which is valid at any position s along the accelerator ring and describes an ellipse with the area $\pi\varepsilon$, called the *beam emittance*. Physically the particle position w and the orbit angle w' lie somewhere on (or inside) the ellipse which changes shape with α, β, and γ, but the shape is repeated after a period. The maximum amplitude is $\sqrt{\varepsilon \cdot \beta_{\text{max}}}$.

Different particles will execute betatron oscillations with different amplitudes, but there will always be a particle with the largest amplitude. If ε_m is the emittance for this particle, the beam as a whole can be characterized by

$$\gamma w^2 + 2\alpha ww' + \beta w'^2 \leq \varepsilon_m. \tag{8.23}$$

The amplitude, or β-function, $\sqrt{\beta(s)}$, does not define the amplitude of a specific particle at position s, but rather specifies the beam diameter in the radial and axial direction respectively, at position s, and thus is a property of magnet lattice design and period.

8.4. PHASE STABILITY

8.4.1. Synchrotron Oscillations in Cyclic Accelerators

In all accelerators, except DC and induction machines, acceleration is achieved by the application of rf or microwave power to a resonant circuit, with the result that the accelerating gap voltages vary sinusoidally. The fundamental equations (Eqs. 8.1 and 8.2) cannot be valid for all particles all the time. Only a vanishing small number of particles will pass an accelerating gap with the correct phase angle and thus increase their energy for the correct amount to satisfy these equations. Oliphant, Veksler, and McMillan realized that particles, which traverse the gap with a phase angle close to the correct phase angle, will oscillate in phase and remain stable. This principle is called *phase stability* and is valid for both cyclic accelerators (cyclotron excepted) and for linacs.

Figure 8.18 shows the phase stability principle for a cyclic accelerator assuming that the revolution time increases with energy. Particle A on the reference orbit passes the gap at the phase stable angle φ_r, and thus the energy gain is exactly that required to remain

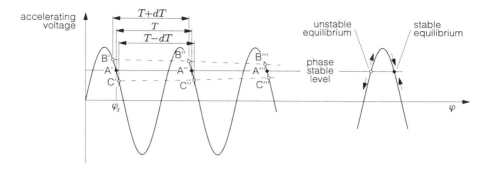

Fig. 8.18. Principle of phase stability ($dT/dp > 0$).

in the reference orbit. In the next gap passage and all which follow, this phase angle will be preserved. Particle B traverses the gap early and thus gains more energy than the reference particle. Since velocity does not change appreciably, but mass does, the radius of the orbit increases and the particle will need longer time to reach the gap the next time. There it will gain less energy (position B″), but still have more energy than particle A, but its phase angle will approach φ_r more closely on each passage. For the particle C, just the opposite is true. It gains less energy, so each revolution time is shorter than that for the particle A, but its phase angle approaches on each subsequent turn φ_r. When B and C reach φ_r, they still have an excess or a deficit of energy, respectively, but continue to move in the same direction, and now their roles are reversed. Over a certain phase angle interval these particles will oscillate stably. Thus the magnetic field can be gradually changed during acceleration or acceleration voltage amplitude can be chosen in relatively wide limits without losing the particles.

Different accelerators show a different phase stability behavior, which depends on how the revolution time, T, changes with energy. The description above is for relativistic particles, but the situation is different for subrelativistic particles. Assume a constant mass, which means that the velocity will increase with energy. Thus a particle with lower velocity takes a longer time to complete a revolution. The phase stable angle must, therefore, lie on the rising side of the sine wave.

To quantitatively describe phase stability we define the momentum compaction factor,

$$\alpha = \frac{dL/L}{dp/p} = \frac{dR/R}{dp/p} \tag{8.24}$$

where $L = 2\pi R$ is the the reference orbit length, and R is its mean radius of curvature. Assuming a circular reference orbit ($r = R$ for simplicity), the particle momentum is $p = qrB$ (Eq. 8.2),

$$\frac{dp}{p} = \frac{dr}{r} + \frac{dB}{B} = (1 - n)\frac{dr}{r}, \tag{8.25}$$

where the momentum compaction factor is

$$\alpha = \frac{1}{1 - n}. \tag{8.26}$$

The revolution time is $T = L/v$, and thus

$$\frac{dT}{T} = \frac{dL}{L} - \frac{dv}{v} = \frac{dL}{L} - \frac{d\beta}{\beta}, \tag{8.27}$$

where $\beta = v/c$. Defining (Appendix G) $\gamma = m/m_0$ and $\eta = p/m_0c = \beta\gamma$, we obtain

$$\frac{d\eta}{\eta} = \frac{d\beta}{\beta}\left(1 + \frac{\beta^2}{1 - \beta^2}\right) = \frac{d\beta}{\beta}\gamma^2 = \frac{dp}{p}. \tag{8.28}$$

Equation 8.27 can thus be written as

$$\frac{dT}{T} = \frac{dL}{L} - \frac{1}{\gamma^2}\frac{dp}{p} = \left(\frac{1}{1-n} - \frac{1}{\gamma^2}\right)\frac{dp}{p} = \left(\alpha - \frac{1}{\gamma^2}\right)\frac{dp}{p}, \tag{8.29}$$

or

$$\frac{dT}{dp} = \left(\frac{1}{1-n} - \frac{1}{\gamma^2}\right)\frac{T}{p},$$

where the sign depends on the values of n and γ. If $|n| < 1$ or if the particles are highly relativistic, $dT/dp > 0$ and the situation corresponds that of Fig. 8.18. If $|n| \gg 1$ and the particles are sub- or moderately relativistic, $dT/dp < 0$, and thus the phase stable angle is on the rising side of the sine wave. Since electrons are highly relativistic even at modest energies, $dT/dp > 0$ is valid for electrons in all practical cases.

The transition energy is

$$W_t = \frac{W_0}{\sqrt{\alpha}}, \tag{8.30}$$

where W_0 is the rest mass energy and α is the momentum compaction factor. At this energy $dT/dp = 0$ and there is no phase stability. In strong focusing proton synchrotrons the revolution time initially decreases with energy because the velocity increase is faster than the increase in radius of curvature, and thus the phase stable angle lies on the rising side of the sine wave, $0 < \varphi_r < 90°$. When the energy reaches the transition energy, the acceleration voltage phase must make a carefully controlled jump. Above the transition energy, the particle velocity changes slowly and the dominant factor is the mass increase. Thus the phase stable angle now lies on the falling side of the sine wave, $90 < \varphi_r < 180°$.

In cyclic accelerators, phase oscillations are called *synchrotron oscillations,* and, in most practical applications, have a frequency, ω_f, which is much lower than the revolution frequency, ω_r. Because the betatron oscillations frequencies, ω_x and ω_y, are of the same order or higher than ω_r, betatron and synchrotron oscillations are not coupled generally and can, therefore, be treated separately.

In most accelerators where the acceleration stability is based on synchrotron oscillations, the power source frequency, ω_{rf}, is a multiple of the revolution frequency, because the resonant circuits, in whose gaps particles are accelerated, can be made smaller, increasing the Q of the circuit due to reduced losses. The harmonic number of ω_{rf} is

$$h = \omega_{rf}/\omega_r.$$

The motion of a particle in the phase-energy space (Fig. 8.18), is described by

$$\frac{d\varphi}{dt} = -h(\omega - \omega_r) = -h\omega_r\frac{\delta\omega}{\omega_r} = h\omega_r\frac{\delta T}{T} = h\omega_r\left(\alpha - \frac{1}{\gamma^2}\right)\frac{\delta p}{p}, \tag{8.31}$$

where Eq. 8.29 was used. The momentum can be written as

$$p = mv = m_0c\gamma\beta = m_0c\sqrt{\gamma^2 - 1},$$

and

$$\frac{dp}{p} = \frac{\gamma^2}{\gamma^2 - 1} \frac{d\gamma}{\gamma} = \frac{1}{\beta^2} \frac{dW}{W},$$

where W is the particle energy, so that Eq. 8.31 takes the form

$$\frac{d\varphi}{dt} = h\omega_r \left(\alpha - \frac{1}{\gamma_r^2}\right) \frac{1}{\beta_r^2} \frac{\delta W}{W}, \tag{8.32}$$

with γ_r and β_r corresponding to the synchronous particle. By defining

$$\Gamma = \left(\alpha - \frac{1}{\gamma_r^2}\right) \frac{1}{\beta_r^2}, \tag{8.33}$$

the variation of the phase angle with time is

$$W_r \frac{d\varphi}{dt} = h\,\omega_r\,\Gamma\,\delta W = h\omega_r\,\Gamma\,(W - W_r)$$

and

$$\frac{dW_r}{dt} \frac{d\varphi}{dt} + W_r \frac{d^2\varphi}{dt^2} = h\,\omega_r\,\Gamma \left(\frac{dW}{dt} - \frac{dW_r}{dt}\right).$$

Because $\omega_f \ll \omega_r$ we can neglect the first term compared to the second,

$$W_r \frac{d^2\varphi}{dt^2} = h\,\omega_r\,\Gamma \left(\frac{dW}{dt} - \frac{dW_r}{dt}\right), \tag{8.34}$$

which expresses the phase dependence on the particle energy.

The increase in energy, dW/dt, depends on the accelerating voltage amplitude and phase. Particles which are accelerated in one or more gaps around the circumference of the accelerator obtain an the energy gain in each gap:

$$\Delta W = qV_{eff} \sin\varphi,$$

where V_{eff} includes the beam coupling coefficient M_B. If the relative energy gain per turn is low, $\sum(\Delta W/W) \ll 1$, we can assume that this energy gain is uniformly distributed along the orbit, and thus instead of differences, differentials can be used:

$$\frac{dW_r}{dt} = \frac{\omega_r}{2\pi} qV_{eff} \sin\varphi_r$$

and

$$\frac{dW}{dt} = \frac{\omega_r}{2\pi} qV_{eff} \sin\varphi,$$

where $\omega \approx \omega_r$ when compared with variation in φ.

Inserting into Eq. 8.34, the equation of motion for synchrotron oscillations becomes

$$\frac{d^2\varphi}{dt^2} = \frac{h\,\omega_r^2\,\Gamma\,qV_{eff}}{2\pi W_r} (\sin\varphi - \sin\varphi_r). \tag{8.35}$$

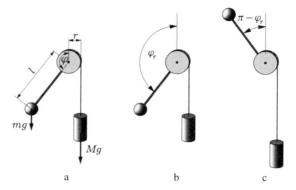

Fig. 8.19. Prestrained pendulum. (**a**) Forces. (**b**) Stable equilibrium. (**c**) Unstable equilibrium.

Equation 8.35 has a mechanical analog in the prestrained pendulum (as seen in Fig. 8.19), whose equation of motion is

$$(Mr^2 + ml^2)\frac{d^2\varphi}{dt^2} = mgl \sin\varphi - Mgr, \tag{8.36}$$

where g is the acceleration of gravity. For equilibrium, $\varphi = \varphi_r$, we have

$$Mr = ml \sin\varphi_r \qquad \text{and} \qquad \sin\varphi_r = \frac{Mr}{ml},$$

which gives

$$\frac{d^2\varphi}{dt^2} = \frac{mgl}{Mr^2 + ml^2}(\sin\varphi - \sin\varphi_r). \tag{8.37}$$

Both Eqs. 8.35 and 8.37 have the form

$$\frac{d^2\varphi}{dt^2} = \text{const}\,(\sin\varphi - \sin\varphi_r).$$

The analysis of the prestrained pendulum helps to determine the stability limits of synchrotron oscillations.

Equilibrium of the prestrained pendulum is obtained if

$$\sin\varphi_r = \frac{Mr}{ml} \le 1, \qquad \text{which implies that} \qquad Mr \le ml.$$

If $Mr = ml$ the deflection angle is $\pi/2$ and the equilibrium is unstable. The inequality $\sin\varphi_r < 1$ has two solutions. The solution $\pi - \varphi_r < \pi/2$ is unstable, and the smallest perturbation starts a motion of the weight M, which cannot be stopped (Fig. 8.19c). $\varphi_r > \pi/2$ corresponds to the stable equilibrium. When the equilibrium point of the pendulum is directly below the suspension point, we obtain $\varphi_r = \pi$. This equilibrium point allows the largest amplitude of oscillations. The stability limits of the pendulum are therefore

$$\pi/2 < \varphi_r < \pi \tag{8.38}$$

In an accelerator with $\Gamma > 0$, the same limits must be valid. When $\varphi_r = \pi/2$, no phase error is tolerated and the beam current quickly becomes zero. When the resonant phase angle is π, there is no acceleration because of $V_{eff} \sin\pi = 0$ and the energy gain is zero.

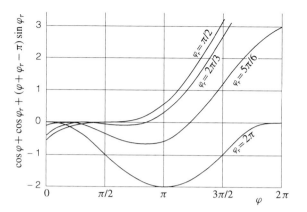

Fig. 8.20. Stabilizing potential $\cos \varphi + \cos \varphi_r + (\varphi + \varphi_r - \pi) \sin \varphi_r$ with synchronous phase angle φ_r as parameter.

In Eq. 8.35, time can be changed for position along the orbit by

$$\frac{d^2\varphi}{dt^2} = \omega_r^2 \frac{d^2\varphi}{d\theta^2},$$

where θ is the azimuthal angle. Multiplying by $d\varphi/d\theta$ and integrating, we obtain

$$\left(\frac{d\varphi}{d\theta}\right)^2 = -\frac{h\,\Gamma\,qV_{eff}}{\pi W_r}(\cos \varphi - \varphi \sin \varphi_r) + \text{const}, \qquad (8.39)$$

with initial conditions of $\varphi = \varphi_0$ and $d\varphi/d\theta = \varphi_0'$:

$$\left(\frac{d\varphi}{d\theta}\right)^2 = -\frac{h\,\Gamma\,qV_{eff}}{\pi W_r}[\cos \varphi - \cos \varphi_0 + (\varphi - \varphi_0) \sin \varphi_r] + (\varphi_0')^2.$$

For a given φ_r the largest amplitude will be obtained for $\varphi_0 = \pi - \varphi_r$ and $\varphi_0' = 0$, as the pendulum analog shows. Synchrotron oscillations are thus governed by

$$\frac{d\varphi}{d\theta} = \sqrt{-\frac{h\,\Gamma\,qV_{eff}\cos \varphi_r}{2\pi W_r}} \cdot F(\varphi, \varphi_r), \qquad (8.40)$$

where

$$F(\varphi, \varphi_r) = \pm \sqrt{\frac{2}{\cos \varphi_r}[\cos \varphi + \cos \varphi_r + (\varphi + \varphi_r - \pi) \sin \varphi_r]}. \qquad (8.41)$$

The parenthetical expression, shown in Fig. 8.20, acts like a *stabilizing potential*. Particles "oscillate" stably in the potential well as long as their energy remains bellow zero. A particle with $\varphi_r = \pi/2$ rolls along the potential hill, without being able to reach a stable position.

$F(\varphi, \varphi_r)$ can be visualized with the so-called "fish diagram" of Fig. 8.21, which shows the phase stable region with φ_r.

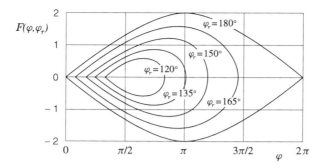

Fig. 8.21. Fish diagram. Normalized phase stable region with synchronous phase angles φ_r as parameter.

From Eq. 8.32, we have

$$\frac{\delta W}{W} \propto F(\varphi, \varphi_r),$$

but since $\delta W/W$ is proportional to $\delta p/p$, and therefore to $\delta r/r$ (Eq. 8.25), we have

$$\frac{\delta r}{r} \propto F(\varphi, \varphi_r),$$

so the particles oscillate with phases and radii in the phase stable region defined by the fish diagram.

8.4.2. Phase Stability in Linacs

In linacs the acceleration takes place in the gaps in series on the electron optical axis, with the acceleration field in each gap varying sinusoidally. Since the gap distance does not change with time, the momentum compaction factor, α, is equal to zero, and Eq. 8.29 becomes

$$\frac{dT}{T} = -\frac{1}{\gamma^2} \frac{dp}{p}. \tag{8.42}$$

where T is the time for a particle to pass two consecutive gaps. The synchronous phase angle must lie on rising voltage, $0 \leq \varphi_r \leq \pi/2$, so the stabilizing potential and phase stable region are inverted when compared with most cyclic accelerators. Figure 8.22a corresponds to Fig 8.20, and Fig. 8.22b corresponds to the fish diagram, Fig. 8.21.

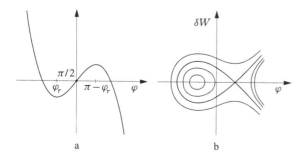

Fig. 8.22. Phase stability in linacs. **(a)** Stabilizing potential as a function of the phase angle φ. **(b)** Normalized phase stable region.

In linear accelerators these phase oscillations are reduced with increasing energy, since $dT/T \propto -dW/W$ decreases and $\gamma \gg 1$. Thus electron linacs have practically no phase stability for energy above a few megaelectronvolts. With $\varphi_r = 90°$ the electron beam is on the rising side of the sine wave because even at $\gamma \gg 1$ the electron velocity $v < c$.

8.5. CYCLIC ACCELERATORS

8.5.1. Cyclotron

One of the first accelerators, the classical cyclotron, consists of a large magnet with circular poles (Fig. 8.23), in whose gap there is a vacuum chamber with two D-shaped electrodes connected to an rf power source. At the center is an ion source. The D-electrode system resonant frequency is adjusted by movable short circuit over the electrode stems, which is placed a distance of $\lambda/4$ from the electrodes. Radio-frequency power feeds the system through an inductive loop near the short circuit, so the voltage between the D-electrodes is much larger (up to 300 kV) than the voltage over the loop feed-through.

The revolution frequency is constant in a constant magnetic field, B, as long as the particle mass does not change, so a cyclotron cannot accelerate electrons. The cyclotron energy is limited by the charge-to-mass ratio of the accelerated particles,

$$W_{max} = \frac{B^2 r_{max}^2}{2} \frac{q^2}{m_0},\tag{8.43}$$

and the magnetic field which is ~ 2 T for iron poles.

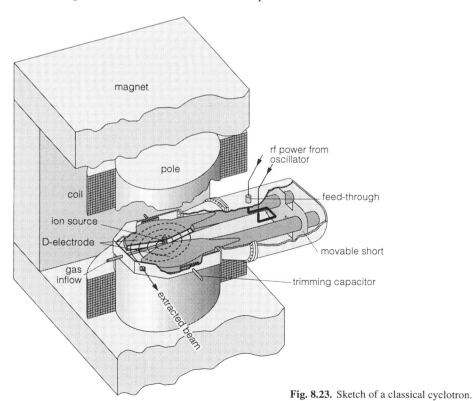

Fig. 8.23. Sketch of a classical cyclotron.

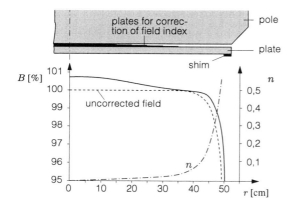

Fig. 8.24. Variation of the magnetic field with radius. (After *Particle Accelerators*, S. M. Livingston and J. P. Blewett, 1962, McGraw-Hill.)

Vertical stability of betatron oscillations requires a magnetic field gradient such that the field decreases with increasing radius (Fig. 8.24), and the field index $n < 1$. The field is tailored by inserting iron shims with different radius between the pole and the plate. A small air gap is introduced to help homogenize the field, and a second shimming plate around the periphery shapes and widens the fringing field.

As the beam velocity approaches that of light, particle mass changes. Thus the D-electrodes rf voltage frequency and the orbital frequency cannot be synchronized during the acceleration. An analysis of the situation in front of the ion source shows that the transit time effects cause the voltage between the D-electrodes to be maximum (i.e., $\varphi = 90°$), at beam injection. With a constant magnetic field (Eq. 8.1), $B_0 = m_0 \omega_{rf}/q$, the orbital phase when crossing the D-electrode gap will increase with increasing particle mass (Fig. 8.25a). Particle acceleration ends when the phase reaches 180°. As the magnetic field must decrease with radius ($n < 1$), the central field must be set higher than B_0, and so the phase will decrease initially, reaching 180° later, as seen in Fig. 8.25b.

Figure 8.25 shows particle performance in the cyclotron during a quarter synchrotron oscillation. Assuming a 20 MeV cyclotron and the magnetic field of Fig. 8.24, the phase changes during the last half-cycle by

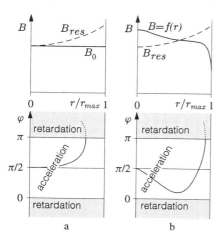

Fig. 8.25. Phase change during the acceleration. (**a**) Constant magnetic field, $B = B_0$ given by Eq. 8.1. (**b**) B changes with radius according to Fig. 8.24. $B_0 = m_0 \omega_{rf}/q$, B_{res} takes into account the relativistic increase of mass. (After *Teilchenbeschleuniger*, R. Kollath, 1962, Friedr. Vieweg, with permission.)

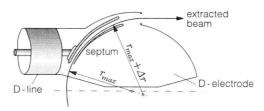

Fig. 8.26. Extraction. (After *Particle Accelerators*, S. M. Livingston and J. P. Blewett, 1962, McGraw-Hill.)

$$\Delta\varphi = \pi\frac{\omega_0 - \omega}{\omega} = \pi\frac{\frac{q}{m_0}B_0 - \frac{q}{m}B}{\frac{q}{m}B} = \pi\left(\frac{mB_0}{m_0 B} - 1\right),$$

where B is the field at the orbit. Since $m/m_0 = W_T/W_0 = 1 + W_k/W_0$, we obtain

$$\Delta\varphi = \pi\left(\frac{B_0 - B}{B} + \frac{B_0 W_k}{BW_0}\right).$$

With $B = 0.985 B_0$, the phase change in the last orbit is

$$\Delta\varphi = 0.12 \quad \text{or} \quad 6.7°.$$

Although the phase change per half-cycle is less during the most of the acceleration, the number of orbits is still limited to less than 100. Thus acceleration voltage between cyclotron D-electrodes is a few hundred kilovolts.

For a final phase angle of 180° the outer orbits are very closely packed, which makes their extraction difficult, so a phase angle of 180° is only used with internal targets. For the beam to be extracted the acceleration must cease at $\sim 90°$, where the orbit spacing is sufficiently large. The beam enters an electrostatic prism (Fig. 8.26) whose electric field deflects the ions and increases their radius of curvature. Since the beam is rather wide because of radial oscillations, the first part of the inner electrode, the septum, must be thin to keep ions in the penultimate orbit from hitting the septum. This limits the extracted beam current density, and thus in most machines the internal beam current must be reduced on extraction to prevent melting the septum. The distance between the extraction electrodes is 5–8 cm, and the voltage is ~ 50 keV.

The classical cyclotron energy limit is circumvented by the isochronous cyclotron based on edge focusing (Thomas [170]; Fig. 8.27a). Here the pole plates have different thickness in different sectors, and so the field varies along the orbit and the reference orbit is no longer circular. The fringe region is traversed at an angle, and thus edge focusing defines the stability. The magnetic field can thus increase with increasing radius, the field index n is less than 0, and the synchronization between the rf field and particles can be maintained even as the particle mass increases.

At the hill and valley edge the magnetic field has an azimuthal component, and the particle velocity has a radial one. This results in axial focusing at every transition (Fig. 8.27c). The focusing is weak, and the energy limit of the Thomas cyclotron is only slightly higher than that of the classical cyclotron. By giving the maxima and minima a spiral shape, the angle between the edge and the orbit becomes larger (Fig. 8.27b). The focusing becomes stronger; but while one edge is focusing, the other is defocusing, similar to strong focusing.

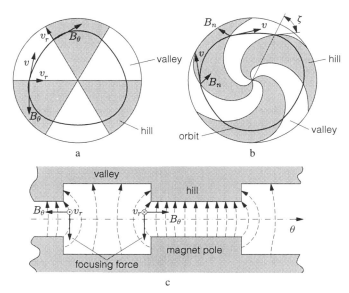

Fig. 8.27. Magnetic field in isochronous cyclotron. (**a**) Thomas cyclotron. (**b**) Isochronous cyclotron. (**c**) Edge focusing as seen from the center along the orbit. (After *Teilchenbeschleuniger*, R. Kollath, 1962, Friedr. Vieweg, with permission.)

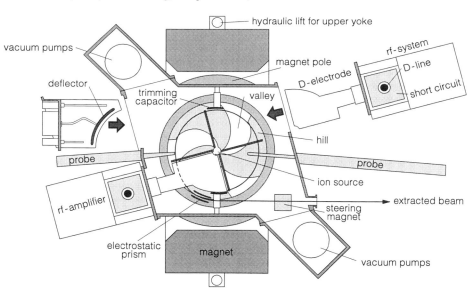

Fig. 8.28. A commercial 20 MeV isochronous cyclotron (Scanditronix).

For a constant revolution frequency, ω, the mean magnetic field must vary with radius as

$$<B> = \frac{B_0}{\sqrt{1 - \beta^2}} = \frac{B_0}{\sqrt{1 - \frac{\omega^2 r^2}{c^2}}}. \tag{8.44}$$

Table 8.1 gives important parameters of a large isochronous cyclotron, the GWI cyclotron at the The Svedberg Laboratory in Uppsala.

Table 8.1

GWI isochronous cyclotron

Magnet	Weight	600 ton
	Pole diameter	2.8 m
	Gap hills (3 times)	20 cm
	Gap valleys (3 times)	36 cm
	Magnetic field	1.54–1.75 T
	Coils, weight	50 ton
	power	300 kW
Vacuum system	Pump capacity	40 000 l/s
	Pressure	10^{-5} Pa
rf system	Electrodes	2
	Power	250 kW
	Frequency	12–24 MHz
	Max. electrode voltage	50 kV
Beam	Particles	p, d, He and other ions
	Energy	max. 200 q^2/A MeV
	Current	1–20 μA
	Extraction efficiency	90 percent

8.5.2. Synchrocyclotron

The relativistic mass increase causes a loss of synchronism which also can be evaded by changing the rf power source frequency with time,

$$f(t) = \frac{qBc^2}{2\pi(W_0 + W_k)},$$ (8.45)

where W_k is the particle kinetic energy. An oscillator circuit with a rotating capacitor accomplishes this. With accelerator phase stable, particles can be accelerated during many thousand of revolutions, with only ~ 10 kV D-electrode voltage.

Since the frequency is not constant, the particles are accelerated intermittently by short pulses, so the duty cycle is low, and only ~ 1 percent of the injected ions are accepted. To obtain stable betatron oscillations the magnetic field index must be $0 < n < 0.2$, and zero near the center (injection conditions), while $n = 0.2$ gives a resonance because $v_x/v_y = \sqrt{1-n}/\sqrt{n} = 2$.

The largest synchrocyclotron was in Dubna with a 6 m diameter, 7200 ton magnet which produced protons with energy up to 680 MeV.

8.5.3. Synchrotron

Instead of changing the rf signal frequency, the magnetic field can be increased with time to keep the particles and the rf signal synchronized. Synchrotrons are either weak focusing ($0 < n < 1$) or strong focusing ($|n| \gg 1$), but both have common properties.

Because the magnetic field increases with particle momentum, the orbit radius is almost constant. The energy increase adjusts itself automatically to the magnetic field increase, and the particles oscillate around the phase stable angle. Consequently, the magnet pole width needs only cover the amplitude corresponding to betatron and synchrotron oscillations (Fig. 8.1).

Figure 8.29 shows several synchrotron dipole bending magnets. The open C-type is used where easy access to the vacuum chamber is important, while the H-type

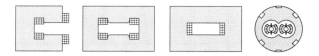

Fig. 8.29. Different types of synchrotron magnet construction.

is magnetically superior. The magnets are iron laminates to decrease eddy current losses. At low field at injection, remanent fields influence the field index; and at high field, saturation influences the field index. Corrections are made with special pole-face windings whose currents are computer-controlled.

The vacuum chamber, made of a low-conductance material, uses stainless steel in large proton machines with a slow magnetic field variation, and ceramic in smaller fast pulsing machines.

A synchrotron ring can contain as many as several thousand dipole magnets in the largest machines (Table 8.2). Between dipoles are straight sections where injection and extraction devices, focusing quadrupoles and multipoles, acceleration cavities, and monitoring equipment are mounted.

Acceleration takes place in one or more cavities, where for electron synchrotrons the rf frequency can be kept constant, because the velocity $v \approx c$, while in proton synchrotrons the frequency is $f_{rf} = hv/L$, where L is the reference orbit length and h is the harmonic number. The cavity resonant frequency in the latter must also vary, and so the cavities are constructed with ferrites which lower the cavity Q-value, but the change of ferrite magnetization allows the cavity stay in resonance. To limit the frequency swing and the remanent field problems the injection energy is chosen as high as possible. This also helps to decrease the transverse beam dimensions, the energy spread, and, hence, the vacuum chamber size.

It is not possible to inject particles into the reference orbit, because an orbit cannot be open and closed at the same time. Two injection methods (Fig. 8.30), both of which use electrostatic prisms, or *inflectors,* are employed. In the first, injection is made under one turn and puts the injected particles directly on the reference orbit. Large synchrotrons, where the long revolution time allows the voltage across the inflector to reach zero before the first injected particles reach the inflector after one revolution, use this method. Smaller machines use multiturn injection (Fig.8.30b), where the inflector deflects the particles and introduces betatron oscillations. The oscillation frequency must be such that during a few turns the particles pass the inflector near the inner vacuum chamber wall. During this time the magnetic field strength is increased and the reference orbit shrinks to about halfway between the inflector septum and the inner vacuum chamber wall. The rf power is switched off during injection and immediately on after injection, to avoid further reference orbit shrinking.

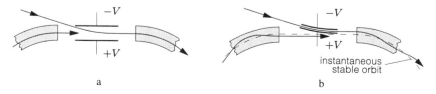

Fig. 8.30. (a) Single-turn injection. (a) Multiturn injection.

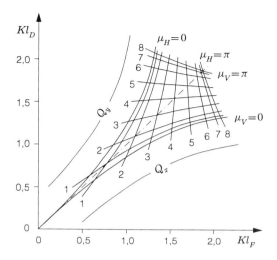

Fig. 8.31. Stability diagram of an AG synchrotron with 16 periods (tiediagram).

Beam extraction from a synchrotron is complicated by the fact that high-energy particles are magnetically stiff and all the particles are on or near the reference orbit. To extract particles the reference orbit is disturbed introducing a betatron oscillation of the whole beam. Usually a very fast kicker magnet containing no iron is used, so that a risetime of a few nanoseconds can be attained, often from EBS tubes. A magnetic prism, a septum magnet, intercepts some beam particles with large-amplitude oscillations to be deflected out of the ring. Depending on the amplitude of the disturbance, extraction can be fast or occur over many turns.

All modern large synchrotrons use strong focusing. Equations 8.15 and 8.16 define the transformation matrices of a magnetic period in a ring. The matrix trace, Eq. 8.18, is $2\cos\mu$, and so the betatron oscillation stability is ensured only when $0 < \mu < \pi$, expressed by the *tiediagram* of Fig. 8.31. The figure is valid for a ring with 16 magnetic periods, under the assumption that $|n| \gg 1$, or $K_F \approx K_D \approx K$. However, for any integer value of betatron oscillations, Q_x and Q_y, a resonance occurs, and thus the stability diagram is crossed by instability lines. Even integer $Q_x + Q_y$ values (coupled x and y oscillations), and half-integer values are dangerous. Thus during the whole acceleration process, particles must lie inside a small region limited by resonance lines. Energy spread is another complication because it causes a dispersion of betatron oscillation frequencies, *tune shift*, which are controlled with nonlinear elements, like sextupole lenses, placed around the ring.

AG synchrotrons are constructed with focusing (F) and defocusing (D) dipole magnets placed in the ring. The order of F- and D-magnets together with spaces (O) constitute the lattice. FODO and FOFDOD combinations are used most often, the latter having the advantage of a large horizontal β-function (Eq. 8.23) in a straight section between two F-halves, which makes injection and extraction easier. A third method, applied in the largest synchrotrons and colliders, is to make gradientless dipole magnets, $n = 0$, where focusing is attained with quadrupoles. The advantage here is that the dipole field is stronger and the frequency of betatron oscillations (Q-values) are controlled by quadrupoles and not fixed by the lattice.

Table 8.2 gives parameters of the two largest proton synchrotrons, the SPS at CERN and the Tevatron at Fermilab, and of the electron-positron collider LEP at CERN.

Table 8.2

Largest synchrotrons

		SPS	Tevatron	LEP
Energy	[GeV]	2×450	2×900	2×50
Injection energy	[GeV]	14	150	20
Mean diameter	[km]	2.2	2.0	8.37
Length	[km]	6.9	6.3	26.7
Magnets: dipoles		744	774	3280
length	[m]	6.26	6.3	5.75
vacuum chamber	[mm]	$152 \times 38/129 \times 53$	61×61	
field strength	[T]	2.03	4.44	0.135
Magnets: quadrupoles		216	224	736
vacuum chamber	[mm]	$143 \times 42/83^{\phi}$	70^{ϕ}	
gradient	[T/m]	22.5	76	10.9
Vacuum: pressure	[Pa]	10^{-7}	10^{-9}	$10^{-11} \times I$
ion pumps		1400	$96 + 130$	2200
turbomolecular pumps		120	48	60 (mobile)
Acceleration: energy gain	[MeV/turn]	7.2	2.8	260
rf cavities		4	8	120
length	[m]	20	2.7	2.2
frequency	[MHz]	200	53.1	352
Cycle time	[s]	14.4	200	continuous
Peak power	[MW]	210		11
Mean power	[MW]	57		11
Price [year]	[M$]	320 [1975]	250 [1975]	~ 1000[1990]

* I, beam current.

These machine lattice consist of constant field dipoles and focusing quadrupoles. Figure 8.32 shows the Tevatron tunnel. In the upper conventional magnet string, particles are accelerated to 150 GeV, while in the lower superconducting string (the hose transports liquid helium) they are accelerated to 1 TeV. All three machines are storage ring colliders — that is, synchrotrons where counterrotating particles and antiparticles are accelerated, and for hours held at the final energy. Each revolution the stored beams are made to collide at a few points along the circumference.

Fig. 8.32. Photograph of the Tevatron tunnel (Fermilab's Visual Media Services).

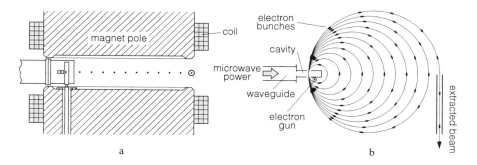

Fig. 8.33. Microtron. (**a**) Cross section and (**b**) electron orbits.

8.5.4. Microtron

If the relativistic particle energy increases in constant steps the revolution time in a constant magnetic field increases by an integer number of rf periods from turn to turn. An accelerator exploiting this fact is the microtron, originally called the electron cyclotron (Veksler [166]). It can accelerate only electrons and the orbits are circular, as shown in Fig. 8.33. A resonant cavity located near the constant magnetic field periphery is provided electrons by a gun, which are accelerated by the cavity electric field on each pass. The orbit radius increases from orbit to orbit, but all orbits are approximately tangent along the cavity axis. The revolution time (Eq. 8.1) is

$$T = 2\pi \frac{m}{eB} = 2\pi \frac{W}{eBc^2},$$

which for the nth revolution is

$$T_n - T_{n-1} = \frac{2\pi}{eBc^2} W_r = \nu\tau, \qquad (8.46)$$

and for the first is

$$T_1 = \frac{2\pi}{eBc^2}(W_0 + W_{inj} + W_r) = \mu\tau, \qquad (8.47)$$

where τ is the rf period, W_0 is the electron rest mass energy, W_{inj} is the injection energy, and μ and ν are integers. The resonant energy gain is

$$W_r = (W_0 + W_{inj})\frac{\nu}{\mu - \nu}, \qquad (8.48)$$

whose corresponding resonant magnetic field is

$$B_r = \frac{2\pi}{ec^2}\frac{W_0 + W_{inj}}{\mu - \nu} f_{rf}, \qquad (8.49)$$

where f_{rf} is the rf power source frequency, $\mu = 2, 3, 4, \ldots$ and $\nu = 1, 2, 3, \ldots$. For the fundamental mode, $\mu = 2, \nu = 1$, the resonant energy gain, $W_0 + W_{inj}$, and corresponding magnetic field are the largest, one reason that most microtrons use the fundamental mode. The injection energy is ~ 50–100 keV, and therefore the resonant energy gain is ~ 550–600 keV. Microwaves with frequency of a few gigahertz are used

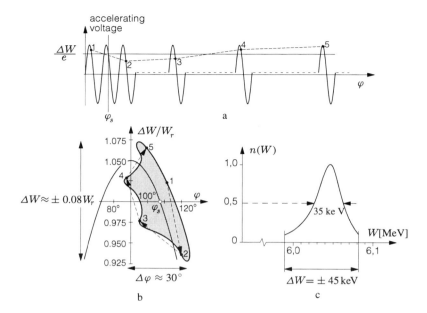

Fig. 8.34. Phase stability in microtrons. (**a**) Phase oscillations. (**b**) Phase stable area for a resonant phase angle of 108°. (**c**) Typical energy distribution in the beam in a microtron with 11 orbits.

to obtain the necessary voltage W_r/e across the cavity gap. B_r is rather small (e.g., at 3 GHz, 0.11 T) for an iron constant field magnet, so the microtron energy is limited by the magnet size.

The microtron phase stability is the same as for a synchrotron, because as the revolution time increases with energy the phase stable angle must lie on the falling sine curve of the gap voltage (Fig. 8.34a). However, in synchrotrons the phase angle changes only slightly from orbit to orbit, while in microtrons the change is large, so instead of a differential equation describing the phase oscillations, difference equations are used. Operating in the fundamental mode the relative energy error in the microtron orbit n is

$$\varepsilon_n = \frac{W_n - (n+1)W_r}{W_r}, \tag{8.50}$$

and the interorbit phase angle at the cavity transit is

$$\varphi_{n+1} = \varphi_n + 2\pi(\varepsilon_n + n + 1),$$

where the factor $2\pi(n+1)$ is disregarded when considering the phase at the transit (e.g., Fig. 8.34b). Thus,

$$\varphi_{n+1} = \varphi_n + 2\pi\varepsilon_n, \tag{8.51}$$

and the energy gain in each transit is

$$W_{n+1} = W_n + eV_{eff}\sin\varphi_{n+1}, \tag{8.52}$$

where the effective gap voltage takes into account the beam coupling coefficient. For resonant electrons, we have

$$W_{n+1,r} = (n + 1)W_r + eV_{eff} \sin \varphi_r. \tag{8.53}$$

Subtracting and dividing by $eV_{eff} \sin \varphi_r = W_r$, the $n + 1$ orbit energy error is

$$\varepsilon_{n+1} = \varepsilon_n + \frac{\sin \varphi_{n+1} - \sin \varphi_r}{\sin \varphi_r}. \tag{8.54}$$

Equations 8.51 and 8.54 give phase and energy recursion formulae which can be solved graphically or numerically to determine the phase stable oscillation. The region bound by the solution is called the *phase stable area* and is shown in Fig. 8.34b for a 108° resonant phase angle, the largest phase stable area for a microtron working in the fundamental mode. The relative energy error in every orbit with $n > 5$, after which most phase unstable electrons are lost, is $\sim \pm 0.08 W_r$.

For other modes Eq. 8.51 must be modified by

$$\varphi_{n+1} = \varphi_n + 2\pi \nu \varepsilon_n, \qquad \bullet \tag{8.55}$$

which leaves Eq. 8.54 unchanged. Considering only small oscillations the phase stability limits are computed, assuming that

$$\delta \varphi_n = \varphi_n - \varphi_r,$$

so that the sine function can be linearized as

$$\sin \varphi_{n+1} = \delta \varphi_{n+1} \cos \varphi_r + \sin \varphi_r.$$

Equations 8.51 and 8.54 are then in matrix form

$$\begin{pmatrix} \delta \varphi_{n+1} \\ \varepsilon_{n+1} \end{pmatrix} = \begin{pmatrix} 1 & 2\pi \nu \\ \cot \varphi_r & 1 + 2\pi \nu \cot \varphi_r \end{pmatrix} \begin{pmatrix} \delta \varphi_n \\ \varepsilon_n \end{pmatrix},$$

whose trace is limited for stable oscillations to

$$-2 \leq 2 + 2\pi \nu \cot \varphi_r \leq 2. \tag{8.56}$$

Thus for

$$\nu = 1 : \quad 90° \leq \varphi_r \leq 122.5°$$
$$\nu = 2 : \quad 90° \leq \varphi_r \leq 107.7°$$
$$\nu = 3 : \quad 90° \leq \varphi_r \leq 102.0°$$

The large phase stability limit for the fundamental mode is the second reason to use this mode.

The demands on magnetic field homogeneity increase with the third power of the orbit number, putting a practical limit on magnet size. Even with pole-face wound correction coils the necessary field homogeneity is difficult to attain for more than ~ 30–40 orbits.

Fig. 8.35. Axial oscillations in a microtron. The photograph was made using strained wires painted with ZnS.

The homogeneous magnetic field of the microtron means that, unlike the cyclotron and synchrocyclotron, there is no field gradient to guarantee the axial and radial beam stability. The zero field index gives radial stability, like in a 180° prism, but there is no axial magnetic force. The axial beam focusing occurs only at the acceleration gap entrance and exit, but since the resonant phase angle lies on the falling voltage sine curve, the electric field is stronger on entrance where it is focusing than on exit where it is defocusing. The large momentum increase during transit through the gap amplifies this effect, and with the small number of orbits it suffices to stabilize the orbits, as seen in Fig. 8.35.

Injection is made from a small electron gun mounted on the side of the cavity (Fig. 8.33), and small magnetic shims compensate for the disturbance caused by the hole in the magnet. Alternately, a thermionic cathode can be mounted directly in the cavity wall, which allows a higher resonant energy gain, but with the drawback that the cathode sputtering contaminates the walls and causes breakdowns in the gap, so that the resonator must be cleaned often.

Extraction, also indicated in Fig. 8.33, uses an iron tube, possibly movable, mounted on the opposite side of the cavity, which screens the magnetic field in an orbit. Electrons enter this field-free region and move tangentially out of the machine. To compensate for the field disturbance in the next-to-last orbit, iron rods are placed parallel to, but above and below, the tube.

If the circular microtron is split along a line crossing the cavity symmetry plane, the halves separated by a distance l, and the single cavity replaced by a small linac, as seen in Fig. 8.36, then we have the race-track microtron. The resonance condition unchanged,

$$T_1 = \frac{2\pi}{eBc^2}(W_0 + W_{inj} + W_r) + \frac{2l}{c} = \mu\tau, \qquad (8.57)$$

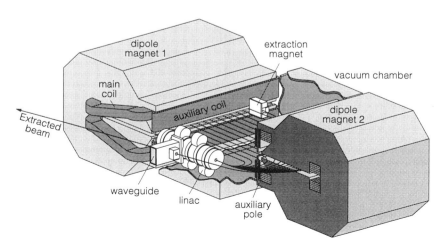

Fig. 8.36. View of a race-track microtron.

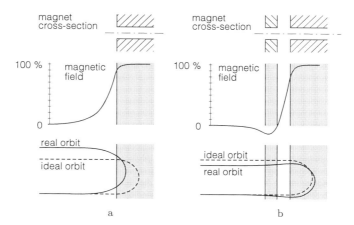

Fig. 8.37. Edge focusing in a race-track microtron. (**a**) Dipole magnet, defocusing in axial direction. (**b**) Main dipole and countering field, focusing.

where $v \approx c$, as in Eqs. 8.46 and 8.49. The resonant energy gain, with $\lambda = \tau c$, is

$$W_r = (W_0 + W_{inj})\frac{v}{\mu - v - 2l/\lambda} \qquad (8.58)$$

and is subject to only technical limits: magnetic field strength (up to ~ 1 T) and magnet size.

The phase stability conditions also remain unchanged, so the phase stable regions have the same shape and size at the same phase stable angle as for the circular microtron. So the relative energy error does not change, which in the fundamental mode is $\Delta W \approx \pm 0.08 W_r$ (Fig. 8.34b).

The focusing properties of the race-track microtron, however, are completely different. In both 180° dipole magnets, fringing fields shape the orbit such that the electrons obliquely cross the dipole edge (Fig. 8.37a), with the result that electrons are axially defocused four times per revolution. Adding auxiliary magnets with countering fields (Fig. 8.37b), the velocity direction is changed, and focusing is obtained at each crossing which ensures axial stability. In the radial direction there is no magnetic focusing except the usual 180° magnetic prism effect. However, a weak focusing in the linac is often enough to keep the beam stable. In race-track microtrons with a large pole separation, additional focusing is achieved by quadrupoles in the common orbit or outer orbit straights.

The two dipole magnets act as two mirrors, so any deviation from the magnetic parallelism in the axial or radial direction deflects the beam in the direction in which the dipoles open, with the deflection increasing as n^2. A deviation of ± 0.1 mrad is sufficient to lose the beam in a 10 to 15 orbit machine. It is practically impossible to obtain such accuracy in machining, setup, and material homogeneity, so special small deflection coils are used in the outer straight orbits to compensate for this drift.

The electrons are injected along the linac axis using an edge focusing deflecting magnet, as seen in Fig. 8.38. Angular deviation in the common orbit is compensated for by another magnet, oppositely excited, with the same edge angle placed further away from the linac. Weak parallel beam displacement can be corrected using other focusing elements.

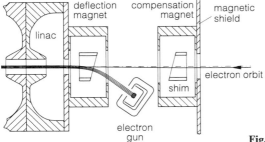

Fig. 8.38. Injection into a race-track microtron.

Table 8.3

Typical microtron parameters

		Circular	Race-track
Electron energy	[MeV]	6–22	2.5–55
Pulsed power in the beam	[MW]	1	1.5
Pulse length	[μs]	4	0.05–5
Duty cycle, maximum	[percent]	0.1	0.1
Energy spread at 6 MeV	[percent]	<1	<2
at full energy	[percent]	0.2	0.3
Mode: μ		2	11
ν		1	1
Number of orbits		11–42	3–15
Diameter of the last orbit	[m]	0.4–1.5	0.47
Acceleration		Cavity	Linac
Energy gain per turn	[MeV]	0.57	2.7–3.7
Power source		Magnetron	Klystron
Pulsed power	[MW]	2	3
Frequency	[MHz]	2998	2998
Injection energy	[keV]	60	25
Magnetic field	[T]	0.11	0.55–0.75
Size: total length	[m]		1.6
magnet height	[m]	0.65	0.4
magnet length	[m]		0.55
magnet diameter	[m]	2.4	

The electron gun can lie inside the vacuum chamber (Fig. 8.38), or outside of it, in which case a beam transport system with lenses guides the beam to the deflecting magnet.

To extract the beam a small magnet is put near the outer edge of the dipole magnet 1 (seen in Fig. 8.36), which can be moved into the orbit from which extraction is to be made. The field deflects the beam by an angle corresponding to the distance between two orbits. After passing the dipole magnet 2, the beam enters the extraction channel in the same orientation independent of the orbit from which it was extracted.

Table 8.3 gives some parameters of two typical microtrons.

8.5.5. Betatron

In the betatron, an induction machine, the magnetic field varies with time, inducing electric field,

$$\oint \mathbf{E}\, d\mathbf{s} = -\frac{d\Psi}{dt},$$

which is used to accelerate electrons along the circular orbit such that

$$\frac{dp}{dt} = -eE = \frac{e}{2\pi r}\frac{d\Psi}{dt}, \tag{8.59}$$

where r is the orbit radius. The electron energy is increased in proportion to the increase of the magnetic field, keeping the orbit radius constant:

$$\frac{mv^2}{r} = \frac{pv}{r} = evB,$$

which gives

$$\frac{dp}{dt} = er\frac{dB}{dt}. \tag{8.60}$$

The betatron condition is obtained by combining Eqs. 8.59 and 8.60,

$$\frac{1}{\pi r^2}\frac{d\Psi}{dt} = 2\frac{dB}{dt},$$

or

$$\frac{d<B>}{dt} = 2\frac{dB_0}{dt}. \tag{8.61}$$

Thus the magnetic field *inside* the orbit, $<B>$, must vary twice as fast as does the magnetic field *on* the orbit, B_0.

A betatron, shown in Fig. 8.39, has the dimension of its central core and short gap adjusted to realize the betatron condition. The higher core field causes an early saturation in the core. Betatron excitation is, therefore, achieved with a combination of two windings, one or both of which can be DC-biased so that the core and gap come into saturation at the same time.

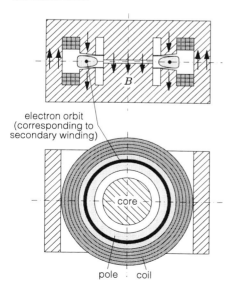

electron orbit
(corresponding to
secondary winding)

core

pole · coil

Fig. 8.39. Sketch of the cross sections of a betatron.

Electrons are injected from a small gun located on the outer vacuum chamber wall at low energy, a few hundred electronvolts. A special winding over the vacuum chamber is pulsed shortly during injection to keep electrons after their first orbit from hitting the back of the gun, while the electron beam moves toward the center of the chamber. For extraction this winding is again pulsed, now with opposite polarity, causing the orbit to expand, and the beam enters the extraction channel. The energy gain per turn is low, ~ 50 eV.

Betatrons are now replaced by electron linacs and microtrons, which give a much higher beam current. The main advantage of betatrons was their very small energy spread which resulted from their low-energy gain per turn. The largest betatron had an energy of 300 MeV, which was the synchrotron radiation limit.

There also exist linear induction accelerators constructed with a number of single-turn toroidal ferrite cores, which, when pulsed, generate an electric field that accelerates the particles. Induction accelerators are used to accelerate very intense electron beams, up to kiloamperes.

8.6. LINEAR ACCELERATORS

Linear accelerators (linacs) are straight accelerators where acceleration is achieved by an rf or microwave field in a number of gaps which beam particles must traverse at a correct phase angle. Between the gaps are field-free spaces, the drift tubes. Electron and ion linacs differ in that electrons behave as highly relativistic particles (i.e., $v \approx c$) when their energies are only a few megaelectronvolts. Thus electron linacs can either use traveling waves or have constant distances between acceleration gaps. For protons and ions the distance between the gaps must vary with their velocity.

8.6.1. Proton Linac

Proton linacs, where drift tubes increase in length, are of three configurations. The Wideröe type, proposed by Ising, has even-numbered drift tubes connected to one pole of an rf source and has odd-numbered ones connected to the other pole. The successive gaps have oppositely directed electric fields and the linac is, therefore, called the half-beta-lambda linac. The Alvarez type [171] has all drift tubes mounted in a cavity which is excited in the TM_{010} mode so the gap fields all point in the same direction, and thus is also known as a beta-lambda linac. The third type, the standing-wave linac, used mostly in small electron linacs, will be described later.

A cell, n, contains a drift tube and two gaps of length

$$L_n = \frac{\beta_n \lambda}{2} \qquad \text{or} \qquad L_n = \beta_n \lambda,$$

for the Wideröe or Alvarez type, respectively. The energy gain per cell is,

$$\Delta W = q M_B \hat{V} \sin \varphi,$$

where M_B is the beam coupling coefficient defined by Eq. 3.33. Since the electric field decreases along the linac as the drift tubes become longer, the Wideröe-type linac is used mainly at low energy.

In the Alvarez linac the drift tube shape is modified with the first tubes being wider and shorter, and those near end longer and slimmer (Fig. 8.41a), so that the electric field

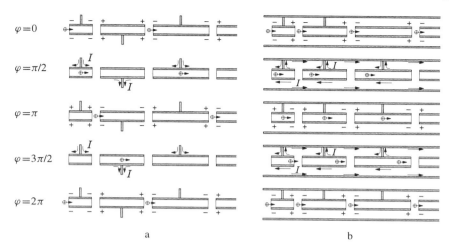

Fig. 8.40. Principle of (a) a Wideröe and (b) an Alvarez-type linac.

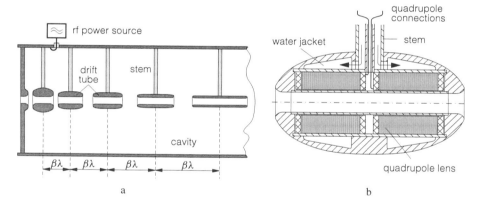

Fig. 8.41. (a) Sketch of the construction of an Alvarez-type linac. (b) Cross section of a drift tube. (After *Particle Accelerators*, S. M. Livingston and J. P. Blewett, 1962, McGraw-Hill.)

strength remains almost constant over the whole length. High-energy linacs are therefore mostly of the Alvarez type with the particle velocity matched to cavity and drift tube construction, and so these linacs have more than one cavity. Figure 8.41b shows a cross section of a drift tube in a proton linac.

The synchronous phase angle must lie on the rising side of the voltage sine wave, $0 \leq \varphi_r \leq \pi/2$, so the electric field is weaker on the gap entrance than on the exit, the gap being defocusing. At relativistic energies the self-magnetic field reduces the defocusing force (Eq. 6.7)

$$F_r = F_e + F_m = F_e(1 - \beta^2).$$

To obtain focusing, especially early in the acceleration, quadrupole lenses are built into the drift tube electrode, Fig. 8.41b.

Table 8.4 gives important parameters of the largest proton linac at the Los Alamos Meson Physics Facility. The accelerator has four Alvarez-type tanks which accelerate protons to 100 MeV, followed by an ~800 m long standing-wave structure.

Table 8.4

Parameters of LAMPF (Los Alamos Meson Physics Facility)

Injection: ion source		30 keV	Duoplasmatron
	injector	750 keV	Pierce system
Drift tube accelerator:	frequency	201.25 MHz	
	structure	4	Alvarez tanks
	I	5.4 MeV	
	II	41.3 MeV	
	III	72.7 MeV	
	IV	100 MeV	
Side-coupled accelerator:	frequency	805 MHz	
	structure	44	$\pi/2$-mode tanks
Beam: energy		800 MeV	
pulse current		17 mA	
mean current		1 mA	
duty cycle		6–12 percent	
secondary beam, mesons		5×10^{10} π^+/second	
Power: copper losses, pulse		38 MW	
beam power, pulse		17 MW	
mean		3 MW	
Length		850 m	

8.6.2. Electron Linac

In a slow-wave structure the electromagnetic wave phase velocity can be adjusted to the particle velocity. Electrons, injected near the crest of the field from a gun with moderate anode voltage $\sim$100–150 keV, will be accelerated. The traveling wave linac, similar in construction to a high-power traveling wave tube, was first built at Stanford 1948 [172]. By connecting a number of resonant cavities with an appropriate coupling a standing-wave linac [173] is obtained, where the electrons are accelerated in the gaps and travel in drift tubes between the gaps, while the electric field changes direction as in an Wideröe-type linac. Figure 8.42 shows an instantaneous picture of the field distribution in both types of electron linacs.

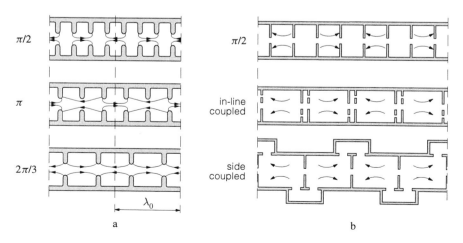

Fig. 8.42. Instantaneous picture of the field distribution (**a**) in a traveling wave and (**b**) in the $\pi/2$ mode in a standing-wave linac.

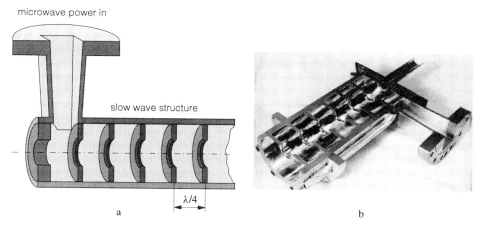

Fig. 8.43. Traveling wave linac (**a**) Construction ($\pi/2$ mode). (**b**) Photograph of a section of the 50 GeV SLAC linac. (Stanford Linear Accelerator Center.)

A traveling wave linac is excited by the microwave power at the beginning of a section (see Fig. 8.43). The electromagnetic wave traveling along the structure is attenuated by losses in the copper structure and power to the electron beam. After a certain length, the attenuation length, the electric field becomes too weak for efficient acceleration and the residual power is removed at the end of a section by a matched load to reduce reflections. Every section of a high-energy linac has its own power source, a klystron, all of which are excited from a common oscillator (Fig. 8.48). The power source of short, one-section linacs, is usually a magnetron.

The largest traveling wave linac, 2 miles long, at the Stanford Linear Accelerator Center (SLAC) (Fig. 8.56), accelerates both electrons and positrons. Table 8.5 gives the important parameters of this machine.

The standing-wave linac consists of a string of coupled cavities excited by microwave power near the central cavity. The resulting electromagnetic wave travels to both ends where it is reflected and builds a standing wave. Such linacs have both acceleration and coupling cavities; the first contain gaps along the optical axis, and the second are constructed as thin cavities either placed between the accelerating cavities (in-line coupled) or moved laterally (side-coupled).

A system of N coupled cavities is equivalent to a system of coupled oscillators and, therefore, has N resonant frequencies,

$$\omega_n = \frac{\omega_0}{\sqrt{1 + k \cos \frac{n\pi}{N}}},$$

where k is the coupling between two contiguous cavities. A dispersion curve for a system of 13 cavities is shown in Fig. 8.45. The $\pi/2$ mode, for $n = N/2$, where N is an odd integer, results in the largest distance to the nearest oscillation mode, has the largest manufacturing tolerances, is insensitive to beam loading, and rapidly responds to fast transient disturbances, and therefore is used in most standing-wave linacs.

In the $\pi/2$ mode the electric field in the two neighboring accelerating or coupling cavities is shifted by π. In coupling cavities the two entering fields from the respective accelerating cavities cancel, if the system is perfectly tuned, and thus there are no losses

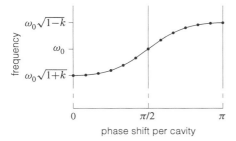

Fig. 8.44. Dispersion curve of a linac with 13 cavities.

in the coupling cavities. The accelerating cavities, $\lambda/2$ long, allow enough space to optimize the gap and cavity shapes to obtain a high shunt impedance. At 3 GHz with $R_{sh} \sim$ 60–80 MΩ/m, Q-values of $\sim$15,000 can be obtained. Figure 8.45 shows a cross section of a small side-coupled standing-wave linac, and Fig 8.46 is a photograph of a similar on-line coupled linac before brazing. The half-length acceleration cavities at both ends make the system electrically symmetric, the injection energy can be lower, the focusing is better, and the output energy is more homogeneous.

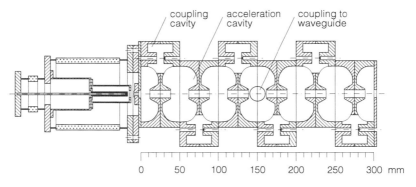

Fig. 8.45. Cross section of a 5 MeV standing-wave side-coupled linac with the electron gun.

Fig. 8.46. Photograph of an on-line coupled linac before brazing. (Royal Institute of Technology, Stockholm.)

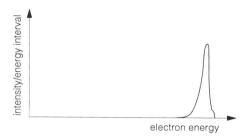

Fig. 8.47. Typical electron energy distribution from a low-energy linac.

In all electron linacs, acceleration takes place near the peak electric field. For the traveling wave linac this occurs when the wave travels at the velocity of the electrons. In standing-wave linacs the beam coupling coefficient must be taken into account when optimizing the gap shapes. As there is little phase focusing, the electron energy distribution is rather broad (Fig. 8.47), so in large electron linacs special provisions are made for adjusting the phase between the sections to narrow the energy spectrum.

There is also little radial focusing, a problem of minor importance since the electrons are highly relativistic. To the laboratory system length, z, and that of the electron rest system, z_e, a Lorentz transformation

$$z_e = \frac{z}{\gamma},$$

where

$$\gamma = \frac{W_{inj} + e <E> z}{W_0},$$

can be applied and results in

$$z_e = z\frac{W_0}{W - W_0} \ln \frac{W}{W_0},$$

under the assumption that $W_{inj} \approx W_0$. Using the SLAC linac as an example, the 2-mile laboratory length gives in the rest system length of 36 cm, and so focusing over this length is not a problem.

Figure 8.48 shows the construction of the electron linac microwave power source: A 10 to 40 kV DC source charges the capacitors in a **P**ulse **F**orming **N**etwork, a lumped delay line with the delay time τ, through a diode. A trigger pulse ignites the hydrogen thyratron, and the PFN input is short-circuited across a pulse transformer. A wave travels along the line, and is reflected at the open end, and the capacitors discharge when the wave returns to the input. A pulse of length 2τ and voltage half that of the DC power source is produced across the pulse transformer primary. The secondary winding is connected to the cathode of a klystron or magnetron, which generates the microwave power. A circulator or ferrite isolator can be inserted between the microwave power source and the accelerator to protect the microwave tube from the reflected power caused by mismatchs in the case of linac or waveguide breakdown.

High-energy linacs have more than one section with an amplifier-type microwave power source, feeding each section, with all tubes being driven by a single oscilla-tor. The input signal phase and amplitude to each tube can be individually adjusted, except in one section which defines the reference phase for the whole accelerator.

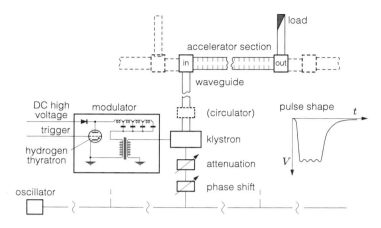

Fig. 8.48. Microwave power distribution.

Table 8.5

Parameters of some electron linacs

		Clinical	Clinical	SLAC
Energy	[MeV]	4–6	30–35	50000
Type		Standing wave	Standing wave	Traveling wave
Pulsed power in the beam	[MW]	≈ 1	2	50
Pulse length	[μs]	1–5	1–5	3
Duty cycle, maximum	percent	0.1	0.1	0.018
Energy spread	percent	<4	<4	<0.01
Peak PFN power	[MW]	2–4	7–8	14,740
Power source		Magnetron	Klystron	Klystron
Number of tubes		1	2	220
Pulsed power	[MW]	2–4	4	67
Length	[m]	0.3–0.6	4–5	3,050

Thus, a small difference between the electron and light velocities which produce phase error can be compensated for.

Cavities in the standing-wave linac and the slow-wave structure in the traveling wave linac are high Q-value resonant devices, and the frequency of each cavity or of a section are adjusted with narrow tolerances by mechanical deformation of the walls while controlling frequency changes. To trim SLAC a device automatically measures a section and exerts force on appropriate positions.

Table 8.5 shows some parameters of two clinical electron linacs and the Stanford linear accelerator.

8.6.3. RFQ — Radio-Frequency Quadrupole

A quadrupole mass spectrometer (see Section 5.2.2) combines DC and AC voltage between the electrodes of a quadrupole to separate and to focus particles. However, an rf voltage can be connected to the electrodes, and as long as the electrodes are flat, they will only focus. But if they have a periodic structure (Fig. 8.49), the electric field will also have a longitudinal component, which can accelerate particles; the field period must be $\beta\lambda$, where $\beta = v/c$ and λ is the electric field wavelength. The electrodes are mounted in a tank

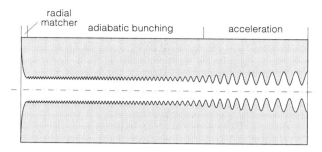

Fig. 8.49. Electrode cross section in an RFQ.

which acts as a resonant circuit, usually operating in the TM_{210} mode. The horizontal and vertical electrode pair are shifted by a half-period which increases the axial field component.

RFQ is an ion accelerator. Ions from a source, possibly accelerated up to ~ 100 keV, enter a radial matching section, where they are initially focused. An adiabatic bunching section follows, where the periodic structure amplitude slowly increases, and the particles are bunched and gradually accelerated. Bunching is effective and more than 90 percent of the injected particles are accelerated. At the end of the RFQ, when bunching is completed, the waviness amplitude increases, but the longitudinal electric field does not increase correspondingly, since the structure period increases because of the $\beta\lambda$ condition which reduces the field strength. This sets the energy limit of a RFQ to ~ 1–2 MeV for protons.

The phase stability, like in a Wideröe-type linac, requires that the phase stable angle be $0 \leq \varphi_r \leq \pi/2$. The periodic structure gives a kind of strong focusing, and the phase angle on the increasing part of the rf sine wave gives the usual defocusing. The total effect is weak-focusing, and thus the RFQ aperture is gradually decreasing toward the end of the structure.

An RFQ can be used as a preaccelerator for large proton linacs, and its cavity is more compact than for proton linacs of the same energy, because it uses higher frequency. RFQ also accepts a larger beam current.

8.7. STORAGE RINGS

8.7.1. Beam Transport

Large accelerators often consist of smaller units (Fig. 8.57), between which the particle beam must be transported with as small loss as possible. This is the task of the beam transport system which includes straight sections (drift space) and a variety of focusing elements like deflection prisms, quadrupoles, sextupoles, and other correcting devices.

A particle beam leaving an accelerator has a diameter, radial and axial velocities, and an energy spread. The beam transport system will transform the diameter and the radial and axial velocities while the energy spread will produce chromatic aberrations. However, all the forces acting on the beam are conservative forces (with some exceptions) and thus Liouville's theorem (Appendix B) can be applied — the phase space area occupied by the beam is constant throughout. For beam transport systems it is useful to define a curved coordinate which follows a reference particle, where s is the logintudinal and x and y are the transverse coordinates. The energy spread is accounted for in the particle momentum, p, distribution.

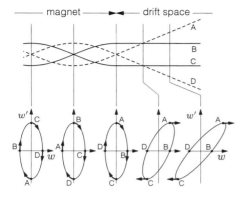

Fig. 8.50. Particle orbits and emittance transformation in a magnetic prism and a field-free region thereafter. (After *Principles of Cyclic Particle Accelerators*, J. J. Livingood, © 1961, Van Nostrand, with permission.)

Individual particles can be followed through the system, whose elements are described by transformation matrices. Appendix H lists the transformation matrices for some of common beam transport elements.

A beam transport system can be viewed in two ways. In one, a number of particles is followed through the system, while the other, more convenient way uses the emittance ellipse, which under linear forces remains an ellipse with constant area (Eq. 8.22). There are two such ellipses, one for each transverse direction, x, x' and y, y', and each ellipse (coordinates w, w') changes shape when the beam passes through an element of the transport system. The maximum value of w determines the beam size.

As an example, consider the passage of a beam through a magnetic prism. The transformation matrix is given by Eq. H.3 inside the homogeneous magnetic field. If the ellipse is upright at the entrance to the magnetic field (Fig. 8.50), it will also be upright at the exit. Different particles (A–D) will change positions and radial velocity components during the transit, but a particle at the boundary of the ellipse initially will remain there. When the particles abruptly leave the field (fringing field, a second-order effect, is discussed in Section 4.7.4) the drift space transformation matrix applies (Eq. H.1). The radial velocity, w', no longer changes, and thus the ellipse expands in the w direction while becoming narrower.

This example beam has no energy spread, $dp = 0$, but such monochromatic beams are not physically realizable. A momentum deviation, $p + dp$, results in chromatic

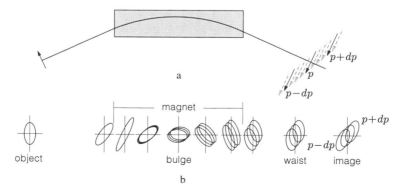

Fig. 8.51. (a) Chromatic aberration. **(b)** Emittance transformation of a beam with energy spread. (After *The Optics of Dipole Magnets*, J. J. Livingood, © 1969, Academic Press, with permission.)

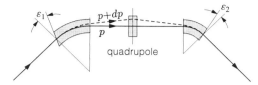

Fig. 8.52. Double-focusing achromatic 90° deflection.

aberrations. Figure 8.51a shows the passage of a multi-momentum beam through a magnetic prism, with different energies giving different images. For each energy its corresponding emittance will be somewhat different, and the common emittance envelope increases, with the beam becoming broader (Fig. 8.51b).

By suitable arrangement of lenses, achromatic imaging is possible, an example of which is given in Fig. 8.52, where a 90° deflection is obtained while the emittance ellipse area remains unchanged.

The emittance ellipse equation (Eq. 8.22) is

$$\gamma w^2 + 2\alpha w w' + \beta w'^2 = \varepsilon,$$

where α, β, and γ are three parameters which vary with position along the beam transport system. Linear elements transform one emittance ellipse to another. If $\mathbf{W}_0 = (w_0, w'_0)$ are the initial position and velocity, after the passage through a linear element the new position and velocity are

$$\mathbf{W}_1 = \mathbf{M}\mathbf{W}_0, \tag{8.62}$$

with

$$\det(\mathbf{M}) = 1. \tag{8.63}$$

Thus a beam transport system can be treated as the successive change in emittance of transited elements.

The emittance reveals the beam behavior as it moves through the transport system. A small emittance implies a thin and parallel beam. However, emittance is based upon the coordinates w, w', while the Liouville's theorem relies on w, p_w. A beam accelerated longitudinally preserves its transverse coordinates w, p_w, but its emittance, defined by w, w', shrinks. This effect must be taken into account by normalization, when particle beams with different energy are compared.

Note that the emittance ellipse area is $\varepsilon\pi$. When emittance is quoted, it is usually just ε, which is normally given in the units "mm·mrad" or "mm·mrad, normalized."

The largest emittance an element can accept is called its *admittance* (note that acceptance is used for the longitudinal phase space). The optimal beam transport system has the beam emittance at the entrance to each element matched to the admittance

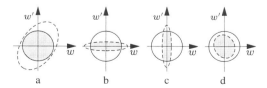

Fig. 8.53. Emittance and admittance. The admittance of the element is shown as a circle, and the beam emittance is shown as an ellipse. The shaded area is admitted. (**a**) The emittance is too large independently of its shape. In **b**, **c**, and **d** the size of the emittance is equal. Only in **d** the whole beam is admitted.

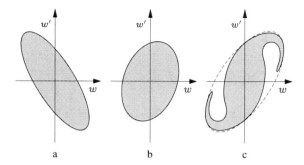

Fig. 8.54. Nonlinear effects. (**a**) Primary emittance. (**b**) After transformation in a linear and (**c**) after a nonlinear element. All shaded surfaces have the same area.

of that element (Fig. 8.53). Since beam transport system can have dozens of elements, such optimization is cumbersome. In the linear case the matrix transformations are of the order 6×6, but with second-order terms included 36×36 matrix elements are added. Second-order (nonlinear) terms deform the emittance ellipse (Fig. 8.54). Although the beam phase space area is not changed (Liouville's theorem), its shape is no longer elliptical, with a central part and tails, reminiscent of a spiral galaxy. This deformed emittance, however, can be circumscribed by a larger "effective" emittance ellipse, which can be used in further computations.

The many computer programs for beam transport computation which exist are of two kinds. One integrates the equation of motion for selected particles through a computed or measured magnetic or electric field and calculates the matrix elements, often including second-order terms. A second type is based on known matrix elements of first and second order, calculating emittance change through a system, accepting conditions on the exit emittance and optimizing (interactively) the beam transport elements. In practice often a combination of both approaches is most useful: The first program computes the matrix elements and a first optimization, and the second only varies the length of the straight sections.

There are cases when the Liouville's theorem is not longer valid, the most important of which is when electron, stochastic, or synchrotron radiation cooling is applied in a storage ring, as discussed in Section 8.7.3. Another is the beam transit through a thin foil, as when stripping injection is used, and the particles are scattered and the emittance is increased.

8.7.2. Storage Rings and Colliders

Storage rings are constructed similar to large synchrotrons, but the particles are kept circulating for a long time (up to some days). If the particles are electrons a storage ring can generate synchrotron light or laser light (Section 8.7.4). Two counterrotating particle beams can be stored and brought to a focus in what is called a *collider*, in this way increasing the collisional energy (Appendix G, Eq. G.9, Fig. 8.3, and Table 8.6). Table 8.7 gives data for the largest existing and projected colliders. Compare also with Fig. 8.4, the Livingston diagram for colliders.

An important parameter in both storage rings and colliders is the luminosity of the beam(s), which is directly proportional to the current density. In colliders it depends on the size of the focus at the point where the beams collide. The current density can be increased by injecting beams with a small emittance and energy spread or by cooling (Section 8.7.3). By choosing an appropriate lattice, the β-function can be adjusted at a focus to give a high luminosity.

Table 8.6

Equivalent energy in collision with stationary target

Laboratory	Accelerator	Particles	Energy [GeV]	γ	Equivalent energy [TeV]
CERN	SPS	p, $\bar{p}$	450	481	432
FNAL	Tevatron	p, $\bar{p}$	1,000	1,067	2,140
CERN	LHC (projected)	p, p	7,000	7,460	104,500
SLAC	SLC	e^-, e^+	50	97,850	9,780
CERN	LEP (upgraded)	e^-, e^+	100	196,000	39,140
CERN	CLIC (proposal)	e^-, e^+	1,000	1,960,000	3,914,000

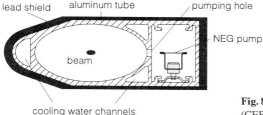

Fig. 8.55. Cross section of the LEP vacuum chamber (CERN).

In circular colliders a single magnet lattice and vacuum chamber can be used if the colliding particles differ in polarity. For pp colliders, like the projected LHC machine (Section 8.8), two separate vacuum chambers placed inside the same superconducting magnet (Fig. 8.68) are to be used.

Storage rings and circular colliders require ultra-high vacuum for a long beam "lifetime" and a low "background," usually between 10^{-10} and 10^{-8} Pa. Figure 8.55 shows the cross section of the LEP vacuum chamber, the length of which is 27 km. The chamber is pumped with NEG pumps, in straight sections with ion pumps, and the initial high-vacuum in the ring is obtained with turbomolecular pumps. The vacuum load is large, because synchrotron light hits the vacuum chamber walls in the deflecting magnets. The lead shield is the radiation screen.

The betatron and synchrotron oscillations create resonance problems, and the long lifetime makes them only worse, as well as some other instabilities, which depend on collective effects. Among them the induced currents (mirror charges) in the vacuum chamber wall lag the beam in phase, which results in positive feedback. Above the transition energy in AG-focusing machines, particles approach as a result of Coulomb repulsion (the "negative mass effect"). Consequently, the charge density becomes nonuniform, which amplifies the effect, and some particles are finally thrown out of the phase stable region. In cooled beams with a very small emittance, intrabeam scattering can limit the current density. In electron machines two electrons can "collide" and scatter with such large angle that the increase of betatron oscillations forces a loss in phase (Touschek effect). In colliders the two high-current-density beams self-focus at high energy, because there is no beam spreading (Eq. 6.8). But in the collision region the Coulomb repulsion predominates and the beams tend to spread. Finally, there is particle loss from scattering with the residual gas in the machine, thus the ultra-high vacuum.

Figure 8.56a shows a layout of a typical medium-energy storage ring, where protons or heavy ions are stripping injected from an isochronous cyclotron (GWI cyclotron, Table 8.1),

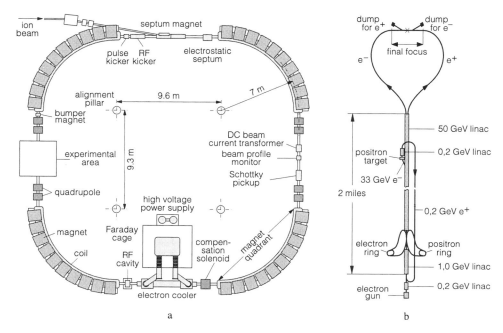

Fig. 8.56. (a) CELSIUS storage ring in Uppsala, Sweden. (b) Layout of the Stanford linear collider (Stanford Linear Accelerator Center).

with proton energy between 48 and 180 MeV. A lattice of four dipole magnets and a number of quadrupoles define the betatron tune of the machine. The machine can accelerate protons in an rf cavity to an energy of 1.4 GeV (with increasing magnetic field). Beam position monitors along the circumference of the machine detect the transverse position of the rf-modulated beam while the beam intensity is measured by a **D**irect **C**urrent **C**urrent **T**ransformer. Without rf voltage the coasting beam expands longitudinally over the whole ring, but can still be followed by a wide-band ferrite loaded cavity, the Schottky pick-up, detecting the revolution frequency and its higher harmonics. To increase the luminosity and to reduce the phase space area the ring is equipped with an electron cooler.

A collider example is the **S**tanford **L**inear **C**ollider, seen in Fig. 8.56b. Electrons from a gun are injected into the SLAC linac (Table 8.5). At an energy of 33 GeV the e⁻ beam is deflected to a target in which positrons are produced which travel in a beam transport system back to the beginning of the linac. From two small accumulation rings, e⁻ and e⁺ are again injected into the linac, accelerated to 50 GeV, and sent along two beam transport systems which terminate in the "final focus," where both beams are ~1 μm wide and collide.

Figure 8.57 shows the central part of the CERN's accelerator complex. Both SPS and LEP beams originate in linacs which feed the PS accelerator. The high-energy protons are either injected into SPS or sent to the antiproton source. The antiprotons are stochastic-cooled in the **A**ntiproton **C**ollector, sent to the **A**ntiproton **A**ccumulator, and from there injected into the SPS, where both p and p̄ are accelerated simultaneously to their final energy and focused in three different straight sections where they collide.

Table 8.7 shows data on the energy and type of particles of the largest existing or projected accelerators.

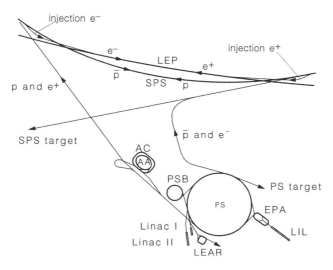

Fig. 8.57. Layout of the central part of the CERN accelerators. LEP, large electron-positron collider; SPS, super proton synchrotron; AC, antiproton collector; AA, antiproton accumulator; PS, proton synchrotron; PSB, proton synchrotron booster; EPA, electron–positron accumulator; LIL, LEP injector linac; LEAR, low-energy antiproton ring; Linac I, injection linac for heavy ions; Linac II, injection linac for protons (CERN).

Table 8.7

Largest high-energy accelerators
(Energy in GeV)

Laboratory		Existing machines (1995)			Projected machines or extensions		
Europe:	CERN	SPS	p	450			
		SPS	pp̄	2×450	LHC	pp	2×7000
		LEP	e^-e^+	2×50		e^-e^+	2×100
	DESY	HERA	ep	$30 + 820$		ep	$32 + 1000$
USA:	FNAL	Tevatron	p	1000			
		Tevatron	pp̄	2×1000			
	SLAC	SLC (linear)	e^-e^+	2×50	B factory	e^-e^+	$9 + 3.1$ $\sim 2 + 2\,A$
Russia:	Serpuchov		p	76	UNK	pp	2×3000
	Novosibirsk		e^-e^+	2×7			
Japan:	KEK	TRISTAN	e^-e^+	2×32	B factory	e^-e^+	$8 + 3.5$ $2.6 + 1.1\,A$
China:	Beijing	BEPC	pp	2×2.8			

CERN - Organisation Européenne pour la Recherche Nucléaire, Genève, Switzerland.
DESY - Deutsches Elektronen-Synchrotron, Hamburg, Germany.
FNAL - Fermi National Accelerator Laboratory, Batavia, Illinois.
SLAC - Stanford Linear Accelerator, Stanford, California.
KEK - National Laboratory for High Energy Physics, Oho, Tsukaba-Shi.

8.7.3. Electron, Stochastic and Synchrotron Cooling

A small emittance and energy spread allow a beam to be better focused and allow the "background" radiation to be smaller. In the accumulation of antiparticles the emittance must be reduced because when antiprotons are created in a high Z-target, the outgoing beam

Fig. 8.58. Aerial view of CERN. (Photograph CERN).

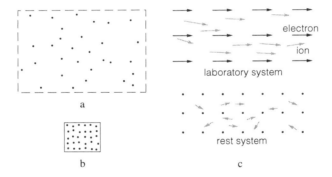

Fig. 8.59. Beam cooling. Emittance (**a**) before and (**b**) after cooling. (**c**) Electron cooling: electron and ion velocities.

is spread in all six coordinates of the phase space, unsuitable for injection into any accelerator or storage ring. Liouville's theorem prohibits emittance change in beam transport elements, so nonconservative forces must be applied. Electron and positrons are automatically cooled in circular machines by their emission of the synchrotron radiation. For particle beams of protons and heavier ions, electron and stochastic cooling are available (Fig. 8.59).

In electron cooling [174] an electron beam with p_x, p_y, $\delta p \approx 0$ is generated using careful electron gun design and confined flow (Section 6.3.3) to reduce the transverse energy to that of the cathode thermal energy, ~ 0.2 eV. Confined flow is ensured by immersing the whole electron beam in a longitudinal magnetic field (Fig. 8.60), while $\delta p \approx 0$ depends on the quality of the high-voltage source for the electron gun.

The ion and electron beams run together in the drift tube, which constitutes a part of the ring straight section (Fig. 8.56a). Coulomb forces between the particles act like friction, reducing the betatron oscillation amplitude and the energy spread of the ion beam, while heating the electrons. Ions circulate in the ring, with new "cool" electrons coming from the gun.

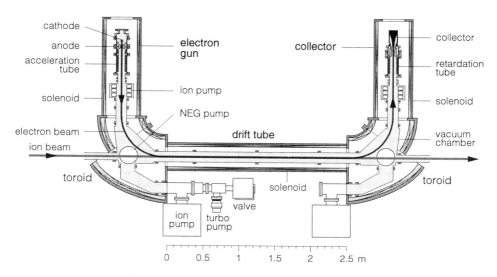

Fig. 8.60. Electron cooler of the CELSIUS storage ring.

The result is a kind of thermal equilibrium, with the ions heated by instabilities in the ring (dominated by intrabeam scattering and residual gas scattering) and cooled by electrons. The emittance and the energy spread can be reduced up to ~ 50 times by electron cooling thus correspondingly increasing the beam lifetime in the ring. Electron cooling is best suited for high-current-density ion beams of not too high energy, because the electrons and the ions must have the same velocity:

$$W_{ke} = \frac{m_{0e}}{m_{0p}} W_{kp} = \frac{1}{1833} W_{kp}. \tag{8.64}$$

The electron cooler for the CELSIUS storage ring has electron-beam currents up to 2.5 A with energies between 10 and 300 keV. At the highest energy the beam power is 0.75 MW, and thus the beam must be retarded before collection. The voltage on the collector is only ~ 5 kV above the cathode voltage for any gun accelerating voltage, and the power loss is a few kilowatts. To drive a beam current of 2.5 A a high-voltage source of only 2 mA is needed.

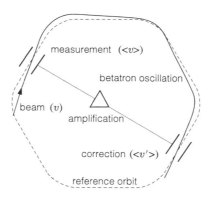

Fig. 8.61. Stochastic cooling principle.

In stochastic cooling [175] a detector with broad-band amplifiers (up to 10 GHz) locates the beam center of gravity in the axial and radial directions. The amplified signal is applied to an electrode pair, with phase such that the center of gravity offset is corrected. The detector and correcting electrodes are on opposite side of the ring, so that the signal can be amplified and reach the electrodes when the bunch to be corrected arrives, Fig. 8.61.

Each correction redistributes the particles with respect to betatron and synchrotron oscillations, so after a time the amplitude of both oscillations is reduced. Stochastic cooling is very effective for low current density beams and so is used in both SPS and Tevatron to decrease the antiproton phase space. In CERN, for example, antiprotons are collected for a whole day, $\sim 8 \times 10^{11}$ $\bar{p}$, before injecting into SPS.

In electron and positron rings, synchrotron radiation damps the amplitude of oscillations in phase space. Emitted photons take energy from both betatron oscillations and the longitudinal motion. A particle radiates

$$-\Delta W = \frac{q^2 \beta^3}{3\varepsilon_0} \frac{\gamma^4}{r} = 8.85 \times 10^{-8} \frac{W^4}{r} \text{ [eV]} \qquad (8.65)$$

per revolution, where W is the beam energy in megaelectronvolts, and r the radius in meters. These losses must be replaced in the acceleration cavity. The attenuation depends on the focusing in the lattice. In AG synchrotrons the attenuation can go over into amplification under certain conditions. Storage rings or electron–positron colliders are therefore usually constructed with gradientless dipoles, and the focusing is achieved entirely with quadrupoles.

8.7.4. Synchrotron Radiation and Free Electron Laser

Any accelerated charged particle emits electromagnetic radiation. In a magnetic dipole the beam is bent and so the particles emit with power that depends on W^4/r — hence the large radii of electron synchrotrons. For LEP, with a circumference of 27 km, the radiation losses per turn are 128 MeV at full energy of 50 GeV.

The high power emitted in synchrotron radiation is a stimulus to build dedicated electron storage rings, an example of which is shown in Fig. 8.62. Here a racetrack microtron injects a 100 MeV electron beam into the ring, where it is accelerated to 550 MeV.

Synchrotron radiation is characterized by high emitted power, narrow light beam, wide spectrum, well-defined polarization, variable time structure, and good stability and acts almost like a point source.

The spectral distribution, opening angle, and polarization can be influenced through the design of the ring. The emitted power is given by Eq. 8.65. The beam, however, is normally bunched because the radiation losses must be compensated, so the emitted power during the passage of the bunch will be larger, because of the q^2 term.

The normalized spectral distribution is shown in Fig. 8.63, where the critical wavelength is

$$\lambda_c = \frac{4\pi r}{3\gamma^3} \approx \frac{1.86}{BW^2} \text{ [nm]}, \qquad (8.66)$$

with B in tesla and W in gigaelectronvolts. The ordinate in Fig. 8.63 is proportional to the number of photons emitted per second. Since photon energy is proportional to $1/\lambda$, the emitted power per wavelength interval rapidly decreases with increasing wavelength.

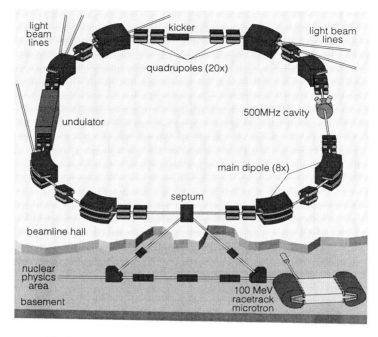

Fig. 8.62. Synchrotron radiation source MAX in Lund, Sweden.

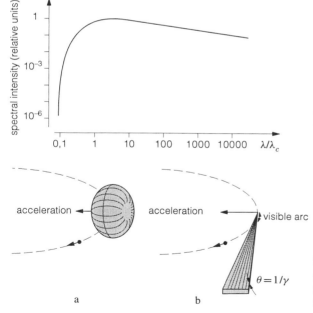

Fig. 8.63. Normalized synchrotron radiation spectrum. (After *European Synchrotron Radiation Facility,* Suplement 1, Y. Farge and P. J. Duke, eds., 1979, European Science Foundation, with permission.)

Fig. 8.64. Angular distribution of synchrotron radiation. **(a)** In the rest mass system of electrons. **(b)** As seen in the laboratory system.

In the center-of-mass coordinate system the emitted radiation power is determined by the dipole radiation distribution per solid angle, whose maximum is perpendicular to the acceleration (Fig. 8.64a). An observer in the laboratory system sees in the axial direction the radiation concentrated into a small cone of angle $\theta \approx 1/\gamma$ in the direction of electron motion (Fig. 8.6b). The light beam width in the radial direction depends on the size

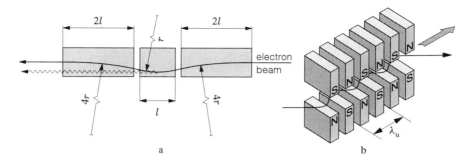

Fig. 8.65. (**a**) Dispersion free wiggler. (**b**) Undulator.

of the azimuth seen by the observer. The light is polarized linearly in the orbit plane, but outside a phase-shifted component is introduced, producing elliptical polarization.

In synchrotron light sources the emitted power can be increased and wavelengths blue-shifted by special wigglers and undulators inserted in the beam line. A wiggler uses a high magnetic field, often from a superconducting magnet. Since the power is proportional to $1/r$ and the wavelength is proportional to $1/B$, an improvement for a factor of 10 can be obtained. Figure 8.65a shows a wiggler in a "chicane" consisting of three magnets, the central one being the radiation source while the other two compensate for the disturbance of the orbit. In a symmetrical structure

$$\int B\,ds = \int Bv\,dt = \frac{1}{e}\int F\,dt = \frac{\Delta p_\perp}{e}$$

can be made zero, in which case the wiggler becomes achromatic.

An undulator has a periodic magnetic structure, and appears as a multiple wiggler (Fig. 8.65b), a situation similar to that of an ubitron (Section 7.6.1), and its description can be applied to the generation of synchrotron light. The electrons passing the undulator oscillate around the axis, their amplitude being small, so the emitted radiation will be axial. The light from all the N periods interferes constructively, with its wavelength determined by the spatial period of the undulator, λ_u, and the energy of the electrons. In addition to plane undulators, as in Fig. 8.65b, there are undulators of two helical magnets with or without iron.

If undulator coherent radiation is trapped between two mirrors, and the losses in the mirrors are smaller than the power emitted in the undulator field, the device, **F**ree **E**lectron **L**aser, will oscillate (Fig. 8.66).

These oscillations result from the coupling of the electric field of the emitted radiation to the transverse motion of the electrons. Some electrons will be accelerated,

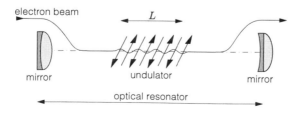

Fig. 8.66. Principle of a FEL.

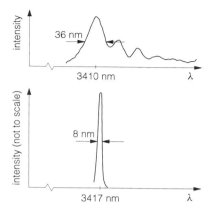

Fig. 8.67. Spectrum of a FEL. (**a**) Below and (**b**) above the limit of oscillations. [After D. A. G. Deacon et al., *Phys. Rev. Lett.*, **38**, 892, (1977), with permission.]

others retarded, depending on the phase relation between the electrons and the field, but on average there will be no energy transfer. However, if the undulator frequency differs slightly from that of the radiation field, the electromagnetic radiation will be amplified, the energy coming from the electron beam, like for the ubitron. A laser constructed according to this principle can be tuned to oscillate from the submillimeter to the UV region with large output power. A FEL can also be made like the proposed microwave tube based on the Čerenkov maser principle of Section 7.6.3.

Figure 8.67 shows a spectrum obtained from a FEL. The upper curve is the undulator output before the parameters were adjusted for oscillations, and the lower curve is the output when the laser action started. The difference in the wavelength points to the fact that the amplification needs a higher electron energy.

8.8. PERSPECTIVE

Probably, the largest synchrotron-type accelerator will be the **Large Hadron Collider** (Table 8.6 and 8.7). LHC will use the LEP tunnel to reduce costs and will be finished at the beginning of the 21st century. Figure 8.68 shows the cross section of a dipole magnet. Larger colliders of the synchrotron type will almost certainly not be built.

The most promising approach to the next energy increase are linear colliders, but not of the conventional type. The energy gain per unit length must be increased from ≤ 10 MeV/m to a much higher value. Different ideas have been proposed how to reach the goal; laser-driven structures, beam-driven plasma waves, and structures driven by

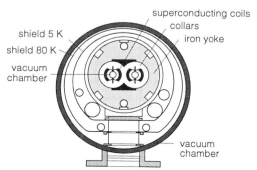

Fig. 8.68. Cross section of the LHC dipole magnet (CERN).

the wake field of a single beam pulse. However, the problems of accelerating field gradient limitation by average power consumption turned the interest toward classical rf structures. One proposal at CERN, the **CERN LI**near **C**ollider (Fig. 8.69), uses a high-current electron beam accelerated in a linac to an average energy of 3–5 GeV. The highly bunched beam delivers energy to 30 GHz traveling-wave transfer structures, which feed power to the main accelerating linac, where an energy gain of up to 80 MeV/m is envisioned. Periodically, the drive beam is reaccelerated in superconducting cavities with a ~ 10 MeV/m field gradient, driven by conventional 1 MW klystrons at 350 MHz, like the one in Fig. 7.21.

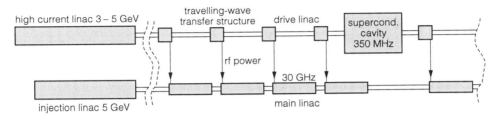

Fig. 8.69. CLIC, a two-beam accelerator (CERN).

Two-beam accelerators for e^+ and e^-, each ~ 15 km long, would reach an energy of 1 TeV. After acceleration, the particles would be deflected with dipoles and chromaticity controlled by sextupoles in long bends of ~ 1 km before the final focus where a beam height of ~ 15 nm is necessary to obtain the design luminosity of 10^{33} cm^{-2}s^{-1}. Many problems remain to be solved; for example, wake fields are induced in all accelerating structures and act on trailing particles within the same bunch, causing beam instability. The beam can be stabilized by rf quadrupoles at the expense of an extremly tight dynamic alignment control of the accelerating structure to within $<1\mu$m, with a time-scale approaching the main linac repetion period of 1700 Hz. A CLIC prototype accelerating section, seen in Fig. 8.70, was built and tested, while in an experiment at SLAC a final focus of ~ 70 nm with the 50 GeV beam was obtained.

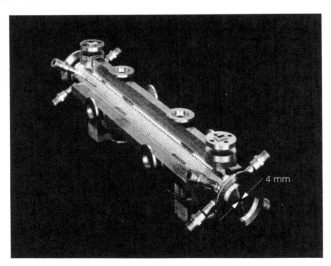

Fig. 8.70. Prototype accelerating section of the CLIC main linac (CERN).

9

Gas Discharges

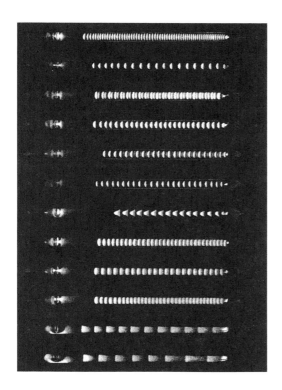

Under certain conditions of gas, pressure, and current density, alternate dark and bright striations apper in the positive column. In the early explorations of the gas discharge physics, striations were a favored experimental topic. [From W. de la Rue and H. W. Müller: Experimental Researches on the Electric Discharge with the Chloride of Silver Battery. Part II. The Discharge in Exhausted Tubes, *Phil. Trans. Roy. Soc.*, **159**, 155, Plate 16 (1878), with permission.]

9.1. INTRODUCTION

HISTORICAL NOTES. Around 1600 Gilbert [1] observed that a charged conductor loses its charge when near a flame. Guericke [176], the constructor of the first electrostatic generator, observed electric sparks from the machine's sulfur sphere to grounded objects. The term "discharge" was coined around 1746 when von Musschenbroek's [177] experiments with "Leiden jars" (discovered in Germany by M. Bose in Wittenberg in 1744 and independently by E. G. Kleist in Kammin in 1745) generated arc discharge between the jar and a grounded wire. At about the same time Franklin [178] investigated atmospheric discharges in hazardous experiments with wires suspended from kites in thunder clouds. He explained St. Elmo's fire as electric discharges and suggested mounting protective lightning conductors on houses. Franklin also named electricity *positive* and *negative*.

Petrov [179] and Davy [180] first demonstrated continuous arc discharges, Davy also showed in 1821 that a magnet attracts or repels the arc with a rotatory motion. Faraday [181] investigated discharges under low pressure between 1831 and 1835 and discovered the glow discharge, with its light and dark bands.

A breakthrough in these investigations came in 1851 when Rühmkorff [182] constructed the induction coil, which generated high voltages with substantially higher currents than could be obtained with electrostatic generators. Another important step was taken in 1858, when Geissler developed a technique for fusing platinum electrodes to glass, and Plücker [8] observed that light emitted by "Geissler tubes" was characteristic of the gas filling the tube. He discovered "cathode rays" emanating from a hole in the anode. Hittorf [9], a student of Geissler, showed these "rays" to spiral along the magnetic field lines. Varley [10] and Hertz [11] tried in the early 1870s to deflect cathode rays by electric field, with and without success respectively. Crookes [81, 183] expressed the opinion that discharges represented the fourth state of matter and that cathode rays are fundamental particles. Goldstein [184] discovered "channel rays," positive ions passing through a hole in the cathode. J. J. Thomson's discovery of the electron in 1897 [13] widened the understanding of gas discharges, but even at the beginning of the century much work was phenomenological. Townsend [185], a student of J. J. Thomson working with dark discharges, tried to assemble the contemporary knowledge and give it an analytical form.

Theoretical work started seriously in the early 1920s, when Langmuir [186] established the first theory of gas discharges and introduced the concept of plasma. Saha [187] made a quantitative analysis of thermal ionization. Their work provided for the rapid development of the field in the decades that followed.

Many electronic components and devices work with high vacuum, which corresponds to a pressure of $\leq 10^{-6}$ Pa, where gas molecules are present at a density of $\leq 10^8/\mathrm{cm}^3$. Collisions between charged particles and gas molecules occur which ionize neutral molecules to form electrons and ions, which interact with any external electric or magnetic fields. Because the number of secondary particles is low, these particles do not disrupt the operation of most vacuum devices and their effects can be neglected. At higher pressure the ionization probability increases and new particles can outnumber the original ones. Processes where ionized particles play an important role in the electric charge transport are called *discharges*.

To introduce discharge physics, we review some properties of gases and vapors. For N gas molecules with mass m in a container of volume V, the mass density ρ and the number density n are defined as

$$\rho = \frac{mN}{V}, \qquad n = \frac{N}{V}. \tag{9.1}$$

If the total gas mass, $M = mN$, is known, the gas number density, or shorter gas density, is

$$n = \frac{N_A}{A}\rho,$$

where N_A is Avogadro's number and A is the mass number, or mass/kilomole. The mean distance between two molecules is

$$d_m = \frac{1}{\sqrt[3]{n}}.$$

The random flux, ν, is the number of particles which cross a surface of unit area per unit time

$$\nu = \frac{1}{4}n v_m, \tag{9.2}$$

where v_m is the mean particle velocity, which for a Maxwellian velocity distribution is

$$v_m = \sqrt{\frac{8kT}{\pi m}}. \tag{9.3}$$

The pressure is defined by the state equation,

$$p = nkT. \tag{9.4}$$

In collisions between gas particles two physical quantities — the gas kinetic cross section, σ, and the mean free path, λ_g — play an important role. The gas kinetic radius, r_k, which is not identical with the physical radius since it decreases with increasing temperature, defines the gas kinetic cross section

$$\sigma = 4\pi r_k^2 \tag{9.5}$$

and the mean free path

$$\lambda_g = \frac{1}{\sqrt{2}\sigma n}. \tag{9.6}$$

The last equation takes into account the movement of the gas molecules. Equation $\lambda = 1/(\sigma n)$, which is based on a particle swarm moving in a stationary gas, is valid for ions, which in an electric field obtain a larger velocity than do neutral gas molecules. However, ion–neutral cross sections may differ greatly from the neutral–neutral ones because of charge transfer. The mean free path for constant cross section is inversely proportional to the gas density:

$$\lambda_g \sim \frac{1}{n}. \tag{9.7}$$

Mean free path determines the number of particles, N, which have not collided in a distance x as

$$N = N_0\, e^{-x/\lambda}. \tag{9.8}$$

The gas pressure, p, is often used to express *density*, usually at 273 K or 293 K, since pressure and density are proportional for constant gas temperature. In many modern applications the temperature can vary over wide limits, so the approximation p proportional to n must be used with caution.

Table 9.1
Important physical properties of gases and vapors

	Noble gases				
	He	Ne	Ar	Kr	Xe
Mass number A [kg/kmol]	4.0	20.2	39.94	83.8	131.3
Atomic mass m [10^{-27} kg]	6.64	33.5	66.31	139.1	218.
Gas kinetic radius r [nm]	0.11	0.13	0.18	0.20	0.24
Mean free path λ_g [μm] at 273 K and 100 Pa	175.	125.	65.5	53.0	36.8
Ionization energy [eV]	24.6	21.5	15.7	14.0	12.1
Lowest excitation energy [eV]	19.8	16.6	11.6	9.9	8.4

	Common gases				
	H_2	N_2	O_2	CO_2	Air
Mass number A [kg/kmol]	2.02	28.0	32.	44.	29.
Molecular mass m [10^{-27} kg]	3.35	46.5	53.12	73.	48.15
Gas kinetic radius r [nm]	0.14	0.19	0.18	0.23	0.19
Mean free path λ_g [μm] at 273 K and 100 Pa	108.	60.	65.5	39.6	59.
Ionization energy [eV]	13.6	15.5	12.5	14.4	
Lowest excitation energy [eV]	10.	6.1	7.9	10.0	

	Metal vapors		Water vapor
	Hg	Cs	
Mass number A [kg/kmol]	200.6	132.9	18.0
Atgomic mass m [10^{-27} kg]	333.	220.6	
Gas kinetic radius r [nm]	0.31	0.18	0.25
Mean free path λ_g [μm] at 273 K and 100 Pa	22.0	65.5	39.4
Ionization energy [eV]	10.4	3.87	12.6
Lowest excitation energy [eV]	4.9	1.4	7.6

Electrons, ions, fast neutral particles, or photons can ionize neutral atoms by collision, with the energy required to free the least bound electron from an atom being the *ionization energy*, W_j, whose corresponding voltage is V_j. If the energy is too low for ionization, excitation of the atom can occur by raising an electron from a lower energy level to a higher one. Excited atoms return to their ground state in a time $\sim 10^{-8}$ s by emitting a photon whose energy is equal to the difference between the two energy levels. Some atoms remain excited for long times in metastable states. In space, when these atoms decay their energy appears in forbidden spectral lines. In gas discharge devices after 10^{-2} to 10^{-4} s they either return to the ground state or gain sufficient energy in a collision to become ionized. An excited atom can also lose energy in collisions with the discharge chamber walls.

Table 9.1 gives a survey of some important physical properties of gases and vapors used in gas discharges.

9.2. IONIZATION

Positive and negative ions can be formed by the passage of decay products of natural radioactivity or cosmic rays, by collisions of fast electrons or ions, and by interaction with photons emitted by excited atoms. Charges can also be generated by secondary electrons from fast ions hitting the cathode and field or thermionic emission from the cathode. Ions and atoms also exchange charges.

Thermal energy causes ionization and excitation by the collision of neutral molecules (e.g., for hydrogen to become 1% ionized a temperature of 7000 K is needed).

An electron under the influence of an electric field, moving in a gas, will be accelerated between the collisions, and when its energy exceeds the gas ionization energy, it can ionize gas particles. At lower energies, electrons can engage either in inelastic collisions which excite gas particles or in elastic collisions. What takes place depends on the particular energy levels of the gas atoms.

In the following, A means a neutral atom, A^* an excited atom, A^+ a positive ion, A^- a negative ion, e an electron, hf a photon, W_e excitation energy, W_j ionization energy, and W_k kinetic energy. In excitation, an atom jumps from a lower to a higher energy level,
$$A + W_e \rightarrow A^*,$$
emitting a photon of energy hf, and the excited electron returns to a lower energy level,
$$A^* \rightarrow A + hf.$$
Ionization, which results in a positive ion and a free electron, is
$$A + W_j \rightarrow A^+ + e,$$
while ionization of an excited atom is
$$A^* + W_k \rightarrow A^+ + e.$$
In recombination a photon is emitted
$$A^+ + e \rightarrow A + hf,$$
while the capture of an electron by a neutral atom produces a negative ion by the process
$$A + e \rightarrow A^- + hf.$$
Charge exchange, especially if resonant, between atoms of the same kind is
$$A_1^+ + A_2 \rightarrow A_1 + A_2^+.$$

For each process a collision cross section can be defined which depends on the energy of the colliding particles. For electrons the ionization cross section is zero for energies below the ionization energy, and then it increases sharply to a maximum at ~ 2–5 times ionization energy, from which its decrease is approximately hyperbolical, as seen in Fig. 9.1.

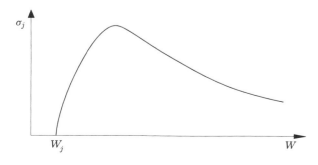

Fig. 9.1. Ionization cross section.

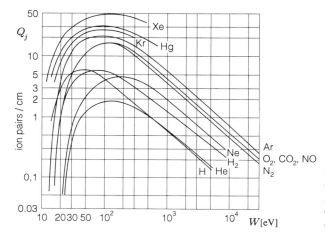

Fig. 9.2. Macroscopic ionization cross-section of some gases. (After *Ionized Gases*, von Engel, © 1955, Clarendon Press, by permission of Oxford Univeristy Press.)

The macroscopic cross section $Q = n_0\sigma\,[\text{m}^{-1}]$, not the microscopic cross section $\sigma\,[\text{m}^2]$, is often quoted in the literature. Here, n_0, the density at 1 torr (133,32 Pa) and $0°$ C, is $3.54 \times 10^{22}\,[\text{m}^{-3}]$. The density at an other pressure or temperature can be computed from Eq. 4.9 as

$$n = n_0 \frac{p}{p_0} \frac{T_0}{T},$$

which for pressures measured in pascals is

$$n = 7.2429 \times 10^{22} \frac{p\,[\text{Pa}]}{T\,[\text{K}]}.$$

The measured macroscopic ionization cross-sections of some gases are shown in Fig. 9.2. For example, the hydrogen microscopic elastic cross section for electrons is a few times $10^{-20}\,\text{m}^2$. Maximum ionization cross section is $\sim 1 \times 10^{-20}\,\text{m}^2$, and maximum excitation cross section corresponding to the Lyman alpha line is $\sim 0.7 \times 10^{-20}\,\text{m}^2$. Other excitation cross sections are much smaller, while the charge exchange cross section is ~ 10 times larger.

The corresponding mean free paths λ_j, λ_{exc}, and so on, mean collision free times τ_j, τ_{exc}, and collision frequencies, ν_j, ν_{exc} are related by

$$\lambda = \frac{1}{\sigma n} = v_e \tau, \qquad \nu = \frac{1}{\tau}, \tag{9.9}$$

where v_e is the mean electron velocity.

If v_e is sufficiently large, some collisions will be elastic, some will result in excitation, and others will result in ionization, all depending on the cross section. The elastic collisions of electrons and atoms and molecules result in small electron energy loss because of the small mass ratio m_e/m_a, the electron losing at most $4m_e/m_a$ of its energy. Inelastic collisions always result in an electron energy loss at least equal to excitation or ionization energy. The effective cross section for electrons includes all collisions and is

$$\sigma_{eff} = \sigma_{elast} + \sum\sigma_j + \sum\sigma_{exc},$$

which corresponds to an effective mean free path of

$$\lambda_{eff} = \frac{1}{n_e \sigma_{eff}}.$$

If monoenergetic electrons with velocity v_e flow through a gas of density n they will form

$$\frac{dn_j}{dt} = \frac{dn_e}{dt} = n_e n \sigma_j(v_e) v_e$$

new electron–ion pairs per unit volume and per unit time. For nonmonoenergetic electrons with a velocity distribution, the mean value must be taken so that

$$\frac{dn_j}{dt} = \frac{dn_e}{dt} = n_e n <\sigma_j(v_e)v_e>, \qquad (9.10)$$

where $<\sigma_j(v_e)v_e>$ is the ionization coefficient. When the mean electron kinetic energy is much smaller than the gas ionization energy, the largest contribution to $<\sigma_j(v_e)v_e>$ is from the "tail" of the electron velocity distribution.

9.3. FIELD-ENHANCED IONIZATION

Two plane electrodes, separated by a distance d, with a DC voltage V_a connected across them are seen in Fig. 9.3.

If $V_a \gg V_j$, and the gas density is such that the mean free path for collision ionization is much smaller than the distance between the electrodes, then $\lambda_{e_j} \ll d$, where λ_{e_j} corresponds to the σ_j maximum. N_0 electrons are emitted per unit time and unit area from the cathode with negligible initial velocity by, for example, photoemission.

Electrons are accelerated by the electric field. If we neglect the energy loss to excitations and elastic collisions with neutral particles, after a distance $x \geq V_j/E$ the electron energy will be large enough to ionize gas particles. $<x>$ is larger than V_j/E because of losses by excitations and elastic collisions, and because the ionization cross section is very small when the electron energy just equals eV_j.

Townsend [185], the first to study these processes, developed an approximate theory to describe the ionization processes, which is valid over a wide range of density and electric fields. When an electron gains enough energy to ionize a neutral gas molecule, an ion–electron pair is formed, in collision in which the projectile electron loses most of its energy. The ion moves toward the cathode, and both the projectile and liberated electron

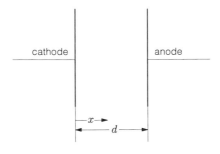

Fig. 9.3. Electrode geometry.

are accelerated toward the anode. When the two free electrons reach the effective ionization energy and collide with the gas, two new electrons are formed, and in this way an electron avalanche is created in which each electron creates α electrons and ions per unit length on its way to the anode. In a layer of thickness dx, through which $N(x)$ electrons flow per unit time,

$$dN = N(x)\,\alpha\,dx$$

electron–ion pairs are formed. Integrating this expression between the cathode and anode, the current multiplication factor, η, between the number of electrons, N_0, or the current, i_0, at the cathode and those at the anode, N_a and i_a, can be computed

$$\eta = \frac{N_a}{N_0} = \frac{i_a}{i_0} = e^{\alpha d}, \qquad (9.11)$$

where α is *Townsend's first coefficient*.

The current transport between the cathode and the anode obeys the continuity equation, and at the anode it is only electrons. At every other position, both electrons and ions transport the current. Although ions can ionize the gas particles on their way to the cathode, the probability is much smaller than that for electrons. Thus the current multiplication factor increases insignificantly when the ionization by ion collision is included.

According to a Townsend theory his first coefficient is

$$\alpha = nA\,e^{-B/(E/n)}. \qquad (9.12)$$

If the gas density changes, other parameters remaining constant, the ion production changes correspondingly. The electron energy gain between two collisions is proportional to E/n; thus the higher the electric field, the shorter the distance an electron must travel before it has enough energy to create an ion. On the other hand, if the gas density is higher, the mean free path will be shorter, and the electron will collide sooner. So

$$\alpha = n\,f\!\left(\frac{E}{n}\right).$$

Table 9.2

Townsend's constants A and B for some gases

Gas	A (ion pairs/torr·cm)	B (V/torr·cm)	Region E/p (V/torr·cm)
He	3	34	20–150
Ne	4	100	100–400
A	12	180	100–600
Kr	17	240	100–1000
Xe	26	350	200–800
H_2	5	130	150–600
N_2	12	342	100–600
CO_2	20	466	500–1000
Air	15	365	100–800
Hg	20	370	200–600

(After *Ionized Gases*, von Engel, © 1955, Clarendon Press, by permission of Oxford Univeristy Press.)

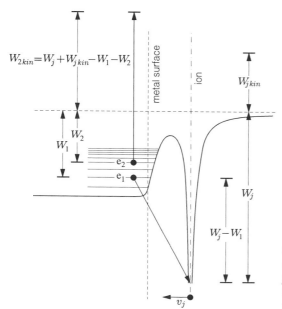

$$W_{2kin} = W_j + W_{jkin} - W_1 - W_2$$

metal surface

ion

W_{jkin}

W_2

W_1

e_2

e_1

W_j

$W_j - W_1$

v_j

Fig. 9.4. Sketch of the secondary emission process caused by the ion impact. (After *Principles of Electron Tubes*, J. W. Gewartowski and H. A. Watson, © 1965, Van Nostrand, with permission.)

Approximate values of A and B at 293 K, with n replaced by p in torr in Eq. 9.12, together with the range of validity, are given in Table 9.2. Tables and diagrams showing α/p as a function of E/p are found in the litterature, but all those data are approximate because even a minute amount of another gas changes the results.

Equation 9.12 agrees well with the measurements over large limits. A and B, characteristic for each gas, are called *Townsend's constants*. Townsend assumed a relationship $B = V_j A$, where V_j is the ionization voltage, but that does not agree with experiment.

When ions hit the cathode, they produce secondary electrons. Each ion must free at least one electron to recombine, but since the ionization energy is larger than the work function of most materials the energy equation can be satisfied even if two electrons are released, thus

$$\frac{m_j}{2} v_j^2 + eV_j = 2e\varphi_c + \frac{m_e}{2} v_e^2, \qquad (9.13)$$

where m_j, v_j and V_j are the mass, velocity, and ionization voltage of the ion, $e\varphi_c$ the cathode work function, and m_e and v_e the mass and velocity of the secondary electrons. Both e_1 and e_2 lie in the valency band, as seen in Fig. 9.4.

As the ion approaches the cathode, there is an interaction between the ion electric field and the electrons on the cathode surface. The potential barrier thickness decreases, allowing one of the electrons, e_1, to tunnel through and recombine with the ion. The excess energy, $\Delta W = W_j + W_{jkin} - W_1$, can release the second electron and give it a kinetic energy, $W_{2kin} = W_j + W_{jkin} - W_1 - W_2$.

Ions with energies greater than a few hundred electronvolts release secondary electrons not only at the surface, but even in deeper regions, as described in Section 2.2.4. In discharges the voltage drop in front of the cathode is a few tens of volts, and the process is as in Fig. 9.4. Low-energy ions move slowly enough so that the electrons in the valency band redistribute, thereby allowing recombination. Only a very small number of secondary electrons are produced at these low kinetic energies.

The secondary emission from the cathode starts a new electron avalanche which generates new electron–ion pairs and new secondary electrons. The first electron in the first avalanche — the first generation — gives $e^{\alpha d}$ electrons which reach the anode, while ($e^{\alpha d} - 1$) ions reach the cathode. If every ion releases γ_j electrons there are

$$\gamma_j(e^{\alpha d} - 1) = K$$

second-generation electrons. Eventually $K e^{\alpha d}$ electrons reach the anode. Correspondingly, ions at the cathode produce $\gamma_j K(e^{\alpha d} - 1) = K^2$, the third generation of secondary electrons. For every electron leaving the cathode,

$$n_a = e^{\alpha d}(1 + K + K^2 + \cdots) = e^{\alpha d}\frac{1 - K^n}{1 - K}$$

electrons reach the anode. If K is greater than 1, the current in the discharge increases are limited only by the power supply. If K is less than 1, and

$$\lim_{n \to \infty} \frac{1 - K^n}{1 - K} = \frac{1}{1 - K},$$

the current multiplication factor is

$$\eta = \frac{e^{\alpha d}}{1 - \gamma_j(e^{\alpha d} - 1)}. \tag{9.14}$$

Many phenomena affect a discharge — photons created in the gas, X-rays at the anode, generation of excited and metastable states, and thermionic emission from the cathode. All create new electrons and contribute a current multiplication factor similar to that of Eq. 9.14. For an arbitrary discharge length with an interelectrode distance d and Townsend coefficient α, the current multiplication factor can be written as

$$\eta_0 = \frac{e^{\alpha d}}{1 - \gamma(e^{\alpha d} - 1)}, \tag{9.15}$$

where γ is the secondary ionization or *Townsend's second coefficient,* which accounts for all secondary processes. η_0 gives the total number of ion–electron pairs which are formed from a single electron which leaves the cathode. γ depends strongly on the cathode material and on the gas, but it is independent on the electric field strength in the discharge region.

As the denominator in Eq. 9.15 goes to zero, the theory becomes invalid and the current must be limited by placing in series a resistor in the external circuit. A *breakdown* occurs and the discharge becomes selfsustained, if

$$e^{\alpha d} \geq 1 + \frac{1}{\gamma},$$

where α again depends on the electric field strength. At the moment of breakdown $E = V_b/d$, where V_b is the breakdown voltage. From the approximate expression 9.12 and Eq. 9.15 we obtain

$$\alpha d = \frac{\alpha/p}{E/p}V_b = A'pd\,e^{-B'pd/V_b} = \ln\left(1 + \frac{1}{\gamma}\right) = M_d = \text{const}, \tag{9.16}$$

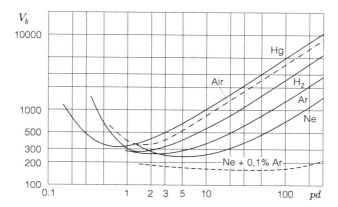

Fig. 9.5. Paschen's curves (After *Ionized Gases*, von Engel, © 1955, Clarendon Press, by permission of Oxford Univeristy Press.)

where $A' = An/p$ and $B' = Bn/p$, and the breakdown voltage is

$$V_b = \frac{B'pd}{\ln(A'pd) - \ln M_d} = f(pd),\qquad (9.17)$$

which only depends on the product of the interelectrode distance and pressure. Constants A', B', and M_d include the influences of ionization voltage, particle mass, electron affinity to make negative ions, and cathode material, geometry, temperature, and work function. Equation 9.17, *Paschen's law* [188], can be visualized graphically in Fig. 9.5, where the Paschen's curves show the lowest breakdown voltage to be roughly an order of magnitude larger than the ionization voltage. Although more correct is the use of density, n, not pressure, p, in Eqs. 9.16–9.17, diagrams in the literature for historical reasons show Paschen's curves with the product pd on the abscissa.

The breakdown voltage reaches infinity twice in Eq. 9.17, once at $pd \to \infty$ and again at $pd = M_d/A'$. Starting from low pressure, the breakdown voltage at first decreases with increasing density because more molecules can be ionized, but at higher pressure the breakdown voltage increases because the mean free path decreases and electrons have difficulty reaching ionization energy.

9.4. GAS DISCHARGES

A tube with two electrodes filled with a low pressure gas is called a *discharge tube*. If the electrodes are connected to a DC source in series with a current-limiting resistor, the voltage drop over the tube will vary with current as seen in Fig. 9.6. Here the gas was helium, the pressure 5160 Pa, the electrodes silver, and the electrode distance 8 mm. The curve can be divided into three regions — *unsustained discharge*, called *Townsend's discharge* (A-C), *glow discharge* (D-E), and *arc discharge* (G); and three transition regions — *breakdown region* (C-D), *anomalous glow discharge* (E-F), and *transition to arc discharge* (F-G).

9.4.1. Unsustained Discharge

Unsustained discharge requires an ionization or electron source. Natural radioactivity and cosmic rays generate enough electrons and ions for a very weak current, but reproducibility requires the cathode be irradiated, for example, by ultraviolet light. Increasing voltage causes the unsustained discharge current to increase until saturation is reached,

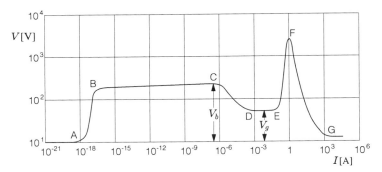

Fig. 9.6. Voltage as a function of the current in a discharge tube (measured up to E). (After *Handbook of Vacuum Physics, Vol. 2, Physical Electronics*, A. H. Beck, ed., 1968, Pergamon Press.)

and all primary charges are collected (region A-B), which in most tubes is $\sim 10^{-18}$ A. When the applied voltage exceeds the gas ionization potential (B), primary electrons create electron–ion pairs according to Eq. 9.11. The current through the tube then increases exponentially with voltage (region B-C). Since Townsend's coefficient α is proportional to $e^{-(Bn/E)}$, only a small voltage increase causes a large current increase, but the discharge is still unsustained, and the process stops with the removal of the primary ion or electron source. In the A-C region the discharge produces no observable light emission.

Chambers to measure ionizing radiation (Section 10.2.1) use the discharge at the beginning of the region A-B since secondary ionization does not occur. Proportional counters (Section 10.2.2) use the region B-C since the number of ions and electrons created is proportional to primary ionization, which depends on the energy loss of the primary radiation and thus can detect single particles or compare energies of particles.

The denominator in Eq. 9.15 determines the upper limit for unsustained discharge. Near zero breakdown occurs, the discharge "ignites" and the current increases two or three orders of magnitude while the tube voltage falls (C-D). Geiger tubes use this region to obtain their high sensitivity (Section 10.2.2), because even if only a single electron–ion pair is created, so many neutral particles become ionized that the discharge spreads over the whole tube volume. Here the current must be limited by a resistor in series.

9.4.2. Glow Discharge

The glow discharge commences when positive ions create so many secondary electrons that an external radiation source is unnecessary to sustain the discharge. Here a breakdown occurs, and the tube "ignites," after which the voltage drops from V_b to the glow voltage, V_g (Fig. 9.6), the breakdown condition given by Eq. 9.17 (or Paschen's curves). The voltage drop after breakdown depends on the positive ion space-charge established in front of the cathode, which creates an electric field that accelerates the emitted secondary electrons. An external primary ionization, like cosmic rays or natural radioactivity, is enough to ignite the discharge.

Visible light is seen when the discharge ignites, and Fig. 9.7a shows schematically the different light and dark bands. In sequence, these bands are: Aston's dark space (1), cathode glow (2), cathode dark space (also called Crooke's or Hittorf's dark space) (3), negative glow (4), Faraday's dark space (5), positive column or plasma (6), and anode glow (7). Each band is related with physical processes in the glow discharge, and Fig. 9.7 shows variation of some quantities along the tube.

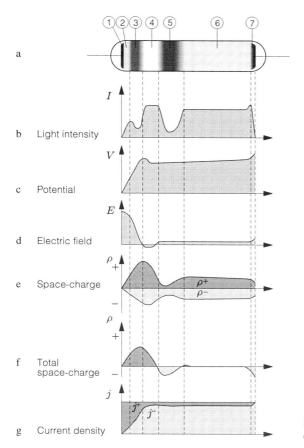

Fig. 9.7. Sketch of the processes in a glow discharge.

The first three bands are called the *cathode drop* and together make ~ 80 percent of the total voltage drop over the tube. The processes in the cathode drop are very complicated and interconnected, but similar to the unsustained discharge near breakdown, the mathematical description is only qualitative. The important of those processes are electron acceleration, ionization, electron multiplication, and secondary electron emission from the cathode. The most important difference is that in the glow discharge the positive ion space-charge dominates in front of the cathode and makes the electric field inhomogeneous. The electron contribution to the current transport near the cathode is small but increases along the region until at the end the current is mainly transported by electrons. To satisfy the continuity equation the current must be transported mainly by positive ions near the cathode, which is possible only if the ions are strongly accelerated and the potential drop must be large. Although the number of ions and electrons created are equal, the ions need a much larger acceleration to attain sufficient energy for secondary emission, while electrons must attain a high energy to create ions. Only then the ions can sustain the secondary emission from the cathode. For most gases and cathode materials $\gamma_j \approx 0.01$–0.2, so 5–100 ions must hit the cathode to release a secondary electron. The electrons move much faster than the ions, but they lose energy in elastic collisions which only heat the gas, and in exciting neutral gas particles, without ionizing them. The cathode voltage drop is therefore ~ 5–10 times the ionization voltage.

We can crudely calculate the ionization degree if we assume a constant positive space-charge inside the cathode drop region and a small current density in the discharge. If V_c is the voltage drop and d_c the cathode drop region length, then if the electric field strength varies linearly in the region (compare Fig. 9.7d), we obtain

$$E = -\frac{2V_c}{d_c}\left(1 - \frac{x}{d_c}\right).$$

In accordance with Fig. 9.7c, $V_c \approx V_a$, which means that the voltage drop across the tube is just the voltage drop across the cathode drop region and

$$\frac{dE}{dx} = \frac{\rho_+}{\varepsilon_0} = \frac{2V_a}{d_c^2}.$$

which expresses the positive space-charge constancy over the whole cathode drop region. Although this is not true, as Fig. 9.7f shows, it allows us to compute the mean ion density

$$n_j - n_e \approx n_j = \frac{\rho_+}{e} = \frac{2\varepsilon_0 V_a}{ed_c^2}.$$

Dividing by the gas density — that is, with $n_0 p/p_0 = 3 \times 10^{25} \times p/p_0$ m^{-3} (n_0 is the density at normal pressure, 10^5 Pa) — and assuming that that the glow discharge voltage drop across the tube is 100 V (see Fig. 9.6), the ratio of ions to neutral molecules is

$$\frac{n_j}{n_0 p/p_0} = \frac{2\varepsilon_0 V_a}{d_c^2 e n_0 p/p_0} = \frac{4 \times 10^{-16}}{d_c^2 p/p_0}.$$

At a pressure of 10^2 Pa, and d_c of a few centimeters, only one in a billion gas particles becomes ionized.

The cathode light coverage is directly proportional to the current density. In Aston's dark space, emitted electrons do not have enough kinetic energy to excite the gas, but when they enter the cathode glow region they are energetic enough to excite and create first ionizations. In the cathode dark space a large number of electrons and ions are produced and the velocity of the incoming electrons is so high that the probability of ionizing is larger than that of exciting the gas. But the velocity of the produced electrons is too low to excite many neutral particles. The cathode dark space thickness, d_k, varies with gas density as

$$d_k = \frac{C}{n}, \qquad C = \text{const},$$

where C is approximately nd for the lowest breakdown voltage of the corresponding Paschen curve. This concordance indicates that important processes take place in the cathode dark space which are necessary for the glow discharge breakdown and ignition.

The electrons leaving the cathode dark space have a broad energy spectrum, and, therefore, a great ability to excite the gas. At low pressures the strongest discharge light comes from the negative glow whose radiation ranges from red to blue lines. Toward the end of the cathode glow region the electrons have lost most of their energy, their mean velocity has decreased, and so has the light, and the Faraday's dark space begins. The space-charge changes the sign to negative, the excitations and ionizations become sporadic, and the electric field strength is very low. By the end of the Faraday's dark space, some electrons have sufficient velocity to excite and to ionize the gas and there begins the positive column or plasma region which emits light and whose processes are described in Section 9.6. If the current density is high, in front of the anode a region with negative space-charge and a positive potential drop appears, the opposite for small discharge currents.

The discharge tube breakdown voltage can be decreased if a small amount, 1 percent or even less, of an auxiliary gas with lower ionization voltage than the main gas is introduced. Often neon and argon are used. With only neon in the tube, electrons must be accelerated up to 21.5 eV to ionize, with argon up to 15.7 eV. The probability for ionizing argon is very low, because the argon partial pressure is small. However, neon has a 16.6 eV metastable level so a large number of neon atoms will be excited to this level. Metastable lifetimes are long compared to the mean time between two collisions, and thus the neon atoms will with high probability collide with an argon atom before decaying. In doing so they transfer their excitation energy to argon atoms and ionize them, considerably reducing the breakdown voltage. Figure 9.5 shows the ignition voltage for such a neon–argon mixture.

If the voltage across the tube is increased or the current-limiting resistance reduced when the entire cathode is covered by light, the voltage across the cathode drop and electric field increase. Such a cathode drop, called anomalous cathode drop, corresponds to the transition (E-F) in Fig. 9.6. Here the cathode is bombarded by more energetic ions, which produce more secondary electrons, which result in a larger discharge current. The more energetic ions cause sputtering which disintegrates the cathode and which can be used to coat surfaces (see Section 10.7).

9.4.3. Arc Discharge

If the voltage across the tube is further increased or the external resistance decreased in the anomalous glow discharge, the current increases. The cathode is heated by the ion bombardment, and at a certain current density (at F in Fig. 9.6) which depends on the cathode electrical and thermal properties, the number of emitted electrons increases sharply. The unstable discharge indicates that electrons liberated by ions cannot be responsible for all the current. The new type of discharge is reached, the arc discharge, whose characteristic is a low voltage across the cathode drop, of the order of the ionization voltage or less. Reducing the current-limiting resistance causes the voltage over the discharge to decrease, so the resistance of an arc discharge is negative. With the current increase the cathode temperature rises until the electron current comes mostly from the thermionic or field emission. This explains the small voltage across the discharge. Thermionic emitted electrons are easily accelerated and because the velocity distribution in plasma is Maxwellian, there are many electrons in the distribution tail, which reach the ionization energy. Arc discharge devices usually work with a higher gas pressure and current densities than do those with glow discharges.

This picture describes well many arcs (e.g., in lamps and welding), but for liquid mercury cathodes, other mechanisms are responsible for electron emission. Liquid mercury is a good heat conductor with a large heat capacity. The emission comes from a small luminous spot which moves across the cathode surface, but thermionic emission alone cannot explain the huge current densities of $\sim 10^5$ A/cm^2. Statistical variation of the mean weak electric field near the cathode surface are supposed to give very high local field strengths, which cause field emission, which is further enhanced by fast local heating. The emitted electrons space-charge disturbs the local field, and the spot moves rapidly to a new position.

Arc discharges are important in both low- and high-voltage switches and circuit break-ers, where they must be suppressed. An appropriate choice of materials, cooling, polishing the surface, and so forth, can reduce cathode emission. A gas or dielectric, which makes the cathode drop and discharge voltage high can be chosen. Recombination is enhanced by electronegative gases (e.g., freon or sulfur hexafluoride). The ions can be swept away either mechanically or electrically. Also gases or liquids, like some kinds of oils, which are difficult to ionize, can be used. Finally, mechanical constructions which quickly increase the distance between the electrodes are effective in reducing arc discharges in switches and breakers.

9.5. DIFFUSION, MOBILITY, CONDUCTIVITY, AND RECOMBINATION

In vacuum, charged particle orbits are directly computed from initial conditions and field distributions. However, in a gas, particles cannot be observed individually, they interact, and are subject to external forces, so the mean values are computed. For example, in a closed system the distribution of the velocities of gas particles is Maxwellian. If an external force acts on the system, the particles will move in the direction of the force with a mean velocity, $\mathbf{v}_D$, superimposed on their random thermal velocities. These random velocities can be neglected in the flow of liquids, or of electrons in a crystal, and only the drift velocity which need not be constant inside the region must be taken into account. Generally, $\mathbf{v}_D$ varies slowly, and thus all particles inside a volume element can be considered to have the same drift velocity.

To connect the drift velocity with the continuity equation we observe a closed volume, $\mathcal{V}$, with particle density n. A surface element, $d\mathbf{S}$, will in time dt be crossed by $n\mathbf{v}_D d\mathbf{S}dt$ particles, as seen in Fig. 9.8. There are $<n>V$ particles inside $\mathcal{V}$, where $<n>$ is the mean density and V the volume. The rate at which the number of particles is reduced is given by the number of particles leaving $\mathcal{V}$:

$$-\frac{\partial(<n>V)}{\partial t} = \int_S n\mathbf{v}_D d\mathbf{S}.$$

Creation or annihilation of particles inside $\mathcal{V}$ is allowed by introducing on the right side a term $<C>V$, with C being negative for annihilation. By Gauss' theorem the integral

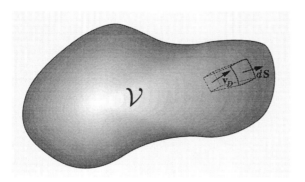

Fig. 9.8. To the derivation of the continuity equation.

over S can be transformed in a volume integral,

$$\frac{\partial(<n>V)}{\partial t} = -\int_S n\mathbf{v}_D d\mathbf{S} + <C>V = -\int_V \nabla \cdot (n\mathbf{v}_D)dV + <C>V.$$

Dividing by V and replacing $<n>$ and $<C>$ by their local values, n and C, we obtain

$$\frac{\partial n}{\partial t} = -\nabla \cdot (n\mathbf{v}_D) + C, \qquad (9.18)$$

which is the *continuity equation* and which applies to hydrodynamics, generation of electrons and holes in semiconductors, particle motion in vacuum, and so forth.

If the particle density is not constant over the volume, there will be a transport of particles from a region with higher to a region with lower concentration in a process called *diffusion*. Experimentally the particle flow $n\mathbf{v}_D$ is found to be proportional to the gradient of the particle concentration

$$n\mathbf{v}_D = -D\nabla n, \qquad (9.19)$$

where D is the diffusion constant and the sign reflects that the particle flow is from higher to lower concentrations, in a direction opposite to the gradient. Combining the continuity and diffusion equations, we obtain

$$\frac{\partial n}{\partial t} = \nabla \cdot (D\nabla n) + C = D\nabla^2 n + C, \qquad (9.20)$$

which is a new form of the continuity equation.

If there are ions in their own gas and the ion density is low, the electric field which an ion generates does not appreciably influence other ions. In the absence of external fields, the ions behave approximately as neutral gas molecules, since they have the same mass and velocity distribution, which is Maxwellian. Thus any drift motion must come from diffusion.

In an electric field the ions are accelerated in the field direction, and they will collide elastically with gas molecules to which they will transfer a part of their kinetic energy heating the gas. In steady state the mean velocity in the electric field direction varies with field strength. In weak electric fields encountered in a low ionized plasma, when $v_D \ll v_T$, the drift velocity is

$$\mathbf{v}_D = \mu\mathbf{E}, \qquad (9.21)$$

where μ is a constant called *mobility*.

An expression for mobility in weak electric field can be derived under the assumption that the ion velocity immediately after collision is the same as if it were a neutral gas molecule, and that the ions are accelerated by the electric force in the field direction. If the mean time interval between two collisions is $<t>$, then ion velocity when it collides is

$$\mathbf{v} = \mathbf{a}<t> = \frac{q\mathbf{E}}{m}<t>.$$

Since the ion does not move with this velocity during $<t>$, the ion drift velocity is the mean velocity over $<t>$,

$$\mathbf{v}_D = \frac{q\mathbf{E}}{2m}<t>, \qquad (9.22)$$

which, according to the definition of mobility, gives

$$\mu = \frac{q <t>}{2m}.$$ (9.23)

Taking account of Maxwellian velocity distribution of ions and neutral gas particles, 2 in the denominator of Eq. 9.23 is replaced by $\beta \approx 1$.

This simple result allows the estimation of the conductivity of an ionized gas where the current is transported by both ions and electrons. Electron collisions with neutral gas molecules can be elastic or inelastic, but in weak field the probability for the former is much higher. Therefore, the total current is

$$\mathbf{J} = \mathbf{J}_+ + \mathbf{J}_- = n_j \mathbf{v}_{Dj} q_j - n_e \mathbf{v}_{De} e = \sigma_c \mathbf{E}.$$ (9.24)

Introducing the drift velocity (Eq. 9.21), the conductivity becomes

$$\sigma_c = n_j \mu_j q_j + n_e \mu_e e = \sum n_\nu \mu_\nu e.$$ (9.25)

where ν represents both the ion and electron subscripts j and e, because $\mathbf{v}_D$ is negative for electrons. The time between two collisions is estimated by using the mean Maxwellian thermal velocity, v_T, and the mean free path

$$<t> = \frac{\lambda_\nu}{v_T},$$

which, when introduced into Eq. 9.23, gives

$$\mu_\nu = \beta \frac{e\lambda_\nu}{m_\nu v_T},$$ (9.26)

and the conductivity becomes

$$\sigma_c = \sum \beta \frac{n_\nu e^2}{m_\nu} \frac{\lambda_\nu}{v_T}.$$ (9.27)

The mean free path, λ_ν, is related to gas kinetic cross section σ_ν (as long as $n_\nu \ll n_0$, Eqs. 9.5 and 9.6),

$$\lambda_\nu \sigma_\nu n_0 \approx 1,$$

where n_0 is the gas density, so the conductivity can also be expressed as

$$\sigma_c = \sum \beta \frac{n_\nu}{n_0} \frac{e^2}{m_\nu \sigma_\nu v_T}.$$

The resistance experienced by ions as they drift through the gas under the influence of an electric field is similar to that experienced by neutral gas molecules drifting under the influence of diffusion, so the diffusion constant and the mobility can be related. Assume an electric field such that $|\mathbf{E}| = E_x = E$. The electric force on ions is

$$F_e = q_j E.$$

A pressure gradient created by the drift velocity will produce diffusion in the opposite direction, which ceases when the force from diffusion compensates the force generating the drift velocity. The gradient of the ion density, n_j, lies in the x direction, and the force on an ion is

$$dF = -dp_j S = -\frac{dp_j}{dx} S dx = -\frac{dp_j}{dx} dV,$$

where dp_j/dx is the pressure gradient. The force on an ion from diffusion is

$$F_D = \frac{dF}{dN} = -\frac{dV}{dN}\frac{dp_j}{dx} = -\frac{1}{n_j}\frac{dp_j}{dx}.$$

From Eqs. 9.19 and 9.4, if temperature is kept constant we obtain

$$v_D = -\frac{D}{n_j}\frac{dn_j}{dx} = -\frac{D}{p_j}\frac{dp_j}{dx}.$$

The electric force and the force from the pressure gradient result in drift velocity $\mathbf{v}_D$. Equating $q_j E$ and $(1/n_j)\, dp_j/dx$ at equilibrium,

$$\frac{q_j v_D}{\mu} = \frac{p_j v_D}{n_j D},$$

and introducing p_j from the equation of state (Eq. 9.4), we obtain

$$\frac{D}{\mu} = \frac{kT}{q_j}. \tag{9.28}$$

This Einstein relation can also be used for systems which are not in equilibrium.

Recombinations of ions and electrons produce neutral atoms or molecules. Electron capture by an ion is very improbable without the involvement of a third particle, since bound atomic electrons can only have specific well-defined energies, while free electrons can have all possible energies. A bound electron easily leaves the atom if sufficient energy is transfered in a collision; however, a free electron seldom possesses exactly the energy and momentum to be captured into one of the vacant levels. Ions and electrons experience, therefore, very many elastic collisions before a recombination occurs.

The presence of a third particle to take up kinetic energy and momentum dramatically increases the recombination probability. Such particles are found on the walls and electrodes in a discharge tube, where the largest number of recombinations takes place. Since the probability of a three-body recombination in the gas is proportional to the square of the density, n^2, it is very low at low pressure. An electron can become attached to a neutral gas molecule creating a negative ion, which can collide with positive ions, easily producing recombination.

The change of the ion density can be calculated if, in a region of space, new charges are created by ionization and are lost by recombination, with the number being proportional to the product of n_j and n_e. The recombination coefficient, R, defines the number of recombinations per unit time and volume, and κ is the rate of the ion–electron pairs production, so the change in the ion and electron density is

$$\frac{dn_j}{dt} = \frac{dn_e}{dt} = \kappa - Rn_j n_e, \tag{9.29}$$

another example of the continuity equation, where the right side is C from Eq. 9.20.

9.6. PLASMA

When proposed in 1928 by Langmuir [189] the name *plasma* was a synonym for the positive column in a gas discharge. Plasma, ionized gas, is the most common form of the matter in the universe: The sun, the stars, the interplanetary and intergalactic space, and the earth ionosphere are all plasma or filled with plasma. On earth the plasma state is found in fluorescent and neon lights, electron flash, gas lasers, welding arcs, and lightning. Plasma is now important in fundamental research of thermonuclear energy generation and direct thermal to electric energy conversion by magnetohydrodynamic generators and in fast growing applications like plasma chemistry, sputtering, and ion etching.

The conditions in a plasma are very complicated because besides electric fields, magnetic fields are often present, so the motion of the electrons and ions are very intricate and have oscillations and instabilities. In the short description of plasma properties, we will restrict the discussions only to electric field.

Macroscopically a plasma is essentially electrically neutral,

$$n_j \simeq n_e, \tag{9.30}$$

and only for brief times and in very small volumes are differences in positive and negative space-charge observed. These small deviations from equilibrium appear because of random particle velocities. Assume a region between two planes at $x = -d$ and $x = d$ in which only electrons exist, so Poisson's equation

$$\frac{\partial^2 V}{\partial x^2} = \frac{e n_e}{\varepsilon_0} \tag{9.31}$$

gives the potential distribution, V. If the electric field is zero at $x = 0$, integrating gives

$$V = \frac{e n_e}{\varepsilon_0} \frac{x^2}{2}.$$

An electron at $x = 0$ and another at $x = d$ see a potential difference, which expressed in energy is

$$\Delta W = \frac{e^2 n_e d^2}{2\varepsilon_0}. \tag{9.32}$$

Electron thermal motion generates small local deviations from the quasineutrality of positive and negative charges. The mean "thermal energy" in the x direction is $kT_e/2$ (Eq. 2.11), where T_e is the temperature which corresponds to the kinetic energy of the electrons, the electron "temperature." The value of d in Eq. 9.32, when the potential and the kinetic energy are equated, determines the maximum dimension of the space-charge region where quasineutrality can be disturbed,

$$\lambda_D = d = \sqrt{\frac{\varepsilon_0 k T_e}{e^2 n_e}}, \tag{9.33}$$

where λ_D is called *Debye length* [190]. An ionized gas is called a plasma if λ_D is smaller than the smallest linear dimension of the region containing the plasma. Since the degree of ionization, n_e/n_0, does not enter the expression for the Debye length, a plasma can be slightly or highly ionized. In a laboratory plasma, electron density can vary between 10^{12} m^{-3} in a low-pressure plasma and 10^{24} m^{-3} in high-pressure arc discharges while the Debye length varies between 10^{-2} m and 10^{-8} m. Cosmic plasma can be much thinner, in the earth magnetosphere $n_e \sim 10^6$ m^{-3} and $\lambda_D \sim 1$ m.

An exactly neutral plasma would have zero electric field, but it is just the electric field which maintains the plasma in a gas discharge. We therefore talk of a *quasineutrality*, which means that $n_j = n_e$ holds except in connection with the Poisson's equation, and that the voltage drop in the direction of current flow must be small, which characterizes a good electric conductor. From this point of view, a highly ionized plasma is similar to a metal.

In the derivation of the Debye length the electron velocity distribution was assumed to be Maxwellian, which agrees with the measurements. There are physical reasons to expect electrons, ions, and neutral particles to have the same velocity distributions. For electrons and ions, elastic collisions with neutral gas particles are similar to collisions between solid spheres. Collisions between charged particles, the Coulomb collisions, are interactions between two electric fields, and they occur at much larger distances. In collisions between electrons and ions the energy transfer is very small because of the large ion–electron mass difference. But in collisions of two electrons the energy transfer can be large, and the cross section is much larger than that for collisions between gas particles. Thus the "electron gas" can be seen as an isolated system, and its velocity distribution is Maxwellian.

In the electric field, electrons attain a much higher velocity than ions, so electrons in a plasma have usually a much higher "temperature" than the heavy particles. In a low-pressure discharge the electrons can reach a temperature $\sim 10^4$–10^5 K, while the gas remains near room temperature. Although the concept of "temperature" strictly has a meaning only for gases with Maxwellian velocity distribution, in plasma physics the "temperature" is used as a measure for the mean kinetic energy of the particles, defined by

$$T_\nu = \frac{2}{3}\frac{W_\nu}{k},$$

where W_ν is the mean kinetic energy of particles ν. It is the strong electron–electron interaction that gives electrons the Maxwellian velocity distribution. The ion temperature is larger than the neutral gas temperature, which is a consequence of the drift velocity in the weak electric field of the plasma.

To the higher electron temperature corresponds a higher drift velocity, and the ratio of the electron and ion drift velocity is $\sim m_j \lambda_j / m_e \lambda_e$, which in a low-pressure plasma can be on the order of a few hundred. Thus the current in a plasma is carried mainly by electrons (Fig. 9.7g), and in Eq. 9.24 the J_+ term can be neglected,

$$\mathbf{J} \approx \mathbf{J}_- = -n_e \mathbf{v}_D e, \tag{9.34}$$

and the conductivity is given by Eq. 9.27:

$$\sigma_c = \beta \frac{n_e e^2}{m_e} \frac{\lambda_e}{v_{Te}}. \tag{9.35}$$

In applications and laboratory a plasma is limited to a region inside a chamber, whose walls often are nonconductive. When the plasma is created the fast electrons reach the wall first and charge them if nonconductive, creating a sheath around the plasma. Under static conditions an electric field is created whose potential difference is such that the net current to the wall becomes zero. The ions which approach the sheath are accelerated toward the wall, and the ion density in the sheath decreases, but not as fast as the electron density, so the space-charge in the sheath is positive. Consequently, the ion current to the wall and the potential in the plasma decrease near the wall. In a low-pressure plasma the potential difference between the plasma axis and the wall can be on the order of several kT_e/e, or a few volts. Similarly, sheaths are formed in a glow discharge between the plasma and the cathode dark space and between the plasma and the anode glow. The Faraday's dark space and the negative glow are plasma, in the context of quasineutrality.

The potential difference between the axis and the walls creates a radial electric field, $E_\perp$. The diameter of the plasma is often much smaller than its length, as in a fluorescent tube. The number of ions reaching the wall is then much larger than the number leaving the plasma in the axial direction. The mean ion lifetime before recombination, τ_j, is within wide limits proportional to the product of the neutral gas density, n, and the tube radius, r, because the mean free path is inversely proportional to density. Since the electron and ion density are almost equal, each electron must ionize an atom in the time interval τ_j, and since the electron and ion currents to the wall are equal in steady state, the number of created electron–ion pairs per unit length must be slightly larger than the number of recombinations at the wall. The axial electric field, $E_\parallel$, accelerates the electrons so that they can reach the energy needed to satisfy these conditions. But the Maxwellian velocity distribution implies a mean velocity too small to provide the ionization energy, and only a few electrons with velocities in the distribution tail ionize neutral gas particles.

The mean electron kinetic energy is a function of $E_\parallel$ and the density; so if the product nr increases, $E_\parallel/n$ must decrease, implying a lower electron temperature. But even the electron axial drift velocity is determined by $E_\parallel/n$, which can be seen by introducing into Eq. 9.23 $<t>$, which is proportional to the mean free path and inversely proportional to n. An increase of nr must be compensated by a lower drift velocity.

This description of plasma processes is qualitative and is based on measurements. The voltage across a discharge tube, the current through it, and the gas pressure are easily measured, but this does not give any information about the electron and ion densities, their energies, drift velocities, and other plasma parameters. Langmuir [186] proposed to measure plasma parameters using a small probe protruding into the plasma. By isolating the probe a process similar to that on the wall occurs, the fast electrons reach the probe first, charging it negative until the potential becomes so high that most of the incoming electrons are repelled. In steady state an equal number of electrons and ions reach the probe, and the probe attains the floating potential, V_f.

By connecting a voltage source between the anode and probe and varying the voltage, the probe potential and the current reaching the probe change. A probe characteristic, shown in Fig. 9.9a, results. The current is taken as positive when it goes from the probe to the plasma. The anode potential can be taken as a reference zero. The probe characteristic has three regions: saturated ion current (A-B), electron current in the retarding field (B-C), and saturated electron current (C-D).

Current to the probe is almost independent on the potential for high negative potentials, since only positive ions which, because of their random thermal motion, happen to penetrate the sheath surrounding the probe carry the current. For a large plane probe the current is practically constant, but such a probe would disturb the plasma. So the probes are small, usually a thin wire. A thin wire, however, introduces problems, since the sheath thickness increases at high negative potential and the probe current varies.

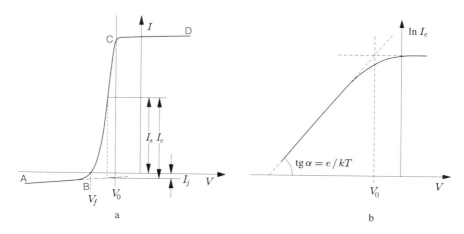

Fig. 9.9. (a) Probe characteristic. **(b)** Electron current vs. probe potential.

As the probe potential is increased, some high kinetic energy electrons reach the probe. Assuming a Maxwellian electron velocity distribution, the current is

$$I_s = I_e - I_j \sim e^{e(V-V_0)/(kT_e)}, \tag{9.36}$$

where V is the probe potential and V_0 is the plasma potential at the probe position. $V - V_0$ is the voltage drop across the sheath, and I_j is the ion current which is often negligibly small. The logarithm of the current with the probe potential, as seen in Fig. 9.9b, is a straight line with the slope e/kT_e, indicating that the velocity distribution is Maxwellian. As the probe potential reaches $V = V_0$, all the electrons are accumulated at the probe, the positive space-charge sheath shrinks to zero, and the probe attains the plasma potential, V_0. As the probe potential is further increased, the probe current has a small increase caused by the negative space-charge sheath which builds up around the probe. The difference between the plasma and the floating potentials, $V_0 - V_f$, is usually a few times higher than the electron temperature, kT_e/e in volts. Using more than one probe the axial electric field can be determined by measuring the plasma potential at different probe positions.

Since a probe must not disturb the plasma too much, it must be small and so the volume of the space-charge sheath must also be small. The sheath thickness depends on the Debye length, so a lower ion density and a lower pressure increase the sheath thickness and a lower limit therefore exists for both the discharge current and pressure. However, if the pressure is too high, the mean free path becomes shorter and can be shorter than the sheath thickness, so ionization can occur inside the sheath and interfere with the measurements. A difference between the work function of the probe and the anode as well as adsorbed gas on the probe surface is important, since the measured potential differences are a few volts. There is a large literature on probe measurements, in which different theories are used for different parameter values.

Other plasma diagnostic methods have therefore been developed, like electromagnetic waves, especially microwave and laser beams, spectroscopic methods, interferometry, and particle beam diagnostic.

An example is the interaction of plasma with microwaves. Plasma is electrically neutral, except in volumes with the Debye length dimensions, where the disturbed neutrality is restored by the space-charge. If all electrons in a plasma move a small distance x

relative to the ions the space-charge produces a surface charge density $en_e x$, whose electric field, $E = en_e x / \varepsilon_0$, subjects all electrons to a force $e(en_e x / \varepsilon_0)$ and thus their equation of motion is

$$m_e \frac{d^2 x}{dt^2} = -e \frac{en_e x}{\varepsilon_0},$$

which is a harmonic oscillation with plasma frequency

$$\omega_{pl} = \sqrt{\frac{n_e e^2}{\varepsilon_0 m_e}}. \tag{9.37}$$

In a plasma with density 10^{18} m^{-3} we have

$$f_{pl} = \frac{\omega_{pl}}{2\pi} \approx 10 \text{ GHz},$$

so in most laboratory plasmas the frequency is in the microwave region.

Plasma oscillations can propagate in plasma, since electrons which move into the neighboring regions transmit the information of what is happening in the oscillating region. Other types of waves can also propagate in a plasma — acoustic, electromagnetic, hydromagnetic (Alfvén), magnetoacoustic, and ion waves, depending on the conditions in plasma and its environment. All these waves can be amplified and create instabilities. This rapidly increasing knowledge of plasmas is of paramount importance for astrophysics and space research, but also for plasma applications in research and technology on the earth.

10

Gas Discharge
Tubes and Devices

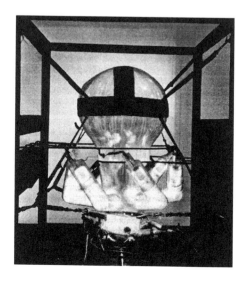

Before the invention of tyristors, large mercury-pool
rectifiers were used to rectify high currents. The
rectifier shown in the figure had a power rating of
500 kW at 600 V. The current density at the cathode
spot was $\sim$10, 000 A/cm^2. (From *Gaseous Conductors*, J. D. Cobine, 1941, McGraw-Hill.)

10.1. INTRODUCTION

HISTORICAL NOTES. The first practical applications of gas discharges were Geissler tubes, often used for decoration because of the beautiful colored effects. In 1859 Becquerel [191] made an attempt to use these tubes as a light source introducing fluorescent material, but the device was too inefficient. Mercury arc lamps ware demonstrated in 1860 [192], when the Hungerford Suspension Bridge in London was illuminated, while the electric arc illuminated Paris in 1863. At the turn of the century an American inventor, Peter Cooper-Hewitt [193], introduced a low-pressure mercury lamp, which was more efficient than contemporary incandescent lamps, but the light was too blue. Attempts to introduce dyes and fluorescent materials to improve spectral distribution failed because of fast detoriation. The high-pressure mercury lamp ("quartz" lamp) [194] was the only discharge lamp used as a source of UV-radiation. When substantial quantities of neon were obtained by Cloude [195] in 1908, the neon- and mercury-vapor-filled sign and decorative lighting tubes became common. Introduction of heated cathodes and development of improved fluorescent powders in the 1920s by General Electric Co. in the United States, and Philips and Osram in Europe, created renewed interest in discharge lighting tubes. The industrial production of sodium [196] and mercury lamps started in the early 1930s, followed by fluorescent lamps, which were first exhibited at the Chicago Centennial Exhibition in 1933.

The rectifying properties of an arc between a mercury cathode and carbon anode [197] were observed in 1882. Introduction of three-phase AC systems as the main electric power source stimulated the use of mercury arc rectifiers [198] everywhere where DC current was necessary (see the chapter-title-page photograph). A control grid, like in a triode, was suggested by Langmuir [199] in 1914, in a tube with thermionic cathode and mercury vapor. However, the cathode disintegrated rapidly because of ion bombardment, and the mechanism of the grid control was only cleared by Langmuir's positive ion sheath theory [186], so the high-current rectifier tubes and thyratrons became feasible when Hull [200] observed disintegration being dependent on the ion energy. The need for short deionization time in radar applications during the second world war triggered the construction of hydrogen thyratrons [201] and TR/ATR tubes [202]. An important advance was the ignition electrode invention [203] in mercury pool tubes; in ignitrons the current could reliably be started at any desired time while the anode was positive.

The laser action principle was proposed in 1958, and two years later the first ruby laser was constructed [204]. Soon laser action on atomic transitions for different gases were reported, the He–Ne laser being the first [205]. High-power and high-efficiency CO_2 lasers followed in 1964 [206], and the excimer lasers were introduced in 1975 [207].

As early as 1852 Grove [208] observed that a gas discharge can etch the surface of the negative electrode, while Plücker [8] showed that small particles were torn off and covered the glass surrounding the cathode. For almost 100 years cathode sputtering was regarded as an undesirable phenomenon, limiting the cathode life and causing difficulties with isolation and secondary electrons. That sputtering can be used to obtain thin films of different metals or to etch the cathode surface was observed at the beginning of the century, but it was mostly used in scientific laboratories (reference 209 cites over 100 early references for the time 1852–1930). Extensive measurements of sputtering rates were also made, but realiable data were obtained by ion beam measurements [210]. The sputtering industrial breakthrough came in the 1950s, induced by the development in semiconductor electronics. Besides improvements in thin film technology, plasma etching ousted the wet chemical methods; the more so after the introduction of reactive gases into plasma [211]. About 1970, ion beams entered the scene, especially as ion beam etching, modification of thin films, and ion-assisted deposition [212]. Ion implantation was suggested by Shockley [213].

In gas discharge devices the current varies from a few electrons per second to many megaamperes, a range of more than 10^{25}. According to Fig. 9.6, three regions are of main interest in technological applications: unsustained discharge (A-C), glow discharge (D-E), and arc discharge (G).

10.2. UNSUSTAINED DISCHARGE DEVICES

10.2.1. Ionization Chamber

A chamber filled with air or a noble gas that has two plane- or cylinder-shaped electrodes (Fig. 10.1), which works in the range A-B in Fig. 9.6, can detect and measure ionizing radiation. The choice of chamber geometry and gas depends on the radiation to be measured. High-energy charged particles ionize gas proportional to the number of electrons per unit volume. Low-energy γ-rays ionize primarily by photoelectric effect, while high-energy photons ionize by Compton scattering and pair production. The first case favors tightly bound electrons to the nucleus, while in the second case heavy nuclei are desirable. Therefore the noble gases are used in ionization chambers not filled with air.

Ionization chambers often have a thin foil window on one side through which α-particles and β-rays can pass.

The saturation current, I_m, is proportional to the primary radiation, j:

$$I_m = jV\rho \frac{p}{10^5} \frac{273}{T}, \tag{10.1}$$

where V is the chamber volume, ρ is the gas density at STP (ρ_{air}=1.293 kg/m³), p is the pressure in Pa, and T is the temperature in K. Since the saturation current increases with pressure, the chambers can use pressures up to 5 MPa and produce saturation currents on the order of 1 pA.

The current produces a voltage drop across the very high resistance, R. This signal is amplified and read out. The chamber RC constant is usually a few seconds.

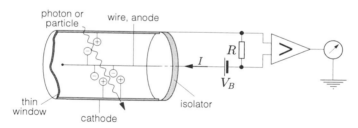

Fig. 10.1. Sketch of an ionization chamber.

10.2.2. Proportional and Geiger Counters

Ionizing radiation detectors, which work in the B-C and C-D region of Fig. 9.6, are called proportional and Geiger [214] counters. The anode of the tube is the central wire, and the cathode is the surrounding cylinder, enclosed in a glass envelope. The primary charged particle or photon initiates the ionization (Fig. 10.2a), and an electron–ion avalanche results in a current in an external circuit. The avalanche starts near the thin wire anode where the electric field is strongest, because the light fast electrons ionize the gas when they are pulled toward the anode. The slower ions form a sheet which drifts to the cathode.

In proportional counters the voltage across the tube is about 400 V and the avalanches are limited to the trajectory of the primary particle. The current is proportional to the primary ionization and, thus, to the energy of the primary particle.

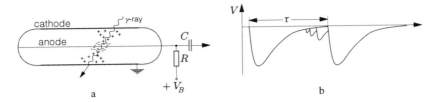

Fig. 10.2. Geiger counter. (**a**) Circuit. (**b**) Output signal.

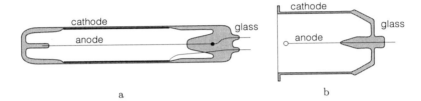

Fig. 10.3. (**a**) γ-Detector, (**b**) β-γ-detector.

Geiger counters operate with voltages up to about 1000 V, so the discharge spreads over the whole tube, and a much higher current pulse is obtained, which is independent of particle energy or type.

The tube is filled with a noble gas or air, along with ethanol or a halogen vapor (bromine) to quench the discharge. The lower ionization energy for these quenching gases allows the noble gas or air ions to be neutralized by collisions with the quenching gas molecules. UV photons emitted in this process are absorbed by the quenching gas, whose molecules are ionized or excited to a higher vibration or rotation state. When the positive quenching gas ions reach the cathode, they dissociate; photoemission, which could initiate a new avalanche, does not occur. Ethanol is consumed in the dissociation, so these tubes are limited to about 10^{10} discharges. Dissociated bromine recombines, and the tube life is not limited by the number of discharges.

An external RC circuit with high resistance and low capacitance helps the quenching, because an ionizing particle causes a voltage drop across R, which decreases the tube voltage and thus causes the avalanche to stop.

The slow ions are neutralized when they hit the cathode; they produce the long dead time τ (0.01–1 ms) (Fig. 10.2b), which is longer for Geiger counters than for proportional counters and increases with the volume in both. Saturation occurs when the typical incoming particles separation becomes $\sim 1/\tau$.

Depending on the particles to be detected, proportional and Geiger counters are constructed with glass envelope (γ-rays) or with a thin mica window (5–100 mg/cm^2 for β-rays and 1.5–2 mg/cm^2 for α-particles; Fig. 10.3).

10.3. GLOW DISCHARGE TUBES

10.3.1. Gas Discharge Light Sources

The positive column in a glow discharge tube emits photons, many in the UV region, whose spectral distribution depends on the filling gas. So many types of discharge lamps have been developed, but the most common is the fluorescent lamp. The main advantage of discharge lamps over the incandescent lamps is their low power consumption per lumen of emitted light, as seen in Fig. 10.4.

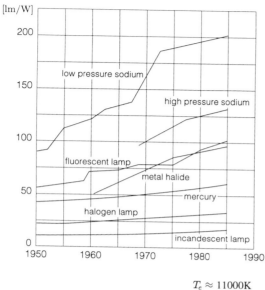

Fig. 10.4. Light yield per watt of glow discharge lamps (Philips).

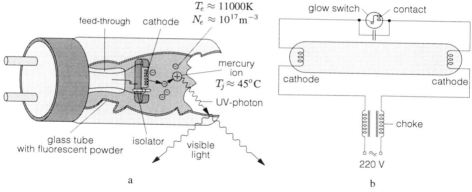

Fig. 10.5. (**a**) Fluorescent lamp. (**b**) Connection.

In fluorescent lamps most of the emission is in the UV region; the discharge occurs in mercury vapor mixed with a noble gas (pressure about 500 Pa). The spectral distribution can be varied over broad limits by choice of the phospor coating of the inside of the tube (Fig. 10.5a). The powder acts as a transformer of light. In standard neutral white lamps a mixture of magnesium tungstenate (blue-white) and zinc beryllium silicate (yellow-green-red) is commonly used.

Common fluorescent lamps are connected to the mains voltage (Fig. 10.5b), but neither 110 V nor 220 V are enough to start the discharge. The lamps are ignited with a glow switch, two bimetal electrodes which bend when heated by the glow discharge, and short-circuit the tube. The current flows then through two spiral-shaped W–BaO-sintered cathodes on each side of the tube. The heated cathodes start to emit electrons. When the bimetal electrodes cool, they open the short-circuiting contact, which results in a high-voltage pulse across the choke, which ignites the main discharge. The choke limits the current. The low discharge voltage across the tube causes the glow switch to remain open and the two cathodes cool down. Electron emission occurs then only by secondary electrons from positive ions bombardment.

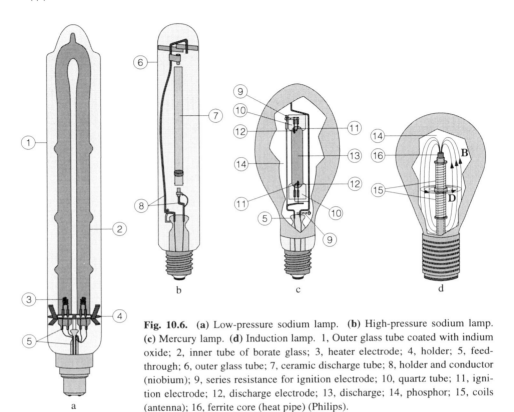

Fig. 10.6. (a) Low-pressure sodium lamp. (b) High-pressure sodium lamp. (c) Mercury lamp. (d) Induction lamp. 1, Outer glass tube coated with indium oxide; 2, inner tube of borate glass; 3, heater electrode; 4, holder; 5, feed-through; 6, outer glass tube; 7, ceramic discharge tube; 8, holder and conductor (niobium); 9, series resistance for ignition electrode; 10, quartz tube; 11, ignition electrode; 12, discharge electrode; 13, discharge; 14, phosphor; 15, coils (antenna); 16, ferrite core (heat pipe) (Philips).

Long discharge lamps used in advertising have two cold cathodes and operate with AC voltage between 700 and 1000 V. The color is determined by the fill gas (He, yellow; Ne, orange-red; Ne–Ar–Hg, blue) or by the tube phosphor coating.

Low-pressure sodium lamps contain two glass cylinders (Fig. 10.6a), and the inner side of boron glass is filled with a neon–argon–sodium mixture. The sodium metal begins to evaporate when the lamp is switched on, and the argon–neon discharge starts. After a few minutes the light comes mainly from the two yellow (589.0 and 589.6 nm) sodium lines. The inner side of the outer tube is coated by indium oxide, and while the visible light passes through, the infrared radiation is almost completely reflected, which keeps the high working temperature. At about 260°C the efficiency is maximum, and about 35 percent of the power results in visible light. The volume between the cylinders is evacuated. The most important use of low-pressure sodium lamps is in public roadway illumination, since the yellow light has a low dispersion in fog. Lamps with power up to 150 W give a light output of 30,000 lumen.

High-pressure sodium lamps have an inner tube of ceramic (Al_2O_3) (Fig. 10.6b), filled with a mixture of xenon– or argon/neon–sodium. At the substantially higher gas pressure, green and blue lines are also emitted (see Fig. 10.7). The evacuated volume between the tubes allows for a drift temperature of about 350°C. Lamps with power of up to 1000 W give 125,000 lumen and are used in sports centers, airports, large squares, and so on, often in combination with mercury or metal-halide lamps to give a more natural spectral distribution. Phosphor coating on the inside of the outer tube gives the same effect.

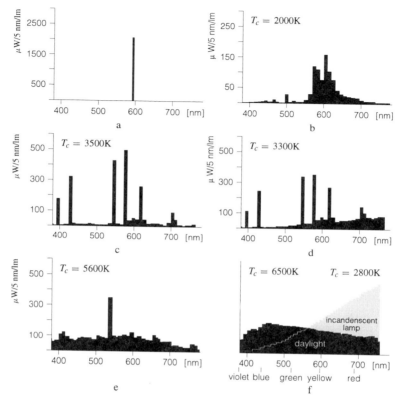

Fig. 10.7. Spectral light distribution of different gas discharge lamps. The distribution varies strongly depending on the construction and use of phosphors. (**a**) Low pressure sodium lamp, (**b**) high-pressure sodium lamp, (**c**) mercury lamp, (**d**) mercury/filament lamp, (**e**) metal halide lamp. T_c is the color temperature (Osram).

Mercury lamps have an inner quartz tube (Fig. 10.6c) filled with liquid mercury and an ignition gas, usually argon/neon. The discharge is started by a faint discharge between an ignition electrode and the main electrode that heats the tube increasing the pressure, and after a few minutes the main discharge ignites. The pressure at the working temperature is between 10^4 and 10^6 Pa, so there are many atoms in a highly excited state which emit spectral lines in the visible region of the spectrum. The high-pressure mercury lamp light is more natural in color; however, red is missing, which can be improved by suitable phosphors. Mercury lamps, used in stores, industrial shops, for the general urban illumination, and so on, have power of up to 2000 W and light flow output of 125,000 lumens. Combination mercury discharge and tungsten filament lamps are useds to compensate for missing the red and infrared part of the spectrum (see Fig. 10.7). The filament is sized to work like an external load, so the tube does not need a choke.

Metal halide lamps, a modern version of high-pressure mercury lamps, use mercury with sodium, thallium, or indium iodide and have a much higher light yield, especially in the green and yellow-red. Low-power lamps up to 150 W produce without phosphors a spectrum with color temperature between 5000°C (neutral white) and 3000°C (warm white). They are used everywhere where high intensity and good color is important (e.g., in shop windows, exhibition halls, color TV transmissions). High-power lamps (power of 2000 W and output of 190,000 lumens) are used in sports illumination, large industry halls, and urban lighting.

In induction lamps an external oscillator generates rf current which feeds a coil inside the lamp (Fig. 10.6d). The frequency is $\sim$ 2.5 MHz. Inside the coil is a ferrite core, and the rf current generates a magnetic field. **D**-lines around the magnetic field lines excite and ionize the gas inside the lamp (mercury vapor and noble gases). The UV radiation is converted into visible light on the phospor coating of the inside of the bulb. The induction lamp strikes immediately without any delay, the high-frequency of the power source prevents any stroboscopic effect, and the expected lifetime is 60,000 hours. The lamps have a light output of 6000 lumens and will dominate in urban areas, shopping centers, public building lighting, and everywhere else where long burning time and difficult relamping are anticipated.

10.3.2. Radar Gas Switching Diodes

Radar gas switching diodes are used in the antenna circuit to switch off the receiver during the transmission; otherwise some of the reflected power could reach the detector and destroy it. The tube (glass or ceramic) is filled with a noble gas – water vapor mixture (the water electronegativity results in a short deionization time), has copper electrodes, and is inserted into the waveguide (Fig. 10.8a).

A preionization electrode creates a weak discharge, so that a fast main discharge ignition is obtained ($\sim$tens of nanoseconds). The waveguide facing the receiver is short-circuited by the TR diode (**T**ransmit-**R**eceive). The addition of a second gas switching diode, the so-called ATR diode (**A**nti-**T**ransmit-**R**eceive), connected at a distance of $\lambda/4$ into a branch of the waveguide, increases the receiver sensitivity. This short-circuits the antenna circuit, and part of the radar signal propagating toward the transmitter is reflected to the receiver (Fig. 10.8b).

Modern high-power radar systems use a circulator to decrease power reaching the receiver during the radar pulse. TR diodes in such systems have as many as three gas switching diodes protection and are terminated with a varactor diode voltage limiter. Some low-power TR diodes contain a β-emitter preionization (e.g., tritium) to decrease the switching time.

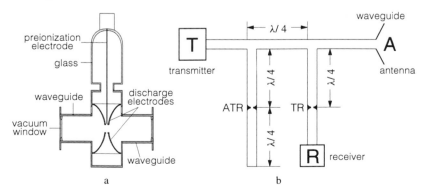

Fig. 10.8. Gas switching diode. (**a**) Construction. (**b**) Connection in low-power radar.

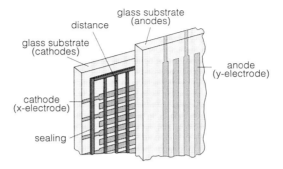

Fig. 10.9. Alphanumeric gas discharge display.

10.3.3. Gas Discharge Displays

Gas discharge displays are multielectrode devices which create alphanumeric characters or halftone pictures. Alphanumeric displays were used mostly in portable computers. A row of cathodes and perpendicularly a column of anodes are made of thin coatings on a glass substrate, using integrated circuit production technique (Fig. 10.9). Between the cathodes and anodes a weak discharge is always burning by connecting to the cathodes a voltage of -200 V, which gives the screen a characteristic orange color. Voltage pulses ($\sim$150 V, $\sim$150 μs) are simultaneously applied on specific cathodes and anodes, on whose crossings the main discharge ignites, to display a dot-matrix representation of an alphanumeric character. Pulse voltage amplitude determines the light intensity.

Industrial prototypes of halftone displays have also been made using both DC and AC, a color example is shown in Fig. 10.10. Here a glass substrate on which rows of auxiliary anodes are etched has rows of thin film anodes on the surface, on the top of which an isolator layer with small holes for the anodes is surrounded by phosphors for the corresponding color. Another glass plate has rows of parallel cavities for the cathodes together with the attendant discharge volume. A discharge ignited in a crossing illuminates the phosphor by UV radiation and creates the desired color. Such 1024×1024 pixels color displays

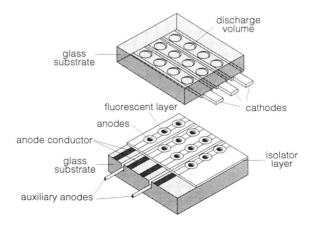

Fig. 10.10. Color display.

were made with ~ 2 pixels/mm. Commercial thin TV and monitor displays will, however, probably be made by either field emission arrays or liquid crystals. All three displays are practically free of aberrations and can be made very thin.

10.4. ARC DISCHARGE TUBES

10.4.1. Overvoltage Spark Gaps

If a spark gap voltage exceeds a limit, in a few nanoseconds an arc discharge begins with high-current and arc voltage of ~ 10 to 25 V, preventing a further voltage increase and damage to the protected device.

Introducing a weak β-emitter in the tube or constructing the electrodes with a strong electric field in a limited volume shortens the ionization time. The electrode surface is usually covered by a low work function material, the isolator is glass or ceramic, and the tube is filled with a noble gas.

Overvoltage spark gaps with voltages between 70 and 50000 V can achieve currents of over 50 kA, but behave differently if DC or pulsed. For DC voltages the ignition is completely determined by the voltage over the tube, while in pulse operation the rate of voltage change determines the ignition time, ~ 1 kV/μs often being sufficient. The discharge is extinguished when the current falls below ~ 0.5 A (Fig. 10.11b).

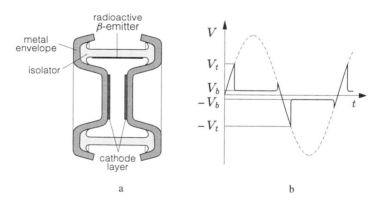

Fig. 10.11. Overvoltage spark gap. **(a)** Construction. **(b)** Voltage as a function of time when limiting an AC voltage. V_t is the ignition voltage, and V_b is the arc voltage.

10.4.2. Electron Flash Tubes

The electron flash tubes are made of hard glass and have a pin-shaped tungsten anode and an Ba-Sr impregnated cathode. The ignition electrode is often mounted on the outside of the tube, and the ignition pulse creates a high electric field around the cathode and ignites the xenon filled tube. The spectral emission from xenon is similar to daylight with color temperature of ~ 6000 K. Xenon ionization voltage is low, so the voltage over the tube is about 400 V, with current of up to 400 A, from the discharge of a large capacitance capacitor, which lasts for ~ 1 ms. Besides the well-known application in photography, the flash tubes are used for optical pumping of lasers, in stroboscopes, and as warning light.

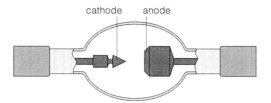

cathode   anode

Fig. 10.12. Arc discharge lamp.

10.4.3. Arc Discharge Lamps

Arc discharge lamps with a short arc are used as strong, point sources. In a quartz glass tube a sharp thoriated tungsten cathode and tungsten anode, Fig. 10.12, are placed a few millimeters apart, and xenon or mercury under high-pressure (a few megapascal) fills the tube. The addition of mercury results in UV emission. The discharge is initiated by a high-voltage pulse, it heats the thoriated cathode to ~ 2000 K which starts electron emission, and the discharge becomes an arc.

Arc lamps with power between 50 W and 30 kW are used with reflectors, as searchlights, for simulation of sunlight, in spectro-photometers, for illumination of photoresist, and in movie projectors.

10.4.4. Ignitron

An electronic switch, the ignitron [203], conducts currents up to hundreds of kiloamperes and has a liquid mercury cathode at the bottom of a metal container, as seen in Fig. 10.13a. Emission from the mercury surface is started by a current pulse through a semiconductor ignition electrode, usually silicon carbide. Alkali metal additives to mercury reduce the arc voltage drop to ~ 4 V. An emitting cathode spot is created on the cathode surface by a large surface electric field, which starts the arc. The container walls, which in high-current tubes are water-cooled, cause the evaporated mercury to condense and run down to the cathode. The anode is heated in large ignitrons to prevent condensation. The arc extinguishes when the anode voltage falls below that of the arc, typically between 15 and 300 V. When working with AC the tube must be ignited during each period. The time jitter is ~ 100 ns, the delay time is $1–10$ μs, anode voltage is up to ~ 100 kV, peak current is ~ 1 MA, and controlled power is up to 1 MW, but the peak voltage and peak current cannot be simultaneously obtained. The tube lifetime depends on the charge per pulse. Large ignitrons of 100 kA, 2 000 C can survive hundred of thousands shots while at 1 MA per pulse these tubes survive only a few shots.

The **L**iquid **M**etal **P**lasma **V**alve is closely related to ignitron. It has a molybdenum cathode with a mercury-filled groove. Heating evaporates mercury which condenses on the walls and runs down into a pump, from where it is pumped back to the cathode. Some LMPV tubes ignite when the anode voltage exceeds the electrical strength, while others have an auxiliary anode which ignites the discharge. Typical LMPV tube operates with anode voltage up to 150 kV, peak current up to 10 kA, pulse repetition frequency of up to 100 Hz, and pulse length of tens of microseconds.

Ignitrons and LMPV tubes are mostly used as switches, closing the circuit in large condenser batteries by starting the discharge current. Applications include driving lasers, processing with shock waves, accelerators, pulsed magnetic field, military interference devices, fusion research, and nuclear explosions simulation.

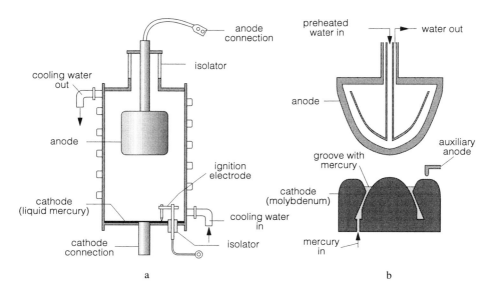

Fig. 10.13. (**a**) Ignitron. (**b**) Liquid metal plasma valve.

10.4.5. Hydrogen Thyratron

Although gas-filled thyratrons are being replaced by thyristors, the hydrogen thyratrons, which can produce short high-power pulses with small jitter, are used in radar modulators, accelerators, and lasers. A high-power thyristor can stand voltages of up to about 5 kV, with peak currents of a few kiloamperes and a current derivative di/dt of 1 kA/μs. Using high-voltage hydrogen thyratrons, anode voltages of up to 250 kV, currents of up to 50 kA, and current derivative of up to 20 kA/μs can be obtained.

In the hydrogen thyratron a large flat anode is situated inside a glass or ceramic cylinder. Opposing it is a indirectly heated cathode (Fig. 10.14). A flat control grid is located between the cathode and the anode, and in some constructions more grids sometimes enclose the anode.

As long as the control grid is held at the cathode potential, the tube does not conduct current, even at anode potentials of a few tens of kilovolts. A short pulse (~ 0.5 μs) to the control grid ignites the tube, over which the voltage drop is only a few volts. The ignition time is ~ 0.5 μs. After the pulse the tube must be driven with low voltage (few hundred volts) for ~ 30 μs to extinguish the discharge by neutralizing all the ions.

Modern high-power tubes have two control electrodes, and one or more gradient electrodes, to smooth out the field distribution near the anode. The auxiliary grid, located near the cathode, can be connected to a low positive potential (~ 100 V) so that a very weak discharge burns continuously. The control grid is connected to a negative potential (~ 100–200 V) which prevents the tube from conducting. With two electrodes the jitter is very small (~ 1 ns), and a lower control grid pulse voltage is needed to start the discharge.

Thyratrons filled with hydrogen have a shorter recovery time, while a deuterium fill allows a higher voltage. The gas pressure (~ 40–60 Pa) is critical and is maintained by a titanium-hydride-filled heated container. Titanium hydride is used because it can either give off or take up hydrogen, depending on the temperature and pressure. The cathode, usually an oxide, in the highest-power thyratrons is of the dispenser type.

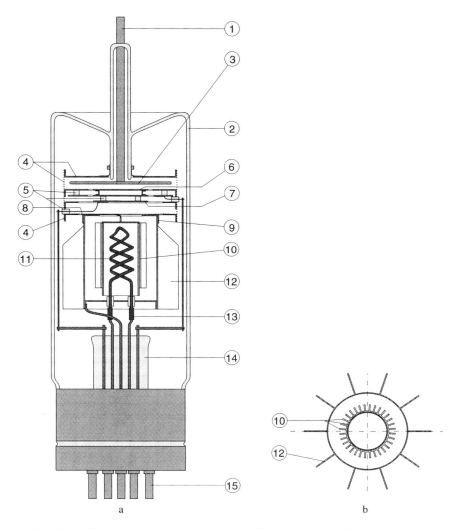

Fig. 10.14. Hydrogen thyratron. **a**. Deuterium filled thyratron CX1159 with glass isolation. **b**. Cross section through the cathode. (EEV.) 1, Anode connection; 2, glass envelope; 3, anode; 4, grid holder (screen and mesh) on cathode potential; 5, isolator and feed-through; 6, control grid; 7, auxiliary grid; 8, mesh which connects cathode and grid holder; 9, cathode holder; 10, cathode; 11, heater; 12, cooling plates (radial leafs); 13, hydrogen or deuterium reservoir with titanium hydride; 14, glass feed-through; 15, connections.

For anode voltages up to ~ 30–50 kV the tube has one voltage gap, with one cathode and one anode. For higher voltages, more gaps and control electrodes are inside the same envelope. A double thyratron, with its two cathodes connected in opposition, can, for example, be used for a kicker magnet where the current changes sign after a current pulse.

Another modern thyratron is the **B**ack **L**ighted **T**hyratron tube, which has no control grid, and the discharge is ignited by UV photons from a laser pulse. Since BLT tubes work at lower pressures than do thyratrons, the cathode–anode gap supports a higher voltage, the ionized gas recombines faster, and the repetition frequency can exceed 100 kHz.

The cathode and anode are made of copper and are separated by ~ 3–5 mm, and the anode has a ~ 5 mm diameter hole through which UV photons from the laser pulse pass and cause a powerful emission from the cathode. For ~ 100 ns, a thin cathode surface layer is heated by ion bombardment to ~ 5000 K, so that the thermionic emission can reach 10 kA/cm² or ~ 1000 times the emission from a dispenser cathode. Since the anode is also heated, it can work as the cathode if the voltage over the tube changes sign. A number of BLT tubes can be connected in series to achieve very high voltages, high di/dt ($> 10^{12}$ A/s), and low jitter (< 1 ns). The hydrogen pressure is again critical and is kept constant with titanium hydride containers. In commercial models, 80 kV and 80 kA per tube is obtained.

10.4.6. Crossatron

In the hydrogen thyratron the tube current is only broken when the anode voltage is ~ 0 V, and thus a tube is needed which can break the discharge while still at high voltage and current. The crossatron is such a tube. It has a cold cylindric cathode, two control grids, and a centered cylindric anode, seen in Fig. 10.15. At the control grids height, two external permanent magnets are mounted. At start, the first control grid is pulsed to $\sim +1$ kV, and between the cold cathode and the grid begins a Penning discharge (Section 10.6), because the electron orbits in the magnetic field are very large. In equilibrium, the first grid potential remains at ~ 200 V and the other control grid is at cathode potential. If pulsed positive with a few hundred volts, the crossatron will begin to conduct.

When the potential on the first grid is reduced to cathode potential the current ceases. Such a procedure does not work with a thyratron, but crossatron is at low pressure (~ 3 Pa) and it is the Penning discharge which provides the necessary preionization. Thus the crossatron functions as a current switch, when provided with the correct control grid mesh size, distance, and pressure.

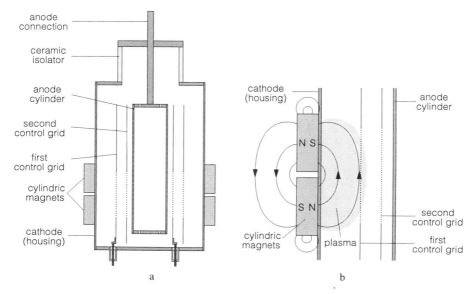

Fig. 10.15. Crossatron. **a**. Cross section. **b**. Region around the cathode with plasma and magnetic field (Hughes Research Laboratory).

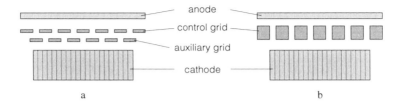

Fig. 10.16. Construction of (**a**) hydrogen thyratron and (**b**) tacitron.

Commercial crossatrons can withstand a 90 kV anode voltage and break a 1 kA current. The voltage drop over the tube when conducting is ~ 200 V, its risetime ~ 20 ns, and the breaking time ~ 50 ns. Crossatrons with anode voltages of 150 kV have been made in the laboratory.

A modern version of thyratron, the tacitron can also break the current. When the thyratron, with its thin mesh control grid, ignites, a thin space-charge layer around the mesh forms, which isolates the grid from the plasma. After ignition the grid voltage does not influence the plasma processes and so it is impossible to cut off the current by changing the control grid potential. The tacitron control grid is thick, as seen in Fig. 10.16, so that the electron mean free path becomes much shorter than the grid channels. When the grid is negative relative to the cathode, the ionized layer covers the whole channel and cuts off the current. Tacitrons with anode voltages up to 20 kV, currents up to 500 A, and repetition frequency as high as 200 kHz are produced. The hydrogen pressure is ~ 5 times lower than that in the thyratron, or ~ 10 Pa. The lifetime is shorter, ~ 1000 hours, to be compared to 5000 hours for thyratrons.

A third tube is under development, capable of cutting off high currents, which exist in vacuum and gas-filled designs. The magnetic field controls the current using the magnetron principle, with the axial magnetic field being generated by a coil loosely wound around the tube (low inductance to obtain a short risetime). The cylindrical dispenser cathode and copper anode have the anode nested in the cathode for the gas version, and vice versa in the vacuum version.

In the vacuum construction the current is established according to the Child–Langmuir law while the magnetic field is zero, but when the axial magnetic field is on, the current is cut off, if the working point is below Hull's parabola. The cathode–anode distance is ~ 1 cm and the anode voltage can be 100 kV. With a 500 cm^2 dispenser cathode the tube can cut off currents of 4 kA. The voltage drop over the tube, while conducting, is ~ 1 kV.

In the low pressure gas construction a control grid is placed between the cathode and anode. With no magnetic field, and the control grid at negative potential with respect to the cathode, the tube current is cut off. With a positive grid voltage the tube starts conducting in accordance with the Child–Langmuir law; but if at the same time the magnetic field is switched on, the electron orbits are increased and strongly ionize the gas. The current greatly increases relative to the space-charge limit. If the magnetic field is switched off and the control grid is set negative relative to the cathode, the tube will no longer conduct. At ~ 0.1 T the tube can cut off currents of a few kiloamperes at 100 kV.

10.4.7. Spark Gap

The spark gap can replace hydrogen thyratrons with a considerable reduction in cost. In a ceramic envelope, with brass or stainless steel metallic base plates, two hemispheric electrodes are placed (Fig. 10.17); at the center of one of them, the ignition electrode

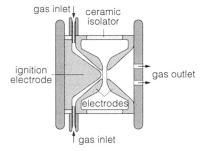

Fig. 10.17. Triggered spark gap.

is located. The tube is filled, or flowed through, with air, hydrogen, or argon/neon; the specific kind influences the the spark behavior rise and deionization times.

At ignition, existing free electrons (~ 500 cm^{-3} STP) form the first avalanche, which results in a large jitter (30–50 ns). UV laser light can be used to increase number of primary charges. The avalanche broadens from ohmic heating, and the discharge envelopes the electrode surface. At high voltages, and with gap lengths of > 5 mm, the accelerated ions sputter and evaporate the cathode, with the result that these evaporated atoms are accelerated toward the anode, where the process is repeated. Such evaporation helps start the discharge. In a gas flow tube, the flow must be appropriately shaped to prevent macroscopic crater formation, which limits the repetition frequency and the mean power and influences the life. Spark gaps with voltages up to a few hundred kilovolts can switch a few hundred kiloamperes, but their life is limited by crater formation to 10^3–10^5 pulses.

Spark gaps are used to switch the circuits in Kerr cells, spark chambers, gas discharge light sources, "smart weapons" modulators, and so on.

10.5. GAS LASERS

At 0 K all electrons in an atom are in their lowest energy levels, but as temperature is increased, some electrons are promoted to a higher energy state and occupy the "tail" of the FD distribution. The number of atoms, N_1 and N_2, with electrons in the levels E_1 and E_2, respectively, is given by

$$\frac{N_2}{N_1} = e^{-(E_2-E_1)/(kT)}$$

This natural equilibrium of Fig. 10.18a is disturbed if energy is supplied from outside.

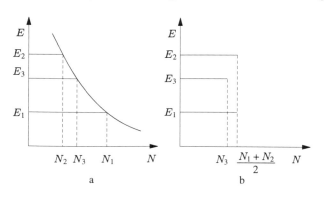

Fig. 10.18. Distribution of electrons on energy levels. **(a)** Natural equilibrium. **(b)** After optical pumping.

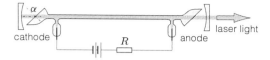

Fig. 10.19. Sketch of a gas laser.

Irradiating the system by incoherent light at frequency f_{12}, or for a gas supplying energy by a gas discharge, many electrons will go over from level 1 to level 2, while other electrons will go from level 2 to level 1. The probability for either transitions is equal, but the number of transitions is proportional to the level population of occupied states. The lower level 1, with a larger population, will therefore lose more electrons to level 2 until both levels have the same number of occupied states (Fig. 10.18b), in a process called *optical pumping*.

If, besides levels 1 and 2, a level 3 with occupation number N_3 exists, then by pumping N_2 can become larger than N_3, and an inversion is said to arise. If the excited atoms are irradiated by light with the frequency f_{23}, more electrons will go from level 2 to level 3 than the reverse. The emitted light becomes more intense than the absorbed light, and light with frequency f_{23} will be amplified. The $3 \rightarrow 1$ transitions will not usually be amplified, because there is no inversion between these levels. If the spontaneous $3 \rightarrow 1$ transitions occur less often than $3 \rightarrow 2$, then an inversion can exist even between these levels.

There are known to be more than ten thousand different transitions which result in stimulated emission. The wavelength extends from 0.1 to over 1000 μm, or from X-rays to long infrared. Commercial **L**ight **A**mplification by **S**timulated **E**mission of **R**adiation devices can be classified by their medium as solid state, semiconductor, dye, gas, molecular, free electron, and X-ray lasers. Here we only discuss gas and molecular lasers.

Gas and molecular lasers, which constitute ~ 80 percent of all industry lasers, are of four kinds. In the first, the laser action is produced by transitions in neutral atoms. In the second, the transition occurs in ionized atoms. In the third, the transitions are between molecules vibrational or rotational states. In the fourth, transitions take place in molecules by a chemical bond in the higher energy level, which breaks when the molecules return to their lowest energy level.

Gas lasers, a sketch of which is shown in Fig. 10.19, are pumped by a gas discharge in a tube filled with a low-pressure gas, closed at both ends by two glass plates. In some lasers these plates are set at the Brewster angle, α, because if $\tan \alpha = n$ (n being the refraction index) the reflections are reduced and polarized light is produced. A mirror is placed on either side of the tube; one is opaque and totally reflects light back into the tube, while the other is semitransparent, and the amplification of the laser determines its transmissivity.

Laser properties vary over wide ranges of power, frequency, and repetition. Output power from a few milliwatts to 100 kW can be reached in continuous operation, and pulsed power is possible up to terawatts. Beam divergence and spectral purity also vary, while a stabilized He–Ne laser has the same order of stability as that of the cesium clock, the primary time standard.

10.5.1. Neutral Atom Laser

He–Ne laser

In a helium–neon laser, whose simplified energy level diagram is shown in Fig. 10.20, a discharge excites the metastable $2^1 S$ and $2^3 S$ helium levels, from which it is improbable

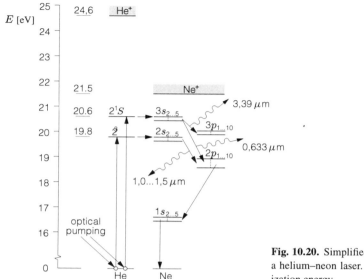

Fig. 10.20. Simplified energy level diagram for a helium–neon laser. He$^+$ and Ne$^+$ is the ionization energy.

that the electrons will return directly to the ground state. Neon has similar 3s and 2s energy levels. A collision between an excited helium atom with a ground state neon atom results in an exchange of energy leaving the helium in the ground state and the neon excited. The lifetimes of neon 2s and 3s levels are ~ 200 ns, and those of the 2p and 3p levels are ~ 10 ns. Pumping is done on helium with inversion of neon at s and p levels. Atoms in 1s level return to their ground state in collisions with the tube walls, and thus the He–Ne laser output power is inversely proportional to the tube diameter.

The infrared 3.39 μm line is the strongest, so to emit the red 0.63 μm line it must be suppressed by mirrors or absorption in the Brewster window. He–Ne laser also emits 0.543, 0.594, 0.730, 1.157, and 1.523 μm light. The output power is usually ~ 1 mW but can be ~ 100 W. The 0.63 μm line efficiency is ~ 0.1 percent.

The laser tube, made of pyrex or quartz glass (Fig. 10.21), is between 0.1 and 1 m long. One window can be of the Brewster type if polarized light is desired. The pumping discharge occurs inside a glass capillary, with diameter of ~ 1 mm. Its length, mirror geometry, and pressure determine the output power, beam diameter and divergence, and spectral line width. The laser needs an ~ 2 kV power supply at a few milliamperes current, with the pressure chosen so $p \cdot d \approx 500$ Pa·mm. For the red 0.63 μm line the helium and neon ratio is 5:1. He–Ne laser is used mostly as a low-power source in measuring, interferometry, and holography.

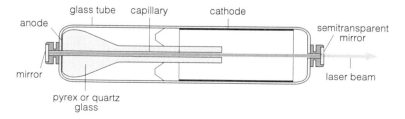

Fig. 10.21. Construction of a low power helium-neon laser.

Metal Vapor Laser

Metal vapors can be used in lasers, the most prevalent being copper and gold, because their melting and evaporation temperatures are low ($\sim 1500°$C for copper and $\sim 1600°$C for gold). Copper laser emits mostly in the yellow-green, while a gold laser emits mostly in the UV and red. Both lasers work only in the pulse mode because the lower levels have longer lifetimes than do the upper levels. They can each reach a mean power of ~ 100 W, pulsed power in excess of ~ 100 kW, at a pulse length of a few tens of nanoseconds and a repetition frequency of up to 20 kHz. Efficiency is about 1 percent.

The laser tube is ceramic, and neon is used as the auxiliary gas in which the discharge starts. The pressure is ~ 3000 Pa. Metal foils are placed near the cathode, which with a 10 kV voltage and 1 kA current evaporates the foil. Since the metal condenses on the tube the lifetime is a few hundred hours. Water or forced air cooling is required.

The copper laser is used to optical pump dye lasers, which results in over 50 percent efficiency. Other applications include high-speed photography and pulsed holography. The gold laser is used in medicine, since its red light affects certain dyes, which when taken up by tumor cells, cause molecule dissociation, and the created free radicals kill the cells.

10.5.2. Ion Lasers

In an argon ion laser the gas is ionized by discharge. An inelastic collision can cause the ionized atom to be put into the $4p$ state. Even if in a higher energy level, it can reach the $4p$ state by photon emission. $4p$ life time is ~ 10 ns, while the lower $4s$ state is ~ 1 ns, so an inversion is obtained for laser emission.

An ~ 30 cm long, ~ 2 mm diameter tube, seen in Fig. 10.22, of Be-O ceramic is used for its high heat conductivity and is filled with pure argon (a special getter keeps the gas clean) at a pressure between 1 and 100 Pa. Current densities of up to 1 kA/cm^2 is used to ionize a large number of argon atoms, of which many will become excited. Some argon lasers have in a tube of Al_2O_3 ceramic a central tungsten insertion, with water-cooled copper cooling flanges. Tungsten insertion temperature near the discharge can reach 5000 K. A low-energy (milliwatts) argon laser has double glass walls between which water circulates. To limit the discharge to the axial region and to decrease the number of particles which hit the isolation wall, high-energy argon lasers often use an axial magnetic field.

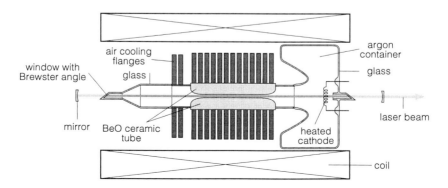

Fig. 10.22. Construction of a argon ion laser.

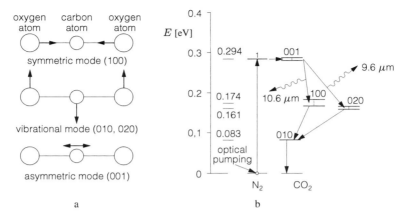

Fig. 10.23. CO₂ molecule. (**a**) Oscillation modes. (**b**) Energy level diagram.

This laser emits many lines from the blue to green, the choice being made by a suitable Brewster angle, which can be changed by a prism. Argon laser output power varies from a few milliwatts to 50 W, and its efficiency is low (< 0.1 percent). It is used mostly in applications where the He–Ne laser power is not high enough, such as "welding" of the retina.

10.5.3. Molecular Lasers

While individual atoms only can be excited to a higher energy level through an electronic transition, molecules can additionally be excited to rotational and vibrational states, where a transition between these states results in infrared photon emission.

The carbon dioxide laser action depends on the transition between oscillatory modes of the CO_2 molecule; vibrational, symmetric, and asymmetric modes. Each vibration mode is associated with a rotational state (Fig. 10.23a).

Besides CO_2 the discharge tube contains He and N_2, the latter in a high rotational state. In collisions this rotational energy is transferred to the CO_2-molecule 001 levels. Because these levels have almost the same energy, the cross section is large. Thus an inversion between the 001 asymmetric and the 100 symmetric mode or the 020 vibration mode occurs. The strongest lines are 9.6 and 10.6 μm, both in the infrared, however more than 100 transition are possible. The efficiency is ∼45 percent. Addition of He to ∼ 80 percent of the gas increases the plasma heat conductivity and the pressure. The number of molecules in the lower 100 states decreases in collisions with the He atoms, thereby increasing the output power.

CO_2 lasers have several cooling system variants: closed for low power, forced gas circulation for high power. In the latter the circulation can be axial and slow (a few liters per second) or transversal and fast (up to 300 m/s). To avoid the CO and O_2 dissociation with time, which decreases the power, a small amount of hydrogen or water vapor is added. With nickel or platinum catalyst electrodes, CO reconverts to CO_2. Glass tube CO_2 lasers produce an output power of ∼ 50 W/m of the active length, forced gas flow commercial units produce an output of 20 kW, and experimental units reach a continuous output of 100 kW. In pulsed mode a few gigawatts can be obtained during 100 ns, with the efficiency of ∼ 20 percent.

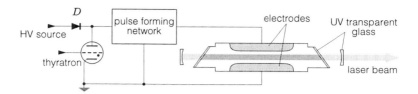

Fig. 10.24. Excimer laser. Construction and electric circuit.

CO_2 lasers are the most widely used industrial gas lasers, mostly in large-scale machining; they are also used in medicine as a surgical knife, with the advantage of fast blood coagulation. However, at 10.6 μm no optical fiber transport can be used, so jointed tubes with mirrors are employed.

10.5.4. Excimer Laser

Excimers, or excited dimers, are molecules which only exist in the excited state (e.g., ArF*, KrF*, KrCl*, XeF*, and XeCl*, where * denotes excitation), because in their ground state they decompose in their constituent atoms. These lasers have become popular since their first construction in 1975, because they emit in the UV: ArF* at 193 nm, KrF* at 248 nm, KrCl* at 222 nm, XeF* at 352 nm, and XeCl* at 308 nm.

Noble gases in an arc discharge are excited to a higher level, where they behave like an alkali atom and bond to a halogen atom (e.g., NaCl bond). However, when the molecule falls to its ground level by photoemission, it dissociates. These molecules cannot remain in the excited state for long; therefore, lasers based on excimers are only used in pulsed mode with a pulse length of less than 50 ns.

Excimer lasers are constructed with parallel nickel electrodes, which are not attacked much by F and Cl, separated by 2–3 cm, as in Fig. 10.24; passive components are Teflon-coated. The mirrors are quartz glass, but as the transparence of quartz decreases with wavelength, MgF_2 or CaF_2 is used at shortest λ. The gas, 5–10 percent noble gas, 0.1–0.5 percent halogen, and the rest He or Ne, circulates driven by a vacuum pump in a closed system, at pressures of $\sim 10^5$ Pa.

Laser action results in 100 MW/liter of gas, and the pulse voltage is 25–35 kV. The pulse generator is a pulse-forming network. The buffer gas, He or Ne, is preionized by a UV light flash tube (not very effective), by an electron beam (complicated because a thin foil is necessary), or by X-rays (effective but expensive), to obtain a homogeneous arc discharge. Commercial excimer lasers use flash tubes.

At pulse power of several hundred megawatts, the electronics limits excimer laser to a few hundred pulses per second. With the short pulses of 5–50 ns the mean power is a few hundred watts, and efficiency is a few percent.

Excimer lasers are used to pump dye lasers, for machining, in production of integrated circuits, in photochemistry, in medicine to peel of thin tissue layers without damaging the tissue below, and for eye surgery.

10.6. ION SOURCES

Ion sources generate ion beams with defined energy, current, and energy distribution. Many different ion sources have been developed; the important ones are sketched in Fig. 10.25.

All have in common a cold or heated cathode in a high-frequency or stationary electric and/or magnetic field in which a gas discharge is created by electrons which ionize gas molecules. Neutral gas molecules can be introduced into the source, a liquid or a solid can be vaporized, or neutral atoms or molecules can be obtained by sputtering. The ions are removed from the discharge by an extraction electrode connected to a desired acceleration voltage, and the ion beam is formed by lenses.

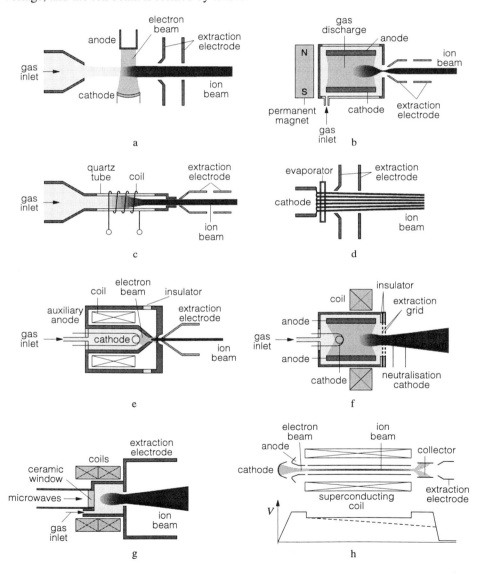

Fig. 10.25. Construction principles for the most important ion sources. (**a**) Ion source with heated cathode, (**b**) Penning source, (**c**) high-frequency ion source, (**d**) ion source with vaporization, (**e**) duoplasmatron, (**f**) ion source with grid extraction, (**g**) ECR ion source, (**h**) EBIS ion source with potential distribution along the axis.

In a heated cathode ion source an electron beam passes the gas flow. Ions are removed by the extraction electrode electric field. These sources are simple and are used when the current density and energy homogeneity requirements are not high.

A Penning source uses a glow discharge between two cold electrodes in a magnetic field. Acceleration takes place between each cathode and anode, and the electron orbits are stretched because the magnetic field forces the electrons to move in spirals, and because of the repelling force of the opposite cathode. The ionization probability is increased, and thus the current is much higher than that from a source with heated cathode. The Penning source principle is used in vacuum technology for both gauges and pumps.

High-frequency ion sources use either a coil or two electrodes, connected to an rf power supply. In either case a glow discharge is ignited in a quartz tube, where neutral molecules are ionized. The ions stream together with gas through the tube where at the end the ions are accelerated by an extraction electrode.

In a vaporized liquid or solid ion source the vaporizer has small holes, and atoms or molecules of the heated vaporized matter hit a heated tungsten cathode, where they are thermally ionized on the cathode surface and focused by a following lens. These ion sources are used when the substance cannot be introduced in gas phase, and they have their most important application in integrated circuit manufacture.

In a duoplasmatron an arc discharge is ignited between the cathode and anode, limited by an auxiliary anode, where the inhomogeneous magnetic field, generated by a cylindric coil near its opening, concentrates the electrons into a small volume. A very strong ionization results with high current densities. The duoplasmatron is mostly used in accelerator technology and spectrometers.

An ion source with grid extraction, known also as Kaufman ion source, was developed by NASA for space-probe ion propulsion. The source, a heated cathode in the center of a chamber through which the working gas flows, is mounted perpendicularly to two anodes. A diverging magnetic field lengthens the electron orbits and high degree of ionization is obtained. Two grids are used for extraction as well as acceleration to the desired energy. Commercial sources giving a few amperes of current in a beam with diameter of up to a few tens of centimeters are mostly used in the manufacture of integrated circuits. The neutralization cathode indicated in the figure emits electrons which neutralize the ion beam space-charge, and it can charge neutral gas atoms without recombination with beam ions.

Electron **C**yclotron **R**esonance sources are used in surface treatment and in accelerator technology. Here microwave power enters a cavity through a cylindric waveguide and a ceramic window. The cavity is pervaded with a working gas. The electric field, generated by the circularly polarized electromagnetic wave, ionizes the gas, while an external magnetic field forces the electrons into circular orbits with cyclotron frequency, ω_c. If the microwave frequency is equal to the cyclotron frequency, the system is in resonance and the energy transfer is very efficient. Usually the industrial frequency of 2.45 GHz, corresponding to magnetic field of 0.0875 T, is used.

Electron **B**eam **I**on **S**ource, exclusively used in accelerator technology, can produce fully stripped ions (e.g., U^{92+}). From a Pierce-type electron gun, with acceleration voltage between 10 and 100 kV, an electron beam enters a high magnetic field region, up to 5 T, generated by an ~ 1 m long superconducting coil. The electron beam of a few amperes is highly compressed, with 1 to 2 mm diameter in the homogeneous part of the magnetic field.

A suitably designed electric field forces the ions to remain in the ion trap, where they interact with the electrons, losing more and more electrons. Finally, the trap is opened by decreasing one of the electron-optical mirror potential (dashed line in Fig. 10.25h), and ions freely leave the trap. Mass analysis selects the ions with the desired Z/A ratio.

10.7. GAS DISCHARGES AND SURFACE PROCESSING

A self-sustaining gas discharge lasts only as long, as enough positive ions hit the cathode and release secondary electrons. Townsend's second coefficient, γ, is ~ 0.01–0.1, so only one in ten to hundred ions releases a secondary electron. The continuity equation requires that the main current be supported by strongly accelerated ions near the cathode, which hit the cathode with high kinetic energy. With suitable pressure this energy can be kiloelectronvolts. Thus these ions can expel individual atoms, molecules, and molecular clusters, which form a thin layer on all surrounding electrodes. This process can be used to produce thin films of some desired material — *sputtering* — or to etch away the cathode outermost layer — *ion etching*. Both processes are affected by the gas, which influences the chemical composition of the film or modifies the cathode etching. Similar effects can be obtained by using an ion source, where the ions are extracted, accelerated, and hit the etched substrate or are the sputtering source.

The semiconductor industry is the most important user of sputtering and ion etching, often a driving force behind the use of gas discharges. Integrated circuits are coated by sputtering and are etched by ion etching. Depending on the process, different plasma-generating methods were developed. DC, AC, rf, and microwaves, combined with magnetic field to increase the efficiency, are used. The discharge, together with other methods, like evaporation, used to make thin films or to etch, are listed in Table 10.1. PECVD and RIE are the most common, and HDLP-P will be necessary when the line width decreases under 0.5 μm.

The yield, the ratio of released atoms to incident ions, depends on ion energy and the cathode material. A threshold, between 5 and 25 eV, must be exceeded for atoms to be released, after which the yield increases sharply and reaches a rather wide maximum at a few kiloelectronvolts. There is a large variation of yield with material, as seen in Fig. 10.27; it varies periodically in concert with the sublimation heat periodicity. The velocity of the released atoms or molecules is about five times larger than the velocity of the atoms released by evaporation. Thus sputtering provides a better substrate adhesion.

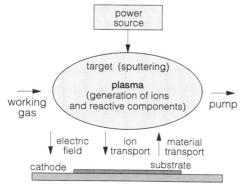

Fig. 10.26. Principle of sputtering and ion etching.

Table 10.1

Sputtering and ion etching methods

ARE	Activated reactive evaporation
CVD	Chemical vapor deposition
IAC	Ion-assisted coating
IAD	Ion-assisted deposition
IBAD	Ion-beam assisted deposition
IBAE	Ion-beam assisted etching
IBED	Ion-beam-enhanced deposition
IBS	Ion-beam sputtering
ICB	Ionized cluster beam deposition
IVD	Ion vapor deposition
PECVD	Plasma-enhanced chemical vapor deposition
RIE	Reactive ion etching
HDLP-P	High-density low-pressure plasma

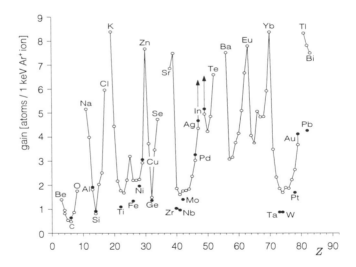

Fig. 10.27. Sputtering yield for different elements. The cathode is hit by 1 keV argon ions. •, Measured; ○, computed. (After *Handbook of Ion Beam Processing Technology*, J. J. Cuomo and S. M. Rossnagel, 1990, Noyes Publications, with permission.)

10.7.1. Sputtering

Integrated circuits require the deposition of thin films of many materials, both of metals (e.g., aluminum and gold) or polycrystals (e.g., silicone oxide and GaAs). Evaporation was the first used and worked well for pure materials, like aluminum for conductors. To produce narrower conductors, modern circuits use alloys (e.g., Al–Si or Al–Si–Cu). However, the vapor pressure differences allow more volatile elements vaporize faster, so the composition of the deposited material is not that of the original alloy. In sputtering the relative yields determine the composition, after a short time equilibrium is reached, and the sputter stream will be that of the alloy, since the vapor pressure of the metals is low. At the low temperature of the substrate, most sputtered atoms will stick to the substrate, a clear advantage of sputtering over evaporation. Noble gases, usually argon, are used in sputtering, though chemically active gases can be added, whose ions react chemically.

Chemical Vapor Deposition, originally used in integrated circuits production, has the disadvantage of high temperatures (e.g., 625°C) for deposition of silicon from SiH_4. Plasma Enhanced Chemical Vapor Deposition uses much lower temperature, and the doping element diffusion is decreased, so that modern integrated circuits with high packing density can be made.

Sputtering is also used to make thin films of refractory metals, like W, Mo, and Ta, which can only be made by sputtering or electron-beam technology. Reactive sputtering is used to make oxide, nitride, sulfide, and carbide films by adding of N_2, O_2, H_2S, and CH_4, respectively, to argon. Sputtering is used to apply antireflection coating to optical glass and to coat isolators. The film properties differ from those of the compact material, and it is possible to obtain a better mechanical stability by using reactive gases.

Sputtering equipment can use both gas discharges and ion beams. The substrate to be coated is mounted on a holder, which usually is the circuit anode. The cathode, called *target,* provides the sputtered atoms. Other positions and configurations are possible, but the bombardment of the substrate with negative ions must be avoided.

A simple sputtering setup, seen in Fig 10.28, uses two electrodes and a DC voltage of $\sim$1–3 kV, a cathode current density of $\sim$1 mA/cm^2, a pressure of $\sim$1 Pa, and an electrode separation of $\sim$5–10 cm. For DC sputtering the substrate must be conducting. An AC system, often driven at main frequency, can have electrodes of different materials, both of which act as targets. In an rf system, where the frequency is a few megahertz, a negative potential drop forms in front of the electrodes, since the electron and ion mobility differ. The plasma and floating potential difference accelerates the ions and electrons. For a symmetrical system the plasma potential can be a few hundred volts. When the substrate electrode capacitance is modified or the electrode is grounded to the vacuum chamber, the plasma potential is $\sim$ 10–20 V. A high electric field is created in front of the target, but there is no current to the target and the electrode can be either a conductor or an insulator. Radio-frequancy sputtering allows a large choice of materials, and is the simplest method to sputter insulators. In a diode-type sputtering system the substrate is enclosed in plasma, and all particles can reach the substrate and react chemically with each other and with the substrate. Since many particles are ionized, the diode sputtering system is chemically very active.

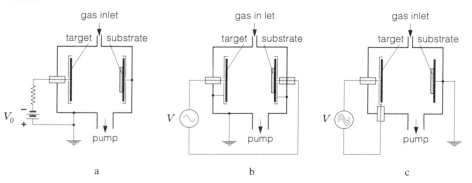

Fig. 10.28. Sputtering equipment of diode type. (**a**) DC system, (**b**) AC system, (**c**) rf system.

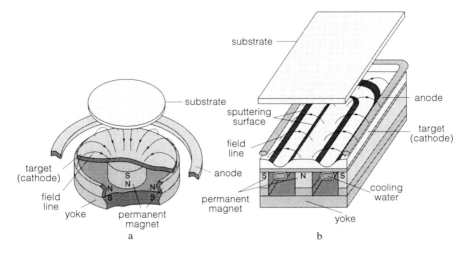

Fig. 10.29. Magnetron-type sputtering system. (**a**) Circular model. (**b**) Rectangular model.

To increase the ion current and sputtering yield a magnetic field can be used. For a DC diode system a coil is wound around the vacuum chamber, and yield increases of ~ 10 can be obtained. A modern magnetron sputtering system is seen in Fig. 10.29. The magnetic field distribution profoundly influences the discharge. Usually the magnetic field is directed parallel to the target surface, so the electrons get a drift velocity, $\mathbf{E} \times \mathbf{B}$, around and along the surface of the target, which dramatically increases the ionization probability. The dense plasma near the cathode allows high current ($\sim$many amperes) with low voltage ($\sim$few hundred volts) and high power (~ 10 kW) at a lower pressure (< 1 Pa) than by using the DC sputtering. Magnetron principle works also with an rf power supply (usually 13.56 MHz). This is surprising, since the electron mobility perpendicularly to the magnetic field lines should be smaller than that of the ions, and thus the target should be positively biased. In reality, a negative bias aproximately equal to peak rf voltage is obtained, indicating that the electron motion is very turbulent, and their drift velocity is much larger than that given by $\mathbf{E} \times \mathbf{B}$. An rf magnetron can also sputter nonconducting materials.

Sputtering can also be done with ion beams, which have the advantage of simpler control of energy, current, incidence angle, or beam divergence when compared to that in plasma sputtering. A more important advantage is that it is possible to work with low pressure, 1–100 mPa, two orders of magnitude less than that with plasma; so the mean free path of primary ions and sputtered atoms exceeds the chamber dimensions. The disadvantage is the higher price. The most commonly used ion sources are grid extraction, rf sources, and ECR.

Ion beams allow direct deposition; the higher the beam energy, the better the adhesion and the more uniform the film quality. The low pressure results in a cleaner deposition as very few foreign atoms reach the film, an effect which benefits from the absence of electric field in front of the substrate. A neutralizing cathode, used with a grid extraction ion source, allows deposition on a substrate which is electrically isolated or of nonconducting material. A working gas can be chosen so the ions react chemically on the substrate surface. Dual ion sources can be used, of which one can be of vaporization type. Among the most successful applications is **I**on-**B**eam-**A**ssisted **D**eposition used in production of thin dielectric films, where optical properties are important.

Thin, high-temperature superconductor and diamond films can be made with plasma or ion-beam deposition. Coating of surfaces with diamond provides new and superior machine tools, but electrical, optical, and thermal properties of such films are of more importance. Diamond can also be doped to make semiconductors. Deposition made at the rate of 60–80 μm/h probably begins with the formation of carbides which seed the growth of diamond crystals.

10.7.2. Ion Etching

In the production of integrated circuits, thin layers must be selectively etched away. For example, silicon oxide must be removed from the silicon substrate, without damaging the latter, because the uppermost silicon layer, less than 100 nm thick, has doping agents which must not be harmed. Wet etching etched even below the mask, limiting the line width. The important practical advantage of ion etching is the ability to control the ion incidence angle. The technique is also much cleaner than the wet etching and can be easily automatized. Equipment for ion etching is similar to the one for sputtering, and both plasma and ion beam can be used.

By the choice of suitable gases it is possible to etch selectively. The most frequently used technology, **R**eactive **I**on **E**tching, uses working gas containing halogens (e.g., chlorine in the form of CCl_4, Cl_2 with possible adding of O_2 when etching aluminum;

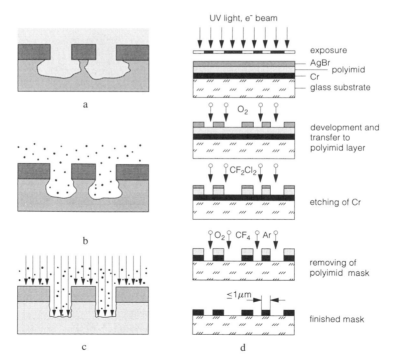

Fig. 10.30. Difference between (**a**) wet etching, (**b**) plasma etching (RIE), and (**c**) etching with the ion beam (IBAE). (**d**) Production of a metal mask for integrated circuit by ion etching (After *Fundamentals of Plasma Chemistry and Technology*, H. V. Boenig, 1988, Technomic Publishing Co., with permission.)

or fluorine in the form of CF_4, SiF_4, CF_nCl_{4-n} when etching silicon). Both ionization and dissociation takes place in the discharge, and free radicals as well as positive and negative ions are produced. Usually an rf discharge is used. The substrate to be etched is attached to a capacitively coupled electrode, which generates a voltage drop from the electron and ion mobility difference, accelerating the ions toward the substrate. The reactive free radicals proceed to the substrate by diffusion, where they are adsorbed on the surface and react chemically. For example a Si or SiO_2 substrate reacts with the F-radicals to form SiF_4 (or its precursors SiF, SiF_2, SiF_3), which is volatile and is pumped away. Ion bombardment of the substrate creates "active sites" by removing already fluoridized Si atoms which passivize the surface (sidewall passivization), so the etching process is much faster and more uniform (Fig. 10.30).

Ion etching is also used for general treatment of metals (the same gases as for Si with O_2 and argon added) or organic compounds (development of photoresist with O_2, of polymers with H_2). The example in Fig. 10.30d shows the production of a metal mask for manufacturing an integrated circuit.

10.7.3. Ion Implantation

The element density increase in modern integrated circuits depends on fabricating precision, so in addition to sputtering and ion etching, precise doping of Si and GaAs is of paramount importance.

A high-energy ion penetrates matter losing energy in elastic and inelastic collisions, until it comes to rest, settling somewhere in the crystal lattice. Radiation damage along its path can be reversed by annealing at high temperatures, but the ion is a permanent part of the crystal structure. By ion implantation any ion can be introduced, with the penetration depth depending on its energy, mass, and charge state as well as on the substrate material. Energies of ~ 10–$10,000$ keV produce penetrations of ~ 10–2000 nm.

An ion implantation device consists of an ion source, often a duoplasmatron, and a Cocroft–Walton- or van de Graaf-type electrostatic accelerator, all in ultrahigh vacuum. The ion beam is energy analyzed with a $90°$ bending magnet or a quadrupole mass filter, followed by a slit, to accuracy often better than 10 eV. Ion current, ~ 10–$10,000$ μA, is measured so the dose can be specified to ~ 1 percent.

Doping with ion implantation is fast. A 100 mm diameter silicon substrate can be given a dose of 10^{13} ions/cm^2 at 1 MeV in ~ 30 sec. Very accurate control of all implantation parameters is possible in a room temperature operation.

10.8. PLASMA METALLURGY AND PLASMA CHEMISTRY

Thermal plasmas are playing an increasingly important role in metallurgy and chemistry, where the high plasma temperature, which can reach 30,000 K, has made a number of new processes possible and others cheaper. Among these are the reduction of iron from ore or recycling scrap iron, reduction of other metals, production of aluminum and different alloys (ferrochrome and ferromanganese), recycling of platina from car catalysts, remelting under low pressure to improve metal properties, and production of acetylene and nitrogen oxides, and so on. Cement, ceramic, and paper industries increase the temperature of cheap waste fuel using plasma torches. Another important application is to render hazardous chemical waste harmless (e.g., dioxin).

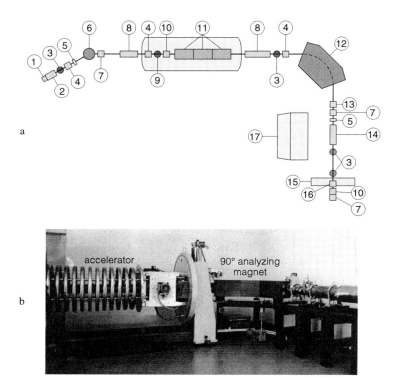

Fig. 10.31. (a) Equipment for ion implantation. 1, Ion source; 2, preacceleration; 3, cryopump; 4, beam guiding; 5, vacuum valve; 6, mass analyzer; 7, ion current monitor; 8, lens; 9, ion pump; 10, beam profile monitor; 11, 1 MeV electrostatic accelerator; 12, energy analyzer with 90° magnet; 13, slit; 14, deflection system; 15, vacuum chamber for substrate exchange; 16, substrate holder; 17, console. **(b)** Photograph of an ion implanting device. The accelerator tank, in drift filled with SF_6, is removed. The rings around the acceleration tube protect from corona discharge. (National Electrostatic Co.)

10.8.1. Plasma Torch

A torch, seen in Fig. 10.32, used to create high-temperature plasma, has a tungsten cathode and a row of water-cooled auxiliary copper anodes between which the working gas streams. The auxiliary anodes start the discharge as the arc is transferred from one anode to the next. In the arc discharge, which burns between the cathode and the main anode (~ 1 kW to 1 MW), the heated gas is partially ionized and streams through the anode opening. A coil wound around the anode generates a magnetic field which concentrates the plasma near the axis until it leaves the torch. By suitably shaped electrodes, which let the gas stream tangentially, or by using a rotating magnetic field in a three-phase system, the plasma is stabilized to avoid its hitting the anode. In a three-phase system the contact point of the arc on the anode rotates along the periphery. Both the cathode and anode are water-cooled, which in a DC-driven system consumes ~ 30–60 percent of the total input power. Large plasma torches are, therefore, AC-driven with two, three, or six torches connected together or by the plasma to the conducting workpiece. Such torches can have an efficiency of ~ 90 percent. Here the plasma originates in a DC source and the arc is transferred to the next torch, which is connected to the AC power source. Some torches use AC and DC simultaneously.

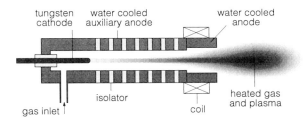

Fig. 10.32. Plasma torch.

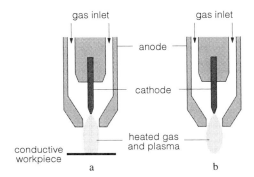

Fig. 10.33. Plasma torch with (**a**) transferred and (**b**) not transferred arc.

The gas-plasma temperature varies between 3000 and 30,000 K, with power density as high as 25 kW/cm^2. A noble gas (argon or helium), diatomic (nitrogen, hydrogen, or oxygen), or mixtures with water vapor constitute the working gas. In the later case the inner side of the anode can be sprayed with water, which cools the anode as it evaporates on its surface. The choice of gas depends on the process for which the plasma torch is used. For diatomic gases the dissociation energy needs a larger power input than that for noble gases, but it increases the power density.

A plasma torch can be used in two ways, seen in Fig. 10.33. In the first the arc is transferred to a conducting workpiece electrically connected to the torch anode. Transferred arc produces a very high temperature where the plasma contacts the workpiece, higher than the melting temperature of any known material. In the second the heated gas-plasma combination only warms up the workpiece so the plasma is neutralized over a larger area of the treated material.

Simple construction plasma torches are most videly used in welding. The arc is transferred to the workpiece and the gas is usually argon, sometimes helium. Some metals and alloys (e.g., Ti) must be welded in a protective atmosphere.

10.8.2. Plasma in Metallurgy

High-temperature, high-power density, and a variety of the working gases makes plasma torches useful in metallurgy. Their use is limited to processes where no practical alternative exists or where the material properties can be improved.

The advantage of a plasma-based plant for iron oxide reduction (Fig. 10.34a) as compared to blast furnaces is a higher efficiency. Prereduced ore and the reduction agent, being coal powder or heavy oil, enter the reduction zone. Crude iron, together with slag, appear almost immediately at the bottom of the shaft, and carbon monoxide, hydrogen,

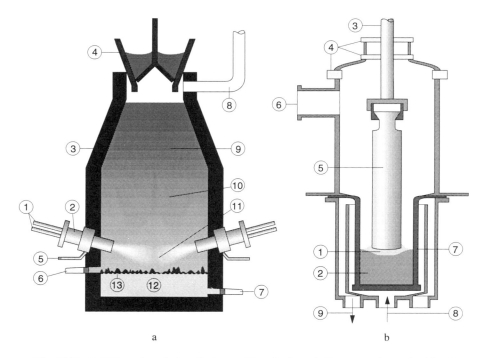

a   b

Fig. 10.34. (a) Plasma-based plant for iron oxide reduction. 1, Gas and coal powder inlet; 2, plasma torch; 3, reduction shaft (ceramic bricks); 4, coke inlet; 5, prereduced iron ore; 6, slag outlet; 7, crude iron outlet; 8, carbon monoxide and hydrogen outlet; 9, coke, 800°C; 10, coke, 950°C; 11, melting and reduction zone, plasma, 1700°C; 12, melt; 13, slag. (b) Arc furnace. 1, Arc discharge; 2, re-melt metal; 3, electric connection and holder; 4, isolator; 5, electrode to be melted; 6, connection to vacuum pump; 7, container (normally copper) for melt and solidified metal; 8, cooling water in; 9, cooling water out.

and a small amount of water evolve. At the shaft top the gas mixture temperature is 800°C, its composition and temperature well adapted for prereduction of the iron ore to ~50–60 percent in a separate unit. The coke in the shaft is preheated by gases and functions as a constantly renewed fireproof furnace interlining. Similar plant is used to produce sponge iron or to remelt scrap iron.

Arc discharge is used to melt refractory metals, like W, Mo, Ti, Ni, Zr, and their alloys, and to produce special steels. Metals can also be remelt to decrease their gas content (e.g., to obtain **O**xygen **F**ree **H**igh **C**onductivity copper). Remelting also improves the mechanical properties by reducing the number of structural defects, especially in refractory materials. Figure 10.34b shows a typical arc furnace. A high-current arc is created between the electrode to be melted, and the water cools remelted material. As the upper electrode surface melts, a feedback circuit controls and keeps constant the distance until the electrode has completely remelted.

10.8.3. Plasma Chemistry

The high temperature generated by a plasma torch can accelerate a number of chemical processes and create new ones. At present, many processes demonstrated on a small scale lay fallow, because the low price of oil makes them uncompetitive with traditional methods.

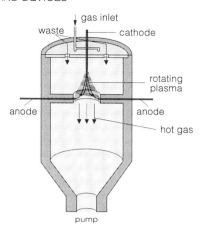

Fig. 10.35. Plasma furnace for hazardous chemical waste.

Among these is the production of acetylene and ethylene from oil, natural gas, or coal; synthesis of nitrogen oxides; manufacturing of pigments (e.g., TiO_2); production of metal carbides, nitrides, and oxides; and ceramic fabrication. Plasma chemistry is also used experimentally in cement and paper industry.

An important application is rendering harmless chemical waste. Ordinary waste, like domestic refuse, is now burned in special furnaces, the heat used in district heating or electricity generation. The use of plasma torches in such plants increases the efficiency and reduces hazardous gases. A chemical waste furnace, shown in Fig. 10.35, has a rotating plasma generated by six symmetrically placed anodes, which produces central temperatures of $\sim 50,000°C$. Rotating the plasma creates a vortex which gasifies the waste by breaking all molecular bonds so that difficult chemical compounds, like dioxins, can be completely destroyed. Only harmless gases (e.g., water vapor, carbon dioxide, etc.) evolve; and solid waste in the form of powder, which includes small amounts of heavy metals in metallic form, remains.

11

Vacuum Technology

Second-generation vacuum pump made by
Otto von Guericke in Magdeburg around
1670. The pump and the two hemispheres
are displayed at Kunskapstivolit, Technical
Museum in Malmö, Sweden. (Photograph
Technical Museum).

11.1. INTRODUCTION

To function satisfactorily, electron tubes and devices must maintain a vacuum. Vacuum is produced and measured in systems which have one or more pumps to remove air or other gases from the chamber. Connecting lines, valves, seals, measuring equipment, and leak detectors are also included in the system. In Latin "vacuum" means "empty," but today we use it for a gas whose pressure is lower than atmospheric. Dry air at the sea level has a pressure of 101,323.2 Pa, 1013 mbar, or 760 torr. The distinction between rough, medium, high (HV), and ultrahigh vacuum (UHV) is made approximately according to Table 11.1. At present the lowest laboratory pressure obtained is $\sim 10^{-15}$ Pa. At the ultra-high vacuum limit (10^{-6} Pa) there are still $\sim 2.5 \times 10^9$ particles/cm^3 at 293 K.

Table 11.1

Vacuum regions

	Pa	mbar
Rough vacuum	$10^5 - 10^2$	$10^3 - 1$
Medium vacuum	$10^2 - 10^{-1}$	$1 - 10^{-3}$
High vacuum	$10^{-1} - 10^{-5}$	$10^{-3} - 10^{-7}$
Ultra-high vacuum	$< 10^{-6}$	$< 10^{-8}$

HISTORICAL NOTES: Galileo, at the beginning of the seventeenth century, generated the first partial vacuum with a simple piston. Then he showed that bodies with different density fall with the same velocity. Torricelli [215] invented the mercury barometer in 1643, and 7 years later Guericke [176] constructed the first vacuum pump with which he showed that 8 pairs of horses were not enough to pull apart an evacuated sphere into hemispheres.

Interest in low pressures was slight until in the second half of the nineteenth century discharge phenomena hastened the development. In 1874 McLeod [216] made the compression manometer and measured pressures down to 10^{-3} Pa. In 1905 Gaede [217] constructed the rotary vane pump, which was sealed by mercury; in 1915 he invented the mercury diffusion pump [218], the first pump able to achieve high vacuum. Dewar [219], the inventor of the thermos bottle, tried to use refrigerated sorption pumping in 1905. The turbomolecular pump, proposed a few years later [220], took almost 50 years to become commercially available [221].

The McLeod manometer was replaced by indirect-reading gauges in about 1920. The Pirani gauge [222], which uses the thermal conductivity change, can measure pressures to ~ 0.01 Pa, while the Penning [223] and the ionization gauge [224] extended the region to $\sim 10^{-5}$ Pa. In 1950 Bayard and Alpert [225] constructed an ionization gauge capable of measuring pressures to $\sim 10^{-9}$ Pa. A modification of the Penning gauge, the magnetron gauge [226], operates to $\sim 10^{-11}$ Pa.

These inventions provided art which made possible the technology from light bulbs to simulation of the conditions in space.

11.2. VACUUM PHYSICS

Vacuum is generated by one or more vacuum pumps, usually connected in series, each characterized by the pressure difference from intake to exhaust. The ultimate pressure, p_g, is the lowest pressure which can be obtained at the pump intake side.

At any cross section on a vacuum line the quantity of pumped gas at a given temperature, called the *suction power,* is given by

$$Q_{pV} = -\frac{d(pV)}{dt} = -kT\frac{dN}{dt} \qquad [\text{Pa} \cdot \text{m}^3/\text{s or mbar} \cdot \text{l/s}]. \qquad (11.1)$$

The gas flow across a vacuum line cross section in a time dt, called the *volume throughput,* is

$$Q_V = -\frac{dV}{dt} \qquad [\text{m}^3/\text{s or l/s (liters/second)}]. \qquad (11.2)$$

The throughput at the pump intake, called the *pumping speed, S,* is defined by the average throughput for a standard test system at a specified temperature and pressure.

In a pumped chamber (recipient), pV, decreases as

$$Q_{pV} = -\frac{d(pV)}{dt} = -\frac{dp}{dt}V - \frac{dV}{dt}p; \qquad (11.3)$$

and for $p \sim \text{const}$ — that is, assuming a negligible flow resistance between the recipient and pump — we obtain

$$Q_{pV} = -\frac{dV}{dt}p = Q_V p_2 = S p_1, \qquad (11.4)$$

where Q_V and p_2 are the volume throughput and pressure at the recipient exhaust, and S and p_1 are the pumping speed and pressure at the pump intake. If $p_1 = p_2 = p$ along the vacuum line of sufficiently large cross section, we obtain

$$Q_V = S, \qquad (11.5)$$

and thus if the pressure drop can be neglected the pumping speed equals volume throughput.

When evacuating a system with constant volume, V, we have

$$Q_{pV} = -\frac{dp}{dt}V = S(p - p_g) \qquad (11.6)$$

and

$$-\frac{dp}{dt} = \frac{S}{V}(p - p_g). \qquad (11.7)$$

For an ideal pump $S = S_0 = \text{const}$ and

$$-\frac{dp}{p - p_g} = \frac{S_0}{V}dt, \qquad (11.8)$$

which after integration is

$$\ln\frac{p_0 - p_g}{p - p_g} = \frac{S_0}{V}t. \qquad (11.9)$$

Here, p_0 is the initial system pressure at $t = 0$, so as $p_g \ll p_0$ we obtain

$$p = p_g + p_0 e^{-(S_0/V)t}, \qquad (11.10)$$

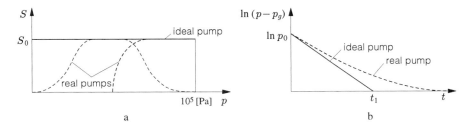

Fig. 11.1. (a) Pumping speed, S, of ideal and real vacuum pumps and (b) $p(t)$ of an ideal and a real pump.

and the system pressure falls exponentially with time. The pumping time needed to reach the pressure p_1 is

$$t = \frac{V}{S_0} \ln \frac{p_0}{p_1 - p_g}. \tag{11.11}$$

Real pumps do not have constant pumping speed.

$$S(p) = -\frac{V}{p - p_g} \frac{dp}{dt}. \tag{11.12}$$

The pressure drop, $-(dp/dt)$, is obtained from a published pumping speed curve. Figure 11.1.a. shows such curves for real and ideal pumps.

For pressures above 0.1 Pa the assumption that the gas volume, V, is that of the system to be evacuated is reasonable. But at HV the desorption of molecules on the recipient surface by thermal motion becomes important. For example, in a 0.1 m^3 vacuum system the inner surface can be ~ 1 m^2, which, covered by a monolayer of say N_2, amounts to $\sim 8 \times 10^{18}$ molecules. At 0.1 Pa there are $\sim 3 \times 10^{18}$ molecules left in gas in the volume. In practice the real surface is, depending on materials and surface structure, 10–100 times larger than the geometrical surface, and thus a much longer time will be needed to obtain a low pressure than can be expected from the pumping speed and the system volume.

The process can be speeded up in several ways. Some molecules like air desorb easily, whereas others like grease, fingerprints, and moisture desorb slowly, so all HV vacuum components must be thoroughly cleaned by rinsing in an organic agent (e.g., alcohol, trichloretylene, or water with detergent), followed by drying in air. UHV surfaces are chemically or electrolytically etched and electropolished, cleaned with trichloretylene vapor, washed in a deionized water ultrasound bath with detergent, rinsed in deionized water, and dried in air at $\sim 150°$C. Gas desorption from a "dirty" stainless steel surface is $\sim 10^{-5}$ Pa l/cm^{2}s, and from a clean one $\sim 10^{-8}$ Pa l/cm^{2}s. Desorption can be accelerated by heating the vacuum system to $\sim 450°$C in a process called "baking." A gas discharge can also be used for outgassing.

The ultimate pressure in a vacuum system is limited by two phenomena: diffusion and permeation. In all materials, foreign atoms slowly diffuse to the surface, and when they are pumped the ultimate pressure is determined by permeation; for metals it is mainly hydrogen, while for glass it is helium which slowly leaks through the vacuum chamber walls. Some organic seals, like O-rings, have such a large permeation that the ultimate pressure remains limited to $\sim 10^{-6}$ Pa. Since for UHV systems diffusion can limit the ultimate pressure, copper, stainless steel, and titanium construction materials are "fired"

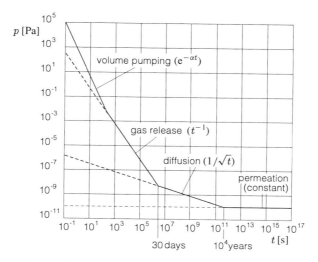

Fig. 11.2. Limiting steps in pumping an unbaked UHV system. (After *A User's Guide to Vacuum Technology*, O'Hanlon, © 1980, John Wiley & Sons, with permission.)

[i.e., heated in vacuum (10^{-4} Pa) to high temperature (700–900°C) for a few hours]. Most of the hydrogen and other gases are pumped out in this process.

Figure 11.2 shows an idealized pumping speed curve of an unbaked vacuum system with metallic seals. Initially, the pressure falls exponentially according to Eq. 11.10 until at $\sim 10^{-3}$ Pa desorption takes over, and the pressure falls more slowly. After a few days of pumping, the pressure is determined by diffusion, and the ultimate pressure will be reached in 10,000 years. It is, therefore, absolutely necessary to clean, bake, and fire a UHV system.

An HV or UHV system should not be vented with atmospheric air. UHV systems use artificial air, nitrogen, or argon; in HV systems, special filters which remove water vapor and small dust particles are recommended.

Connecting elements between the pump and the recipient — such as tubes, valves, bellows, and so on — cause pressure drop. In analogy with electric circuits the flow resistance, R, is introduced in

$$Q_{pV} = \frac{p_2 - p_1}{R} = G(p_2 - p_1), \tag{11.13}$$

where $G = 1/R$ is the flow conductivity, p_2 is the pressure at the recipient, and p_1 is the pressure at the pump. For two vacuum lines connected in series we have

$$R = R_1 + R_2, \tag{11.14}$$

and for two vacuum lines connected in parallel we obtain

$$G = G_1 + G_2. \tag{11.15}$$

According to the continuity equation we have

$$Q_{pV} = S_1 p_1 = S_2 p_2 = G(p_2 - p_1), \tag{11.16}$$

or

$$S_2 = \frac{S_1}{S_1/G + 1}. \tag{11.17}$$

Here S_2 is the throughput at the end of the vacuum line with flow conductivity G for pump speed S_1.

Although flow conductivity can be computed for some simple geometries (e.g., a cylindrical tube or a hole in a wall), it is not solely defined by geometry. It depends to a high degree on the pressure and, hence, the mean free path, λ, since the pressure influences the motion of the molecules inside the tube. If the mean free path is small compared to the tube diameter, the internal frictional forces dominate and thus G will be pressure-dependent. Such flow, called *viscous,* can be turbulent or laminar depending on the Reynolds number. In vacuum technology, mostly laminar flow is observed. A special case of laminar flow, the *Poiseuille flow,* in a cylindrical tube with a circular cross section and a parabolic velocity distribution has

$$G = \frac{r^4 \pi}{8\eta l} \frac{p_1 + p_2}{2}, \tag{11.18}$$

where η is the dynamic coefficient of viscosity (for air at STP $\eta = 1.81 \times 10^{-5}$ kg/ms), r and l are the tube radius and length ($l \gg r$), and p_1 and p_2 are the pressures at both ends. The viscous forces govern the flow when $\lambda/d < 0.01$.

At low pressure the mean free path becomes longer than the tube diameter ($\lambda/d \geq 1$), the friction against the tube walls dominates, and the flow is called *molecular.* A particle which hits the tube stops, and then it leaves in a direction which is independent of its arrival. Such a particle may not necessarily pass through the tube; it can hit the side and leave in the direction of its arrival.

For molecular flow the conductivity of a cylindric tube of circular cross section is

$$G = \frac{r^3 \pi}{l} \sqrt{\frac{8kT}{\pi M}}, \tag{11.19}$$

where M is the molecular mass. For air at 22°C, in a tube with diameter d and length l cm we have

$$G = 12.1 \frac{d^3}{l} \quad [\text{l/s}].$$

An aperture with the area A in a thin wall (e.g., a valve) has

$$G = A \sqrt{\frac{kT}{2\pi M}} \tag{11.20}$$

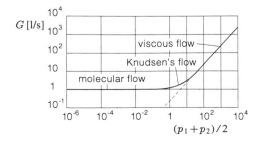

Fig. 11.3. Conductivity of a 1 m long tube with 1 cm radius (air at 22°C).

and for air at $22°C$ with A in cm^2 we have

$$G = 11.6A \quad [l/s]$$

The region between viscous and molecular flow is called *Knudsen flow* [227]. To estimate the flow type in a vacuum system the following rule of thumb is used: If $pd > 10$ it is laminar, while if $pd < 1$ it is molecular, where p is the pressure in pascals and d is the tube diameter in meters.

11.3. VACUUM PUMPS

No vacuum pump can cover the entire pressure region from atmospheric to UHV, so two or more pumps must be connected in series, and for UHV possibly also a few pumps in parallel. Vacuum pump performance is determined by the pumping speed, S (m^3/h or l/s), which for many pumps is constant over many decades of pressure. The quantitative measure, especially at low pressure, is, however, the suction power, Q_{pV} (Pa·m^3/h or mbar·l/s), the amount of gas per unit time transported from the intake to the exhaust at a given pressure.

The following factors are of interest: exhaust pressure, ultimate pressure, and transmission of undesirable vapors back into the vacuum system. Depending on the exhaust pressure the main categories of pumps are:

- Roughing pumps, which exhaust to atmospheric pressure, work alone or in combination with other pumps, and have a lowest obtainable pressure that is limited.
- High-vacuum pumps, which only pump effectively if exhausted to a volume at a pressure lower than atmospheric.
- Entrainment pumps, where the pumped gas is bound in electrodes, in a getter (e.g., chemically active surface), or on a low-temperature surface.

In HV and UHV vacuum systems a roughing pump is connected in series with one or more pumps of the other two types.

11.3.1. Roughing Pumps

Most roughing (also called backing) pumps are mechanical, while if only a small volume is pumped a sorption or a membrane pump is used. Sometimes a water jet pump is adequate, which at $10°C$ can give $\sim 10^3$ Pa. The most common roughing pump is the rotary vane pump.

11.3.1.1. Rotary Vane Pump

The rotary vane pump, seen in Fig. 11.4, consists of a casing, with a cylindrical working chamber, and an eccentrically mounted rotor, a point on the circumference of which always remains in contact with the casing. Two vanes glide in the rotor slits, pressed against the working chamber walls by springs, and sweep out a crescent-shaped volume. The rotor rotates, compresses the gas, and forces it out through the exhaust valve. The working chamber is immersed in oil, which seals the clearance between different moving parts of the pump, lubricates, transports the heat to the casing and fills the space below the exhaust valve.

A rotary vane pump has a high compression, $\sim 10^5$ at ~ 1 Pa. When pumping vapor, it compresses only to the vapor saturation pressure corresponding to pump working temperature. If water vapor is pumped at $70°C$, it can be compressed only to $\sim 2.5 \times 10^4$ Pa,

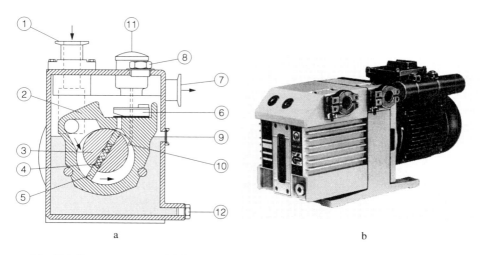

Fig. 11.4. Rotary vane pump. (**a**) Cross section. 1, Intake; 2, working chamber; 3, rotor; 4, spring; 5, vane; 6, exhaust valve; 7, exhaust; 8, oil filling; 9, window to control oil level; 10, gas ballast intake; 11, gas ballast valve; 12, oil drain plug. (**b**) A 4 l/s rotary vane pump (Leybold).

because as the compression continues the vapor condenses without changing the pressure. The exhaust valve does not open, so water vapor remains in the pump and emulsifies with the oil, detoriating the lubricating properties and decreasing the suction power. Water vapor condensation can be avoided with a gas ballast valve. Here, before starting the compression, a small amount of air is admitted to the compression space through a valve, so that the compression ratio is kept below 1:10. The pumped vapor is compressed together with the gas ballast air and exhausted.

Rotary vane pumps are either single- or two-stage units, where the single-stage unit exhausts at atmospheric pressure and its oil is in direct contact with the air. Some air is absorbed, partially released inside the working chamber, and expelled, until equilibrium is obtained. In a two-stage pump, the exhaust of the first stage is at a much lower pressure and the ultimate pressure is decreased. Single-stage and two-stage pumps have, respectively, an ultimate pressure of $\sim$1 Pa and $\sim$0.1–0.01 Pa. The pumping speed, S, varies between 0.25 and 500 l/s for commercial units. Different mineral-, hydrocarbon- or polyester-based oils are used, which have a rather high vapor pressure (10^{-3}–10^{-1} Pa). For a very clean vacuum a trap is installed to prevent the oil vapor from entering the vacuum system. For environmental security an oil mist filter or a tube conducting pumped gases outside the building should be used.

11.3.1.2. Root's Blower

Root's blower, seen in Fig. 11.5, is an intermediary between roughing and high-vacuum pumps. Pumping speed is largest between 5×10^2 and 10^{-2} Pa, where rotary vane pump speed decreases. A rotary vane pump and a Root's blower combination has a large pumping capacity from atmospheric pressure to less than 0.01 Pa.

In Root's blower, two rotors rotate synchronously at high velocity (1500–3000 turns/minute), without lubricating oil, inside a working chamber while maintaining a small distance ($\sim$ 0.1–0.2 mm) to each other and to the chamber. The backstreaming is negligible in the narrow slits because the flow resistance increases with decreasing pressure.

Fig. 11.5. Root's pump. (**a**) Cross section. 1, Overflow valve; 2, intake; 3, casing; 4, rotor; 5, exhaust. (**b**) A 500 l/s Root's pump (Leybold).

These pumps are constructed with an overflow valve, which connects the exhaust to the intake, and is loaded by a spring that opens when the pressure difference between the exhaust and intake exceeds a given limit. The valve returns a certain amount of gas to the vacuum side. In larger pumps the gas is cooled before entering back to the vacuum side. This helps to cool the pump.

The compression ratio is $\sim 1{:}4$ at atmospheric pressure and reaches $\sim 1{:}60$ at 1 Pa. The ultimate pressure is lowered by a similar factor as compared to a rotary vane pump. Commercial Root's blowers have pumping speeds ranging from 30 to 25,000 l/s.

11.3.1.3. Sorption Pump

The sorption pump is based on the absorption of gases in solids and adsorption on the surface, respectively. A gas, which is in thermodynamic equilibrium with the surface, is adsorbed by weak van der Waals electrostatic forces. The adsorption on a surface is strongly enhanced by cooling to the condensation temperature of the gas to be pumped, so the equilibrium gas pressure remains low, even if the surface has adsorbed a gas layer many molecules thick.

A stainless steel canister is filled with a zeolite (aluminum calcium silicate) molecular sieve, an extremely porous material ($800 \text{ m}^2/\text{cm}^3$). The pump size is limited to ~ 1 kg of the adsorbing material, enough to evacuate ~ 40 l, but for larger systems more than one pump can be used, resulting in faster pump-down, as seen in Fig. 11.6. With 500 g of zeolite a 15 l vacuum system can be pumped to 1 Pa in ~ 10 minutes.

The pump is immersed in liquid nitrogen, and to speed up the zeolite cooling there are radial copper fins inside the canister. When the ultimate pressure is attained, the pump is isolated from the vacuum system by a valve. The pump is then heated to free the adsorbed gas. However, to desorb the bound water it should be baked at $200–300°C$.

The important sorption pump properties are:

- Selectivity, pumping only gases with condensation temperatures higher than the temperature to which the pump is cooled down. When pumping air the residual atmosphere is $\sim 70\%$ neon.
- Limited capacity, so adsorbed gas layer thickness must not be large to obtain low ultimate pressure.

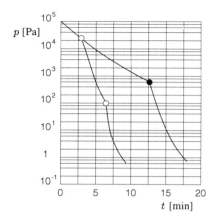

Fig. 11.6. Pressure as function of the pumping time with two (●) and three (○) sorption pumps (Leybold).

- Clean vacuum is obtained since the pump only contains solid materials with low vapor pressure, and thus is used in small UHV systems, where oil contamination must be avoided, in combination with ion or cryopumps.
- Water absorption is detrimental, as little as 2% by weight reduces the sorption capacity for all gases.

11.3.1.4. Diaphragm Pump

The diaphragm pump is a multistage pump normally used in combination with the turbomolecular pump. Each stage has moderate compression and the ultimate pressure is ∼ 200 Pa. In the small pump chambers, each with ∼ 0.2–0.4 l, a viton- or teflon-covered neoprene diaphragm moves back and forth, mounted on a holder outside the vacuum system. Two chambers connected in parallel on the vacuum side and two more connected in series result in pumping speeds of a few cubic meters per hour with a very clean vacuum. These pumps can pump chemically active gases, like halogen gases in sputtering and ion etching devices.

11.3.2. High-Vacuum Pumps

11.3.2.1. Diffusion Pump

The oil diffusion pump, seen in Fig. 11.7, is widely used and has no movable parts. Pump fluid is a boiler-heated vapor which flows up and out through a number of jet nozzles, after which it expands and its velocity distribution is changed from thermal to supersonic in the downwards expansion direction. The gas molecules' random motion causes them to be caught in the vapor stream, compressed in the lower part of the pump, and pumped out by a side duct where a roughing pump maintains low pressure. Modern pumps have several compression stages. The pump walls are water-cooled, so the vapor condenses and is returned as a thin film of liquid to the boiler. The pumping fluid is oil with very low vapor pressure.

With a 1 to 10 Pa rough vacuum an ultimate pressure less than 10^{-6} Pa can be obtained. In some pumps the oil is continuously cleaned by fractional distillation obtaining ultimate pressures of 10^{-9} Pa. Oil molecules backstreaming from the pump into the vacuum system can be avoided by a baffle (in UHV systems it is necessary), which is mounted directly above the intake flange. Baffles are cooled by water, freon, Peltier effect, or liquid nitrogen.

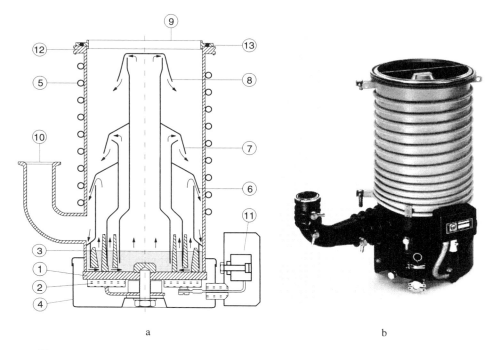

Fig. 11.7. Diffusion pump. (**a**) Cross section. 1, Boiler; 2, heating element; 3, oil level; 4, protection cage; 5, cooling water envelope; 6, lower nozzle; 7, central nozzle; 8, top nozzle; 9, intake; 10, exhaust; 11, electric connection; 12, flange; 13, sealing. (**b**) A 400 l/s diffusion pump (Leybold).

Diffusion pumps have a very large pumping speed depending on the flange diameter and the velocity of the gas which enters the pump, 100–1000 l/s, but some commercial units have 100,000 l/s with a 1.25 m flange diameter.

11.3.2.2. Turbomolecular Pump

Turbomolecular pumps produce a very clean vacuum (no oil or lubrication on the high-vacuum side) with an ultimate pressure of $\sim 10^{-8}$–10^{-9} Pa. The principle, already known at the beginning of the century, could because of mechanical tolerances not be realized in commercial hardware until the 1950s.

The pump has a bladed turbine (Fig. 11.8) with a stator-rotor blade distance of ~ 1 mm, producing a low compression in each stage. Thus many stages are required for large compression. These pumps are manufactured with both horizontal and vertical axis, and modern units have a second, molecular stage, a cylinder which rotates between two helicoidal grooves. The pump works in the molecular flow region, where the mean free path is larger than the distance between the stator and rotor blades. When the rotor blade peripheral velocity (up to 60,000 turns/minute) reaches the mean velocity of the pumped gas molecules the compression begins.

The high-vacuum side intake flange has the same diameter as the rotor blades for high pumping speed. The pumps can be baked at $\sim 150°$C or higher for pumps with ceramic ball bearings. The motor is driven by a power source with frequency 400–1500 Hz. Through a special port, dry air or nitrogen venting gas can be admitted when the rotation has decreased to $\sim 50\%$.

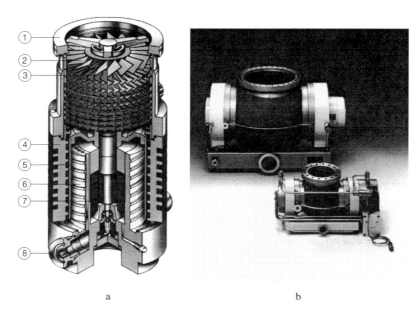

Fig. 11.8. Turbomolecular pump. **(a)** Cross section. 1, High-vacuum side; 2, rotor disk; 3, stator disk; 4, molecular stator 1; 5, molecular rotor; 6, molecular stator 2; 7, motor; 8, backing vacuum. **(b)** Two turbomolecular pumps with horizontal axis (Balzers-Pfeiffer).

Fig. 11.9. Rotor of a turbomolecular pump with horizontal axis (Balzers-Pfeiffer).

Pumping speed is large, in commercial units of 20–5000 l/s, and is practically constant between 0.1 and 10^{-8} Pa. The pump has a large compression ratio which depends on the kind of pumped gas, for N_2 up to $\sim 10^{12}$. Hydrogen and helium have a larger thermal velocity than do the heavier gases like oxygen, nitrogen, argon, and organic compounds. The hydrogen and helium partial pressures on the high-vacuum side are therefore higher, and at the ultimate pressure ($< 10^{-10}$ Pa) the residual gas atmosphere contains $> 95\%$ hydrogen (see Fig. 11.24).

11.3.2.3. Ion Pump

In its simplest form the ion pump has a cold cathode and an anode (Fig. 11.10), between which burns a Penning discharge. Because there is no pumping liquid or movable parts, it creates a very clean vacuum and the ultimate pressure is $\sim 10^{-9}$ Pa.

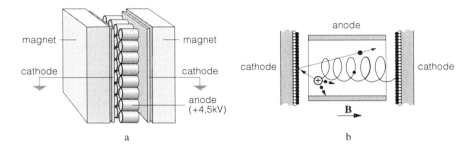

Fig. 11.10. (**a**) Ion pump.   (**b**) Principle.   ⊕, Ion; ●, titanium atom; ., electron.

The cathodes are two titanium plates at ground potential, between which is a hollow anode. The potential difference between the cathodes and the anode is 4–6 kV. A permanent magnet generates an ~ 0.1 T magnetic field in cathode–anode direction. The cold cathodes emit secondary electrons which are trapped in the potential well. The magnetic field causes the electrons to travel over long distances in spiral paths before reaching the anode, so the probability that an electron will ionize a gas molecule is very high. The ions, accelerated along the magnetic field lines, hit the cathode with a large kinetic energy. The cathode material, titanium, is sputtered toward the anode and the inner pump walls and makes there an active getter film. The ions penetrate into the cathode and remain there. An ion pump does not pump noble gases well. Argon, ~ 1% in air, does not react chemically and is only adsorbed on, or buried below, the surface. When a cluster of argon atoms is freed, the pressure periodically increases which is called the argon instability. Triode pumps (Fig. 11.11b), have hollow titanium cathodes which by sputtering cover the noble gas atoms with a thin layer of titanium on the outer walls.

An ion pump can be started below ~ 0.1 Pa, and the backing pump can be disconnected from the system by a valve or by a permanent seal as in microwave tubes. When an ion pump pumps a system, no separate vacuum-measuring instrument is necessary, because within wide limits the discharge current is proportional to pressure, the calibration depending on the system residual gas.

Pumping speed, ~ 2–500 l/s in commercial units, decreases at low pressure because S is proportional to the discharge current and, thus, to pressure. The maximum, around 10^{-3} to 10^{-5} Pa, is defined as the nominal pumping speed. Ion pumps, intended to be used in HV and UHV systems, can be baked at ~ 400°C.

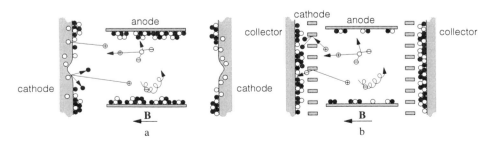

Fig. 11.11. Schematic presentation of (**a**) a diode and (**b**) a triode ion pump. ●, Titanium atom; ○, gas particle; ⊕, ion; ⊖, electron. (Leybold.)

Fig. 11.12. Ion pumps (Leybold).

11.3.2.4. Sublimation Pump

A sublimation pump is a UHV pump which pumps chemically active gases (H_2, O_2, CO, CO_2, H_2O, etc.) and therefore is especially useful in pumping, together with a turbomolecular pump, hydrogen. Titanium atoms evaporate from a directly heated Ti–Mo–alloy wire (Fig. 11.13a), and sublime on the cold (often LN_2) chamber walls where they make a thin, clean layer which binds active gases by adsorption, diffusion, and formation of chemical compounds. Pumped gases cannot be desorbed by heating, and the layer must be cyclically renewed to ensure continuous pumping. The wire evaporates at a rate of ~ 0.1 g/h, so a 1.2 g Ti wire can be used for ~ 12 hours; therefore a sublimation pump can only be used

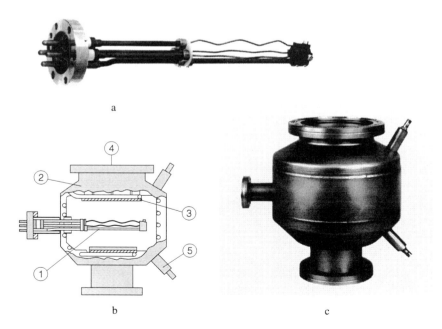

Fig. 11.13. (a) Titanium sublimator with flange. (b) Sublimation pump, cross section. (c) Sublimation pump housing. 1, Titanium sublimator; 2, chamber; 3, optical screen (prevents Ti-atoms to enter the UHV system); 4, intake; 5, connection for water cooling or liquid nitrogen (Balzers).

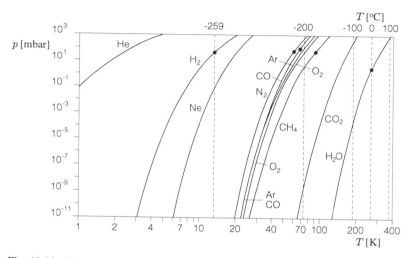

Fig. 11.14. Vapor pressure of common gases. •, Melting point. (Source: Balzers, Vacuum Components catalog.)

in a UHV system to complement other pumps. For example, the pump of Fig. 11.18c is activated at 10^{-8} Pa for ~ 3 minutes per day, so the wire lasts ~ 8 months.

The active gas pumping speed is 3–10 l/s cm^2 at room temperature, and 9–14 l/s cm^2 at the liquid nitrogen temperature. The pump can be made in any desired geometry, but must be constructed to allow baking at 400°C.

11.3.2.5. Cryogenic Pump

A cryogenic pump uses the fact that van der Waals forces keep gas molecules on a container surface cooled to a sufficiently low temperature. Cryogenic pumps, most of which use helium as working gas, are clean HV or UHV pumps and, unlike ion pumps, do retain no condensed gases after heating and must be vented periodically. Two helium cryogenic pump models have evolved; in the first a gas refrigerator reduces the temperaure to within a few degrees of liquid helium, and the second uses liquid helium directly.

In the first, called a *cryopump*, helium is externally compressed to ~ 2 MPa and transported through flexible, high-pressure hoses to a two-step expansion valve (Fig. 11.15). In the first the gas is cooled to ~ 70 K, while in the second it is cooled to ~ 12–20 K, and all gases except H$_2$, He, and Ne are pumped effectively at these low temperatures. The pumping surface, the condensor, is covered with charcoal or zeolite in which the above-mentioned gases are adsorbed and trapped, so an ultimate pressure of $<10^{-9}$ Pa can be obtained. Between the housing and condensor is a radiation screen connected to the first expansion step, which optically screens the condensor and in large units is cooled by liquid nitrogen. Contaminating lubricating oils in the compressor are removed in a separator and adsorber. The rest of the oils and contaminations are trapped at the warm end of the expansion valve and flushed by the returning helium. Cryopumps are built for pumping speeds from 400 l/s up to 100,000 l/s (in the latter for water vapor $\sim 350,000$ l/s, for argon $\sim 50,000$ l/s), which is almost constant from 1 Pa to the ultimate pressure.

Liquid helium pumps are used mainly in scientific laboratories which have helium liquefaction equipment. These pumps are mounted directly in the vacuum system. The stainless steel canister surface is covered with a thick silver layer or a bonded molecular sieve.

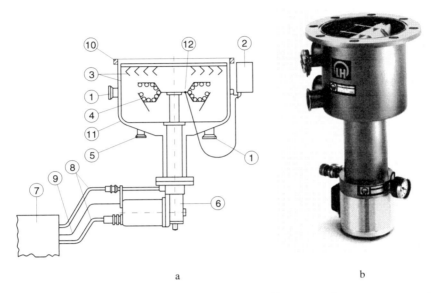

a   b

Fig. 11.15. Cryopump with refrigerator. (**a**) Sketch. 1, Backing vacuum; 2, temperature measuring instrument; 3, radiation screen (~ 70 K); 4, condensor with charcoal (~ 20 K); 5, vacuum gauge connection; 6, expansion valve; 7, compressor; 8, high-pressure hose; 9, electric connection; 10, intake; 11, pump housing; 12, temperature sensor (Balzers). (**b**) A 1500 l/s cryopump (Leybold).

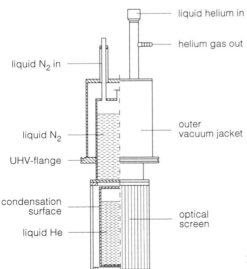

Fig. 11.16. Cryogenic pump with liquid helium (Leybold).

Helium evaporating slowly inside the container keeps the temperature at ~ 4.2 K and leaves the pump through an exhaust, to be liquefied anew. A level sensor controls the liquid helium inflow. The helium canister is enclosed in an optically tight system of blackened radial copper plates, cooled by liquid nitrogen as protection from heat radiation. Liquid helium pump pumps at a higher speed than do refrigerator pumps because of its lower, more stable temperature and has ultimate pressure of $< 10^{-11}$ Pa.

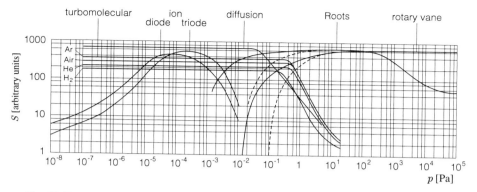

Fig. 11.17. Typical pumping speed: - - - - with, ——— without gas ballast. (Source: Balzers and Leybold catalogs.)

Table 11.2

Vacuum pumps

Pump type	Initial pressure (Pa)	Ultimate pressure (Pa)	Pumping speed (l/s)
Rotary vane pump	10^5	10^{-1}	0.25–500
Root's pump	10^3	10^{-2}	30–25,000
Sorption pump	10^5	10^{-1}	(100 l/1.5 kg)
Diffusion pump	10^1	10^{-6}	10–100,000
Turbomolecular pump	10^2	10^{-8}	4–10,000
Ion pump	10^0	10^{-9}	1–5000
Titanium sublimation pump	10^{-6}	10^{-10}	(3–10 l/s cm^2)
NEG pump	10^{-1}	10^{-10}	(1–3 l/s cm^2)
Cryopump	10^{-1}	10^{-10}	1000–100,000

The evaporated helium can be pumped by a backing pump to a pressure of $\sim 5 \times 10^3$ Pa with a temperature decrease to 2.3 K where all gases except helium are effectively pumped (Fig. 11.14). The helium consumption here is rather large.

11.3.2.6. Nonevaporable Getter Pump

NEG pump, also a UHV pump, works by surface adsorption and by gas diffusion through the material (zirconium alloyed with aluminum and a small amount of iron). At atmospheric pressure or when the pumping speed decreases, an NEG pump must be activated at $\sim 800°C$, or $\sim 450°C$, depending on the alloy type. During activation the pressure should be below 1 Pa and the diffused gases are released. Pumping speed and pumping capacity are largest for hydrogen, ~ 3 l/s cm^2, but even N_2, O_2, CO, and CO_2 are pumped well. The material in steel sheets can be piled up or folded for pumping speeds of >1000 l/s.

11.4. MEASUREMENT

Vacuum gauges cover the range from atmospheric pressure to the interplanetary vacuum, but only a few types are commonly used: those which measure backing vacuum to ~ 0.1 Pa and those used in the HV and UHV region. Common vacuum gauges record a change in a physical property of the residual gas which depends on the gas density. Heat conduction

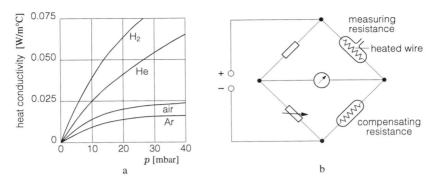

Fig. 11.18. (a) Heat conduction vs. pressure. **(b)** Pirani gauge.

can be used down to ~ 0.1 Pa, ionization at lower pressure. These manometers depend on the residual gas composition and, thus, *do not measure absolute pressure*.

11.4.1. Pirani Gauge

The heat transfer between two surfaces at different temperatures depends on the pressure and on the nature of the gas. Figure 11.18a shows this dependence for the Knudsen flow pressure range where the mean free path is comparable with the recipient size. At sufficiently low pressure the heat is transported mostly by radiation.

The gauge based on heat conduction is the Pirani gauge and is used at pressures of down to ~ 0.1 Pa from the atmosphere, where the heat transfer changes only slowly with pressure.

The Pirani gauge forms one arm of a Wheatstone bridge (Fig. 11.18b). The high-resistance temperature coefficient gauge wire is compared with a compensating resistance. Any pressure variation changes the wire temperature and, thus, its resistance, which causes the bridge to go out of balance. The unbalance can directly be used for pressure measurement, or a feedback circuit can change the voltage over the bridge until a balance is obtained, by keeping the wire temperature at $\sim 130^\circ$C. The voltage change is directly calibrated in pressure. The bridge must be zeroed at low pressure by adjusting a variable resistor in the other branch. A Pirani gauge, which can be used at pressures of down to $\sim 10^{-3}$ Pa, has its compensating resistance in an evacuated tube and works at constant temperature.

11.4.2. Piezoresistive Pressure Transducer

Piezoresistive pressure transducer measuring elements are two silicon plates, of ~ 10 mm^2, as seen in Fig. 11.19, with gap evacuated to high vacuum. One of the plates is an $\sim 25\ \mu$m thick membrane with four piezoresistors diffused by evaporation, making a bridge connected to a constant current source and balanced at low pressure. Pressure change

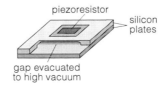

Fig. 11.19. Piezoresistive pressure transducer.

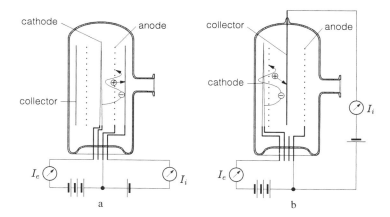

Fig. 11.20. (a) Ionization gauge. (b) Bayard–Alpert gauge.

results in an unbalance creating a signal proportional to pressure. The piezoresistive pressure transducer measures absolute pressure in a range between atmospheric and ~100 Pa.

11.4.3. Ionization Vacuum Gauge

Ionization vacuum gauge relies on the fact that ionization rate in a gas is proportional to the gas density, but depends also on the gas composition and electron energy. For an electron current emitted from a heated or cold cathode the ion current for pressures < 0.1 Pa (space-charge limit) is

$$I_j = CI_e p.$$

C depends on tube geometry and determines the gauge sensitivity.

Hot Cathode Ionization Gauge

The hot cathode ionization gauge is similar to a triode (Fig. 11.20a). It uses heater current regulation so that the filament emits a constant electron stream. The electrons are accelerated toward the cylindrical mesh of the tube anode, which is connected to a stabilized power supply. Most electrons miss the anode, and make a few oscillations around it, being trapped in the potential well. The positive ions, produced in electron–gas molecule collisions, reach the collector, which is negative with respect to cathode, and are conducted to an electrometer amplifier. The tube can measure pressure to ~ 10^{-5} Pa.

The electrons, which hit the anode with large kinetic energy, produce photons, which cause photoemission from the collector. It is not possible to separate this undesired photoemission current which flows *from*, and the ion current which flows *to*, the collector. The Bayard–Alpert gauge, seen in Fig. 11.20b, places the cathode outside the mesh anode, and the collector is a thin wire at the anode center. The small collector surface makes it less vulnerable to photons; therefore this gauge can measure pressure down to 10^{-9} Pa or, with two collectors, using voltage-modulation, down to 10^{-11} Pa.

Cold Cathode Ionization Gauge

The cold cathode gauge, which uses the Penning discharge, has a magnetic field that stretches the electron orbits (Fig. 11.21) and increases the ionization probability, resulting in enhanced sensitivity compared to the hot cathode gauge. The discharge takes place between grounded cathodes and a circular anode held at a potential of a few kilovolts and produces a current which is the sum of the electron and ion current.

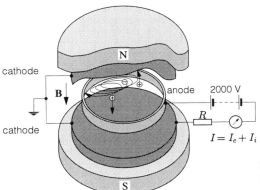

Fig. 11.21. Penning gauge. **B**, magnetic field
(~ 0.04 T); I_e, electron current; I_i, ion current;
I, total current.

The Penning gauge range is 1–10^{-4} Pa, limited at lower pressure where it has difficulty maintaining the discharge. However, auxiliary electrodes facilitate ignition and reduce photoemission problems, so modern magnetron-type Penning gauges can operate down to 10^{-11} Pa.

11.4.4. Residual Gas Analysis

The device uses residual gas ions from a source, separated by electric and magnetic fields. After separation the ion current hits the collector where it is a measure of the residual gas partial pressure, so by scanning over different masses the composition is determined.

Modern residual gas analyzers (Fig. 11.22) separate ions with an electric quadrupole, similar to that used in a mass spectrometer (Section 5.2.2) but with simpler construction, since the ions are analyzed directly and the sensitivity is lower. The ion current is measured by a Faraday cage or by a secondary electron multiplier in which the smallest partial pressure is $\sim 10^{-12}$ Pa. The ions make many periodic oscillations,

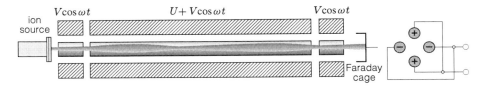

Fig. 11.22. Residual gas analyzer with electric quadrupole (Fisons Instruments).

Table 11.3

Vacuum gauges

Type	Pressure [Pa]														
	10^5	10^4	10^3	10^2	10^1	10^0	10^{-1}	10^{-2}	10^{-3}	10^{-4}	10^{-5}	10^{-6}	10^{-7}	10^{-8}	10^{-9}
Mechanical	—	—	—												
Compression		—	—	—	—	—	—								
Piezoresistive	—	—	—	—											
Pirani		—	—	—	—	—	—								
Penning						—	—	—	—	—	—				
Ionization			—	—	—	—	—	—	—	—	—	—	—		

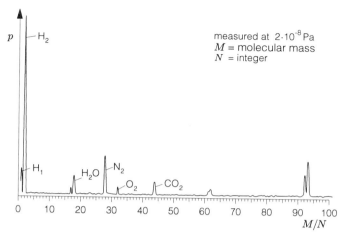

Fig. 11.23. Typical residual gas mass scan in a high-vacuum system (Balzers).

and only those with a mass selected by the applied voltages pass through the quadrupole. By simultaneously varying the voltages U and V, their mutual relationship is maintained, and a complete mass scan is obtained (Fig. 11.23).

11.5. LEAKS AND LEAK DETECTORS

Chambers, pumps, connection elements, valves, baffles, traps, filters, and gauges in a vacuum system all have surfaces covered by adsorbed gas molecules which deteriorate the vacuum, but also may have small holes or leaks which do the same. For example, a 10 liter vacuum chamber at 293 K and 0.001 Pa has $\sim 10^{18}$ molecules in a monomolecular surface layer, with only $\sim 10^{15}$ molecules in the volume. Desorption from the walls, which for a clean metal surface is $\sim 10^{-8}$–10^{-9} Pa l/s cm^2, slows down the pumping. For pressures below 0.1 Pa, desorption mainly determines the pumping time.

In an UHV system all surfaces must be clean, smooth, polished if possible, and baked at $\sim 400°C$, and in a closed vacuum system getter materials (e.g., Ba, Sr, Mg) which both absorb and adsorb many residual gases are indicated.

A leak of magnitude q_l is defined as air in Pa l/s which enters the system,

$$q_l = V\frac{dp}{dt}.$$

For an HV system q_l should be less than 10^{-7} Pa l/s, and for UHV we must have $q_l < 10^{-10}$ Pa l/s.

Since the size of the leak determines the ultimate pressure, leaks must be located and sealed. To distinguish between gas desorption and leaks, the chamber is disconnected from the pump, and the pressure increase with time is noted. If the pressure increases approximately linearly for many minutes a leak is indicated while if the derivative decreases with time the gas desorption is the probable cause.

Large leaks are detected by overpressure in the chamber, and painting the outer surface by a low surface tension liquid. Leaks of $>10^{-8}$ Pa l/s produce "bubbles." When overpressure is not possible, the evacuated system can be sprayed by a test gas (freon, CF_2Cl_2, carbon dioxide, or hydrogen), and any gauge response with spray position is noted.

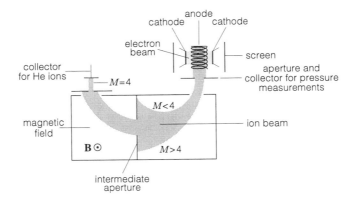

Fig. 11.24. Helium leak detector.

For smaller leaks, detectors with halogen-containing molecules, flammable gases or helium are used. The halogen detector uses a substance like freon which strikes a hot ($\sim 800°C$) platinum anode surface and suddenly increases the ionization. Halogen partial pressure to $\sim 10^{-6}$ Pa can be detected, which corresponds to $\sim 10^{-4}$ Pa l/s. A flammable gas detector uses an n-type semiconductor surface to adsorb oxygen, whose resistance, after equilibrium is reached, becomes constant. When a flammable gas, like hydrogen, carbon monoxide, or methane, hits the element, positive charges are produced in the semiconductor, momentarily decreasing its resistance. The detection limit is $\sim 10^{-2}$ Pa l/s.

Helium leak detector (Fig. 11.24), works like a mass spectrometer (Section 5.2.1.4) tuned to detect only helium ions. Its sensitivity, which results from helium easily penetrating very small leaks, is $q_l < 10^{-15}$ Pa l/s.

11.6. VACUUM COMPONENTS

11.6.1. Filters, Baffles and Traps

Filters, baffles, and traps are inserted to prevent backstreaming of oil vapors into the chamber. In a baffle or trap the oil partial pressure falls to a value which corresponds to the temperature of the device, and thus they are often cooled with liquid nitrogen (LN_2).

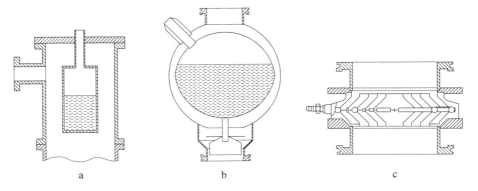

Fig. 11.25. (a) Simple trap with LN_2. (b) Optically tight trap with LN_2. (c) Cooled Chevron baffle.

The simplest trap consists of two stainless steel coaxial cylinders and is placed between the pump and the chamber. The inner cylinder is filled with LN_2 (Figs. 11.25a and 11.25b), where organic compounds condense on the cold walls but evaporate if the temperature is allowed to rise. A Chevron baffle (Fig. 11.25c) has optically tight capture plates cooled by water, freon, or LN_2 on which vapor particles condense. Plates are often coated with Teflon to prevent capillary creeping along the surface.

Oil filters, used between diffusion and backing pumps to prevent harmful additives in the diffusion pump oil from contacting rotary vane pump oil, contain charcoal and zeolite or Fuller's earth. Similar filters on the rotary vane pump exhaust prevent air pollution. Filters must be regularly regenerated by heating or by changing the active material.

11.6.2. Construction Components and Valves

Pumps, chambers, and gauges are joined by flanges, tubes, bellows, and valves, all of which must be vacuum tight, made with low partial pressure materials, and have clean, smooth surfaces. Each is tightened with seals made of elastic and deformable material, whose choice is governed by partial pressure and permeation.

In medium- to high-vacuum systems the O-ring is the common seal, which has elastic properties and fills small irregularities. Systems with neoprene O-rings cannot be heated above $90°C$, with viton above $150°C$. O-rings are used in systems whose ultimate pressure is $> 10^{-5}$ Pa, since their permeation is $\sim 10^{-6}$ Pa l/s. They are placed in rectangular or trapezoidal cross-section grooves (Figs. 11.26a and 11.26b) while the construction of Fig. 11.26c reduces the leakage; the space between the O-rings is evacuated by the roughing pump.

Detachable connections use standard elements like the PNEUROP system, seen in Fig. 11.27a. Tubes of different lengths, T-pieces, cross-pieces, elbows, transitions, reducing elements, and gauges can be obtained. Larger flanges can be connected with claw grips around the O-ring seal periphery (Fig. 11.27b) (ISO-K flanges in the PNEUROP system). O-rings are smeared with (apiezon or silicon) vacuum grease, which has a low partial vapor pressure. Glass seals, a pair of conically ground tubes, sealed with vacuum grease are sometimes used. Vacuum hoses are connected by nipples.

O-ring permeation is large because of voids between the polymer chains; and thus for UHV systems aluminum, indium, or gold ring seals (Fig. 11.28), or standardized flanges ("conflat," CF flanges) with flat copper rings (Fig. 11.29), are used. Sealing results from plastic deformation when the flanges are screwed together. O-rings (Helicoflex) with aluminum, stainless steel, or gold envelope and special spring alloy retain elasticity to $500°C$.

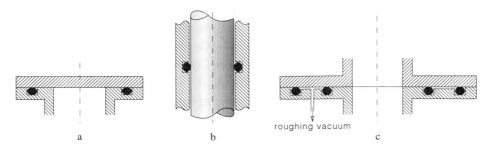

roughing vacuum

a   b   c

Fig. 11.26. Connections with O-rings. **(a)** Axial seal. **(b)** Radial seal. **(c)** Double seal with pumping.

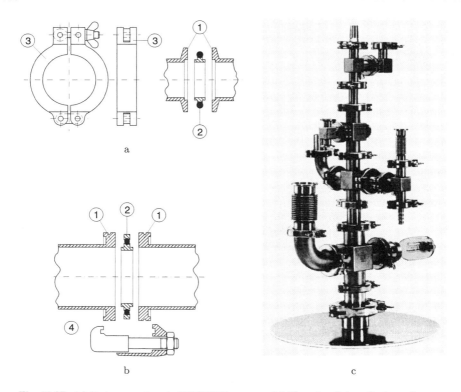

Fig. 11.27. **(a)** Fast connections in PNEUROP system. **(b)** Clamping fittings for large flanges. **(c)** Linked-up PNEUROP components (Leybold). 1, Flange; 2, seal with centering ring (with O-ring); 3, clamping ring; 4, claw grip.

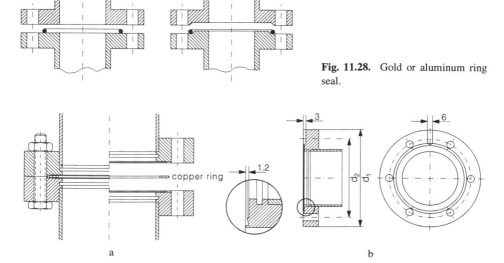

Fig. 11.28. Gold or aluminum ring seal.

Fig. 11.29. CF-flange ("conflat"-flange). **(a)** Two flanges with copper seal. **(b)** Construction drawing.

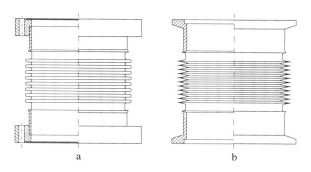

Fig. 11.30. (**a**) Folded tubes bellow, (**b**) membrane bellow.

Bellows (Fig. 11.30) are either (a) welded folded stainless steel tubes which allow moderate motion in axial and radial direction or (b) thin S-formed stainless steel disks which allow moderate radial and large axial motion.

Glass or plexiglass sight glasses can be used in HV systems, but UHV systems require Pyrex, quartz, or sapphire because of baking. Feedthroughs transfer liquid and electric current through the chamber wall, and they allow rotation or linear motion to be brought into a vacuum system. Seals are made with O-rings in HV, with bellows in UHV systems.

Valves, which allow parts of a vacuum system to be isolated, are attached to all pumps, traps, and baffles to avoid contamination when the system is vented. Most HV and UHV valves only allow air pressure on one side and should be opened only when similar pressures exist on both sides of the seal. A pump valve should have a diameter of pump intake. The HV valve (Fig. 11.31a) is sealed with O-rings, and the UHV valve (Fig. 11.31b) is sealed with metal seals. A gas dosing valve (Fig. 11.31c) controls gas inlet.

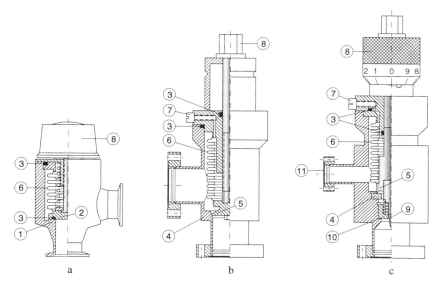

Fig. 11.31. (**a**) HV valve. (**b**) UHV valve. (**c**) Needle valve. 1, Seat; 2, plate; 3, seal with O-ring; 4, seat(stainless steel, gold plated); 5, plate (hard copper alloy or sapphire); 6, bellow for opening/closing mechanism; 7, leak detector connection; 8, opening/closing mechanism; 9, needle; 10, seat; 11, gas inlet (Balzers).

11.7. VACUUM MATERIALS

Vacuum materials must satisfy sometimes conflicting conditions, like strength (pressure difference is 10^5 N/m^2) and low vapor pressure, and should easily be machined, joined, polished, and cleaned. In UHV systems, baking and low outgassing rate are of paramount importance.

11.7.1. Metals

Metals are used for most chamber constructions and components since they must have a low vapor pressure at high temperature, be easily machined and joined to each other as well as to ceramic and glass, and be nonporous and nonpermeable for gases. Although partial vapor pressure does not set the limit for the use of most metals in vacuum systems, zinc (and thus brass), lead, cadmium, selenium, and sulfur should be avoided. Sulfur also limits the use of certain stainless steels in UHV.

The material choice shrinks with decreasing pressure required. In UHV, only some stainless steels (e.g., L304 and LN316), aluminum, and copper can be used, and for electrodes and cathodes we can use tungsten, molybdenum, tantalum, and titanium.

Table 11.4

Metals in vacuum technology

Metal	Density (g/cm^3)	Fusing point (°C)	Application
Stainless steel	7.86	1530	Chambers, flanges, valves, pumps
Fe-Co-Ni, Fe-Ni alloys	8.3	1450	Melt-in material for glass and ceramic
W	19.3	3407	Cathodes, anodes, arc discharge electrodes, filaments, parts to be used at high working temperature
Mo	10.3	2617	Cathodes, anodes, parts to be used at high working temperature
Ta	16.8	3014	Anodes, grids, cathodes, parts to be used at high working temperature
Ti	4.52	1670	Anodes, grids, getter material
Pt	21.4	1772	Melt-in material for glass, grids
Au	19.3	1064	Brazing material, gold plating
Ag	10.5	960	Brazing material, silver plating of copper electrodes and stainless steel details in UHV
Ni	8.9	1453	Anodes, grids, screens, support for oxide cathodes
Cu	8.9	1084	Anodes, cooling, collectors, seals, conductors, melt-in material for glass
Al	2.7	660	Chambers, deflection plates, electrodes, electron windows
Be	1.85	1280	Windows for X-ray radiation or particles
Zr	13.1	1852	Getter material

Even these materials require "firing" — that is, heating in high vacuum ($\sim 10^{-4}$ Pa) to 700–950°C a few hours to drive out dissolved gas. Some of the most commonly used metals and their applications are given in Table 11.4.

11.7.2. Glass and Ceramic

Glass behaves as a supercooled liquid; that is it solidifies without crystallizing, which is a consequence of the viscosity increasing quickly with decreasing temperature. After heating, the glass must be annealed. Glasses used in vacuum technology are silicon oxides with the addition of sodium, calcium, or lead oxides (soft glasses), boric oxide (borosilicate glass), or fused silica (quartz glass). Glass has high compression strength, but it is brittle, sensitive to blows, and can shatter if unequally heated. It tolerates most chemicals and heat up to ~ 400°C. It is a good electric isolator, is transparent to light, and can easily be joined with suitable metals, but it is important to choose the right glass and the thermal expansion coefficients must match to within $\sim 10^{-6}/$°C.

Ceramic is a sintered material, and the most widely used are aluminum oxide (Al_2O_3), beryllium oxide (BeO), and boron nitride (BN). Ceramics have high compression and tensile strength, are chemically resistant, are good electric isolators, are vacuum tight even without glazing, have low vapor pressures, and can easily be joined to metals. Aluminum oxide ceramic can only be ground but recrystallized mica is machinable. Ceramic is used for electric feedthroughs, for isolators between the electrodes, and as vacuum chambers in accelerators.

11.7.3. Organic Materials

Organic materials used in vacuum are elastomers (like rubber, neoprene, viton, and teflon) and thermoplastics (like PVC and polyeten). Elastomers are used in O-rings and other seals and as electric isolators and feedthroughs because of the high dielectric breakdown strength. But because of the large desorption and permeation they are limited to medium and HV systems. It is enough to have one O-ring to limit the ultimate pressure to $>10^{-6}$ Pa. Plastic gloves must be avoided when touching UHV components.

Vacuum grease is used for sealing conical connections and hoses, as well as to smear the O-rings. For temporary leak seals and for joining elastomers with other materials, epoxy resins like Araldit can be employed, but not in UHV because of their high vapor pressure.

Carbon- and silicon-based oils are used for lubrication and as a propellant in diffusion pumps but not in UHV systems where modern turbomolecular pumps with unlubricated magnetic bearings can be used. Oils for diffusion pumps must have a low vapor pressure at room temperature and must be resistant to oxygen degradation at working temperature.

11.8. JOINING TECHNIQUE

Rigid joints are created by gluing, soldering, brazing, welding, melting, or mechanical pressuring. Materials with *matched* thermal dilatation coefficients, heat conductivity, and mechanic strength should be used. Gluing by epoxy resin can be used only in connection with medium vacuum and only exceptionally in HV, when materials which cannot be joined in another way (like plastics or rubber) must be joined. Special glue, which evaporates when heated, can be applied for temporary joints during assembly.

Table 11.5

Brazing alloys

Alloy (composition in percent)						Fusing temperature in °C
Ag	Au	Cu	Ni	In	Pd	
60		27		13		720
72		28				840
58		32			10	840
	80	20				910
	82		18			950
	35	65				1020
	100					1083

Metals and metallized surfaces can be joined by soldering or brazing when the materials to be joined have a higher fusing temperature than the soldering or brazing alloy. During the joining a new alloy is formed as alloy metals and the materials to be joined diffuse into each other, with the resulting alloy solidifying more slowly than pure metals. Soldering below $\sim 600°$C should not be used in vacuum technology because the reduction agent results in a "dirty" joint. Brazing is carried out either in vacuum furnaces or in furnaces with reducing hydrogen atmosphere at temperatures between 700°C and 1100°C. Of the many alloys with different fusing temperatures and components, some are specified in Table 11.5.

When joining metal to ceramic, the latter must first be metallized to obtain a strong joint. The metallization compound must react chemically with ceramic, so a powder mixture of suitable metals (e.g., molybdenum and manganese) is applied to the ceramic and sintered at high temperature after which suitable metals can be brazed. Ceramic can be metallized also by reactive ion plating.

Metal joints, most frequently stainless steel, are made by plasma, electron-beam, or laser welding to guarantee a clean welding seam. In plasma welding a protecting gas (argon or helium) is employed. Electron-beam welding gives a superior seam, but at high cost. For smaller details, laser welding is the most convenient method. For UHV it is important to use appropriate electrodes when welding stainless steel because "firing" is normally made at 950°C and a wrong choice of electrodes increases the granularity (i.e., carbon inclusions), which weakens the seam and increases the magnetic permeability.

Glass and metals are fused by melting at $\sim 800°$C, depending on the type of glass and metal. The molten glass penetrates into the small irregularities at the metal surface while metal oxides partially resolve in glass and a continuous transition between metal and glass is formed. Cooling the joint slowly results in an almost strain-free transition if the metal and glass have similar thermal expansion coefficients.

Pure copper or copper–ceramic joints can also be made with high mechanic pressure where copper becomes liquid and UHV tight connections can be obtained. For example, a copper evacuation tube for modern metal–ceramic electron tubes is sealed by pressing together tools with circular punches until the tube is cut.

Appendices, Symbols Used in Text, List of Tables, Common Abbreviations, Physical Constants, Bibliography, Name Index, Subject Index

Appendix A

Energy Distribution of Fermions

Observe a three-dimensional potential box (Fig. A.1). From de Broglie's wavelength

$$\lambda = \frac{h}{p}$$

and with quantum numbers n_1, n_2, n_3, the momentum components are

$$p_x = \frac{n_1 h}{2a}, \qquad p_y = \frac{n_2 h}{2b}, \qquad p_z = \frac{n_3 h}{2c}.$$

Schrödinger's equation allows real solutions if $n_i \lambda / 2 =$ length of the box side. The energy of the particle is

$$E = \frac{1}{2m}(p_x^2 + p_y^2 + p_z^2) = \frac{h^2}{8m}\left(\frac{n_1^2}{a^2} + \frac{n_2^2}{b^2} + \frac{n_3^2}{c^2}\right).$$

Assume that the box is a cube with $a = b = c$. Then

$$E = \frac{h^2}{8ma^2}(n_1^2 + n_2^2 + n_3^2),$$

for example, $E_{211} = E_{121} = E_{112} = 6h^2/(8ma^2)$. All three quantum numbers must be positive:

$$n_1 > 0, \quad n_2 > 0, \quad n_3 > 0.$$

Put

$$n_1^2 + n_2^2 + n_3^2 = \rho^2.$$

The number of states between 0 and E, with $n > 0$ in a sphere octant, as seen in Fig. A.2, is

$$S(E) = \frac{1}{8}\frac{4\pi}{3}\rho^3, \qquad \rho = \frac{2a}{h}\sqrt{2mE}.$$

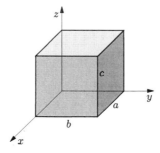

Fig. A.1. Three-dimensional potential box

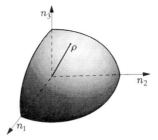

Fig. A.2. Three-dimensional momentum space.

Therefore

$$S(E) = \frac{1}{8}\frac{4\pi}{3}\frac{8a^3}{h^3}\sqrt{(2mE)^3} = \frac{4\pi V}{3h^3}\sqrt{(2mE)^3},$$

with the box volume $V = a^3$.

The electrons have two spin directions, so the number of states in the energy interval dE is

$$dS(E) = \frac{V}{h^3}4\pi\sqrt{(2m)^3}\sqrt{E}dE.$$

Appendix B

Liouville's Theorem

Under the influence of conservative forces the volume of a phase space element does not change [96].

Six-dimensional phase space coordinates are: x, y, z, p_x, p_y, p_z.
Consider only two dimensions: x and $p_x = p$. Particle motion is described by Hamilton's function:

$$H = E_k + E_p,$$

where E_k is its kinetic and E_p its potential energy. The potential energy does not depend on the momentum, p, and the kinetic energy on the coordinate, x, so the equations of motion are

$$\dot{p} = -\frac{\partial H}{\partial x}, \qquad\qquad \dot{x} = \frac{\partial H}{\partial p}.$$

Under the influence of conservative forces the sum of the kinetic and potential energy is constant, and

$$\frac{dH}{dt} = \frac{\partial H}{\partial x}\dot{x} + \frac{\partial H}{\partial p}\dot{p} = \frac{\partial H}{\partial x}\frac{\partial H}{\partial p} - \frac{\partial H}{\partial p}\frac{\partial H}{\partial x} = 0$$

A volume element is

$$\Delta V = (x_2 - x_1)(p_2 - p_1),$$

as seen in Fig. B.1. After time dt the coordinates x_1, x_2, p_1 and p_2 have changed to $x_1 + \dot{x}_1 dt, \cdots p_2 + \dot{p}_2 dt$ and the volume of the displaced element is

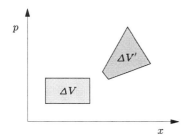

Fig. B.1. A volume element change in two-dimensional phase space.

$$\Delta V' = (x_2 + \dot{x}_2 dt - x_1 - \dot{x}_1 dt)(p_2 + \dot{p}_2 dt - p_1 - \dot{p}_1 dt)$$
$$= (x_2 - x_1)(p_2 - p_1) + (p_2 - p_1)(\dot{x}_2 - \dot{x}_1)dt + (x_2 - x_1)(\dot{p}_2 - \dot{p}_1)dt + \cdots$$

Neglecting the quadratic terms in dt and replacing the differences with differentials, we obtain

$$\Delta V' = dx\,dp + dp\frac{\partial \dot{x}}{\partial x}dx\,dt + dx\frac{\partial \dot{p}}{\partial p}dp\,dt = dx\,dp\left\{1 + \left(\frac{\partial \dot{x}}{\partial x} + \frac{\partial \dot{p}}{\partial p}\right)dt\right\}.$$

But

$$\frac{\partial \dot{x}}{\partial x} + \frac{\partial \dot{p}}{\partial p} = \frac{\partial}{\partial x}\frac{\partial H}{\partial p} - \frac{\partial}{\partial p}\frac{\partial H}{\partial x} = 0$$

and

$$\Delta V' = \Delta V.$$

In six dimensions we have

$$\Delta V = dx\,dy\,dz\,dp_x\,dp_y\,dp_z = \text{invariant}.$$

Appendix C

Beam Loading Admittance

An electron beam from an electron gun with voltage V_0 and beam current I_0 passes through a gridded cavity, and the distance between the grids is d. The cavity is connected to a microwave source, and the voltage across the gap is $V_1 \ll V_0$. The electron beam is velocity-modulated in the gap (Eq. 3.35):

$$v_z = \frac{dz}{dt} = v_0 - \frac{eV_1}{\omega md}(\cos \omega t - \cos \omega t_1) = v_0 - \frac{v_0 V_1}{2\omega V_0 T_0}(\cos \omega t - \cos \omega t_1), \quad \text{(C.1)}$$

where $v_0 = \sqrt{2\eta V_0}$, $T_0 = d/v_0$ and t_1 is the time when the electron passes the first grid. Integrating Eq. C.1 we obtain

$$z = v_0(t - t_1) - \frac{v_0 V_1}{2\omega^2 V_0 T_0}[\sin \omega t - \sin \omega t_1 - \omega(t - t_1)\cos \omega t_1]. \quad \text{(C.2)}$$

The electron passes the second grid, $z = d$, at t_2, and the transit time is $\tau = t_2 - t_1$. Note that $\tau \neq T_0$ because of the modulation, but is neglected in Chapter 7.

Then

$$d = v_0\tau - \frac{v_0 V_1}{2\omega^2 V_0 T_0}(\sin \omega t_2 - \sin \omega t_1 - \omega\tau \cos \omega t_1).$$

Division with v_0 gives

$$\tau = T_0 + \frac{V_1}{2\omega^2 V_0 T_0}(\sin \omega t_2 - \sin \omega t_1 - \omega\tau \cos \omega t_1). \tag{C.3}$$

The second term on the right side is small, so t_1 can be approximated by $t_2 - T_0$ and $\tau \approx T_0$. Transit time τ is then

$$\tau = T_0 + \frac{V_1}{2\omega^2 V_0 T_0}[\sin \omega t_2 - \sin \omega(t_2 - T_0) - \omega T_0 \cos \omega(t_2 - T_0)]. \tag{C.4}$$

The current in the cavity at t is the sum of all electron charges between the grids. Each electron induces in cavity walls the current

$$i = \frac{e}{d}\frac{dz}{dt},$$

so during time dt_1 charge $I_0 dt_1$ passes through the first grid. At time t the induced current is

$$i = \frac{I_0 dt_1}{d}\frac{dz}{dt},$$

where dz/dt is the electron velocity computed at t. The total current is

$$i(t) = \int_{t_2-\tau}^{t_2} \frac{I_0}{d}\frac{dz}{dt}dt_1. \tag{C.5}$$

Substituting dz/dt from Eq. C.1 we obtain

$$i(t_2) = I_0\frac{\tau}{T_0} - \frac{I_0 V_1}{2(\omega T_0)^2 V_0}[\omega\tau \cos \omega t_2 - \sin \omega t_2 + \sin \omega(t_2 - \tau)]. \tag{C.6}$$

In the first term on the right side, τ from Eq. C.4 is entered, and in the second term the approximation $t_1 = t_2 - T_0$ and $\tau = T_0$ can be used because the second term is small, giving

$$i(t_2) = I_0 + \frac{I_0 V_1}{2V_0}\left[\frac{2(1 - \cos \omega T_0) - \omega T_0 \sin \omega T_0}{(\omega T_0)^2}\sin \omega T_0\right.$$

$$\left. + \frac{2\sin \omega T_0 - \omega T_0(1 + \cos \omega T_0)}{(\omega T_0)^2}\cos \omega T_0\right]. \tag{C.7}$$

The hf current at gap exit is then

$$i(t_2) = g_0 V_0 + g V_1 \sin \omega T_0 + b V_1 \cos \omega T_0,$$

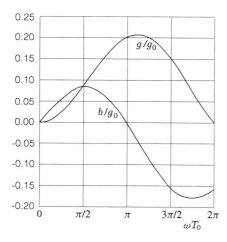

Fig. C.1. Normalized high-frequency beam conductance, g/g_0, and beam susceptance, b/g_0, vs. transit angle ωT_0.

where

$$g_0 = \frac{I_0}{V_0},$$

$$g = \frac{g_0}{2} \frac{2(1 - \cos \omega T_0) - \omega T_0 \sin \omega T_0}{(\omega T_0)^2}, \quad \text{and} \quad (C.8)$$

$$b = \frac{g_0}{2} \frac{2 \sin \omega T_0 - \omega T_0(1 + \cos \omega T_0)}{(\omega T_0)^2}.$$

g and b vary as sine functions as seen in Eq. C.7: g varies in phase with the modulating voltage, and b is phase-shifted 90°. $Y = g + jb$ is called the *beam loading admittance* or *beam loading* and it looks like a shunt over the gap.

These results are valid for small signal amplitude and a gridded gap. At large signal amplitude higher harmonic components must be included, and the expression for the beam loading admittance takes the form of integrals, which only can be solved numerically. In cavities without grids, numerical results will differ, but the general conclusions remain. Because the transit angle must be smaller than $\pi/2$, the beam loading fundamental frequency component is always capacitive.

Appendix D

Slow-Wave Structures

A strong interaction between an electron beam and electromagnetic wave can be achieved only if the beam velocity and the wave phase velocity are almost equal. In waveguides the electromagnetic wave phase velocity is greater than the light velocity, while in microwave tubes the electrons are accelerated only to a fraction of light velocity. Slow-wave structures are circuits where the phase velocity can be reduced [228].

The simplest slow-wave structure is the helix (Fig. D.1). The electromagnetic wave phase velocity in the TEM mode for a straight wire above a conductive plane is equal to the light velocity, $v_f = c$. If the wire is formed as a helix and the conductive plane as a cylinder around the helix, the approximation

$$v_f = c \frac{p}{\sqrt{p^2 + (\pi d)^2}} = c \sin \psi < c \quad (D.1)$$

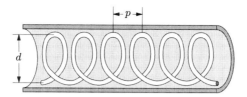

Fig. D.1. Helix as a slow-wave structure.

is adequate, where p is the pitch of the helix, d its diameter, and ψ the pitch angle. In a general case the helix can be filled with dielectric material,

$$v_f = c \frac{p}{\sqrt{\epsilon_r [p^2 + (\pi d)^2]}};$$

ϵ_r must not be large, because the losses increase and the efficiency of the structure is impaired.

In a rectangular waveguide the distribution of the electromagnetic field is determined by the solution of the wave equation, for example, in the TE mode the solution is

$$H_z = A \cos \frac{m\pi x}{a} \cos \frac{n\pi y}{b} e^{j(\omega t - \beta z)}, \qquad (D.2)$$

with

$$\beta^2 + \left(\frac{m\pi}{a}\right)^2 + \left(\frac{n\pi}{b}\right)^2 = k^2 = \frac{\omega^2}{c^2}. \qquad (D.3)$$

The propagation constant, β, is a function of frequency,

$$\omega = f(\beta), \qquad (D.4)$$

and the phase velocity is

$$v_f = \frac{\omega}{\beta}. \qquad (D.5)$$

For a rectangular waveguide in the TE mode, Eq. D.4 can be visualized in the form of a *Brillouin diagram,* shown in Fig. D.2.

Note that all solid curves represent different propagation modes in the waveguide, and all have a phase velocity greater than the light velocity, approaching the asymptote $\omega = \beta c$ near the waveguide cutoff frequency.

The slow-wave structures are, by nature, periodic; translation for a period does not change the appearance of the structure. For the helix the period, L, is equal to pitch. For all waves transported in media with periodic boundary conditions, Floquet's theorem [229] is valid:

> *The stationary solution for a propagation mode in a structure with periodic bound-ary conditions has the property that the field in two neighboring cells differs only by a complex constant.*

Mathematically, Floquet's theorem can be expressed — for example, for the electric field — by

$$\mathbf{E}(x, y, z - L) = \Gamma \mathbf{E}(x, y, z), \qquad (D.6)$$

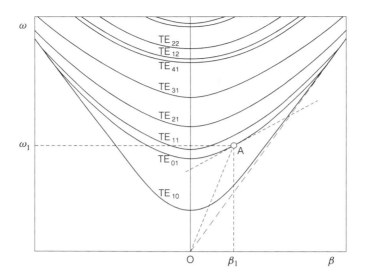

Fig. D.2. Brillouin diagram, $\omega = f(\beta)$, for TE modes in a rectangular waveguide. All curves have a common asymptote, with the slope being given by the light velocity. In point A the slope of the line OA gives the phase velocity, the slope of the curve tangent in A the group velocity, $v_g = \partial\omega/\partial\beta$ (see Appendix F). (After *Principles of Electron Tubes*, J. W. Gewartowski and A. H. Watson, © 1965, Van Nostrand, with permission.)

where

$$\Gamma = e^{j\beta_0 L},$$

and β_0 is the phase constant. In a periodic structure **E** is a periodic function of the axial coordinate z with the period L, so

$$\mathbf{E}(x, y, z) = \mathbf{E}_p(x, y, z) e^{-j\beta_0 z}, \tag{D.7}$$

and $\mathbf{E}_p$ can be expressed by a Fourier series

$$\mathbf{E}_p(x, y, z) = \sum_{-\infty}^{\infty} E_n(x, y) e^{j(2\pi n/L)z}, \tag{D.8}$$

where

$$E_n(x, y) = \frac{1}{L} \int_0^L \mathbf{E}(x, y, z) e^{-j(2\pi n/L)z} dz.$$

The electric field in the slow-wave structure can thus be described by

$$\mathbf{E}(x, y, z) = \sum_{-\infty}^{\infty} E_n(x, y) e^{-j(2\pi n/L)z} e^{-j\beta_0 z} = \sum_{-\infty}^{\infty} E_n(x, y) e^{-j\beta_n z}, \tag{D.9}$$

where

$$\beta_n = \beta_0 + \frac{2\pi n}{L} \tag{D.10}$$

defines the phase constant for the nth mode, $n = -\infty, \cdots, -1, 0, 1, \cdots, \infty$. Eq. D.9 satisfies both the wave equation and Floquet's theorem.

The functions $E_n(x, y)\,e^{-j\beta_n z}$ are called *space harmonic waves,* similarly to the Fourier series in the time domain. Every space harmonic wave, for every n, is the solution of the wave equation, so their sum must satisfy the wave equation. However, every space harmonic wave alone does not satisfy the boundary conditions; only the sum of all waves does. It is therefore impossible that the electromagnetic wave transports power along the structure by only one space harmonic wave, because all must be present. The consequence is that an "observer" (e.g., an electron) never can be in phase with the total field; when it moves along the structure, it can be in phase only with one of the space harmonic waves.

Space harmonic waves in a structure have the same frequency but different phase velocities

$$v_{fn} = \frac{\omega}{\beta_n} = \frac{\omega}{\beta_0 + \frac{2\pi n}{L}}. \tag{D.11}$$

The phase velocity decreases when $|n|$ or β_0 increases, and in a slow-wave structure it is possible to achieve phase velocities lower than the light velocity. A strong interaction between the electrons and the electromagnetic wave in the structure is possible, if the phase velocity of *one space harmonic wave* is adjusted so that it is almost equal to the electron velocity. Electrons will feel the phase of the other waves change; and if they move in the field long enough, the mean value of the field will correspond only to the space harmonic component, with which the electron is synchronous, and the contribution from all other components can be neglected.

Solution of the wave equation for a slow-wave structure allows calculation of $\omega = \mathrm{f}(\beta_n)$ and of v_{fn}. Figure D.4 shows a Brillouin diagram for a slow-wave structure made of a rectangular waveguide with thin metal plates, according to Fig. D.3.

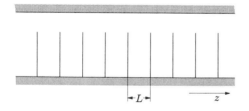

Fig. D.3. Slow-wave structure made of a rectangular waveguide with thin metal plates. (After *Principles of Electron Tubes,* J. W. Gewartowski and A. H. Watson, © 1965, Van Nostrand, with permission.)

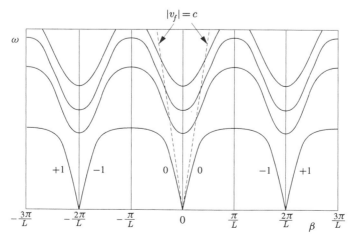

Fig. D.4. Brillouin diagram for the structure in Fig. D.3. (After *Principles of Electron Tubes,* J. W. Gewartowski and A. H. Watson, © 1965, Van Nostrand, with permission.)

Infinitely many propagation modes exist, one for each value of n. From Eq. D.10 it can be seen that a complete ω–β space harmonic wave diagram can be obtained by a parallel translation of the $n = 0$ branch by $2\pi n/L$ along the β axis.

According to Eq. D.9 for both positive and negative values of n there exist solutions of the wave equation. The energy transport in the structure occurs with the group velocity, but, for example for $n = -1$, the group velocity can be negative, even if the phase velocity is positive. This means that an electron beam can move in the positive z direction and have the same velocity as a space harmonic wave which propagates in this direction, which energy transport occurs in the negative direction (see also Section 7.3.4 and Appendix F).

Appendix E

Small-Signal TWT Theory

According to Appendix D the phase constant for the nth space harmonic mode in a slow-wave structure is

$$\beta_n = \beta_0 + \frac{2\pi n}{L}, \tag{D.10}$$

the propagation constant is

$$\beta_e = \frac{\omega}{v_0},$$

and by a suitable choice of the acceleration voltage or of circuit parameters we have

$$\beta_e \approx \beta_n \qquad \text{or} \qquad v_0 \approx v_{fn},$$

where v_{fn} is the phase velocity of the nth space harmonic wave, and $n = 0$ is the normal choice for a TWT. The phase of the other waves, as viewed by the electrons, will change continuously. The TWT slow-wave structure length has a number of hf periods, and the electrons move over many wavelengths of the electromagnetic field when traveling along the tube. The mean value of the field thus corresponds only to a single harmonic component, the nth space harmonic component with which the electrons are almost synchronous, so the contribution from the other components can be neglected.

We consider only the high-frequency velocity, current, space-charge, and so on, components and assume that the radial forces caused by the space-charge are compensated by an external axial magnetic field. We assume that the electrons move in the positive z direction and in one dimension, and that the signal amplitude is small, so as to avoid nonlinear effects. Thus all high-frequency components vary as

$$\mathbf{A} = \Re[a\, e^{j\omega t - \Gamma z}], \tag{E.1}$$

where the amplitude a depends neither on time nor on the space coordinate and Γ is a complex propagation constant. Differentiating by t and z we get

$$\frac{\partial \mathbf{A}}{\partial t} = \Re[j\omega a\, e^{j\omega t - \Gamma z}]$$

and

$$\frac{\partial \mathbf{A}}{\partial z} = \Re[-\Gamma a\, e^{j\omega t - \Gamma z}].$$

The partial derivative with respect to time is just a multiplication by $j\omega$, while with respect to the z coordinate it is just a multiplication by $-\Gamma$, so in all subsequent equations the exponential term can be omitted.

The Electronic Equation

The velocity, current density, and space-charge have time constant and high-frequency components:

$$v_{tot} = v_0 + v,$$
$$i_{tot} = -I_0 + i, \tag{E.2}$$
$$\rho_{tot} = -\rho_0 + \rho.$$

The high-frequency current is, in first order,

$$i = (v_0\rho - \rho_0 v)S, \tag{E.3}$$

where S is the beam cross section. The continuity equation,

$$\nabla \cdot \mathbf{J} + \frac{\partial \rho}{\partial t} = 0,$$

gives

$$\frac{\partial i}{\partial z} = -\frac{\partial \rho}{\partial t}S \tag{E.4}$$

and

$$-\Gamma i = -j\omega\rho S. \tag{E.5}$$

By introducing ρ of Eq. E.5 into Eq. E.3 the high-frequency velocity is

$$v = -\frac{j\omega - v_0\Gamma}{j\omega\rho_0 S}i. \tag{E.6}$$

The electric field influencing the beam electron motion has components caused by the space-charge and the circuit. By Newton's second law the total field is $\mathcal{E}_{zT}$:

$$\frac{d\mathbf{v}}{dt} = \eta\mathcal{E}_{zT} = \frac{\partial \mathbf{v}}{\partial t} + \frac{\partial \mathbf{v}}{\partial z}\frac{\partial z}{\partial t}, \tag{E.7}$$

where $\mathbf{v}$ includes the *exponential term* and η includes the *sign of the negative electron charge*. For small-signal amplitude the axial electron velocity changes little and thus can be approximated by the DC beam velocity,

$$\frac{\partial z}{\partial t} = v_0.$$

Now Eq. E.7 becomes

$$(j\omega - v_0\Gamma)v = \eta E_{zT} = \eta(E_{zn} + E_{zr}), \tag{E.8}$$

where E_{zn} is the axial space harmonic wave component, which is approximately synchronous with the electron mean velocity, and E_{zr} is the space-charge-generated electric field.

E_{zr}, given by the Poisson's equation for an infinitely wide electron beam and taking the sign of the hf space-charge component in Eq. E.2, is

$$-\Gamma E_{zr} = \frac{\rho}{\epsilon_0}. \tag{E.9}$$

Substituting ρ of Eq. E.5 we obtain

$$E_{zr} = j\frac{i}{\omega\epsilon_0 S}. \tag{E.10}$$

The electron beam is not infinitely wide, so the same reasoning as for klystrons can be applied. The reduced plasma frequency, ω_q, must be used and a factor F is introduced such that

$$F = \frac{\omega_q}{\omega_p},$$

and then

$$E_{zr} = jF^2\frac{i}{\omega\epsilon_0 S}.$$

The plasma frequency is

$$\omega_p^2 = -\eta\frac{\rho_0}{\epsilon_0},$$

and

$$E_{zr} = j\frac{\omega_q^2}{\eta\rho_0\omega S}i. \tag{E.11}$$

Equations E.6, E.8, and E.11, with $I_0 = v_0\rho_0 S$, are combined to give the high-frequency beam current

$$i = \frac{j\beta_e I_0 E_{zn}}{2V_0\left[(\Gamma - j\beta_e)^2 + \dfrac{\omega_q^2}{v_0^2}\right]}. \tag{E.12}$$

The equation is called *electronic* because it describes the interaction between the high-frequency beam current component and the space harmonic wave with the amplitude E_{zn}, which travels along the slow-wave structure.

The Circuit Equation

The high-frequency beam current component induces a current in the circuit, which adds to the existing one, generated by the input signal, so the total current changes along the tube. The power transported at any point z along the TWT depends on electric field of the nth space harmonic wave. The equivalent circuit of a slow-wave structure can be approximated by a line of capacitances and inductances with characteristic impedance

$$Z_0 = \sqrt{\frac{L}{C}}.$$

If line losses are neglected, the impedance can be defined as the ratio of the voltage across the line squared and the delivered power,

$$Z_0 = \frac{|V|^2}{2P}.$$

The voltage in a slow-wave structure can be related to the electric field, allowing for the presence of the electron beam. The beam coupling impedance includes the interaction of the slow-wave structure with the beam,

$$Z_n = \frac{\int |E_{zn}|^2 \, dS}{2PS}. \tag{E.13}$$

Here, P is the mean value of the power flow at z, S is the beam cross-section area, and the integral is over the whole beam cross section. Solving Maxwell's equations for the TWT slow-wave structure, the impedance can be computed. For manufactured structures the impedance is obtained by measuring the electric field along the structure as a function of the delivered power. The beam coupling impedance is usually a few kiloohms. A high impedance corresponds to a strong electric field at a constant power flow and produces a strong interaction with the beam, so high impedance means high amplification. From Eq. E.13 it is seen that for a high impedance the electric field must be concentrated in the region where the electron beam is situated, which means near the tube axis.

The instantaneous power flow along the structure is

$$p = \frac{\mathcal{E}_{zn}^2}{\beta_n^2 Z_n},$$

where $\mathcal{E}_{zn}$ is instantaneous amplitude of the nth space harmonic wave taken as a mean across the beam cross section. The power increases because of the interaction with the beam,

$$dp = \frac{2\mathcal{E}_{zn} d\mathcal{E}_{zn}}{\beta_n^2 Z_n}. \tag{E.14}$$

The change in $\mathcal{E}_{zn}$ over dz is caused by new electromagnetic waves which spread in both directions from the point on the slow-wave structure in front of the current segment dz (Fig. E.1). Thus

$$d\mathcal{E}_{zn} = d\mathcal{E}_{zn+} + d\mathcal{E}_{zn-},$$

but by symmetry we have

$$d\mathcal{E}_{zn-} = d\mathcal{E}_{zn+}. \tag{E.15}$$

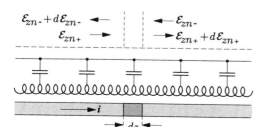

Fig. E.1. Interaction between the high-frequency current and the electric field in the slow-wave structure.

The total beam power transported to the circuit is then

$$
\begin{aligned}
dp &= dp_- + dp_+ \\
&= \frac{2}{\beta_n^2 Z_n} (\mathcal{E}_{zn-} d\mathcal{E}_{zn-} + \mathcal{E}_{zn+} d\mathcal{E}_{zn+}) \\
&= \frac{2}{\beta_n^2 Z_n} (\mathcal{E}_{zn-} + \mathcal{E}_{zn+}) \, d\mathcal{E}_{zn-} \\
&= \frac{2}{\beta_n^2 Z_n} (\mathcal{E}_{zn-} + \mathcal{E}_{zn+}) \, d\mathcal{E}_{zn+},
\end{aligned}
\tag{E.16}
$$

which in terms of the high-frequency beam current i is

$$
dp = -i \mathcal{E}_z \, dz,
\tag{E.17}
$$

where $\mathcal{E}_z$ is the total electric field from all waves in the slow-wave structure. However, only the nth space harmonic wave component must be considered, that wave most synchronous with the beam. Interaction with other components can be neglected, since phase differences along the whole TWT counteract and cancel each other. Equation E.17 can be written as

$$
dp = -i \, (\mathcal{E}_{zn-} + \mathcal{E}_{zn+}) \, dz.
\tag{E.18}
$$

Comparing Eqs. E.18 and E.16 the increase of the hf field over dz is

$$
d\mathcal{E}_{zn-} = d\mathcal{E}_{zn+} = -\frac{1}{2} \beta_n^2 Z_n i \, dz,
\tag{E.19}
$$

so the total electric field in a point, z_0, has three components:

$$
E_{zn1}(z_0) = E_z(0) \, e^{-\Gamma_0 z_0}, \text{ generated by the input signal;}
$$

$$
E_{zn2}(z_0) = \int_0^{z_0} e^{-\Gamma_0(z_0 - z)} \, dE_{zn+}, \text{ for all waves coming from left; and}
$$

$$
E_{zn2}(z_0) = \int_{z_0}^{L} e^{-\Gamma_0(z - z_0)} \, dE_{zn-}, \text{ for all waves coming from right.}
$$

Here, L is the slow-wave structure length, and $\Gamma_0 = \alpha + j\beta_n$ the propagation constant of the nth space harmonic wave with no electron beam. We sum these three components using Eq. E.19:

$$
E_{zn}(z_0) = E_z(0) \, e^{-\Gamma_0 z_0} - \frac{1}{2} \beta_n^2 Z_n \int_0^{z_0} i \, e^{-\Gamma_0(z_0 - z)} \, dz - \frac{1}{2} \beta_n^2 Z_n \int_{z_0}^{L} i \, e^{-\Gamma_0(z - z_0)} \, dz,
$$

which must be valid for every point, z, along the whole slow-wave structure, so

$$
E_{zn}(z) = E_z(0) \, e^{-\Gamma_0 z} - \frac{1}{2} \beta_n^2 Z_n \int_0^{z} i(\zeta) \, e^{-\Gamma_0(z - \zeta)} \, d\zeta - \frac{1}{2} \beta_n^2 Z_n \int_{z}^{L} i(\zeta) \, e^{\Gamma_0(z - \zeta)} \, d\zeta.
$$

$$
\tag{E.20}
$$

This integral equation connects the electric field in every point along the tube with the high-frequency beam current, i.

We transform Eq. E.20 into a differential equation, but this must be done with care because the integral limits are themselves functions of z. If

$$f(z) = \int\limits_{\alpha(z)}^{\beta(z)} F(\zeta, z)\, d\zeta,$$

then (Leibnitz's rule; see, for example, Korn and Korn, *Mathematical Handbook for Scientist and Engineers*, McGraw-Hill, 1968, p. 103)

$$\frac{df}{dz} = \int\limits_{\alpha}^{\beta} \frac{\partial F}{\partial z}\, d\zeta + F(\beta, z)\frac{d\beta}{dz} - F(\alpha, z)\frac{d\alpha}{dz}. \tag{E.21}$$

By differentiating Eq. E.20 twice using this rule we obtain

$$\frac{d^2 E_{zn}}{dz^2} = \Gamma_0^2 E_z(0)\, e^{-\Gamma_0 z} - \frac{1}{2}\Gamma_0^2 \beta_n^2 Z_n \int\limits_{0}^{z} i(\zeta)\, e^{-\Gamma_0(z-\zeta)}\, d\zeta$$

$$- \frac{1}{2}\Gamma_0^2 \beta_n^2 Z_n \int\limits_{z}^{L} i(\zeta)\, e^{\Gamma_0(z-\zeta)}\, d\zeta + \Gamma_0 \beta_n^2 Z_n i, \tag{E.22}$$

but differentiating $\mathcal{E}_{zn}$ twice gives

$$\frac{d^2 E_{zn}}{dz^2} = \Gamma^2 E_{zn}. \tag{E.23}$$

Multiplying Eq. E.20 by Γ_0^2 and subtracting from Eq. E.22, taking into account Eq. E.23, the resulting equation becomes

$$\Gamma^2 E_{zn} - \Gamma_0^2 E_{zn} = \Gamma_0 \beta_n^2 Z_n i, \tag{E.24}$$

and so the amplitude of the nth space harmonic wave is

$$E_{zn} = \frac{\Gamma_0 \beta_n^2 Z_n}{\Gamma^2 - \Gamma_0^2} i. \tag{E.25}$$

Rearranging, the high-frequency beam current is

$$i = \frac{\Gamma^2 - \Gamma_0^2}{\Gamma_0 \beta_n^2 Z_n} E_{zn}. \tag{E.26}$$

Equation E.26 is called the *circuit equation*, because it describes the influence of the high-frequency beam current on the circuit field.

The Solution

We derived the electronic and circuit equation assuming that the interaction is a result of the modulation with small signal amplitude and that all equations are linear. This means that all field and beam quantities have the form

$$e^{j\omega t - \Gamma z}.$$

The ratio E_{zn}/i from the circuit equation (Eq. E.26) and from the electronic equation (Eq. E.12) can be equated:

$$\frac{\Gamma_0 \beta_n^2 Z_n}{\Gamma^2 - \Gamma_0^2} = \frac{2V_0 \left[(\Gamma - j\beta_e)^2 + \frac{\omega_q^2}{v_0^2} \right]}{j\beta_e I_0},$$

or

$$(\Gamma^2 - \Gamma_0^2)\left[(\Gamma - j\beta_e)^2 + \frac{\omega_q^2}{v_0^2} \right] = \frac{j\beta_e \beta_n^2 \Gamma_0 Z_n I_0}{2V_0}, \tag{E.27}$$

the solution of which gives four values of the propagation constant Γ, and so there are four different waves traveling along a TWT.

Pierce [230] introduced some parameters, which are helpful in analyzing the solution:

$$C^3 = \frac{Z_n I_0}{4V_0}, \qquad\qquad R = \frac{\omega_q^2}{4C^2\omega^2}, \tag{E.28}$$

$$\Gamma_0 = j\beta_e(1 + Cb - jCd) = \alpha + j\beta_n, \quad \text{and} \quad \Gamma = j\beta_e(1 + jC\delta).$$

C, dimensionless parameter with values between 0.01 and 0.1, is called the amplification parameter. To keep the synchronism between the motion of the electron beam and the phase velocity of the electromagnetic wave, $\beta_e \gtrsim \beta_n$; thus

$$b = \frac{\beta_n - \beta_e}{\beta_e C} = \frac{v_0 - v_{fn}}{v_{fn} C} > 0. \tag{E.29}$$

d is a measure of the attenuation in the circuit,

$$d = \frac{\alpha}{\beta_e C}. \tag{E.30}$$

Neglecting the fourth-order products we obtain

$$\Gamma^2 - \Gamma_0^2 = -2\beta_e^2 C(-b + j\delta + jd),$$

$$(\Gamma - j\beta_e)^2 = \beta_e^2 C^2 \delta^2,$$

and Eq. E.27 becomes

$$-\beta_e^2 C(-b + j\delta + jd)\left(\beta_e^2 C^2 \delta^2 + \frac{\omega_q^2}{v_0^2} \right) = j\beta_e \beta_n^2 \Gamma_0 C^3. \tag{E.31}$$

Dividing by $\beta_e^4 C^3(-b + j\delta + jd)$, observing that $\beta_e^2 v_0^2 = \omega^2$, and introducing R, we obtain

$$\delta^2 - 4R = -\frac{j\beta_n^2 \Gamma_0}{\beta_e^3} \frac{1}{(-b + j\delta + jd)}.$$

Let $\beta_e \approx \beta_n$, introduce Γ_0 according to Eq. E.28, and neglect Cd and Cb in comparison to 1,

$$\frac{j\beta_n^2 \Gamma_0}{\beta_e^3} \approx -1.$$

What remains is

$$\delta^2 = \frac{1}{(-b + j\delta + jd)} + 4R, \tag{E.32}$$

which is a third-order equation, the fourth root having disappeared with the approximation $(C\delta)^2 \ll 1$.

The propagation constant depends on all four parameters, $\Gamma = f(b, d, R, C)$, which are small in the linear theory and thus are justified to be approximated:

$$b = d = R = 0.$$

What remains,

$$\delta^3 = -j, \tag{E.33}$$

has solutions

$$\delta_1 = \frac{\sqrt{3}}{2} - \frac{j}{2}, \quad \delta_2 = -\frac{\sqrt{3}}{2} - \frac{j}{2}, \quad \text{and} \quad \delta_3 = j, \tag{E.34}$$

and hence

$$\Gamma_1 = -\frac{\sqrt{3}}{2}\beta_e C + j\beta_e\left(1 + \frac{C}{2}\right),$$

$$\Gamma_2 = \frac{\sqrt{3}}{2}\beta_e C + j\beta_e\left(1 + \frac{C}{2}\right), \quad \text{and} \tag{E.35}$$

$$\Gamma_3 = j\beta_e(1 - C).$$

Since all field and beam components vary as $e^{j\omega t - \Gamma z}$, Γ_1 corresponds to a wave whose amplitude increases with z, Γ_2 to a wave whose amplitude decreases with z, and Γ_3 to a wave with constant amplitude. The first wave, responsible for the amplification in a TWT, travels slower than the electrons as the $\beta_e(1 + C/2)$ term shows, and thus is the reason electrons wander in the retarding electric field. The signal amplitude increases exponentially along the tube.

The fourth wave has a propagation constant,

$$\Gamma_4 = -j\beta_e\left(1 - \frac{C^3}{4}\right),$$

travels in the negative z direction, and has a very small amplitude, so it is usually neglected in the analysis of the traveling wave tube, except for reflection problems (Large Signal Problems in Section 7.3.3).

For b, d, and R different from zero the propagation of the waves is unchanged but values of the amplification factor and other parameters are different and are a question of large signal theory, which only can be solved numerically.

The input signal must start three waves, which are characterized by their electric field, velocity, and current as

$$E_{zT} = E_{zT1} + E_{zT2} + E_{zT3},$$
$$v = v_1\,e^{-\Gamma_1 z} + v_2\,e^{-\Gamma_2 z} + v_3\,e^{-\Gamma_3 z}, \quad \text{and} \qquad \text{(E.36)}$$
$$i = i_1\,e^{-\Gamma_1 z} + i_2\,e^{-\Gamma_2 z} + i_3\,e^{-\Gamma_3 z}.$$

The amplitudes can be computed from the initial conditions at the beginning of the slow-wave structure, $z = 0$, where the beam is not yet modulated:

$$v(0) = 0, \qquad i(0) = 0, \qquad E_{zr}(0) = 0, \quad \text{and} \quad E_{zT}(0) = E_{zn}(0).$$

E_{zn} can be obtained from Eq. E.13, with Z_n either computed or measured. Assuming E_{zn} constant across the beam, Eq. E.13 gives

$$E_{zn} = \sqrt{2\beta_e^2 Z_n P},$$

where P is the input power and $\beta_n \approx \beta_e$.

To solve the linear system E.36 we introduce Γ from Eq. E.28, and the velocity and the current from Eqs. E.8 and E.6, where

$$j\omega - v_0\Gamma = j\omega - j\beta_e v_0(1 + jC\delta) = \omega C\delta.$$

Denote the three waves by subscript ν for the hf velocity and current:

$$v_\nu = \frac{\eta}{\omega C\delta_\nu} E_{zT\nu} \quad \text{and}$$
$$i_\nu = -\frac{j\rho_0 S}{C\delta_\nu} v_\nu = \frac{j\eta\rho_0 S}{\omega C^2\delta_\nu^2} E_{zT\nu}. \qquad \text{(E.37)}$$

With these initial conditions Eq. E.36 becomes

$$E_{zT} = E_{zT1} + E_{zT2} + E_{zT3} = E_{zn}(0),$$
$$v \Rightarrow \frac{1}{\delta_1} E_{zT1} + \frac{1}{\delta_2} E_{zT2} + \frac{1}{\delta_3} E_{zT3} = 0, \quad \text{and} \qquad \text{(E.38)}$$
$$i \Rightarrow \frac{1}{\delta_1^2} E_{zT1} + \frac{1}{\delta_2^2} E_{zT2} + \frac{1}{\delta_3^2} E_{zT3} = 0.$$

The solution for the growing wave is thus

$$E_{zT1} = \frac{\delta_1^2}{(\delta_1 - \delta_2)(\delta_1 - \delta_3)} E_{zn}(0) = \frac{1}{3} E_{zn}(0), \qquad \text{(E.39)}$$

with amplitude at the beginning of the tube a third of the amplitude of the input signal, but increasing exponentially according to

$$e^{\frac{\sqrt{3}}{2}\beta_e C z}.$$

If N is the number of electronic wavelengths, λ, with the length L, along the slow-wave structure, we obtain

$$L = N\lambda, \qquad \omega = 2\pi f, \qquad f\lambda = v_0, \qquad \text{and} \qquad \beta_e = \frac{\omega}{v_0} = \frac{2\pi N}{L},$$

so

$$E_{zn}(L) \approx E_{zT1}(0)\, e^{\frac{\sqrt{3}}{2}\beta_e CL} = \frac{1}{3} e^{\sqrt{3}\pi CN} E_{zn}(0), \tag{E.40}$$

and the traveling wave tube amplification is

$$A = 20 \log \frac{E_{zn}(L)}{E_{zn}(0)} = 20 \log \frac{1}{3} + \sqrt{3}\pi CN \log e = -9.54 + 47.3CN. \tag{E.41}$$

The amplification parameter is given by Eq. E.28:

$$C = \sqrt[3]{\frac{Z_n I_0}{4V_0}},$$

where the beam coupling impedance must be known.

Appendix F

Phase and Group Velocity

In a Brillouin diagram, $\omega = f(\beta)$ (Fig. D.4), the slope of a line from the origin to a point on the dispersion curve gives the phase velocity, $v_f = \omega/\beta$; the slope of the curve tangent in the same point gives the group velocity, $v_g = \partial\omega/\partial\beta$. The transport of the energy along the structure occurs with the group velocity, and the interaction with the beam accounts for the phase velocity. There are infinitely many wave equation solutions in the form of space harmonic waves, and any space harmonic wave ($-\infty \leq n \leq \infty$) is a solution, but the boundary conditions can be satisfied only if all of them are present. Therefore, there are solutions where the phase velocity is positive, while the group velocity is negative.

This circumstances are visualized in Fig. F.1. In the left figure an observer walks and observes the waves surging the shore. The waves are coming at an oblique direction and the observer tries to walk with such a velocity as to have the wave crest next to him. The observer is then in phase with the waves, walking with the wave phase velocity. The wave propagation and energy transport occurs with group velocity. The ratio between the group and phase velocity, if the waves make an angle θ with the shore, is

$$v_g = v_f \sin\theta.$$

The right figure shows a similar situation; the difference is the breakwater pier along the shore with equidistant openings, and the observer can see the waves only through them. The group velocity of the waves has a component parallel with the shore, but with the opposite direction as in the left figure, so the energy transport occurs from right to left. Assume that the observer stands at the point a when the crest passes the opening, while the next crest is at c. He walks to the right, and when he passes the opening at b,

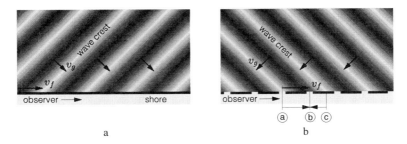

Fig. F.2. Phase and of group velocity.

the wave front has moved so that the crest is there. The observer and the wave are traveling toward each other, and if the observer continues to walk with the same velocity, he will see a wave crest, when he passes by an opening. He is even then in phase with the waves, and observes a phase velocity equal to his walking velocity. But the group velocity has a component that has the opposite direction.

Similarly, the space harmonic wave in a slow-wave structure can have a phase and group velocity that have opposite directions. If the electron beam travels in the positive z direction, it can be in phase with a space harmonic wave, but the power transport can be in the negative direction. Microwave tubes, which use this kind of power flow, work with *backward waves*.

Appendix G

Relativistic Formulae

A collection of some relativistic mechanics formulae are given below. If v is the particle and c is the light velocity, then

$$\beta = v/c, \tag{G.1}$$

$$\gamma = \frac{1}{\sqrt{1 - v^2/c^2}} = \frac{1}{\sqrt{1 - \beta^2}}. \tag{G.2}$$

The relativistic length contraction is

$$L = L_0/\gamma, \tag{G.3}$$

and the time dilatation is

$$t = \gamma t_0. \tag{G.4}$$

The relativistic mass of the particle is

$$m = \frac{m_0}{\sqrt{1 - v^2/c^2}} = \gamma\, m_0, \tag{G.5}$$

and the relativistic energy equation is

$$m^2 c^4 - m^2 v^2 c^2 = m_0^2 c^4, \tag{G.6}$$

where $W = mc^2$ is the total and $W_0 = m_0c^2$ is the rest mass energy. Introducing the relativistic momentum $\mathbf{p} = m\mathbf{v}$, Eq. G.6 becomes

$$W^2 - \mathbf{p}^2c^2 = W_0^2. \tag{G.7}$$

Kinetic energy of a particle is

$$W_k = W - W_0 = m_0c^2(\gamma - 1). \tag{G.8}$$

Collision of a high-energy particle (subscript L) with an in-the-laboratory-system still-standing particle of equal rest mass is described by

$$(W_L + W_0)^2 - p^2c^2 = (2W_S)^2,$$

where the subscript S denotes the energy in the center of mass system. Taking into account Eq. G.7, we have

$$4W_S^2 = 2W_0^2\left(\frac{W_L}{W_0} + 1\right),$$

and the available kinetic energy in the center-of-mass system becomes

$$W_{Sk} = W_0\left[\sqrt{\frac{1}{2}\left(\frac{W_{Lk} + W_0}{W_0} + 1\right)} - 1\right], \tag{G.9}$$

which is much less than the kinetic energy W_{Lk} of the projectile particle in the laboratory system.

The relativistic Doppler effect shifted frequency is

$$f = \frac{f_0}{\gamma(1 \pm \beta)} = f_0\sqrt{\frac{1 \mp \beta}{1 \pm \beta}} = \frac{f_0}{\gamma \pm \sqrt{\gamma^2 - 1}}, \tag{G.10}$$

where the upper sign accounts for receding and the lower sign accounts for approaching source and receiver.

Appendix H

Beam Transport Matrices

Defining the position, angle, and momentum deviation of a particle with respect to a reference particle by a vector

$$\mathbf{P} = \begin{pmatrix} x \\ x' \\ y \\ y' \\ \ell \\ \delta \end{pmatrix},$$

where x is the radial displacement, x' is the angle in the radial plane the orbit makes with respect to the assumed reference particle orbit, y is the transverse displacement,

y' is the transverse angle, ℓ is the path length difference between the particle orbit and the assumed reference particle orbit, and $\delta = \Delta p/p$ the momentum error, the passage of a particle through a beam transport element can be described by a matrix equation

$$\mathbf{P} = \mathbf{M} \cdot \mathbf{P}(0),$$

where $\mathbf{P}(0)$ is the initial coordinate vector.

1. Drift Space
L = length of the drift space.

$$\mathbf{M} = \begin{pmatrix} 1 & L & 0 & 0 & 0 & 0 \\ 0 & 1 & 0 & 0 & 0 & 0 \\ 0 & 0 & 1 & L & 0 & 0 \\ 0 & 0 & 0 & 1 & 0 & 0 \\ 0 & 0 & 0 & 0 & 1 & 0 \\ 0 & 0 & 0 & 0 & 0 & 1 \end{pmatrix}. \tag{H.1}$$

2. Weak Rotationally Symmetric Lens
Electric potential is equal on both sides of the lens, $V_1 = V_2$.

$$\mathbf{M} = \begin{pmatrix} 1 & z_{H2} \\ 0 & 1 \end{pmatrix} \begin{pmatrix} 1 & 0 \\ -\frac{1}{f} & 1 \end{pmatrix} \begin{pmatrix} 1 & z_{H1} \\ 0 & 1 \end{pmatrix}, \tag{H.2}$$

z_{H1} and z_{H2} are the principal plane distances from the lens center. Radial and transverse matrices are equal.

3. Magnetic Prism with Homogeneous Field
θ = deflection angle, r = radius of curvature.

$$\mathbf{M} = \begin{pmatrix} \cos\theta & r\sin\theta & 0 & 0 & 0 & r(1-\cos\theta) \\ -\frac{1}{r}\sin\theta & \cos\theta & 0 & 0 & 0 & \sin\theta \\ 0 & 0 & 1 & r\theta & 0 & 0 \\ 0 & 0 & 0 & 1 & 0 & 0 \\ -\sin\theta & -r(1-\cos\theta) & 0 & 0 & 1 & r(1-\sin\theta) \\ 0 & 0 & 0 & 0 & 0 & 1 \end{pmatrix}. \tag{H.3}$$

4. Edge Focusing
ε = edge angle, r = radius of curvature of the reference orbit.

$$\mathbf{M} = \begin{pmatrix} 1 & 0 & 0 & 0 & 0 & 0 \\ \tan\varepsilon/r & 1 & 0 & 0 & 0 & 0 \\ 0 & 0 & 1 & 0 & 0 & 0 \\ 0 & 0 & -\tan\varepsilon/r & 1 & 0 & 0 \\ 0 & 0 & 0 & 0 & 1 & 0 \\ 0 & 0 & 0 & 0 & 0 & 1 \end{pmatrix}. \tag{H.4}$$

5. Magnetic Prism with Gradient

$$n = -\frac{dB/B_0}{dr/r_0}, \qquad K_r = \sqrt{|1-n|}/r, \qquad K_t = \sqrt{|n|}/r, \qquad L = r\theta.$$

For $0 < n < 1$

$$\mathbf{M} = \begin{pmatrix} \cos K_r L & \frac{1}{K_r}\sin K_r L & 0 & 0 & 0 & \frac{1}{rK_r^2}(1-\cos K_r L) \\ -K_r \sin K_r L & \cos K_r L & 0 & 0 & 0 & \frac{1}{rK_r}\sin K_r L \\ 0 & 0 & \cos K_t L & \frac{1}{K_t}\sin K_t L & 0 & 0 \\ 0 & 0 & K_t \sin K_t L & \cos K_t L & 0 & 0 \\ -\frac{1}{rK_r}\sin K_r L & -\frac{1}{rK_r^2}(1-\cos K_t L) & 0 & 0 & 1 & -\frac{1}{r^2 K_r^3}(K_r L - \sin K_r L) \\ 0 & 0 & 0 & 0 & 0 & 1 \end{pmatrix}.$$

(H.5a)

For $n > 1$

$$\mathbf{M} = \begin{pmatrix} \cosh K_r L & \frac{1}{K_r}\sinh K_r L & 0 & 0 & 0 & \frac{1}{rK_r^2}(1-\cosh K_r L) \\ -K_r \sinh K_r L & \cosh K_r L & 0 & 0 & 0 & \frac{1}{rK_r}\sinh K_r L \\ 0 & 0 & \cos K_t L & \frac{1}{K_t}\sin K_t L & 0 & 0 \\ 0 & 0 & K_t \sin K_t L & \cos K_t L & 0 & 0 \\ -\frac{1}{rK_r}\sinh K_r L & -\frac{1}{rK_r^2}(1-\cosh K_t L) & 0 & 0 & 1 & -\frac{1}{r^2 K_r^3}(K_r L - \sinh K_r L) \\ 0 & 0 & 0 & 0 & 0 & 1 \end{pmatrix}.$$

(H.5b)

6. Magnetic Quadrupole

$$K^2 = G/B\rho, \qquad G = B_{top}/a, \qquad B\rho = \text{magnetic rigidity}.$$

$$\mathbf{M} = \begin{pmatrix} \cos KL & \frac{1}{K}\sin KL & 0 & 0 & 0 & 0 \\ -K\sin KL & \cos KL & 0 & 0 & 0 & 0 \\ 0 & 0 & \cosh KL & \frac{1}{K}\sinh KL & 0 & 0 \\ 0 & 0 & K\sinh KL & \cosh KL & 0 & 0 \\ 0 & 0 & 0 & 0 & 1 & 0 \\ 0 & 0 & 0 & 0 & 0 & 1 \end{pmatrix}. \qquad (\text{H.6})$$

7. Solenoid

$L = $ effective length and $K = B(0)/(2B\rho)$, where $B(0)$ is the field inside the solenoid and $B\rho$ is the magnetic rigidity.

$$\mathbf{M} = \begin{pmatrix} \cos^2 KL & \frac{1}{K}\sin KL\cos KL & \sin KL\cos KL & \frac{1}{K}\sin^2 KL & 0 & 0 \\ -K\sin KL\cos KL & \cos^2 KL & -K\sin^2 KL & \sin KL\cos KL & 0 & 0 \\ -\sin KL\cos KL & -\frac{1}{K}\sin^2 KL & \cos^2 KL & \frac{1}{K}\sin KL\cos KL & 0 & 0 \\ K\sin^2 KL & -\sin KL\cos KL & K\sin KL\cos KL & \cos^2 KL & 0 & 0 \\ 0 & 0 & 0 & 0 & 1 & 0 \\ 0 & 0 & 0 & 0 & 0 & 1 \end{pmatrix}.$$

(H.7)

Appendix K

Historical References

Any discrepancy between the years mentioned in Historical Notes and their corresponding references is a result of the fact that many inventions were documented after their conceptions. For example, new devices and ideas used during World War II were published after the war ended.

[1] Gilbert, W., **De magnete, magneticisque corporibus, et de magno magnete tellure; Physiologia nova, plurimis & argumentis, & experimentis demonstrata**, Londini, 1600

[2] Browne, Sir T., **Pseudoxia epidemica: or, enquiries into very many received tenents, and commonly presumed truths**, London, 1646

[3] Watson, W., "A sequel to the experiments and observations tending to illustrate the nature and properties of electricity," *Phil. Trans. Roy. Soc.*, **44**, 704 (1746-47, printed 1748)

[4] Faraday, M., "Experimental researches in electricity, VII, §869 ff," *Phil. Trans. Roy. Soc.*, 77 (1834)

[5] Faraday, M., "Experimental researches in electricity, XIV, §1679 ff," *Phil. Trans. Roy. Soc.*, 265 (1838)

[6] Stoney, J. G., "On the physical units of nature," *Trans. Roy. Dublin Soc.*, **3**, 51 (1881)

[7] Stoney, J. G., "On the cause of double lines and of equidistant satellites in the spectra of gases," *Trans. Roy. Dublin Soc.*, **4**, 563 (1891)

[8] Plücker, J., "Über die Einwirkung des Magneten auf die electrischen Entladungen in verdünten Gasen," *Pogg. Ann. Phys.*, **103**, 88 (1858);
"Über die Einwirkung des Magnets auf die electrische Entladung," *Pogg. Ann. Phys.*, **113**, 249 (1861)

[9] Hittorf, J., "Über die Electrizitätsleitung der Gase, Teil I und II," *Pogg. Ann. Phys.*, **136**, 1 & 197 (1869)

[10] Varley, C. F., "Some experiments on the discharge of electricity through rarefied media and the atmosphere," *Proc. Roy. Soc.*, **19**, 246 (1871)

[11] Hertz, H. R., "Versuche über die Glimmentladung," *Wied. Ann. Phys.*, **19**, 782 (1883);
"Über den Durchgang von Katodenstrahlen durch dünne Metallschichten," *Wied. Ann. Phys.*, **45**, 28 (1892)

[12] Perrin, J. B., "Nouvelles propriétés des rayons cathodiques," *C. R. Acad. Sci. Paris*, **121**, 1130 (1895)

[13] Thomson, J. J., "Cathode rays," *Phil. Mag.*, **44**, 293 (1897)

[14] Kaufmann, W., "Die magnetische Ablenkbarkeit der Katodenstrahlen und ihre Abhängigkeit von Entladunspotential," *Wied. Ann. Phys.*, **61**, 544 (1897)

[15] Fitzgerald, G. F., "Dissociation of atoms," *The Electrician*, 103 (Mai 21, 1897)

[16] Guthrie, F., "On a relation between heat and electricity," *Phil. Mag.*, **46**, 257 (1873)

[17] Houston, E. J., "Notes on phenomena in incandescent lamps," *Trans. AIEE*, **I**, 1 (1884) (paper presented at the AIEE meeting in Philadelphia concerning Edison effect)

[18] Preece, W. H., "On a peculiar behaviour of glow-lamps when raised to high incandescence," *Proc. Roy. Soc.*, **38**, 219 (1885)

[19] Braun, K. F., "Über ein Verfahren zur Demonstration und zum Studium des zeitliches Verlaufes variabler Ströme," *Wied. Ann. Phys.*, **60**, 552, (1897)

[20] Fleming, J. A., British Patent, 24 850, 1904

[21] Wehnelt, A., "Über den Austritt negativer Ionen aus glühenden Metallverbindungen und damit zusammenhängende Erscheinungen," *Ann. Phys.*, **14**, 425 (1904)

[22] de Forest, L., "The audion. A new receiver for wireless telegraphy," *Trans. IEEE*, **25**, 719 (1906)

[23] Boot, H. A. H., Randall, J. T., "The cavity magnetron," *J. Inst. Electr. Eng.*, *3A*, **93**, 928 (1946)

[24] Bardeen, J., Brattain, W. H., "The transistor – A semiconductor triode," *Phys. Rev.*, **74**, 230 (1948);
Shockley, W., "The theory of p–n junctions in semiconductors and p–n junction transistors," *Bell System Techn. J.*, **28**, 435 (1949)

[25] Busch, H. W., "Berechnung der Bahn von Katodenstrahlen in axialsymmetrischen elektromagnetischen Felde," *Ann. Phys.*, **81**, 974 (1926);
"Über die Wirkungsweise der Konzentrierungsspule bei der Braunscher Röhre," *Arch. Electrotech.*, **18**, 585 (1927)

[26] Maxwell, J. C., **A treatise on electricity and magnetism**, Oxford, 1873

[27] Gauss, K. F., **Theoria motus corporum coelestium in sectionibus conicis solem ambientum**, Hamburg, 1809

[28] Seidel, P. L., "Über ein Verfahren die Gleichungen, auf welche die methode der klensten Quadraten führt, sowie lineäre Gleichungen überhaupt, durch successive Annäherung aufzulösen," *Münch. Akad. Abh., 3 Abt.*, **11**, 81 (1874)

[29] Courant, R., Friedrichs, K., Lewy, H., "Über die partiellen Differenzengleichungen der mathematishen Physik," *Math. Ann.*, **100**, 32 (1928)

[30] Bargmann, V., Montgomery, D., von Neumann, J. L., "Solution of linear systems of high order," *Report for the Bureau of Ordnance*, Priceton, 1946, reprinted in von Neumann Collected Works, vol. V, p. 421

[31] Young, D. M., "Iterative methods for solving partial differential equations of elliptic type," *Trans. Am. Math. Soc.*, **76**, 92 (1954)

[32] Courant, R., "Variational methods for the solution of problems of equilibrium and vibrationes," *Bull. Am. Math. Soc.*, **49**, 1 (1943)

[33] Frankel, S. P., "Convergence rates of iterative treatments of partial differential equations," *Math. Tab. Aid. Comput.*, **4**, 65 (1940)

[34] Brandt, A., "Multi-level adaptive technique (MLAT) for fast numerical solution to boundary value problems," *Proceedings of the 3rd international confonference on numerical methods in fluid mechanics*, Paris 1972;
"Multi-level adaptive solutions to boundary-value problems," *Math. Comput.*, **31**, 333 (1977)

[35] Boris, J., "Relativistic plasma simulation – Optimization of a hybrid code," *Proceedings of the 4th conference on numerical simulation of plasmas*, Office of Naval Research, 1970

[36] Newton, Sir I., **Philosophiae naturalis principia mathematica**, London, 1687

[37] Boyle, R., **A continuation of new experiments, physico-mechanical, touching the spring and weight of air and their effects** (second edition), London, 1662

[38] Mariotte, E., **Seconde essai de physique, de la nature de l'air**, Paris, 1679

[39] Gay-Lussac, L. J., "Sur la dilatation des gaz et vapeurs," *Ann. de Chimie*, **43**, 137 (1802)

[40] Avogadro, A., "D'un manière de déterminer les masses relatives des molécules élémentaires des corps, et les proportions selon lesquelles elles entrant dans ces combinaisons," *J. Phys.*, **73**, 58 (1811)

[41] Carnot, S., **Réflexions sur la puissance motrice du feu et sur les machines propres à déveloper cette puissance**, Paris, 1824

[42] Clausius, R. J. E., "Über einen Grundsats der mechanischer Wärmetheorie," *Pogg. Ann. Phys.*, **120**, 426 (1863);
"Über verschiedene für die Anwendung bequeme Formen der Hauptgleichung der mechanischer Wärmetheorie," *Pogg. Ann. Phys.*, **125**, 353 (1865)

[43] Boltzmann, L., "Mechanische Bedeutung des zweiten Hauptsatses der Wärmetheorie," *Wien Akad. Sitzungsber.*, **53**, 195 (1866)

[44] Maxwell, J. C., **Theory of Heat**, London, 1870

[45] Gibbs, J. W., " On the equilibrium of heterogeneous substances," *Trans. Connecticut Acad.*, **III**, 108 (1875), 343 (1878)

[46] Mach, E., **Die Prinzipien der Wärmelehre**, Leipzig, 1896

[47] Planck, M., "Über irreversible Strahlungsvorgänge," *Ann. Phys.*, **1**, 69 (1900);
"Über das Gesetz der Energieverteilung in Normalspektrum," *Ann. Phys.*, **4**, 553 (1901)

[48] Einstein, A., "Über einen die Erzeugung und Verwandlung des Lichtes betreffenden heuristischen Gesichtspunkt," *Ann. Phys.*, **17**, 132 (1905)

[49] Richardson, O. W., "The electrical conductivity imparted to a vacuum by hot conductors," *Phil. Trans. Roy. Soc.*, **A 201**, 497 (1903)

[50] Langmuir, I., "The electron emission from tungsten filaments, containing thorium," *Phys. Rev.*, **4**, 544 (1914)

[51] Armstrong, E. H., "Some recent development in the audion receiver," *Proc. IRE*, **3**, 215 (1915)

[52] Schottky, W., "Über Hochvakuumverstärker. III. Teil: Mehrgitterröhren," *Arch. Elektrotechn.*, **8**, 299 (1919)

[53] Holst, G., Tellegen, B. D. H., US patent 1 945 040, 1927

[54] Smith, W., "Effect of light on selenium during passage of electric current," *Am. J. Sci.*, **5**, 301 (1873)

[55] Hertz, H. R., "Sehr schnelle electrische Schwingungen," *Wied. Ann. Phys.*, **31**, 421 (1887); "Über einen Einfluss des ultravioletten Lichtes auf die electrische Entladung," *Wied. Ann. Phys.*, **31**, 983 (1887)

[56] Hallwachs, W., "Über den Einfluss des Lichtes auf electrostatisch geladene Körper," *Wied. Ann. Phys.*, **33**, 301 (1888)

[57] Elster, J., Geitel, H., "Über die Verwendung des Natriumamalgames zu lichtelectrischen Versuchen," *Wied. Ann. Phys.*, **41**, 161, 166 (1890)

[58] Pauli, W., "Zusammenhang des Abschlusses der Elektronengruppen im Atom mit der Komplexstruktur der Spektren," *Zeit. Phys.*, **31**, 765 (1925)

[59] de Broglie, L., "A tentative theory of light quanta," *Phil. Mag.*, **47**, 466 (1924)

[60] Drude, P., "Zur Elektronentheori der Metalle I," *Ann. Phys.*, **1**, 566 (1900); "Zur Elektronentheori der Metalle II," *Ann. Phys.*, **3**, 369 (1900)

[61] Lorentz, H. A., **Ergebnisse und Probleme der Elektronentheorie**, Berlin, 1905

[62] Schrödinger, E., "Quantisierung als Eigenwertproblem," *Ann. Phys.*, **79**, 361 (1926)

[63] Sommerfeld, A., "Zur Elektronentheori der Metalle auf Grund der Fermischer Statistik," *Zeit. Phys.*, **47**, 1 (1928)

[64] Seebeck, T. J., "Magnetische Polarisation der Metalle und Erze durch Temperatur-Differenz," *Preuss. Akad. d. Wiss. (Berlin) Abh.*, **1822/23**, 265 (1825), *Pogg. Ann. Phys.*, **6**, 1, 133, 253 (1826)

[65] Dushman, S., "Electron emission from metals as a function of temperature," *Phys. Rev.*, **21**, 623 (1923)

[66] Nottingham, W. B., "Thermionic emission from tungsten and thoriated tungsten filaments," *Phys. Rev.*, **49**, 78 (1936); Hutson, A. R., "Velocity analysis of thermionic emission from single-crystal tungsten," *Phys. Rev.*, **98**, 889 (1955)

[67] Child, D. C., "Discharge from hot CaO," *Phys. Rev.*, **32**, 492 (1911); Langmuir, I., "The effect of space charge and residual gases on thermionic currents in high vacuum," *Phys. Rev.*, **2**, 450 (1913)

[68] Langmuir, I., Blodgett, K., "Currents limited by space charge between coaxial cylinders," *Phys. Rev.*, **22**, 347 (1923)

[69] Langmuir, I., Blodgett, K., "Currents limited by space charge between concentric spheres," *Phys. Rev.*, **24**, 49 (1924)

[70] Schottky, W., "Über den Einfluss von Strukturwirkungen, besonders bei der Thomsonscher Bildkraft, auf die Elektronenemission der Metalle," *Phys. Zeit.*, **15**, 872 (1914)

[71] Fowler, R. H., "The analysis of photoelectric sensitivity curves for clean metals at various temperatures," *Phys. Rev.*, **38**, 45 (1931)

[72] Fowler, R. H., Nordheim, L., "Electron emission in intense electric fields," *Proc. Roy. Soc.*, **A 119**, 173 (1928)

[73] Lemmens, H. J., Jansen, M. J., Loosjes, R., "A new thermionic cathode for heavy loads," *Philips Tech. Rev.*, **11**, 341 (1950)

[74] Fermat, P., memoirs of his son Samuel

[75] Snell, van R. W., lectures ~1610

[76] Newton, Sir I., **Opticks: or a treatise of the reflections, refractions, inflections and colours of light**, London, 1704

[77] Huygens, C., **Traité de la lumière**, Leide (Leyden), 1690

[78] Young, T., "On the theory of light and colours," *Phil. Trans. Roy. Soc.*, **20**, 12 (1802)

[79] Fresnel, A. J., "Mémoire sur la diffraction de la lumière, où l'on examine particulierment le phénomène des franges colorées que présentent les ombres des corps éclairés par un point lumineux," *Ann. chim. phys.*, **1**, 239 (1816); "Mémoire sur la diffraction de le lumière," *Ann. chim. phys.*, **11**, 246 (1819)

[80] Germer, L. H., Davisson, C. J., "The scattering of electrons by a single crystal of nickel," *Nature*, **119**, 558 (1927); "The scattering of electrons by a nickel crystal," *Phys. Rev.*, **30**, 585 (1927)

[81] Crookes, Sir W., "On the illumination of lines of molecular pressure and the trajectory of molecules," *Phil. Trans. Roy. Soc.*, **170**, 135 (1879)

[82] Rogowski, W., Flegler, E., "Eine neue Bauart des Katodenoszillographen," *Arch. Elektrotech.*, **18**, 513 (1927)

[83] Davisson, C. J., Calbick, C. J., "Electron lenses," *Phys. Rev.*, **38**, 585 (1931)

[84] Ruska, E. A., Knoll, M., "Die magnetische Sammelspule für schnelle Elektronenstrahlen," *Z. Techn. Phys.*, **12**, 389 (1931);
"Nachtrag zur Mitteilung," *Z. Techn. Phys.*, **12**, 488 (1931)

[85] Holst, J. H. de Boer, Teves, M. C., Veenmans, C. F, "An apparatus for the transformation of light of long wavelength into light of short wavelenght," *Physica*, **1**, 297 (1934)

[86] Lipich, F., "Über Brechung und Reflexion unendlich dünner Strahlensysteme an Kugelflächen," *Akad. d. Wiss. (Wien) Denkschrift*, **38**, **2**, 163 (1877)

[87] Picht, J., "Beiträge zur Theorie der geometrischen Elektronenoptik," *Ann. Phys.*, **15**, 926 (1932)

[88] Glaser, W., "Strenge berechnung magnetischer Linsen der Feldform $H = H_0/[1 + (z/a)^2]$," *Z. Phys.*, **117**, 285 (1940)

[89] Scherzer, O., "Zur Theorie elektronenoptischen Linsenfehler," *Z. Phys.*, **80**, 193 (1933);
"Die schwache elektrische Einzellinse geringster sphärischen Aberration," *Z. Phys.*, **101**, 23 (1936);
"Über einige Fehler von Elektronenlinsen," *Z. Phys.*, **101**, 593 (1936)

[90] Gabor, D., "A space-charge lens for the focusing of ion beams," *Electronic Eng.*, **15**, 295, 328, 372 (1942);
The electron microscope, Hulton Press, London, 1945

[91] Seeliger, R., "Über die Justierung sphärisch korrigierter elektronoptischer Systeme," *Optik*, **10**, 29 (1953)

[92] Christofilos, N. C., US Patent, 2 736 799, (1950)

[93] Courant, E. D., Livingston, M. S., Snyder, H. S., "The strong-focusing synchrotron – A new high energy accelerator," *Phys. Rev.*, **88**, 1190, (1952)

[94] Barber, N. F., "Note on the shape of an electron beam in a magnetic field," *Proc. Leeds Phil. Lit. Soc., Sci. Sect.*, **2**, 427 (1933)

[95] Wehnelt, A., "Verteiligung des Stromes an der Oberflächen von Katoden in Entladungsröhren," *Ann. Phys.*, **7**, 237 (1902)

[96] Liouville, J., "Expression remarquable de la quantité qui, dans le movement d'un système de points matériels a liaison quelconques, est un minimum au vertu de principle de la moindre action," *J. math. pures et appliquées, ser 2*, **1**, 297 (1856)

[97] Langmuir, D. B., "Theoretical limits of cathode-ray tubes," *Proc. IRE*, **25**, 977 (1937);
Pierce, J. R., "Limiting current densities in electron beams," *J. Appl. Phys.*, **10**, 715 (1939)

[98] Shoulders, R. K., "Microelectronics using electron-beam-activated machinig techniques," *Adv. Computers*, **2**, 135 (1961)

[99] Spindt, C. A., "A thinn-film field-emission cathode," *J. Appl. Phys.*, **39**, 3504 (1968);
Spindt, C. A., Shoulders, K. R., Heynich, L. N., US Patent 3 755 704 (1973);
Spindt, C. A., Brodie, I., Humprey, L., Westerberg, E. R., "Physical properties of thin-film field emission cathodes with molybdenum cone," *J. Appl. Phys.*, **47**, 5248 (1976)

[100] Iams, H. E., Salzberg, B., "The secondary emission phototube," *Proc. IRE*, **23**, 55 (1935)

[101] Weiss, G., "Über Sekundärelektronen-Vervielfacher," *Z. Tech. Phys.*, **17**, 623 (1936)

[102] Zworykin, V. K., Morton, G. A., Malter, L., "The secondary emission multiplier — A new electronic device," *Proc. IRE*, **24**, 351 (1936)

[103] Zworykin, V .K., Rajchman, J. A., "The electrostatic electron multiplier," *Proc. IRE*, **27**, 558 (1939)

[104] Mc Gee, J. D., Flinn, E. A., Evans, M. D., "An electron image multiplier," in *Advances in Electronics and Electron Physics*, **12**, 87 (1960);
Burns, J., Neumann, M. J., "The channeled image intensifier," in *Advances in Electronics and Electron Physics*, **12**, 97 (1960)

[105] Brüche, E., "Elektronmikroskopische Abbildung mit lichtelektrischen Elektronen," *Z. Phys.*, **86**, 448 (1933);
Schaffernicht, W., "Über die Umwandlung von Lichtbilder in Elektronenbilder," *Z. Phys.*, **93**, 762 (1935)

[106] Zworykin, V. K., Morton, G. A., "Applied electron optics," *J. Opt. Soc. Amer.*, **26**, 181, (1936)

[107] Morton, G. A., Ramberg, E. G., "Electron optics of an image tube," *Physics*, **7**, 451 (1936)

[108] Krizek, V., Vand, V., "The development of infra-red technique in Germany," *Electron. Eng.*, **18**, 316, (1946)

[109] Nipkow, P., German patent 30 105, 1884

[110] Zworykin, V. K., US patent 2 141 059, 1923

[111] Farnsworth, P. T., "Television by electron image scanning," *J.Franklin Inst.*, **218**, 411 (1934)

[112] Zworykin, V. K., "The iconoscope — A modern version of the electric eye," *Proc. IRE*, **22**, 16 (1934)

[113] Ardenne, M. von, "Das Elektronen-Rastermikroskop. Theoretische Grundlagen," *Z. Phys.*, **109**, 553 (1938);
      "Das Elektronen-Rastermikroskop. Praktische Ausführung," *Z. Techn. Phys.*, **19**, 407 (1938)

[114] Müller, E. W., "Elektronmikroskopische Beobachtungen von Feldkatoden," *Z. Phys.*, **106**, 541 (1937)

[115] Binning, G., Rohrer, H., Gerber, Ch., Weibel, E., "Tunneling through a controlled vacuum gap," *Appl. Phys. Lett.*, **40**, 178 (1982);
      Binning, G., Rohrer, H., "Scanning tunneling microscopy," *Helv .Phys. Acta*, **55**, 727 (1982)

[116] Binning, G., Gerber, C., Quate, C. F., "Atomic force microscope," *Phys. Rev. Lett.*, **56**, 930 (1986)

[117] Farnsworth, P. T., "Bombardment of metal surfaces by slow moving electrons," *Phys. Rev.*, **20**, 358 (1922)

[118] Auger, P., "Sur les rayons β secondaries produit dans un gaz par les rayons X," *C. R. Acad. Sci. Paris*, **180**, 65 (1925)

[119] Lander, J. J., "Auger peaks in the energy spectra of secondary electrons from various materials," *Phys. Rev.*, **91**, 1382 (1952)

[120] Harris, L. A., "Analysis of materials by electron-excited Auger electrons," *J. Appl. Phys.*, **39**, 1419, 1428 (1968)

[121] Rutherford, E., "The connection between β and γ spectra," *Phil. Mag.*, **28**, 305 (1914)

[122] Steinhardt, R. G. Jr., "An X-ray photoelectron spectrometer for chemical analysis," Ph. D. dissertation, Lehig University, 1950;
      Steinhardt, R. G. Jr., Serfass, E. J., "X-ray photoelectron spectrometer for chemical analysis," *Analyt. Chem.*, **23**, 1585 (1951);
      Siegbahn, K., in **Beta- and gamma-ray spectroscopy**, Amsterdam, 1956

[123] Thomson, J. J., "Rays of positive electricity," *Phil. Mag.*, **20**, 252 (1910)

[124] Dempster, A., "A new method for positive ray analysis," *Phys. Rev.*, **11**, 316 (1918)

[125] Aston, F., Lindemann, F. A., "The possibility of separating isotopes," *Phil. Mag.*, **37**, 523 (1919);
      Aston, F., "A positive ray spectrograph," *Phil. Mag.*, **38**, 709 (1919)

[126] Viehböck, F., Herzog, R. F. K., "Ion source for mass spectrography," *Phys. Rev.*, **76**, 855 (1949)

[127] Herzog, R. F. K., Leibl, H. J., "Sputtering ion source for solids," *J. Appl. Phys.*, **34**, 2893 (1963)

[128] Paul, W., Steinwedel, H., "Ein neues Massenspektrometer ohne Magnetfeld," *Z. Naturforschung*, **8 a**, 448 (1953);
      Paul, W., Reinhard, H. P., von Zahn, U., "Das elektrische Massenfilter als Massenspektrometer und Isotoptrenner," *Z. Phys.*, **152**, 143 (1958)

[129] Pierce, J. R., "Rectilinear electron flow in beams," *J. Appl. Phys.*, **11**, 548 (1940)

[130] Brillouin, L., "A theorem of Larmor and its importance for electrons in magnetic field," *Phys. Rev.*, **67**, 260 (1945)

[131] Wang, C. C., "Electron beams in axially symmetric electric and magnetic fields," *Proc. IRE*, **38**, 135 (1950)

[132] Pierce, J. R., "Spatially alternating magnetic fields for focusing low-voltage electron beams," *J. Appl. Phys.*, **24**, 1247 (1953)

[133] Vibrans, G. E., "Computation of spreading of an electron beam under acceleration and space-charge repulsion," *Tech. Rept. No. 308*, Lincoln Lab. (1953)

[134] Busch, H. W. H., ref. 25; the article gives a solution for electron motion in a general electric and magnetic field, implicitly including Busch's theorem

[135] Arsenjewa-Heil, A., Heil, O., "Eine neue Methode zur Erzeugung kurzer, ungedämpfter, elektromagnetischen Wellen grosser Intensität," *Z. Phys.*, **95**, 752 (1935)

[136] Brüche, E., Recknagel, A., "Über die Phasenfokussierung bei der Elektronenbewegung in schnell-veränderlichen elektrischen Feldern," *Z. Phys.*, **108**, 459 (1938)

[137] Varian, R. H., Varian, S. F., "A high frequency osillator and amplifier," *J. Appl. Phys.*, **10**, 321 (1939)

[138] Pierce, J. R., Shepherd, W. G., "Reflex oscillators," *Bell Syst. Techn. J.*, **26**, 460 (1947)

[139] Kompfner, R., "Traveling wave valve," *Wireless World*, **52**, 369 (1946);
"The traveling-wave tube as amplifier at microwaves," *Proc. IRE*, **35**, 124 (1947)

[140] Hull, A. W., "The effect of a uniform magnetic field on the motion of electrons between coaxial cylinders," *Phys. Rev.*, **18**, 31 (1921);
"The path of electrons in the magnetron," *Phys. Rev.*, **23**, 112 (1924)

[141] Posthumus, K., "Oscillations in a split-anode magnetron. Mechanism of generation," *Wireless Eng.*, **12**, 126 (1935)

[142] Samuel, A. L., US patent 2 036 342, 1934

[143] Alekseiev, N. F., Malairov, D. D., "Получение мощных колебаний магнетроном в сантиметровом диапазоне волн (Generation of high-power oscillations with the magnetron in the centimeter band)," *Ž. Tehn. Fiz.*, **10**, 1297 (1938)

[144] Hartree, D. R., *CVD* (Coordination of Valve Development Committee, Ministry of Defence) *Report No. 1536*, 1941

[145] Rieke, F. F., "Analysis of magnetron performance," *MIT Rad. Lab. Report 52-10*, Sept. 1943

[146] Smith, P. H., "Transmission line calculator," *Electronics*, **12**, 29 (1939)

[147] Syers, J., Brit. patent 588 916, 1941;
Spencer, P. L., US patent 2 546 870, 1941

[148] Gaponov, A. V., "Взаимодействие непрямолинейых электроннйх потоков с електромагнитиым волнами в линиях передачи (Interaction between nonlinear electron beams and electromagnetic waves in transmission lines)," *Izv. VUZ Radiofiz.*, **2**, 837 (1959)

[149] Twiss, R. Q., "Radiation transfer and the possibility of negative absorption in radio astronomy," *Aust. J. Phys.*, **11**, 564 (1958)

[150] Schneider, J., "Stimulated emission of radiation by relativistic electrons in a magnetic field," *Phys. Rev. Lett.*, **2**, 504 (1959)

[151] Philips, R. M., "The ubitron, a high-power traveling-wave tube based on periodic beam interaction in unloaded waveguide," *IRE Trans. Electron Dev.*, **ED-7**, 231 (1960)

[152] Ono, S., Yamanouchi, Y., Shibata, Y., Koike, Y., *Proceedings of the 4th international conference on microwave tubes*, 355 (1962)

[153] Nation, J.J., "On the coupling of an high-current relativistic electron beam to a slow wave structure," *Appl. Phys. Lett.*, **17**, 491 (1970)

[154] Bekefi, G., Orzechowski, T., "Giant microwave burst from a field-emission, relativistic-electron-beam magnetron," *Phys. Rev. Lett.*, **37**, 379 (1976)

[155] Mahaffey, R. A., Sprangle, P., Golden, J., Kapetanakos, C. A., "High-power microwaves from a nonisochronic reflecting electron system," *Phys. Rev. Lett.*, **39**, 843 (1977)

[156] Rutherford, E., "Collision of α particles with light atoms. An anomalous effect in nitrogen," *Phil. Mag.*, **37**, 581 (1919)

[157] Slepian, J., US patent 1 645 304 (1922)

[158] Ising, E., "Prizip einer Methode zur herstellung von Kanalstrahlen hoher Voltzahl," *Ark. mat., astronomi och fysik*, **18**, No. 30, 1 (1924)

[159] Wideröe, R., "Über ein neues Prinzip zur Herstellung hoher Spannungen," *Arch. Elektrotech.*, **21**, 387 (1928)

[160] Lawrence, E. O., Edlefsen, N. F., "On the production of high speed protons," *Science*, **72**, 376 (1930)

[161] Lawrence, E. O., Livingston, M. S., "A method of producing high speed hydrogen ions without the use of high voltages," *Phys. Rev.*, **37**, 1707 (1931)

[162] Cockroft, J. D., Walton, E. T. S., "Experiments with high velocity positive ions," *Proc. Roy. Soc.*, **A 129**, 47 (1930)

[163] Van de Graaff, R. J., "A 1,500,000 volt electrostatic generator," *Phys. Rev.*, **38**, 1919 (1931)

[164] Kerst, D., "Acceleration of electrons by magnetic induction," *Phys. Rev.*, **58**, 841 (1940);
"Induction electron accelerator," *Phys. Rev.*, **59**, 110 (1941);
"The accelerator of electrons by magnetic induction," *Phys. Rev.*, **60**, 47 (1941)

[165] Bethe, H. A., Rose, M. E., "The maximum energy obtainable from the cyclotron," *Phys. Rev.*, **52**, 1254 (1937)

[166] Veksler, W. J., "A new method for acceleration of relativistic particles," *Doklady Ak. Nauk*, **43**, 329 (1944);
       "On a new method of acceleration of relativistic particles," *Doklady Ak. Nauk*, **44**, 365 (1944)
[167] McMillan, E. M., "The synchrotron — A proposed high energy particle accelerator," *Phys. Rev.*, **68**, 143 (1945)
[168] O'Neill, G. K., "Design of a quantum electrodynamics limit experiment," *Bull. Am. Phys. Soc., Ser. 2*, **3**, 158 (1958);
       "Component design and testing for the Princeton-Stanford colliding-beam experiment," *Proceedings of the international conference on high energy accelerators,* Brookhaven, 1961, **USAEC**, 247 (1961)
[169] Kerst, D., Serber, R., "Electron orbits in the induction accelerator," *Phys. Rev.*, **60**, 53 (1941)
[170] Thomas, L. H., "The paths of the ions in the cyclotron. I. Orbits in the magnetic field," *Phys. Rev.*, **54**, 580 (1938);
       "The paths of the ions in the cyclotron. II. Path in the combined electric and magnetic field," *Phys. Rev.*, **54**, 588 (1938)
[171] Alvarez, L. W., "The design of a proton linear accelerator," *Phys. Rev.*, **70**, 799 (1946)
[172] Woodyard, J. R., "A comparison of the high frequency accelerator and betatron as a source of high energy electrons," *Phys. Rev.*, **69**, 50 (1946);
       Ginzton, E. L., Hansen, W. W., Kennedy, W. R., "A linear electron accelerator", *Rev. Sci. Instr.*, **19**, 89 (1948)
[173] Knapp, E., "Accelerating structure research at Los Alamos," *Minutes of the conference on proton linear accelerators at Yale University*, 131 (1963);
       "Design, construction, and testing of rf structures for a proton linear accelerator," *IEEE Trans. Nucl. Sci.*, **NS-12**, 118 (1965)
[174] Budker, G. I., "Status report of works on storage rings at Novosibirsk," *Proceedings of the international symposium on electron and positron storage rings*, Saclay (1966)
[175] Meer, S. van der, "Stochastic dampning of betatron oscillations in the ISR," *CERN/ISR-PO/72-31* (1972)
[176] Guericke, O. von, **Experimenta nova** (ut vocantur) **Magdenburgica de vacuo spatio**, Amstelodami (Amsterdam), 1672
[177] Musschenbroek, P. van, **Institutiones physicae**, Leiden, 1748
[178] Franklin, B., "New Experiments and Observations on Electricity," *letters to Collinson, secretary of The Royal Society, fourth letter, 29.07.1750.*
[179] Petrov, V. V., "Notice sur les expériences galvano-voltaïques," *Akad. med. chir.*, St. Petersburg (1803)
[180] Davy, H., "On some new electrochemical researches, on various objects, particularly the metallic bodies, from the alkalies and earths, and on some combinations of hydrogen," *Phil. Trans. Roy. Soc.*, **25**, 16 (1810);
       "Further researches on the magnetic phenomena produced by electricity; with some new experiments on the properties of electrified bodies in their relation to conducting powers and temperature," *Phil. Trans. Roy. Soc.*, **35**, 425 (1821)
[181] Faraday, M., "A speculative touching electric conduction and the nature of matter," *Phil. Mag.*, **24**, 136 (1844)
[182] Rühmkorff, H. D., "Sur quelques expériences d'induction magneto-electrique," *C. R. Acad. Sci. Paris*, **73**, 922 (1871), (inductor invented in 1851)
[183] Crookes, W., "On a fourth state of matter," *Proc. Roy. Soc.*, **30**, 469 (1880)
[184] Goldstein, E., "Über die Entladung der Elecriticität in verdünnten Gasen," *Wied. Ann. Phys.*, **11**, 832, (1881), *Wied. Ann. Phys.*, **12**, 249 (1881)
[185] Townsend, J. S. E., **Electricity in Gases**, Oxford, 1915
[186] Langmuir, I., "Positive ion currents in the positive column of the mercury arc," *Gen. Elect. Rev.*, **26**, 731 (1923);
       Langmuir, I., Mott-Smith, H., "Studies of electric discharges in gases at low pressures, Part I — V," *Gen. Elect. Rev.*, **27**, 449, 538, 616, 762, 810 (1924)
[187] Saha, M. N., "Ionization in the solar chromosphere," *Phil. Mag.*, **40**, 472 (1920);
       "On the temperature ionization of the higher groups in the periodic classification," *Phil. Mag.*, **44**, 1128 (1922)

[188] Paschen, F., "Über die zum Funkübergang in Luft, Wasserstoff und Kohlsäure bei verschiedenen Drucken erforderliche Potentialdifferenz," *Wied. Ann. Phys.*, **37**, 69 (1889)

[189] Tonks, L., "The birth of 'plasma'," *Am. J. Phys.*, **35**, 857 (1967)

[190] Debye, P., Hückel, E., "Zur Theorie der Elektrolyte," *Phys. Zeit.*, **24**, 185 (1923)

[191] Becquerel, E., "Phosphorence de gaz par l'électricité," *C. R. Acad. Sci. Paris*, **48**, 404 (1859)

[192] Way, J. T., Hungerford suspension bridge in London, Sept. 3, 1860, in Cobine, J. D., "The development of gas discharge tubes," *Proc. IRE*, **50**, 970 (1962)

[193] Cooper-Hewitt, P., Mercury arc lamp, 1901, in Buttolph, L. J., "The Cooper-Hewitt Lamp. Part II. Development and application," *Gen. Elect. Rev.*, **23**, 858 (1920)

[194] Küch, R., Retschinsky, T., "Photometrische und spektralphotometrische Messungen am Quecksilber-lichtbogen bei hohem Dampfdruck," *Ann. Phys.*, **20**, 563 (1906)

[195] Cloude, G., "Les tubes luminiscent au néon," *C. R. Acad. Sci. Paris.*, **151**, 1122 (1908), **152**, 1377 (1908)

[196] Mac Kay, G. M. J., Charlton, E. E., "The luminous electrical discharge in sodium vapor," *Phys. Rev.*, **21**, 209 (1923)

[197] Jamin, M. M., "Sur le courant de réaction de l'arc électrique," *C. R. Acad. Sci. Paris*, **94**, 1615 (1882)

[198] Cooper-Hewitt, P., "Mercury arc rectifier," *Electric World and Engineer*, 121 (Jan 1903)

[199] Langmuir, I., US patent 1 289 823, 1914

[200] Hull, A. W.,"Gas-filled thermionic valves," *Trans. AIEE*, **47**, 753 (1928)

[201] Drewell, P., "Über die Erzeugung und Anwendung kurzer Stromstöse mittels Röhrenschaltung," *Z. Techn. Phys.*, **16**, 614 (1935); "Die Wirkungsweise der gittergesteuerten Gasentladungsröhre bei ihrer Verwendung als Schwingungs-erzeuger," *Z. Techn. Phys.*, **17**, 249 (1936)

[202] Lawson, J. L., *MIT Rad. Lab. Report* (1941); Samuel, A. L., McCrae, J. W., Mumford, W. W., "Gas discharge TR switch," *Bell Tel. Lab*, MM-42-140-26 (1942)

[203] Slepian, J., Ludwig, L. R., "A new method for initiating the cathode of an arc," *Trans. AIEE*, **52**, 643 (1933)

[204] Maiman, T.H., "Optical and microwave-optical experiments in ruby," *Phys. Rev. Lett.*, **4**, 564 (1960)

[205] Javan, A., Bennett, W. R. Jr., Heriott, D. R., "He–Ne laser," *Phys. Rev. Lett.*, **6**, 106 (1961)

[206] Patel, C. K. N., "Interpretation of CO_2 optical maser experiments," *Phys. Rev. Lett.*, **12**, 588 (1964)

[207] Searles, S. K., Hart, G. A., "Stimulated emission at 281.8 nm from XeBr," *Appl. Phys. Lett.*, **27**, 243 (1975)

[208] Grove, W. R., "On the electro-chemical polarity of gases," *Phil. Trans. Roy. Soc.*, **142**, 87 (1852).

[209] Fruth, H. F., "Cathode sputtering. A commercial application," *Physics*, **2**, 280 (1932)

[210] Harrison, D. E. Jr., Magnuson, G. D., "Sputtering tresholds," *Phys. Rev.*, **122**, 1421 (1961); Stuart, R. V., Wehner, G. K., "Sputtering yields at very low bombarding ion energies," *J. Appl. Phys.*, **33**, 2345 (1962)

[211] Irving, S. M., "A dry photoresist removal method," National Meeting ECS, Abstract 180, 1967; "A plasma oxidation process for removing photoresist films," *Solid State Techn.*, **14** (**6**), 47 (1971)

[212] Chopra, K. L., Randlett, M. R., "Duoplasmatron ion beam source for vacuum sputtering of thin films," *Rev. Sci. Instr.*, **38**, 1147 (1967); Wasa, K., Hayakawa, S., "Low pressure sputtering system of the magnetron type," *Rev. Sci. Instr.*, **40**, 693 (1969)

[213] Shockley, W., US patent 2 787 564, 1954

[214] Rutherford, E., Geiger, H., "An electrical method of counting the number of α-particles from radio-active sources," *Proc. Roy. Soc.*, **81**, 141 (1908); Geiger, H., "Einfache Methode zur Zahlung von α- und β-Teilchen," *Verh. d. Dtsch. Phys. Ges.*, **15**, 534 (1913)

[215] Torricelli, E., his friend Viviani made the glass tube and performed the first experiment in 1643

[216] McLeod, H., "Apparatus for measurements of low pressure of gas," *Phil. Mag.*, **48**, 110 (1874)

[217] Gaede, W., "Demonstration einer rotierenden Quecksilberluftpumpe," *Phys. Zeit.*, **6**, 758 (1905);
      "Demonstration einer neuen Verbesserung an der rotierenden Quecksilber-Luftpumpe," *Phys. Zeit.*, **8**, 852 (1907)

[218] Gaede, W., "Die Diffusion der Gase durch Quecksilberdampf bei niederen Drucken und die Diffusions-
      luftpumpe," *Ann. Phys.*, **46**, 357 (1915)

[219] Dewar, J., "Studies of charcoal and liquid air," *Roy. Inst. Proc.*, **18**, 433 (1906)

[220] Gaede, W., "Die äusere Reibung der Gase und ein neues Prinzip für Luftpumpen: die Molekularluft-
      pumpe," *Phys. Zeit.*, **13**, 864 (1912);
      "Die molekularluftpumpe," *Ann. Phys.*, **41**, 337 (1913)

[221] Becker, W., "Eine neue Molekularpumpe," *Vakuum-Technik*, **7**, 149 (1958)

[222] Pirani, M. S., "Selbstzeigendes Vakuum-Messinstrument," *Verh. d. Deut. Phys. Ges.*, **8**, 686 (1906)

[223] Penning, F. M., "Ein neues Manometer für niedrige Gasdrucke, insbesondere zwischen 10^{-3} und 10^{-5}
      mm," *Physica*, **4**, 71 (1937)

[224] Buckley, O. E, "An ionisation manometer," *Proc. Natl. Acad. Sci.*, **2**, 683 (1916);
      Dushman, S., Found, C. G., "Studies with the ionization gauge," *Phys. Rev.*, **17**, 7 (1921)

[225] Bayard, R. T., Alpert, D. A., "Extension of the low pressure range of the ionization gauge," *Rev. Sci.
      Instr.*, **21**, 571 (1950)

[226] Redhead, P. A., "The magnetron gauge: A cold cathode vacuum gauge," *Can. J. Phys.*, **37**, 1260 (1959);
      Lafferty, J. M., "Hot-cathode magnetron ionization gauge for measurement of ultrahigh vacua," *J. Appl.
      Phys.*, **32**, 424 (1961)

[227] Knudsen, M., "Die Gesetze der Molekularströmung und der inneren Reibungsströmung der Gase durch
      Röhren," *Ann. Phys.*, **28**, 75 (1909)

[228] Brillouin, L., "Wave guides for slow waves," *Phys. Rev*, **71**, 483 (1947)

[229] Floquet, G. A. M., "Sur les équations différentielles linéaires à coefficients périodiques," *C. R. Acad.
      Sci. Paris*, **91**, 880 (1880)

[230] Pierce, J. R., "Theory of the beam-type traveling-wave tube," *Proc. IRE*, **35**, 111 (1947);
      Traveling-wave tube, Toronto, 1950

Appendix L

Nobel Prizes Relevant to Electron Physics

1901 W. C. RÖNTGEN
      "in recognition of the extraordinary services he has rendered by discovery of the remarkable rays
      subsequently named after him"

1902 H. A. LORENTZ and P. ZEEMAN
      "in recognition of the extraordinary services they have rendered by their researches into the influence
      of magnetism upon radiation phenomena"

1903 A. H. BECQUEREL
      "in recognition of the extraordinary services he has rendered by his discovery of spontaneous ra-
      dioactivity"

      P. CURIE and M. CURIE SKLODOWSKA
      "in recognition of the extraordinary services they have rendered by their joint researches on the
      radiation phenomena discovered by Professor Henri Becquerel"

1905 P. E. LENARD
      "for his work on cathode rays"

1906 J. J. THOMSON
      "in recognition of the great merits of his theoretical and experimental investigations on the conduction
      of electricity by gases"

1908 Lord E. RUTHERFORD
"for his investigations into the disintegration of the elements, and the chemistry of radioactive substances"

1914 M. VON LAUE
"for his discovery of the diffraction of X-rays by crystals"

1919 M. PLANCK
"in recognition of the services he rendered to the advancement of Physics by his discovery of energy quanta"

1921 A. EINSTEIN
"for his services to Theoretical Physics, and especially for his discovery of the law of the photoelectric effect"

1922 N. BOHR
"for his services in the investigation of the structure of atoms and of the radiation emanating from them"

1922 F. W. ASTON
"for his discovery, by means of his mass spectrograph, of isotopes, in a large number of non-radioactive elements, and for his enunciation of the whole-number rule"

1924 M. SIEGBAHN
"for his discoveries and research in the field of X-ray spectroscopy"

1925 J. FRANCK and G. HERTZ
"for their discoveries of the laws governing the impact of an electron upon an atom"

1927 A. H. COMPTON
"for his discovery of the effect named after him"

C. T. WILSON
"for his method of making the paths of electrically charged particles visible by condensation of vapour"

1928 O. W. RICHARDSON
"for his work on the thermionic phenomenon and especially for the discovery of the law named after him"

1929 L. DE BROGLIE
"for his discovery of the wave nature of electrons"

1932 W. HEISENBERG
"for the creation of quantum mechanics, the application of which has, *inter alia*, led to the discovery of the allotropic forms of hydrogen"

1933 E. SCHRÖDINGER and P. A. M. DIRAC
"for the new discovery of new productive forms of atomic theory"

1935 J. CHADWICK
"for the discovery of the neutron"

1936 V. F. HESS
"for his discovery of cosmic radiation"

C. D. ANDERSON
"for his discovery of the positron"

1936 P. J. W. DEBYE
"for his contributions to our knowledge of molecular structure through his investigation on dipole moments and on the diffraction of X-rays and electrons in gases"

1937 C. J. DAVISSON and G. P. THOMSON
"for their experimental discovery of the diffraction of electrons by crystals"

1938 E. FERMI
"for his demonstration of the existence of new radioactive elements produced by neutron irradiation, and for his related discovery of nuclear reactions brought about by slow neutrons"

1939 E. O. LAWRENCE
"for the invention and development of the cyclotron and for results obtained with it, especially with regard to artificial radioactive elements"

1945 W. PAULI
"for the discovery of the Exclusion Principle, also called the Pauli Principle"

1948 P. M. BLACKETT
"for his development of the Wilson cloud chamber method, and his discoveries therewith in the fields of nuclear physics and cosmic radiation"

1951 J. D. COCKROFT and E. T. WALTON
"for their pioneer work on the transmutation of atomic nuclei by artificially accelerated atomic particles"

1956 W. SHOCKLEY, J. BARDEEN, and W. H. BRATTAIN
"for their researches on semiconductors and their discovery of the transistor effect"

1964 C. H. TOWNES, N. G. BASOV, and A. M. PROCHOROV
"for fundamental work in the field of quantum electronics, which has led to the construction of oscillators and amplifiers based on the maser–laser principle"

1968 L. W. ALVAREZ
"for his decisive contribution to elementary particle physics, in particular the discovery of large number of resonance states, made possible through his development of the technique of using hydrogen bubble chamber and data analysis"

1970 H. ALFVÉN (divided)
"for fundamental work and discoveries in magneto-hydrodynamics with fruitful applications in different parts of plasma physics"

1981 K. M. SIEGBAHN (divided)
"for his contribution to the development of high-resolution electron spectroscopy"

1984 C. RUBBIA and S. VAN DER MEER
"for their decisive contributions to the large project, which led to the discovery of the field particles W and Z, communication of weak interaction"

1986 E. RUSKA
"for his fundamental work in electron optics, and for the design of the first electron microscope"

G. BINNIG and H. ROHRER
"for their design of the scanning tunneling microscope"

1989 N. F. RAMSEY (divided)
"for the invention of the separated oscillatory fields method and its use in the hydrogen maser and other atomic clocks"

Appendix M

Langmuir–Blodgett Factor

β^2 as a function of $\frac{r}{r_c}$ for cylindric electrode geometry
r_c is the cathode radius and r the radius of an arbitrary point
$(\beta^2$ for $r > r_c, -\beta^2$ for $r < r_c)$

$\dfrac{r}{r_c}$ or $\dfrac{r_c}{r}$	β^2	$-\beta^2$	$\dfrac{r}{r_c}$ or $\dfrac{r_c}{r}$	β^2	$-\beta^2$
1.00	0.0000	0.0000	3.8	0.6420	5.3795
1.01	0.00010	0.00010	4.0	0.6671	6.0601
1.02	0.00039	0.00040	4.2	0.6902	6.7705
1.04	0.00149	0.00159	4.4	0.7115	7.5096
1.06	0.00324	0.00356	4.6	0.7313	8.2763
1.08	0.00557	0.00630	4.8	0.7496	9.0696
1.10	0.00842	0.00980	5.0	0.7666	10.733
1.15	0.01747	0.02186	5.2	0.7825	11.061
1.20	0.02815	0.03849	5.4	0.7973	11.601
1.30	0.05589	0.08504	5.6	0.8111	12.493
1.40	0.08672	0.14856	5.8	0.8241	13.407
1.50	0.11934	0.2282	6.0	0.8362	14.343
1.60	0.1525	0.3233	6.5	0.8635	16.777
1.70	0.1854	0.4332	7.0	0.8870	19.337
1.80	0.2177	0.5572	7.5	0.9074	22.015
1.90	0.2491	0.6947	8.0	0.9253	24.805
2.0	0.2793	0.8454	8.5	0.9410	27.701
2.1	0.3083	1.0086	9.0	0.9548	30.698
2.2	0.3361	1.1840	9.5	0.9672	33.791
2.3	0.3626	1.3712	10.0	0.9782	36.976
2.4	0.3879	1.5697	12.0	1.0122	50.559
2.5	0.4121	1.7792	16.0	1.0513	81.203
2.6	0.4351	1.9995	20.0	1.0715	115.64
2.7	0.4571	2.2301	40.0	1.0946	327.01
2.8	0.4780	2.4708	80.0	1.0845	867.11
2.9	0.4980	2.7214	100.0	1.0782	1174.9
3.0	0.5170	2.9814	200.0	1.0562	2946.1
3.2	0.5526	3.5293	500.0	1.0307	9502.2
3.4	0.5581	4.1126	∞	1.0000	∞
3.6	0.6148	4.7298			

If $w = \ln r/r_c \ll 1$ the following expressison is valid:

$$\beta^2 = w^2(1 - 0.8w + 0.344w^2 + \cdots)$$

$$\frac{d\beta^2}{dw} = 2w - 2.4w^2 + 1.374w^3 - 0.509w^4 + \cdots$$

$(-\alpha)^2$ as a function of $\frac{r}{r_c}$ for spherical electrode geometry

r_c is the cathode radius and r the radius of an arbitray point

$\dfrac{r}{r_c}$ or $\dfrac{r_c}{r}$	$(-\alpha)^2$	$\dfrac{r}{r_c}$ or $\dfrac{r_c}{r}$	$(-\alpha)^2$
1.00	0.00000	2.12	0.91729
1.01	0.00010	2.15	0.96124
1.02	0.000397	2.17	0.99098
1.05	0.00245	2.20	1.03627
1.10	0.00962	2.25	1.1135
1.15	0.02127	2.30	1.1929
1.20	0.03716	2.35	1.2745
1.25	0.05710	2.40	1.3581
1.30	0.08091	2.45	1.4438
1.35	0.10842	2.50	1.5314
1.40	0.13949	2.60	1.7127
1.45	0.17399	2.70	1.9016
1.50	0.21178	2.80	2.0979
1.55	0.25276	2.90	2.3015
1.60	0.29682	3.0	2.5120
1.65	0.34386	3.2	2.9533
1.70	0.39380	3.5	3.6634
1.75	0.44655	3.7	4.1673
1.80	0.50202	4.0	4.9662
1.85	0.56016	4.5	6.4057
1.90	0.62089	5.0	7.9708
1.92	0.64589	6.0	11.46
1.95	0.68414	7.0	15.35
1.97	0.71013	10.0	29.19
2.00	0.74985	30.0	178.2
2.02	0.77682	50.0	395.3
2.05	0.81798	100.0	1144
2.07	0.84589	300.0	6031
2.10	0.88846	500.0	13015

If $w = \ln r/r_c \ll 1$ the following expression is valid:

$$\alpha = w - 0.3w^2 + 0.075w^3 - 0.001432w^4 + 0.002161w^5 - 0.0002697w^6 + \cdots$$

Appendix N

Frequency Bands

Band No	Name	Frequency	Wavelength
2	ELF (extreme low frequency)	30–300 Hz	10000–1000 km
3	VF (voice frequency)	300–3000 Hz	1000–100 km
4	VLF (very low frequency)	3–30 kHz	100–10 km
5	LF (low frequency)	30–300 kHz	10–1 km
6	MF (medium frequency)	300–3000 kHz	1000–100 m
7	HF (high frequency)	3–30 MHz	100–10 m
8	VHF (very high frequency)	30–300 MHz	10–1 m
9	UHF (ultrahigh frequency)	300–3000 MHz	100–10 cm
10	SHF (superhigh frequency)	3–30 GHz	10–1 cm
11	EHF (extreme high frequency)	30–300 GHz	10–1 mm
12	Decimillimeter	300–3000 GHz	1–0.1 mm
	P band	0.23–1 GHz	130–30 cm
	L band	1–2 GHz	30–15 cm
	S band	2–4 GHz	15–7.5 cm
	C band	4–8 GHz	7.5–3.75 cm
	X band	8–12.5 GHz	3.75–2.4 cm
	Ku band	12.5–18 GHz	2.4–1.67 cm
	K band	18–26.5 GHz	1.67–1.13 cm
	Ka band	26.5–40 GHz	1.13–0.75 cm
	Millimeter	40–300 GHz	7.5–1 mm
	Submillimeter	300–3000 GHz	1–0.1 mm

SYMBOLS USED IN TEXT

$\mathbf{A} \ldots \mathbf{Z}$	matrices	F	focus
$\mathbf{A} \ldots \mathbf{z}$	vectors	$F, \mathbf{F}$	force
$\mathcal{A} \ldots z$	rf components	F_e	electric force
a	anode (subscript)	F_m	magnetic force
	object distance	FD	Fermi–Dirac (subscript)
a_{ij}	matrix coefficient	g	beam loading conductance
$a, \mathbf{a}$	acceleration		grid (subscript)
A	amplification	g_m	transconductance
	amplitude	G	flow conductivity
	area		termodynamic potential
	mass number	G_L	load conductance
	thermionic constant	G_r	cavity conductance
	work	h	mesh size (FDM)
$\mathbf{A}$	magnetic vector potential		Planck's constant
$\mathbf{A}$	matrix		step length
b	image distance	H	Hamilton's function
	beam loading susceptance		principal plane (subscript)
$B, \mathbf{B}$	magnetic induction	$H, \mathbf{H}, \mathcal{H}$	magnetic field
B_B	Brillouin field	i	time-varying current
c	heat capacitivity	i, I	electric current
	light velocity	$\mathbf{i}, \mathbf{j}, \mathbf{k}$	unit vectors in Cartesian
	cathode (subscript)		coordinates
c, C	constant, integration constant	$\Im$	imaginary part of a
C	amplification parameter (TWT)		complex number
	capacitance	j	imaginary unit, $\sqrt{-1}$
d	diameter		ion (subscript)
	distance	$J, \mathbf{J}$	current density
D	diffusion constant	J_0	DC current density
$\mathbf{D}$	dielectric displacement vector	J_b	initial current density
e	electron (subscript)	J_n	Bessel function
	electron charge	J_r	space-charge current density
E	energy (Chapter 2)	J_{sat}	saturation current density
$E, \mathbf{E}, \mathcal{E}$	electric field	k	Boltzmann's constant
E_F	Fermi energy	k, K	constant
E_k	kinetic energy (Chapter 2)	$\mathbf{K}$	surface current
E_p	potential energy (Chapter 2)	l, L	length
f	frequency	L	attenuation, power loss
	focal length		inductance
F	Faraday's constant	m	molar weight
	reduction factor for space-charge	m, M	mass

m_e	electron mass		R	resistance
m_0	electron rest mass		R_i	inner resistance
m_p	proton mass		R_a	dynamic plate resistance
M	molecular mass		$\Re$	real part of a complex
M_B	beam coupling coefficient			number
$M_l, \mathcal{M}_l$	lateral magnification		$s, \mathbf{s}$	path length
M_θ	angular magnification		S	energy state
MB	Maxwell–Boltzmann (subscript)			entropy
$\mathbf{M}$	magnetic auxiliary vector			optical path
n	field index			pumping velocity
	gas density		$S, \mathbf{S}$	area
	refraction index		t	time
n, N	integer		t_0	transit time, initial value
	number of particles		T	temperature
$\mathbf{n}$	unit outward pointing vector		$u, \mathbf{u}$	velocity (central motion)
N_A	Avogadro's number		u, w	principal orbits (lens)
N_M	number of kilomoles		U	energy (Chapter 2)
p	perveance		v, V	volume
	pressure		$v, \mathbf{v}$	velocity
$p, \mathbf{p}$	momentum		v_0	constant velocity, mean
p_g	ultimate pressure			velocity
P	power		$v_D, \mathbf{v}_D$	drift velocity
	distribution function		v_f	phase velocity
P_{FD}	Fermi–Dirac distribution		v_T	mean thermal velocity
P_{MB}	Maxwell–Boltzman distribution		v_z	axial velocity
q, Q	electric charge		V	potential, voltage
Q	heat		V_0	DC potential
	Q-value		V_b	breakdown voltage
	potential ratio (lens)		V_B	DC source voltage
Q_e	external Q-value		V_m	virtual cathode potential
Q_L	loaded Q-value		V_{st}	control potential
Q_o	unloaded "cold" Q-value		W	work, energy
Q_w	unloaded "varm" Q-value		W_f	work function
Q_{pV}	suction power		W_k	kinetic energy
Q_V	volume flow		W_i	injection energy
Q_x, Q_y	betatron tune		W_p	potential energy
r	reflexion coefficient		W_T	total relativistic energy
r, R	radius		$\mathcal{W}$	probability
r, z, θ	cylindrical coordinates		x, y, z	Cartesian coordinates
r, θ, φ	spherical coordinatess		$\dot{x}, \dot{y}, \dot{z}$	time derivatives
$\mathbf{r}_0, \mathbf{z}_0, \theta_0$	unit vectors in cylindrical		x', y'	space derivatives
	coordinates		X	bunching parameter
R	flow resistance		Y	beam loading impedance
	gas constant		Z	impedance, shunt
	recombination coefficient			impedance
	residuum		Z_0	characteristic impedance

Z_n	beam coupling impedance	ρ	radius of curvature
α	angle		reflection coefficient
	attenuation factor		space-charge
	constant	ρ_0	time constant space-charge
	momentum compaction factor	σ	conductivity
	Townsend's first coefficient		gas kinetic cross section
α_m	limiting angle for total reflexion		surface charge
β	constant	τ	rf period
	phase constant		time constant
	propagation constant		transit time
	ratio v/c	τ_j	ion life-time
	underrelaxation coefficient	φ	phase angle
$\beta_e, \beta_n, \beta_q$	propagation constant		potential
γ	relativistic factor, $\sqrt{1 - \beta^2}$	φ_f	work function potential
	secondary ionization factor	φ_r	phase stable angle
	(Townsend's second coefficient)	ω	angular frequency
δ	secondary emission factor		angular velocity
ε	emittance		overrelaxation coefficient
	error	ω_c	cyclotron frequency
	permitivity	ω_f	phase oscillation frequency
ε_0	permitivity of vacuum	ω_p	plasma frequency
η	current amplification factor	ω_q	reduced plasma frequency
	dynamic viscosity coefficient	ω_r	revolution frequency
	ratio between electron charge	Γ	complex propagation constant
	and mass	Ψ	magnetic flux
η_q	ratio between particle charge	Δ	residual
	and mass (q/m_0)		
θ	angle		
	transit time angle		
κ	constant		
λ	mean free path		
	wavelength		
λ_D	Debye length		
μ	amplification factor		
	chemical potential		
	mobility		
	ratio between particle mass and		
	restmass (m/m_0)		
	permeability		
μ_0	permeability of vacuum		
μ_e	electrochemical potential		
ν	constant, number of particles per		
	unit volume		
	collision frequency		
ρ	density		
	mass number		

LIST OF TABLES

COMMON ABBREVIATIONS

AC	alternating current
AES	Auger electron spectrometer
AFC	automatic frequency control
AFM	atomic force microscope
AM	amplitude modulation
ARE	activated reactive evaporation
BWO	backward wave oscillator
CAD	computer-aided design
CCD	charge-coupled device
CERN	Organisation Europiéenne pour la Recherche Nucléaire
CFA	cross-field amplifier
CLIC	CERN linear collider
CRT	cathode ray tube
CVD	chemical vapor deposition
CW	continuous wave
DC	direct current
DMD	digital mirror device
EBASLM	electron-beam addressed spatial light modulator
EEM	electron emission microscope
EMA	electron microprobe analysis
EMP	electromagnetic pulse (nuclear explosion)
ESCA	electron spectroscopy for chemical analysis
FD	Fermi–Dirac
FEA	field emission array cathodes
FEE	field emission emitter
FEL	free electron laser
FIM	field ion microscope
FM	frequency modulation
HDTV	high-definition television
HPM	high-power microwaves
HV	high-voltage source
IAC	ion-assisted coating
IAD	ion-assisted deposition
IBAD	ion-beam-assisted deposition
IBAE	ion-beam-assisted etching
IBED	ion-beam-enhanced deposition
IBS	ion-beam sputtering
ICB	ion-cluster-beam deposition
IDT	image dissector tube
IEM	ion emission microscope
IIXS	ion-induced X-ray spectroscopy

ITS	ion trap spectrometer
IVD	ion vapor deposition
LEED	low-energy electron diffraction
LEP	large electron–positron collider (CERN)
MB	Maxwell–Boltzmann
MCP	microchannel plate
MS	mass spectrometer
MSLM	microchannel spatial light modulator
PECVD	plasma-enhanced chemical vapor deposition
PFN	pulse-forming network
PIXE	proton-induced X-ray emission
PMT	photomultiplier
RBS	Rutherford backscattering
RIE	reactive ion etching
RMEED	reflection medium-energy electron diffraction
rf	high frequency (radio frequency)
SEM	scanning electron microscope
SIMS	secondary ion mass spectrometer
SNMS	sputtering neutral mass spectrometer
STEM	scanning transmission electron microscope
STM	scanning tunneling microscope
STP	standard pressure and temperature (10^5 Pa, $0°$C)
TEM	transmission electron microscope
TOF	time-of-flight spectrometer
TV	television
TWT	traveling wave tube
UPS	UV photoelectron spectroscopy
VSWR	voltage standing-wave ratio
XPS	X-ray photoelectron spectroscopy

PHYSICAL CONSTANTS

Light velocity in vacuum	c	$= 2.99792458 \times 10^8$ m/s
Permeability of vacuum	μ_0	$= 1.25663706 \times 10^{-7}$ Vs/Am
Permittivity of vacuum	ϵ_0	$= 8.85418782 \times 10^{-12}$ As/Vm
Atomic mass unit	m_u	$= 1.6605402 \times 10^{-27}$ kg
Proton rest mass	m_p	$= 1.6726485 \times 10^{-27}$ kg
Neutron rest mass	m_n	$= 1.6749543 \times 10^{-27}$ kg
Electron rest mass	m_e	$= 9.1093897 \times 10^{-31}$ kg
Ratio proton/electron rest mass	m_p/m_e	$= 1.83615152 \times 10^3$
Electron charge	e	$= 1.60217733 \times 10^{-19}$ As
Electron charge/mass ratio	e/m_e	$= 1.7588196 \times 10^{11}$ As/kg
Planck's constant	h	$= 6.6260755 \times 10^{-34}$ Js
Avogadro's number	N_A	$= 6.0221367 \times 10^{26}$ kmol^{-1}
Molar gas constant	R	$= 8.314510 \times 10^3$ J/(kmol K)
Boltzmann's constant	k	$= 1.380658 \times 10^{-23}$ J/K
Mole volume (STP)	V_m	$= 22.41410$ m^3/kmol
Faraday constant	F	$= 9.6485309 \times 10^7$ As/kmol

1 electronvolt	$1.60217733 \times 10^{-19}$ J
1 electronvolt/k	1.1604445×10^4 K
1 Kelvin$\cdot k$	8.617386×10^{-5} eV
Electron rest mass energy	5.110034×10^5 eV
Proton rest mass energy	9.382796×10^8 eV

BIBLIOGRAPHY

Comprehensive Works

Hemenway C. L., Henry R. W., Caulton M., *Physical Electronics*
  John Wiley & Sons, New York, 1967

Klemperer O., *Electron Physics*
  Butterworths, London, 1972

Beck A. H., Ahmed H., *An Introduction to Physical Electronics*
  Edward Arnold Ltd., London, 1968

Eichmeier J., *Moderne Vakuumelektronik*
  Springer, Berlin, 1981

Lonngren K. E., *Introduction to Physical Electronics*
  Allyn and Bacon, Boston, 1988

Numerical Methods

Southwell R. V., *Relaxation Methods in Theoretical Physics*
  Clarendon Press, Oxford, 1946

Goldstine H. H., *The Computer from Pascal to von Neumann*
  Princeton University Press, Princeton, NJ, 1973

Goldstine H. H., *A History of Numerical Analysis from the 16th through the 19th Century*
  Springer, Berlin, 1977

Colonias J. S., *Particle Accelerator Design: Computer Programs*
  Academic Press, New York, 1974

Birkhoff G., Lynch R. E., *Numerical Solution of Elliptic Problems*
  SIAM, Philadelphia, 1984

Birdsall C. K., Langdon A. B., *Plasma Physics via Computer Simulation*
  McGraw-Hill, New York, 1985

Hockney R. W., Eastwood J. W., *Computer Simulation Using Particles*
  Adam Hilger, Bristol and Philadelphia, 1988

Electron Optics with Applications

Spangenberg K. R., *Vacuum Tubes*
  McGraw-Hill, New York, 1948

Glaser W., *Grundlagen der elektronenoptik*
  Springer, Wien, 1952

Flugge S., Ed., *Handbuch der Physik, Band 33, Korpuskularoptik*
  Springer, Berlin, 1956

Bakish R., Ed., *Introduction to Electron Beam Technology*
    John Wiley & Sons, New York, 1962

Livingood J. J., *The Optics of Dipole Magnets*
    Academic Press, New York, 1964

Steffen K. G., *High Energy Beam Optics*
    Interscience Publishers, New York, 1965

Banford A. P., *The Transport of Charged Particle Beams*
    Spon Ltd., London, 1966

Septier A., Ed., *Focusing of Charged Particles*
    Academic Press, New York, 1967

Kirnstein P. T., Kino G. S., Waters W. E., *Space-Charge Flow*
    McGraw-Hill, New York, 1967

Beck A. H., Ed., *Handbook of Vacuum Physics, Vol. 2, Physical electronics*
    Pergamon Press, London, 1968

Grivet P., *Electron Optics*
    Pergamon Press, Oxford, 1972

Dahl P., *Introduction to Electron and Ion Optics*
    Academic Press, New York, 1973

Hawkes P. W., *Image Processing and Computer Aided Design in Electron
    Optics*
    Academic Press, New York, 1973

Septier A., Ed., *Applied Charged Particle Optics, Part A*
    Academic Press, New York, 1980

Septier A., Ed., *Applied Charged Particle Optics, Part B*
    Academic Press, New York, 1980

Septier A., Ed., *Applied Charged Particle Optics, Part C*
    Academic Press, New York, 1983

Briggs D., Seah M. P., *Practical Surface Analysis*
    John Wiley & Sons, New York, 1983

Loretto M. H., *Electron Beam Analysis of Materials*
    Chapman and Hall, London, 1984

Szilagyi M., *Electron and Ion Optics*
    Plenum Press, New York, 1988

Hawkes P. W., Ed., *Aspects of Charged Particle Optics*
    Academic Press, Boston, 1989

Vickerman J. C., Brown A., Reed N. M., *Secondary Ion Mass Spectroscopy*
    Clarendon Press, Oxford, 1989

de Wolf D. A., *Basics of Electron Optics*
    John Wiley & Sons, New York, 1990

Humphries S. Jr., *Charged Particle Beams*
    John Wiley & Sons, New York, 1990

Briggs D., Seah M. P., *Practical Surface Analysis, Vols. 1 & 2*
    John Wiley & Sons, New York, 1990, 1992

Microwave tubes

Sims G. D., Stephenson M. I., *Microwave Tubes and Semiconductor Devices*
Blackie & Son Ltd, London, 1963

Slater J. C., *Microwave Electronics*
Van Nostrand, New York, 1963

Schevchick V. N., *Fundamentals of Microwave Electronics*
Pergamon Press, London, 1963

Gewartowski J. W., Watson A. H., *Principles of Electron Tubes*
Van Nostrand, Princeton, 1965

Sutton G., *Power Traveling-Wave Tubes*
The English Universities' Press Ltd., London, 1965

Okress E. C., Ed., *Microwave Power Engineering*
Academic Press, New York, 1968

Coleman J. T., *Microwave Devices*
Reston Publishing Co. Inc., Reston, 1982

Gilmour A. S. Jr., *Microwave Tubes*
Artech House, Dedham, 1986

Granatstein V. L., Alexeff I., *High Power Microwave Sources*
Artech House, Boston, 1987

Liao S. Y., *Microwave Electron-Tube Devices*
Prentice-Hall, New York, 1988

Benford J., Swegle J. *High Power Microwaves*
Artech House, Boston, 1992

Particle Accelerators

Livingood J.J., *Principles of Cyclic Particle Accelerators*
Van Nostrand, Princeton, 1961

Kollath R., *Particle Accelerators*
Pitman and Sons, London, 1967

Persico E., Ferrari E., Segre S. E., *Principles of Particle Accelerators*
Benjamin, New York, 1968

Livingston S. M., *Particle Accelerators: A Brief History*
Harvard University Press, Cambridge, 1969

Puglisi M., Stipcich S., Torelli G., Eds., *New Techniques for Future Accelerators*
Plenum Press, New York, 1986

Gas Discharges and Applications

von Engel A., *Ionized Gases*
Clarendon Press, Oxford, 1955

Hirsh M. N., Oskam, H. J., *Gaseous Electronics*
Academic Press, New York, 1978

von Engel A., *Electric Plasmas: Their Nature and Uses*
Taylor & Francis Ltd., London, 1983

Wilson J., Hawkes J. F. B., *Lasers, Principles and Applications*
    Prentice-Hall, New York, 1987

Feinman J., Ed., *Plasma Technology in Metalurgical Processes*
    Iton and Steel Society, Warrendale, 1987

Orfeuil M., Ed., *Arc Plasma Processes*
    Int. Union for Electroheat, Paris-La-Défense, 1988

Rossnagel S. M., Cuomo J. J., Westwood W. D., Ed., *Handbook of Plasma
    Processing Technology*
    Noyes Publications, Park Ridge, NJ, 1989

Cuomo J. J., Rossnagel S. M., Ed., *Handbook of Ion Beam Processing Technology*
    Noyes Publications, Park Ridge, NJ, 1990

Eichler J., Eichler H.-J., *Laser, Grundlagen, Systeme, Anwendungen*
    Springer, Berlin, 1990

Wasa K., Hayakawa S., *Handbook of Sputter Deposition Technology*
    Noyes Publications, Park Ridge, NJ, 1992

Vacuum Technique and Vacuum Materials

Barrington E., *High Vacuum Engineering*
    Prentice-Hall, New York, 1963

Rosebury F., *Handbook of Electron Tubes and Vacuum Techniques*
    Addison-Wesley, New York, 1965

Kohl W. E., *Handbook of Materials and Techniques for Vacuum Physics*
    Reinhold, New York, 1967

Weisler G. L., Carlson R. W., *Vacuum Physics and Technology*
    Academic Press, New York, 1979

O'Hanlon J. F., *A User's Guide to Vacuum Technology*
    John Wiley & Sons, New York, 1980

Journals

Applied Physics
Applied Physics Letters
CERN publications
Doklady Akademii Nauk (continues Doklady Akademii Nauk SSSR)
Electronics
IEEE Transactions on Component Parts
IEEE Transactions on Electron Devices
IEEE Transactions on Image Processing
IEEE Transactions on Microwave Theory and Techniques
IEEE Transactions on Nuclear Science
IEEE Transactions on Plasma Science
Journal of Applied Physics
Journal of Computational Physics
Journal of Electron Microscopy
Journal of Electron Spectroscopy and Related Phenomena
Journal of Mass Spectrometry and Ion Processes
Journal of Physics D. Applied Physics
Journal of Physics E. Scientific Instruments
Journal of Scientific Instruments
Journal of Vacuum Science and Technology
Nuclear Instruments and Methods
Numerical Methods for Partial Differential Equations
Optical and Quantum Electronics
Optics and Laser Technology
Optics and Photonic News
Optik: Zeitschrift für Licht- und Elektronenoptik
Particle Accelerators
Philips Technical Review
Physical Review
Physical Review Letters
Physics of Fluids B. Plasma Physics
Physics Today
Plasma Physics
SLAC publications
Tehničeska Fizika (continues Žurnal Tehničeskoj Fiziki)
Zeitschrift für Physik
Žurnal Eksperimentaljnoj i Teoretičeskoj Fiziki

Book series

Advances in Electronics and Electron Physics
Nuclear Instruments and Methods in Physic Research A.
    Accelerators, Spectrometers, Detectors and Associated Equipment
Nuclear Instruments and Methods in Physic Research B.
    Beam Interactions with Materials and Atoms

NAME INDEX

SUBJECT INDEX